The Complete BR Diesel & Electric LOCOMOTIVE DIRECTORY

Details every Diesel, Electric and Gas Turbine Locomotive of British Rail and its Predecessors

By Colin J. Marsden

Oxford Publishing Co.

Second Edition
© 1993 C. J. Marsden

All rights reserved. No part of this book may be reproduced or transmitted in any form or by any means, electronic or mechanical, including photocopying, recording or by any information storage or retrieval system, without written permission from the copyright owner.

A catalogue record for this book is available from the British Library

ISBN 0-86093-508-6

Oxford Publishing Co. is part of the
Haynes Publishing Group PLC
Sparkford, Near Yeovil, Somerset, BA22 7JJ

Haynes Publications Inc.
861 Lawrence Drive, Newbury Park, California 91320 USA

Printed by: J. H. Haynes & Co. Ltd.
Typeset in Univers Roman Medium

Photograph credits (Plate section, between pages 176 and 177):
Author's collection: 1, 2, 3, 4, 5, 8, 9, 10, 13, 14, 17, 19, 20, 23, 24, 30, 31, 32, 33, 36, 37, 38, 42, 67, 68, 69, 70, 71, 72, 73 and 74.
Colin J. Marsden: 18, 21, 22, 25, 26, 28, 34, 39, 40, 41, 43, 44, 45, 46, 47, 48, 49, 51, 52, 53, 54, 55, 56, 57, 58, 59, 60, 61, 62, 64, 65, 66, 75, 76, 77, 78, 79 and 80.
British Rail: 11, 29, 50 and 63.
Brush Electrical Machines: 7.
Les Elsey: 27. A. R. Goult: 12. Derek Porter: 6.
Alex Swain: 15 and 16.

Cover illustrations:

Front: Representing Classes 47, 22, 52, 35 and 47 are Nos D1719, D6353, D1000, D7065 and D1740, seen outside the repair factory at Old Oak Common in 1966.

Front below, left: Nameplate and coat of arms of No. 73128.
Middle: Combined number and nameplate of No. 31430.
Right: Nameplate of No. 37229, as carried 5/84–10/88.

Rear: Classes 37, 31 and 40 at Stratford on 26th March 1991, with members of each class restored to former identities; Nos D6700 (37119), D5583 (31165) and D200 (40122).
(All Colin J. Marsden)

Network SouthEast liveried Class 47/7 No. 47701 *Old Oak Common Traction & Rolling Stock Depot* hurries away from Yeovil Junction past Coker Wood on 10th June 1992 with the 11.15 Waterloo–Exeter St Davids. *Colin J. Marsden*

Contents

Introduction	4
Abbreviations	5
Motive Power Depots	
GWR (WR) Codes	6
BR 1948 Codes	6
BR 1973 Alphabetic Codes	11
Diesel and Gas Turbine Locomotives	
Pre-Nationalisation Numbers	14
Experimental and Trials Locomotives	16
BR 1948 Numbers	18
BR 1957 Numbers	26
TOPS Numbers	162
Electric Locomotives	
Pre-Nationalisation Numbers	186
BR 1948 Numbers	188
BR 1957 Numbers	190
TOPS Numbers	200
Departmental Stock	
Pre-Grouping	206
Pre-Nationalisation	206
BR	208

Painted in Trainload Coal livery, Class 37/7 No. 37798, originally allocated the number D6706, pauses at Aberdare on 14th April 1992 with an mgr coal working from Tower Colliery to Aberthaw. *Colin J. Marsden*

INTRODUCTION

Welcome to the second edition of *The Complete BR Diesel & Electric Locomotive Directory*, up-dated to 31st December 1992. This is a major work detailing all diesel and electric locomotives operated by BR or its constituent companies, showing in one book, all numbers carried, dates of renumbering, allocated and carried names, together with dates of namings and denamings, details of constructors, introduction dates, works numbers (if allocated), depots of original allocation, dates of withdrawal, depots of final allocation, and disposal details, for each locomotive.

The book is divided into sections, covering different number ranges, and for the majority of locomotives, data is given in the BR 1957 number sequence, ie D1-D9999 for diesel, and E1-E9999 for electric locomotives. Pre-Nationalisation sequences, prototype locomotives, and of course those allocated BR TOPS numbers from new, are all given in separate sections. To assist readers in the location of information, especially with the present fleets, a full TOPS - BR 1957 number conversion table is included.

Information concerning introduction, works numbers, allocations and withdrawals has been obtained from official records. However, in some cases the introduction dates are only available for the original allocation, and the locomotive *may* have been completed some time before the date quoted. Data on disposals has again been checked with official BR records and where possible, those of the major disposal houses. Much information has also been provided by personal observations. Dates given for renumbering indicate the actual month/year that changes were recorded, this also applies to the dates of namings and denamings, the latter not always accurately recorded by the railways.

Details of locomotive storage is given in the 'notes' column, but only where such official records have been made by BR. The code [U] in the storage information indicates that a locomotive was/is stored in an unserviceable condition, while the letter [S] shows the locomotive stored in a serviceable condition, and could be returned to traffic with little attention.

Readers of this book are able to update the information by reference to the topical railway press, for example the monthly publication *Railway Magazine* gives regular reports on locomotive introduction, withdrawals and disposals.

The author would be very interested to receive any comments on the information provided, or any additional information available, this should be sent to 'The Editor', Locomotive Directory, c/o Oxford Publishing Co., Haynes Publishing Co. Ltd, Sparkford, Near Yeovil, Somerset, BA22 7JJ.

The author would like to thank numerous people who have assisted with information for this title, but, a special thank you must go to Roger Wood for general information, Jim Sheldon for naming and denaming data, and Ashley Butlin for disposal records.

Colin J. Marsden
February 1993

Posed inside the now closed depot at Eastfield on 21st August 1992 is, on the right, Class 37/4 No. 37430 *Cwmbran*, while on the left is withdrawn Class 26/1 No. 26028.
Colin J. Marsden

ABBREVIATIONS

A.Barclay	Andrew Barclay
A.Whitworth	Armstrong Whitworth
APCM	Associated Portland Cement [Blue Circle Industries]
B.Boveri	Brown Boveri
BC	British Coal
BOCE	British Open Cast Executive
BR	British Railways [Rail]
BRB	British Railways Board
BRCW	Birmingham Railway Carriage and Wagon Company
BREL	British Rail Engineering Ltd
BRML	British Rail Maintenance Ltd
B.Peacock	Beyer Peacock Ltd
BSC	British Steel Corporation
BTH	British Thomson Houston
CCD	Coal Concentration Depot
CEGB	Central Electricity Generating Board
C-WKS	Crewe Works
DEPG	Diesel Electric Preservation Group
DTG	Diesel Traction Group
EE	English Electric
EE.DK	English Electric Dick Kerr
EE.VF	English Electric Vulcan Foundry
EE.RSH	English Electric Robert Stephenson & Hawthorns
GEC	General Electric Company
GM-DD	General Motors - Diesel Division (Canada)
GM-EMD	General Motors - Electro Motive Division (USA)
GWR	Great Western Railway
H.Clarke	Hudswell Clarke
H.Leslie	Hawthorn, Leslie
H.Wolfe	Harland and Wolfe
H-WKS	Horwich Works
KWVR	Keighley and Worth Valley Railway
LMSR	London Midland and Scottish Railway
LNER	London and North Eastern Railway
LSWR	London & South Western Railway
L & Y	Lancashire & Yorkshire Railway
M.Vickers	Metropolitan Vickers
NBL	North British Locomotive Company
NCB	National Coal Board
NER	North Eastern Railway
NS	Nederlands Spoorwegen
NSF	National Smokeless Fuels
NRM	National Railway Museum, York
NYMR	North Yorkshire Moors Railway
RFS	RFS Industries
R.Hornsby	Ruston and Hornsby
R/I	Reinstated
RSH	Robert Stephenson and Hawthorns Ltd
RTC	Railway Technical Centre
SR	Southern Railway (Region)
TOPS	Total Operations Processing System
WD	War Department
WLA	Western Locomotive Association
YEC	Yorkshire Engine Company
[S]	Serviceable
[U]	Unserviceable

* Number allocated but not carried

Disposal Code Abbreviations

A	Awaiting removal or scrapping
C	Cut up
D	To Departmental stock
E	Exported
I	To internal user fleet or returned to capital stock from departmental stock
P	Preserved
R	Rebuilt
S	Sold for further use
X	Scrapped (details given)

One of the most unusual trains to be recorded on film travelling over the Paignton branch in recent years was this formation of Class 33/0 No. 33052 *Ashford* and Class 37/0 No. 37141 hauling preserved Class 55 'Deltics' Nos D9000 *Royal Scots Grey* and D9016 *Gordon Highlander* from Old Oak Common to Paignton on 16th June 1992, prior to taking part in the Torbay & Dartmouth Railway's gala weekend. *Colin J. Marsden*

MOTIVE POWER DEPOTS

GWR (WR) DEPOT CODES

Code	Location
ADR	Aberbeeg
ADR	Aberdare
AYN	Abercynon
BGP	Bath Green Park
BMJ	Bristol Marsh Junction
BND	Barnwood
BPL	Barnstaple Junction
BRO	Bromsgrove
BRW	Bristol Barrow Road
BRY	Barry
BTD	Bristol Bath Road
CAT	Cathays
CED	Cardiff East Dock 01/61-03/62
CRM	Carmarthen
CTN	Canton
DID	Didcot
DG	Danygraig 01/61-03/62
DNG	Danygraig
DYD	Duffryn Yard
EBW	Ebbw Junction
EXJ	Exmouth Junction
EXR	Exeter
GLO	Gloucester
HED	Hereford
LA	Laira
LDR	Landore

Code	Location
LLY	Llanelly
LTS	Llantrisant
MTR	Merthyr
NA	Newton Abbot
NEA	Neath
OOC	Old Oak Common
OXF	Oxford
PPL	Pontypool Road
PZ	Penzance
RDG	Reading
RYR	Radyr
SBZ	St Blazey
SDN	Swindon
SED	Swansea East Dock
SHL	Southall
SLO	Slough
SPM	St Philips Marsh - Bristol
STJ	Seven Tunnel Junction
TDU	Tondu
TEM	Templecombe
THT	Treherbert
TN	Taunton
TRO	Truro
WAD	Wadebridge
WBY	Westbury
WOS	Worcester
YEO	Yeovil

BR 1948 CODES

Code	Location	Date From	Date Closed or Code Changed *	Notes
1A	Willesden	1948	05/73*	To WN
1B	Camden	1948	01/66	
1C	Watford	1948	03/65	To S.P
1D	Devons Road	1948	09/63*	To 1J
1D	Marylebone	09/63	05/73*	To ME
1E	Bletchley	03/52	05/73*	To BY
1F	Rugby	09/63	12/68	To S.P
1G	Woodford Halse	09/63	06/65	
1H	Northampton	09/63	09/65	
1J	Devons Road	09/63	02/64	
2A	Rugby	1948	09/63*	To 1F
2A	Tyseley	09/63	05/73*	To TY
2B	Bletchley	1948	07/50*	To 4A
2B	Nuneaton	07/50	09/63*	To 5E
2B	Oxley	09/63	03/67	To S.P
2C	Northampton	1948	07/50*	To 4B
2C	Warwick	07/50	11/58	
2C	Stourbridge Junction	09/63	04/67	
2D	Nuneaton	1948	07/50*	To 2B
2D	Coventry	07/50	11/58	
2D	Banbury	09/63	10/66	To S.P
2E	Warwick	1948	07/50*	To 2C
2E	Northampton	03/52	09/63*	To 1H
2E	Saltley	09/63	05/73*	To SY
2F	Coventry	1948	07/50*	To 2D
2F	Market Harborough	10/55	04/58*	To 15F
2F	Woodford Halse	04/58	09/63*	To 1G
2F	Bescot	09/63	05/73*	To BS
2G	Woodford Halse	02/58	04/58*	To 2F
2G	Ryecroft	09/63	04/67	

Code	Location	Date From	Date Closed or Code Changed *	Notes
2H	Monument Lane	09/63	01/67	
2J	Aston	09/63	10/65	
2K	Bushbury	09/63	04/65	
2L	Leamington	09/63	06/65	
2M	Wellington	09/63	08/64	
2P	Kidderminster	09/63	08/64	
3A	Bescot	1948	06/60*	To 21B
3B	Bushbury	1948	06/60*	To 21C
3C	Ryecroft	1948	06/60*	To 21F
3D	Aston	1948	06/60*	To 21D
3E	Monument Lane	1948	06/60*	To 21E
4A	Bletchley	07/50	03/52*	To 1E
4B	Northampton	07/50	03/52*	To 2E
5A	Crewe North	1948	05/65	
5A	Crewe Diesel	05/65	05/73*	To CD
5B	Crewe South	1948	11/67	
5C	Stafford	1948	07/65	To S.P
5D	Stoke	1948	11/67	To S.P
5E	Alsager	1948	06/62	
5E	Nuneaton	09/63	06/66	To S.P
5F	Uttoxeter	1948	12/64	
5H	WCML electric traction (Crewe Electric)	10/63	01/66	
6A	Chester	1948	05/73*	To CH
6B	Mold Junction	1948	04/66	
6C	Birkenhead Mollington St	1948	09/63*	To 8H
6C	Croes Newydd	09/63	06/67	

6

Code	Location	Date From	Date Closed or Code Changed *	Notes	Code	Location	Date From	Date Closed or Code Changed *	Notes
6D	Chester Northgate	1948	01/60		11A	Carnforth	1948	04/58*	To 24L
6D	Shrewsbury	09/63	06/71		11A	Barrow in Furness	04/58	06/60*	To 12E
6E	Wrexham Rhosddu	1948	02/58*	To 84K	11B	Barrow in Furness	1948	04/58*	To 11A
6E	Chester (WR)	02/58	04/60		11B	Workington	04/58	06/60*	To 12F
6E	Oswestry	09/63	01/65		11C	Oxenholme	1948	06/60*	To 12G
6F	Bidston	1948	02/63		11D	Tebay	1948	06/60*	To 12H
6F	Machynlleth	09/63	12/66		11E	Lancaster (Green Ayre)	10/51	02/57*	To 24J
6G	Llandudno Junction	03/52	05/73*	To LJ					
6H	Bangor	03/52	06/65		12A	Carlisle Upperby	06/50	02/58*	To 12B
6J	Holyhead	03/52	01/67	To S.P	12A	Carlisle Kingmoor	02/58	05/73*	To KD
6K	Rhyl	03/52	02/63		12B	Carlisle Upperby	1948	06/50*	To 12A
					12B	Carlisle Canal	06/50	07/51*	To 68E
7A	Llandudno Junction	1948	03/52*	To 6G	12B	Penrith	10/55	02/58	
7B	Bangor	1948	03/52*	To 6H	12B	Carlisle Upperby	02/58	01/68	To S.P
7C	Holyhead	1948	03/52*	To 6J	12C	Penrith	1948	10/55*	To 12B
7D	Rhyl	1948	03/52*	To 6K	12C	Workington	10/55	04/58*	To 11B
					12C	Carlisle Canal	04/58	06/63	
8A	Edge Hill	1948	05/68	To S.P	12C	Barrow in Furness	09/63	05/73*	To BW
8B	Warrington (Dallam)	1948	10/67		12D	Workington	1948	10/55*	To 12C
8C	Speke Junction	1948	05/68	To S.P	12D	Carlisle Canal	02/58	04/58*	To 12C
8D	Widnes	1948	04/64		12D	Kirkby Stephen	04/58	11/61	
8E	Brunswick	05/50	04/58*	To 27F	12D	Workington	09/63	01/68	
8E	Northwich	04/58	03/68		12E	Moor Row	1948	07/54	
8F	Springs Branch	04/58	05/73*	To SP	12E	Kirkby Stephen	02/58	04/58*	To 12D
8G	Sutton Oak	04/58	03/69		12E	Barrow in Furness	06/60	09/63*	To 12C
8H	Allerton	07/60	09/63*	To 8J	12E	Tebay	09/63	01/68	
8H	Birkenhead M St	09/63	05/73*	To BC	12F	Workington	06/60	09/63*	To 12D
8J	Allerton	09/63	05/73*	To AN	12G	Oxenholme	06/60	01/63	
8K	Bank Hall	09/63	10/66		12H	Tebay	06/60	09/63*	To 12E
8L	Aintree	09/63	06/67						
8M	Southport	09/63	06/66		13A	Trafford Park	1948	05/50*	To 9E
8P	Wigan Central (L&Y)	09/63	04/64		13B	Belle Vue	1948	05/50*	To 26G
8R	Walton	09/63	12/63		13C	Heaton Mersey	1948	05/50*	To 9F
					13D	Northwich	1948	05/50*	To 9G
9A	Longsight	1948	05/73*	To LO	13E	Brunswick	1948	05/50*	To 8E
9B	Edgely	1948	05/68		13F	Walton	1948	05/50*	To 27E
9C	Macclesfield	1948	06/61		13G	Lower Ince Wigan	1948	05/50*	To 10F
9C	Reddish	09/63	05/73*	To RS					
9D	Buxton	1948	09/63*	To 9L	14A	Cricklewood East	1948	05/73*	To CW
9D	Newton Heath	09/63	05/73*	To NH	14B	Kentish Town	1948	04/63	
9E	Trafford Park	05/50	01/57*	To 17F	14B	Cricklewood West	09/63	03/67	
9E	Trafford Park	04/58	03/68		14C	St Albans	1948	01/60	
9F	Heaton Mersey	05/50	01/57*	To 17E	14C	Bedford	09/63	11/71	To S.P
9F	Heaton Mersey	04/58	05/68		14D	Neasden	02/58	06/62	
9G	Northwich	05/50	04/58*	To 8E	14E	Bedford	04/58	09/63*	To 14C
9G	Gorton	04/58	06/65		14F	Marylebone	06/61	09/63*	To 1D
9H	Gorton	02/58	04/58*	To 9G					
9H	Patricroft	09/63	07/68		15A	Wellingborough	1948	09/63*	To 15B
9J	Agecroft	09/63	10/66		15A	Leicester Midland	09/63	06/66	
9K	Bolton	09/63	07/68		15B	Kettering	1948	09/63*	To 15C
9L	Buxton	09/63	03/68	To S.P	15B	Wellingborough	09/63	10/68	To S.P
9M	Bury	09/63	04/65		15C	Leicester Midland	1948	09/63*	To 15A
9P	Lees	09/63	04/64		15C	Kettering	09/63	06/65	
					15D	Bedford	1948	04/58*	To 14E
10A	Springs Branch	1948	04/58*	To 8F	15D	Coalville	04/58	09/63*	To 15E
10A	Carnforth	09/63	08/68	To S.P	15D	Leicester (GC)	09/63	07/64	
10B	Preston	1948	02/58*	To 24K	15E	Leicester (GC)	02/58	09/63*	To 15D
10B	Blackpool Central	09/63	02/64		15E	Coalville	09/63	10/65	To S.P
10C	Patricroft	1948	02/58*	To 26F	15F	Market Harborough	04/58	05/65	
10C	Fleetwood	09/63	02/66						
10D	Plodder Lane	1948	10/54		16A	Nottingham	1948	09/63*	To 16D
10D	Sutton Oak	10/55	04/58*	To 8G	16A	Toton	09/63	05/73*	To TO
10D	Lostock Hall	09/63	05/72		16B	Spital Bridge	1948	08/60*	To 35C
10E	Sutton Oak	1948	10/55*	To 10D	16B	Kirby in Ashfield	10/55	09/63*	To 16E
10E	Accrington	09/63	10/72		16B	Annesley	09/63	01/66	
10F	Lower Ince	05/50	03/52		16B	Colwick	01/66	04/70	
10F	Rose Grove	09/63	08/68		16C	Kirby in Ashfield	1948	10/55*	To 16B
10G	Skipton	09/63	04/67	To S.P	16C	Mansfield	10/55	04/60	
10H	Lower Darwen	09/63	02/66		16C	Derby	09/63	03/67	To S.P
10J	Lancaster (Green Ayre)	09/63	04/66		16D	Mansfield	1948	10/55*	To 16C

7

Code	Location	Date From	Date Closed or Code Changed *	Notes
16D	Annesley	02/58	09/63*	To 16B
16D	Nottingham	09/63	11/67	
16E	Kirkby in Ashfield	09/63	10/70	
16F	Burton	09/63	05/73*	To BU
16G	Westhouses	09/63	10/66	To S.P
16J	Rowsley	09/63	03/67	
17A	Derby	1948	09/63*	To 16C
17B	Burton	1948	09/63*	To 16F
17C	Coalville	1948	04/58*	To 15D
17C	Rowsley	04/58	09/63*	To 16J
17D	Rowsley	1948	04/58*	To 17C
17E	Heaton Mersey	01/57	04/58*	To 9F
17F	Trafford Park	01/57	04/58*	To 9E
18A	Toton	1948	09/63*	To 16A
18B	Westhouses	1948	09/63*	To 16G
18C	Hasland	1948	09/63*	To 16H
18D	Staveley (Barrow Hill)	1948	02/58*	To 41E
19A	Sheffield (Brightside/Grimethorpe)	1948	02/58*	To 41B
19B	Sheffield (Millhouses)	1948	02/58*	To 41C
19C	Canklow	1948	02/58*	To 41D
20A	Holbeck	1948	02/57*	To 55A
20B	Stourton	1948	02/57*	To 55B
20C	Royston	1948	02/57*	To 55D
20D	Normanton	1948	02/57*	To 55E
20E	Manningham	1948	02/57*	To 55F
20F	Skipton	1948	06/50*	To 23A
20F	Skipton	10/51	02/57*	To 24G
20G	Hellifield	1948	06/50*	To 23B
20G	Hellifield	10/51	02/57*	To 24H
20H	Lancaster (Green Ayre)	1948	06/50*	To 23C
21A	Saltley	1948	09/63*	To 2E
21B	Bournville	1948	02/60	
21B	Bescot	06/60	09/63*	To 2F
21C	Bromsgrove	1948	02/58*	To 85F
21D	Stratford upon Avon	1948	02/53	To S.P
21D	Aston	06/60	09/63*	To 2J
21E	Monument Lane	06/60	09/63*	To 2H
22A	Bristol Barrow Rd	1948	02/58*	To 82E
22B	Gloucester Barnwood	1948	02/58*	To 85E
23A	Bank Hall	1948	06/50*	To 27A
23A	Skipton	06/50	10/51*	To 20F
23B	Aintree	1948	06/50*	To 27B
23B	Hellifield	06/50	10/51*	To 20G
23C	Southport	1948	06/50*	To 27C
23C	Lancaster (Green Ayre)	06/50	10/51*	To 11E
23D	Wigan Central (L&Y)	1948	06/50*	To 27D
24A	Accrington	1948	09/63*	To 10E
24B	Rose Grove	1948	09/63*	To 10F
24C	Lostock Hall	1948	09/63*	To 10D
24D	Lower Darwen	1948	09/63*	To 10H
24E	Blackpool Central	1948	06/50*	To 28A
24E	Blackpool Central	04/52	09/63*	To 10B
24F	Fleetwood	1948	06/50*	To 28B
24F	Fleetwood	04/52	09/63*	To 10C
24G	Skipton	02/57	09/63*	To 10G
24H	Hellifield	02/57	06/63	
24J	Lancaster (Green Ayre)	02/57	09/63*	To 10J
24K	Preston	02/58	09/61	
24L	Carnforth	04/58	09/63*	To 10A
25A	Wakefield	1948	09/56*	To 56A
25B	Huddersfield	1948	02/57*	To 55G
25C	Goole	1948	09/56*	To 53E
25D	Mirfield	1948	09/56*	To 56D
25E	Sowerby Bridge	1948	09/56*	To 56E
25F	Low Moor	1948	09/56*	To 56F
25G	Farnley Junction	1948	09/56*	To 55C
26A	Newton Heath	1948	09/63*	To 9D
26B	Agecroft	1948	09/63*	To 9J
26C	Bolton	1948	09/63*	To 9K
26D	Bury	1948	09/63*	To 9M
26E	Bacup	1948	10/54	
26E	Lees	10/55	09/63*	To 9P
26F	Lees	1948	10/55*	To 26E
26F	Belle Vue	10/55	04/56	
26F	Patricroft	02/58	09/63*	To 9H
26G	Belle Vue	05/50	10/55*	To 26F
27A	Bank Hall	06/50	09/63*	To 8K
27B	Aintree	06/50	09/63*	To 8L
27C	Southport	06/50	09/63*	To 8M
27D	Wigan Central (L&Y)	06/50	09/63*	To 8P
27E	Walton	05/50	09/63*	To 8R
27F	Brunswick	04/58	09/61	
28A	Blackpool Central	06/50	04/52*	To 24E
28B	Fleetwood	06/50	04/52*	To 24F
30A	Stratford	1948	05/73*	To SF
30B	Hertford East	1948	11/60	
30C	Bishops Stortford	1948	11/60	
30D	Southend on Sea	1948	02/59	
30E	Colchester	1948	05/73*	To CR
30F	Parkeston Quay	1948	01/67	To S.P
31A	Cambridge	1948	05/73*	To CA
31B	March	1948	05/73*	To MR
31C	Kings Lynn	1948	07/62	
31D	South Lynn	1948	07/61	
31E	Bury St Edmunds	1948	01/59	To S.P
31F	Spital Bridge	07/58	02/60	
32A	Norwich	1948	05/73*	To NR
32B	Ipswich	1948	05/68	To S.P
32C	Lowestoft	1948	07/62	
32D	Yarmouth South Town	1948	07/62	
32E	Yarmouth Vauxhall	1948	01/59	
32F	Yarmouth Beach	1948	03/59	
32G	Melton Constable	1948	03/59	
33A	Plaistow	1948	06/62	
33B	Tilbury	1948	09/62	
33C	Shoeburyness	1948	06/62	
34A	Kings Cross	1948	06/63	
34B	Hornsey	1948	05/71	
34C	Hatfield	1948	01/61	
34D	Hitchin	1948	05/73*	To HI
34E	Neasden	1948	02/58*	To 14D
34E	New England	07/58	10/68	
34F	Grantham	07/58	09/63	
34G	Finsbury Park	04/60	05/73*	To FP
35A	New England	1948	07/58*	To 34E
35B	Grantham	1948	07/58*	To 34F
35C	Spital Bridge	08/50	07/58*	To 31F
36A	Doncaster	1948	05/73*	To DR
36B	Mexborough	1948	07/58*	To 41F
36C	Frodingham	1948	05/73*	To FH
36D	Barnsley	1948	07/58*	To 41G
36E	Retford (GC)	1948	06/65	

Code	Location	Date From	Date Closed or Code Changed *	Notes	Code	Location	Date From	Date Closed or Code Changed *	Notes
37A	Ardsley	1948	07/56*	To 56B	53A	Dairycoates	1948	01/60*	To 50B
37B	Copley Hill	1948	07/56*	To 56C	53B	Botanic Gardens	1948	01/60*	To 50C
37C	Bradford Hammerton St	1948	07/56*	To 56G	53C	Springhead	1948	07/61	
					53C	Alexandra Dock (Hull)	12/58	11/63	
38A	Colwick	1948	07/58*	To 40E	53D	Bridlington	1948	06/58	To S.P
38B	Annesley	1948	02/58*	To 16D	53E	Cudworth	1948	07/51	
38C	Leicester (GC)	1948	02/58*	To 15E	53E	Goole	09/56	01/60*	To 50D
38D	Staveley (GC)	1948	07/58*	To 41H					
38E	Woodford Halse	1948	02/58*	To 2G	54A	Sunderland	1948	10/58*	To 52G
					54B	Tyne Dock	1948	10/58*	To 52H
39A	Gorton	1948	02/58*	To 9H	54C	Borough Gardens	1948	10/58*	To 52J
39B	Darnall	1948	06/55*	To 41A	54D	Consett	1948	10/58*	To 52K
40A	Lincoln	1948	05/73*	To LN	55A	Holbeck	02/57	05/73*	To HO
40B	Immingham	1948	05/73*	To IM	55B	Stourton	02/57	08/67	
40C	Louth	1948	12/56		55B	York	12/67	05/73*	To YK
40D	Tuxford	1948	07/58*	To 41K	55C	Farnley Junction	09/56	11/66	
40E	Langwith Junction	1948	07/58*	To 41J	55C	Healey Mills	12/67	05/73*	To HM
40E	Colwick	07/58	01/66*	To 16B	55D	Royston	02/57	09/71	
40F	Boston	1948	01/64		55E	Normanton	02/57	11/67	To S.P
					55F	Manningham	02/57	04/67	
41A	Darnall	06/55	04/64*	To 41B	55F	Hammerton Street	12/67	05/73*	To HS
41A	Tinsley	04/64	05/73*	To TI	55G	Huddersfield	02/57	01/67	
41B	Sheffield (Brightside/Grimethorpe)	02/58	04/64		55G	Knottingley	12/67	05/73*	To KY
					55H	Neville Hill	01/60	05/73*	To NL
41B	Darnall	04/64	10/65		55J	Low Moor	08/67	09/67	
41C	Sheffield (Millhouses)	02/58	12/61						
41C	Wath	11/63	05/73*	To WH	56A	Wakefield	09/56	06/67	
41D	Canklow	02/58	06/65		56A	Knottingley	07/67	12/67*	To 55G
41E	Staveley (Barrow Hill)	02/58	05/73*	To BH	56B	Ardsley	07/56	10/65	
41F	Mexborough	07/58	03/64		56B	Healey Mills	09/66	12/67*	To 55C
41G	Barnsley	07/58	01/60		56C	Copley Hill	07/56	09/64	
41H	Staveley (GC)	07/58	06/65		56D	Mirfield	09/56	04/67	
41J	Langwith Junction	07/58	11/66		56E	Sowerby Bridge	09/56	01/64	
41J	Shirebrook	11/66	05/73*	To SB	56F	Low Moor	09/56	08/67*	To 55J
41K	Tuxford	07/58	02/59		56G	Hammerton Street	07/56	12/67*	To 55F
50A	York	1948	12/67*	To 55B	60A	Inverness	1948	05/73*	To IS
50B	Neville Hill	1948	01/60*	To 55H	60B	Aviemore	1948	08/66	
50B	Dairycoates	01/60	09/70		60C	Helmsdale	1948	09/64	
50C	Selby	1948	09/59		60D	Wick	1948	08/62	
50C	Hull (Botanic Gardens)	01/60	05/73*	To BG	60E	Forres	1948	01/64	
50D	Starbeck	1948	09/59						
50D	Goole	01/60	02/73		61A	Kittybrewster	1948	08/67	To S.P
50E	Scarborough	1948	04/63		61B	Aberdeen	1948	05/73*	To AB
50F	Malton	1948	04/63		61C	Keith	1948	09/66	
50G	Whitby	1948	04/59						
					62A	Thornton Junction	1948	09/69	
51A	Darlington	1948	05/73*	To DN	62B	Dundee Tay Bridge	1948	04/67	
51B	Newport	1948	06/58		62B	Dundee West	04/67	05/73*	To DE
51C	West Hartlepool	1948	09/67	To S.P	62C	Dunfermline	1948	09/69	
51D	Middlesbrough	1948	06/58		62C	Dunfermline Townhill	07/70	05/73*	To DT
51E	Stockton	1948	06/59						
51F	West Auckland	1948	02/65		63A	Perth	1948	07/70	
51G	Haverton Hill	1948	06/59		63B	Stirling	1948	06/60*	To 65J
51H	Kirby Stephen	1948	02/58*	To 12E	63B	Fort William	06/60	07/70*	To 65H
51J	Northallerton	1948	03/63		63C	Forfar	1948	11/59	
51K	Saltburn	1948	01/58		63C	Oban	11/59	05/63	
51L	Thornaby	06/58	05/73*	To TE	63D	Fort William	1948	05/55*	To 65J
					63D	Oban	05/55	11/59*	To 63C
52A	Gateshead	1948	05/73*	To GD	63E	Oban	1948	05/55*	To 63D
52B	Heaton	1948	06/67						
52C	Blaydon	1948	03/65		64A	St Margarets	1948	02/67	
52D	Tweedmouth	1948	01/69		64A	Millerhill	04/67	05/73*	To MH
52E	Percy Main	1948	02/66		64B	Haymarket	1948	05/73*	To HA
52F	North Blyth	1948	01/68		64C	Dalry Road	1948	10/65	
52G	Sunderland	10/58	09/67		64D	Carstairs	1948	06/60*	To 66E
52H	Tyne Dock	10/58	09/67	To S.P	64E	Polmont	1948	06/60*	To 65K
52J	Borough Gardens	10/58	06/59		64F	Bathgate	1948	12/66	
52J	South Gosforth	02/64	05/73*	To GF	64G	Hawick	1948	01/66	
52K	Consett	10/58	05/65	To S.P	64H	Leith Central	11/59	04/72	

9

Code	Location	Date From	Date Closed or Code Changed *	Notes	Code	Location	Date From	Date Closed or Code Changed *	Notes
65A	Eastfield	1948	05/73*	To ED	73C	Hither Green	1948	05/73*	To HG
65B	St Rollox	1948	11/66		73D	Gillingham	1948	06/59	
65C	Parkhead	1948	12/62		73D	St Leonards	07/63	05/73*	To SE
65D	Dawsholm	1948	10/64		73E	Faversham	1948	01/64	To S.P
65E	Kipps	1948	12/62		73F	Ashford	10/58	05/73*	To AF
65F	Grangemouth	1948	05/73*	To GM	73G	Ramsgate	10/58	05/73*	To RM
65G	Yoker	1948	10/64		73H	Dover	10/58	06/61	To S.P
65H	Helensburgh	1948	11/61		73J	Tonbridge	10/58	01/65	To S.P
65H	Fort William	07/70	05/73*	To FW					
65I	Balloch	1948	03/61		74A	Ashford	1948	10/58*	To 73F
65J	Fort William	05/55	06/60*	To 63B	74B	Ramsgate	1948	10/58*	To 73G
65J	Stirling	06/60	12/66		74C	Dover	1948	10/58*	To 73H
65K	Polmont	06/60	05/64		74D	Tonbridge	1948	10/58*	To 73J
					74E	St Leonards	1948	06/58	
66A	Polmadie	1948	03/72	To S.P					
66B	Motherwell	1948	05/67	To S.P	75A	Brighton	1948	05/73*	To BI
66C	Hamilton	1948	05/73*	To HN	75B	Redhill	1948	06/65	To S.P
66D	Greenock	1948	11/66	To S.P	75C	Norwood Junction	1948	08/66	To S.P
66E	Carstairs	06/60	03/67	To S.P	75C	Selhurst	08/66	05/73*	To SU
66F	Beattock	07/62	05/67	To S.P	75D	Horsham	1948	07/59	
					75D	Stewarts Lane	06/62	05/73*	To SL
67A	Corkerhill	1948	05/73*	To CK	75E	Three Bridges	1948	06/65	To S.P
67B	Hurlford	1948	12/66		75F	Tunbridge Wells West	1948	09/63	
67C	Ayr	1948	05/73*	To AY	75G	Eastbourne	1948	06/65	
67D	Ardrossan	1948	09/69						
67D	Largs	09/69	05/73	To S.P	81A	Old Oak Common	1948	05/73*	To OC
67E	Dumfries	07/62	11/66		81B	Slough	1948	06/64	
67F	Stranraer	07/62	03/68		81C	Southall	1948	05/73*	To SZ
					81D	Reading WR	1948	05/73*	To RG
68A	Carlisle Kingmoor	1948	02/58*	To 12A	81E	Didcot	1948	06/65	
68B	Dumfries	1948	07/62*	To 67E	81F	Oxford	1948	05/73	To OX
68C	Stranraer	1948	07/62*	To 67F					
68D	Beattock	1948	07/62*	To 66F	82A	Bristol Bath Road	1948	05/73*	To BR
68E	Carlisle Canal	07/51	02/58*	To 12D	82B	St Philips Marsh	1948	06/64	
					82C	Swindon	1948	05/73*	To SW
70A	Nine Elms	1948	07/67		82D	Westbury	1948	10/63*	To 83C
70B	Feltham	1948	08/70		82E	Yeovil Pen Mill	1948	02/58*	To 71H
70C	Guildford	1948	07/67		82E	Bristol Barrow Road	02/58	11/65	
70D	Basingstoke	1948	09/63	To S.P	82F	Weymouth	1948	02/58*	To 71G
70D	Eastleigh	09/63	05/73*	To EH	82F	Bath (S & D)	02/58	03/66	
70E	Reading (SR)	1948	12/62	To S.P	82G	Templecombe	02/58	10/63*	To 83G
70E	Salisbury	12/62	07/67						
70F	Fratton	09/54	09/63	To S.P	83A	Newton Abbot	1948	05/73*	To NA
70F	Bournemouth	09/63	05/73*	To BM	83B	Taunton	1948	06/68	To S.P
70G	Newport (I O W)	09/54	11/57		83C	Exeter	1948	10/63	To S.P
70G	Weymouth	09/63	07/67		83C	Westbury	10/63	06/68	To S.P
70H	Ryde (I O W)	09/54	05/73*	To RY	83D	Laira	1948	09/63*	To 84A
70I	Southampton Docks	09/63	02/66		83D	Exmouth Junction	09/63	03/67	
					83E	St Blazey	1948	09/63*	To 84B
71A	Eastleigh	1948	09/63*	To 70D	83E	Yeovil	09/63	06/65	
71B	Bournemouth	1948	09/63*	To 70F	83F	Truro	1948	09/63*	To 84C
71C	Dorchester	1948	06/57		83F	Barnstaple	09/63	09/64	
71D	Fratton	1948	09/54*	To 70F	83G	Penzance	1948	09/63*	To 84D
71E	Newport (I O W)	1948	09/54*	To 70G	83G	Templecombe (S & D)	10/63	03/66	
71F	Ryde (I O W)	1948	09/54*	To 70H	83H	Plymouth Friary	02/58	05/63	
71G	Bath (S & D)	1948	02/58*	To 82F					
71G	Weymouth	02/58	09/63*	To 70G	84A	Stafford Road	1948	09/63	
71H	Templecombe (S & D)	1948	02/58*	To 82G	84A	Laira	09/63	05/73*	To LA
71H	Yeovil Pen Mill	02/58	01/59		84B	Oxley	1948	09/63*	To 2B
71I	Southampton Docks	1948	09/63*	To 70I	84B	St Blazey	09/63	05/73*	To BZ
71J	Highbridge	09/54	02/58		84C	Banbury	1948	09/63*	To 2D
					84C	Truro	09/63	10/65	
72A	Exmouth Junction	1948	09/63*	To 83D	84D	Leamington	1948	09/63*	To 2L
72B	Salisbury	1948	12/62*	To 70E	84D	Penzance	09/63	05/73*	To PZ
72C	Yeovil	1948	09/63*	To 83E	84E	Tyseley	1948	09/63*	To 2A
72D	Plymouth Friary	1948	02/58*	To 83H	84E	Wadebridge	09/63	10/64	
72E	Barnstaple	1948	09/63*	To 83F	84F	Stourbridge Junction	1948	09/63*	To 2C
72F	Wadebridge	1948	09/63*	To 84E	84G	Shrewsbury	1948	01/61*	To 89A
					84G	Kidderminster	01/61	09/63*	To 2P
73A	Stewarts Lane	1948	06/62*	To 75D	84H	Wellington	1948	09/63*	To 2M
73B	Bricklayers Arms	1948	06/62	To S.P	84J	Cross Newydd	1948	01/61*	To 89B

Code	Location	Date From	Date Closed or Code Changed *	Notes
84K	Chester (WR)	1948	02/58*	To 6E
84K	Wrexham Rhosddu	02/58	01/60	
85A	Worcester	1948	05/73*	To WS
85B	Gloucester Horton Road	1948	05/73*	To GL
85C	Hereford	1948	01/61*	To 86C
85C	Gloucester Barnwood	01/61	05/64	
85D	Kidderminster	1948	01/61*	To 84G
85D	Bromsgrove	01/61	09/64	
85E	Gloucester Barnwood	02/58	01/61*	To 85C
85F	Bromsgrove	02/58	01/61*	To 85D
86A	Ebbw Junction	1948	09/63*	To 86B
86A	Canton	09/63	05/73*	To CF
86B	Newport Pill	1948	06/63	
86B	Ebbw Junction	09/63	05/73*	To EJ
86C	Canton	1948	01/61*	To 86A
86C	Hereford	01/61	11/64	To S.P
86D	Llantrisant	1948	01/61*	To 88G
86E	Severn Tunnel Junction	1948	06/68	To S.P
86F	Tondu	1948	01/61*	To 88H
86F	Aberbeeg	01/61	12/64	
86G	Pontypool Road	1948	10/67	
86H	Aberbeeg	1948	01/61*	To 86F
86J	Aberdare	1948	01/61*	To 88J
86K	Abergavenny	1948	11/54	
86K	Tredegar	11/54	06/60	
87A	Neath (Court Sart)	1948	06/65	
87A	Landore	11/69	05/73*	To LE
87B	Duffryn Yard	1948	03/64	
87B	Margam	03/64	05/73*	To MG
87C	Danygraig	1948	01/61*	To DG
87C	Danygraig	03/62	03/64	
87D	Swansea (East Dock)	1948	07/64	
87E	Landore	1948	11/69*	To 87A
87F	Llanelly	1948	07/64	To S.P
87G	Carmarthen	1948	04/64	
87H	Neyland	1948	09/63	
87H	Whitland	10/63	03/69	To S.P
87J	Fishguard	1948	09/63	
87K	Swansea (Victoria)	1948	08/59	
88A	Cathays	1948	12/57	
88A	Radyr	12/57	01/61*	To 88B
88A	Canton	01/61	09/63*	To 86A
88B	Cardiff East Dock	1948	01/61*	To CED
88B	Cathays	01/61	03/62*	To 88M
88B	Radyr	01/61	10/68	
88C	Barry	1948	09/64	To S.P
88D	Merthyr Tydfil	1948	10/64	
88D	Rhymney	11/64	04/65	To S.P
88E	Abercynon	1948	10/64	To S.P
88F	Treherbert	1948	10/67	To S.P
88G	Llantrisant	01/61	10/64	
88H	Tondu	01/61	03/64	
88J	Aberdare	01/61	03/65	
88K	Brecon	01/61	12/62	
88L	Cardiff East Dock	03/62	09/63	
88M	Cathays	03/62	11/64	
89A	Oswestry	1948	01/61*	To 89D
89A	Shrewsbury	01/61	09/63*	To 6D
89B	Brecon	1948	11/59	
89B	Cross Newydd	01/61	09/63*	To 6C
89C	Machynlleth	1948	09/63*	To 6F
89D	Oswestry	01/61	09/63*	To 6E

LONDON MIDLAND AREA CODES

Code	Area	From	to
D01	LM - London (Western) Division	04/66	05/73
D02	LM - Birmingham Division	04/66	05/73
D05	LM - Stoke Division	04/66	05/73
D08	LM - Liverpool Division	06/68	05/73
D09	LM - Manchester Division	06/68	05/73
D10	LM - Preston Division	06/68	05/73
D14	LM - London (Midland) Division	01/65	06/68
D15	LM - Leicester Division	01/65	11/67
D16	LM - Nottingham Division	01/65	05/73
LMML	Line Power Controller - Derby	01/65	06/68
LMWL	London Midland (Western Lines)	01/65	06/68

BR 1973 ALPHABETIC CODES

Code	Location and Notes
AB	Aberdeen Ferryhill
AC	Aberdeen Clayhills
AF	Ashford Chart Leacon
AL	Aylesbury
AN	Allerton
AY	Ayr
BC	Birkenhead Mollington Street
BD	Birkenhead North
BE	Bedford Midland Station
BG	Botanic Gardens
BH	Barrow Hill
BI	Brighton
BJ	Bristol Marsh Junction
BK	Birkenhead Central
BL	Blyth
BM	Bournemouth
BN	Birkenhead North (To BD 12/76)
BN	Bounds Green
BP	Beattock
BP	Blackpool North Carriage Sidings
BQ	Bury (From 01/75)
BR	Bristol Bath Road
BS	Bescot
BU	Burton
BV	Bury (To BQ 01/75)
BW	Barrow
BX	Buxton
BY	Bletchley
BZ	St Blazey
CA	Cambridge
CC	Clacton
CD	Cricklewood (To CW from 12/73)
CD	Crewe Diesel
CE	Crewe Diesel (To CD from 12/73)
CE	Crewe Electric
CF	Cardiff Canton
CG	Croxley Green
CH	Chester
CJ	Clapham Junction Carriage Sidings
CK	Corkerhill
CL	Clacton (To CC 12/76)
CL	Glasgow Cowlairs (To GC from 10/83)

Code	Location and Notes	Code	Location and Notes
CL	Carlisle Upperby	LL	Liverpool Edge Hill Carriage Sidings
CP	Crewe Carriage Sidings	LN	Lincoln
CR	Colchester	LO	Longsight Diesel
CW	Crewe Electric (To CE from 12/73)	LR	Leicester
CW	Cricklewood	LV	Liverpool Street
CY	Clapham Yard stabling point		
		MA	Manchester Longsight Carriage Sidings
DA	Darnall	MC	March (To MR 09/73)
DE	Dundee	ME	Marylebone
DL	Doncaster Major (Level 5)	MG	Margam
DN	Darlington	MH	Millerhill
DR	Doncaster	ML	Motherwell
DT	Dunfermline Townhill	MN	Machynlleth
DY	Derby Etches Park	MR	March
EC	Edinburgh Craigentinny	NA	Newton Abbot
ED	Eastfield	NC	Norwich Crown Point
EG	Edge Hill	NH	Newton Heath
EH	Eastleigh	NL	Neville Hill
EJ	Ebbw Junction	NM	Nottingham Carriage Sidings
EM	East Ham	NO	Norwich (To NR 09/73)
EN	Euston Downside	NP	North Pole International
EU	Euston Station	NR	Norwich (To NC 12/83)
EX	Exeter St Davids	NW	Northwich
FH	Frodingham	OC	Old Oak Common
FP	Finsbury Park	OM	Old Oak Common Carriage Sidings
FR	Fratton	ON	Orpington
FW	Fort William	OO	Old Oak Common HST
		OX	Oxford
GC	Glasgow Cowlairs	OY	Oxley Carriage Sidings
GD	Gateshead		
GF	South Gosforth	PA	Cambridge Street
GI	Gillingham	PB	Peterborough
GL	Gloucester	PC	Polmadie Carriage Sidings
GM	Grangemouth	PE	Peterborough Nene
GO	Goole	PH	Perth
GS	Glasgow Shields Road (To GW 03/75)	PI	Pwllheli (Temporary location only in 1980)
GU	Guide Bridge	PM	St Philips Marsh
GW	Glasgow Shields Road (From 03/75)	PO	Polmadie
		PQ	Parkeston Quay
HA	Haymarket	PY	Bristol Pylle Hill
HC	Holyhead Carriage Sidings	PZ	Penzance
HD	Holyhead		
HE	Hornsey	RB	Ranelagh Bridge
HF	Hereford	RE	Ramsgate
HG	Hither Green	RG	Reading
HI	Hitchin	RL	Ripple Lane
HM	Healey Mills	RM	Ramsgate
HN	Hamilton	RS	Reddish
HO	Holbeck	RY	Ryde
HQ	BRB Headquarters (Paper allocation)		
HR	Hall Road	SA	Salisbury
HS	Hammerton Street, Bradford	SB	Shirebrook
HT	Heaton	SE	St Leonards
HY	Hyndland	SF	Stratford
		SG	Slade Green
IL	Ilford	SH	Strawberry Hill
IM	Immingham	SI	Soho
IP	Ipswich	SL	Battersea, Stewarts Lane
IS	Inverness	SP	Springs Branch
		SR	Stratford Level 5
KD	Carlisle New Yard (To KM 01/75)	ST	Severn Tunnel Junction
KD	Carlisle Kingmoor	SU	Selhurst
KM	Carlisle New Yard	SW	Swindon
KM	Carlisle Kingmoor (To KD 01/75)	SX	Stratford (To SF from 09/73)
KX	Kings Cross	SY	Saltley
KY	Knottingley	SZ	Southall
LA	Laira	TE	Thornaby
LE	Landore	TF	Thornton Fields Carriage Sidings
LG	Longsight Electric	TI	Tinsley
LH	Lostock Hall	TJ	Thornton Junction
LJ	Llandudno Junction	TO	Toton

12

Code	Location and Notes	Code	Location and Notes
TS	Tyseley	WK	Workington
TW	Tunbridge Wells West	WM	Wimbledon (To WD 12/76)
TY	Tyne Yard	WN	Willesden
		WO	Wellingborough
VR	Aberystwyth (Vale of Rheidol)	WS	Worcester
		WT	Westhouses (From 01/75)
WB	Wembley Carriage Sidings	WU	Westhouses (To WT 01/75)
WC	Waterloo & City	WY	Westbury
WD	Wimbledon (From 12/76)		
WH	Wath	YC	York Clifton
WJ	Watford	YK	York

Green liveried Class 47/4 No. 47484 *Isambard Kingdom Brunel* is seen inside the now closed 'Factory' at Old Oak Common on 20th March 1993.
Colin J. Marsden

DIESEL AND GAS TURBINE LOCOMOTIVES
PRE NATIONALISATION NUMBERS

SOUTHERN RAILWAY 0-6-0 DIESEL SHUNTERS

Original SR No.	BR 1948 No.	Date Re No.	Built By	Works No.	Date Introduced	Depot of First Allocation	Date Withdrawn	Depot of Final Allocation
1	15201	06/51	SR Ashford	-	09/37	75C	11/64	70D
2	15202	10/48	SR Ashford	-	09/37	75C	12/64	73F
3	15203	05/50	SR Ashford	-	10/37	75C	11/64	75C

GREAT WESTERN RAILWAY 0-6-0 DIESEL SHUNTER

Original GWR No.	BR 1948 No.	Date Re No.	Built By	Works No.	Date Introduced	Depot of First Allocation	Date Withdrawn	Depot of Final Allocation
2	15100	03/48	H.Leslie	3853	04/36	82C	04/65	82C

LONDON & NORTH EASTERN RAILWAY 0-6-0 DIESEL SHUNTERS

Original LNER No.	BR 1948 No.	Date Re No.	Built By	Works No.	Date Introduced	Depot of First Allocation	Date Withdrawn	Depot of Final Allocation
8000	15000	06/52	LNER Doncaster	1960	08/44	30A	08/67	5B
8001	15001	05/52	LNER Doncaster	1963	08/44	30A	04/67	5B
8002	15002	06/50	LNER Doncaster	1973	12/44	30A	08/67	5B
8003	15003	11/51	LNER Doncaster	1978	03/45	30A	05/67	5B

LONDON, MIDLAND & SCOTTISH RAILWAY DIESEL SHUNTERS

LMS No.	Original LMS No.	Date Re No.	BR 1948 No.	Date Re No.	Built By	Works No. and Year of Build	Date Introduced	Depot of First Allocation	Date Withdrawn	Depot of Final Allocation
1831	-	-	-	-	LMSR Derby	8071 1931	05/34	Derby	09/39	Derby
7050	7400	11/34	-	-	DK Preston	2047 1934	11/34	Agecroft	03/43	Agecroft
7051	7401	11/34	-	-	Hunslet	1697 1932	05/33	Leeds	12/45	Chester
7052	7402	11/34	-	-	Hunslet	1721 1933	01/34	Leeds	12/43	Nottingham
7053	7403	12/34	-	-	Hunslet	1723 1934	11/34	Leeds	12/42	Eastleigh (SR)
7054	7404*	-	-	-	Hunslet	1724 1934	11/34	Leeds	05/43	Speke Jn
7055	7405*	-	-	-	H/Clarke	D580 1934	12/34	Speke Jn	04/39	Speke Jn
7056	7406*	-	-	-	H/Clarke	D581 1935	10/35	Speke Jn	05/39	Speke Jn
7057	7407*	-	-	-	H/Wolff	-	02/35	Bolton	01/44	Heysham
7058	7408	11/34	13000*	-	A/Whitworth	D20 1933	02/34	Willesden	11/49	Toton
7059	-	-	-	-	A/Whitworth	D54 1935	05/36	Crewe South	11/44	Willesden
7060	-	-	-	-	A/Whitworth	D55 1935	12/36	Crewe South	12/42	W Department
7061	-	-	-	-	A/Whitworth	D56 1935	06/36	Crewe South	11/44	W Department
7062	-	-	-	-	A/Whitworth	D57 1935	06/36	Crewe South	11/44	W Department
7063	-	-	-	-	A/Whitworth	D58 1935	07/36	Crewe South	11/44	W Department
7064	-	-	-	-	A/Whitworth	D59 1935	12/36	Carlisle	11/44	Willesden
7065	-	-	-	-	A/Whitworth	D60 1935	07/36	Carlisle	12/42	W Department
7066	-	-	-	-	A/Whitworth	D61 1935	08/36	Carlisle	12/42	W Department
7067	-	-	-	-	A/Whitworth	D62 1935	09/36	Carlisle	11/44	W Department
7068	-	-	-	-	A/Whitworth	D63 1935	10/36	Carlisle	12/42	W Department
7069	-	-	-	-	H/Leslie	3841	01/36	Crewe South	12/40	W Department
7070	-	-	-	-	H/Leslie	3842	01/36	Crewe South	12/40	W Department
7071	-	-	-	-	H/Leslie	3843	01/36	Crewe South	12/40	W Department
7072	-	-	-	-	H/Leslie	3844	01/36	Crewe South	12/40	W Department
7073	-	-	-	-	H/Leslie	3845	04/36	Crewe South	12/40	W Department
7074	-	-	12000	07/49	H/Leslie	3846	04/36	Crewe North	04/61	5B
7075	-	-	-	-	H/Leslie	3847	06/36	Willesden	12/40	W Department
7076	-	-	12001	09/49	H/Leslie	3848	09/36	Crewe South	02/62	5B
7077	-	-	-	-	H/Leslie	3849	09/36	Preston	12/40	W Department
7078	-	-	-	-	H/Leslie	3850	12/36	Willesden	12/40	W Department
7079	-	-	12002	09/49	H/Leslie	3816	04/34	Crewe South	06/56	5B
7080	-	-	12003	01/50	LMSR Derby	-	05/39	18A	11/67	8F
7081	-	-	12004	04/49	LMSR Derby	-	05/39	1A	12/67	8F
7082	-	-	12005	08/49	LMSR Derby	-	05/39	18A	09/67	8C
7083	-	-	12006	04/48	LMSR Derby	-	06/39	18A	09/67	8C
7084	-	-	12007	10/49	LMSR Derby	-	06/39	18A	10/67	8C
7085	-	-	12008	07/49	LMSR Derby	-	09/39	18A	07/67	8C

14

Original SR No.	Disposal Code	Disposal Detail	Date Cut Up	Notes
1	C	G Cohen, Morriston	10/69	Loan to WD 1941-45. After withdrawal used as generator
2	C	J Cashmore, Newport	11/66	Loan to WD 1941-45.
3	C	J Cashmore, Newport	08/66	Loan to WD 1941-45.

Original GWR No.	Disposal Code	Disposal Detail	Date Cut Up	Notes
2	C	G Cohen, Morriston	01/66	

Original LNER No.	Disposal Code	Disposal Detail	Date Cut Up	Notes
8000	C	A King, Norwich	08/68	
8001	C	J Cashmore, Great Bridge	08/67	
8002	C	A King, Norwich	08/68	
8003	C	Slag Reduction, Ickles	06/68	

LMS No.	Disposal Code	Disposal Detail	Date Cut Up	Notes
1831	C	BR Crewe Works	08/51	After withdrawal used as generator
7050	P	Museum of Army Transport, Beverley	-	Built: 1934, WD use: 08/40, To LMS: 08/41
7051	P	Middleton Railway	-	Built: 01/32, WD use: 08/40, To LMS: 06/41, WD use: 08/44, To LMS: 06/45
7052	C	Birds, Long Marston	05/69	Built: 05/33, WD use: 08/40, To LMS: 02/42
7053	C	Hunslet Co, Leeds	07/55	Built: 1933, WD use: 10/39, To LMS: 01/41
7054	C	NCB Hickleton Main	10/74	WD use: 10/39, To LMS: 03/40, WD use: 10/44, To LMS: 07/41, WD use: 03/42, To LMS: 07/42
7055	C	BR Thornaby	02/64	Used as Departmental No 953
7056	C	T Muir, Thornton	01/56	Used as Departmental MPU No. 1
7057	E	Northern Counties Committee, Ireland	-	Built: 07/34, WD use: 03/41, To LMS: 06/43
7058	C	BR Derby Works	01/50	Built: 06/33, WD use: 08/41, To LMS: 06/43
7059	E	WD in France, later sold to SNCB	-	Built: 07/35, WD use: 08/40, To LMS: 05/41, WD use: 08/41, To LMS: 01/44
7060	E	WD in Egypt	-	Built: 1935, WD use: 01/41
7061	E	WD in France, later sold to SNCB	-	Built: 1935, WD use: 10/40
7062	S	WD in UK	-	Built: 1935, WD use: 09/40
7063	C	E L Pitts, Brackley	03/67	Built: 1935, WD use: 09/40, To LMS: 03/41, WD use: 11/42, To LMS: 12/42, WD use: 11/44
7064	E	WD in France, later sold to SNCB	-	Built: 1935, WD use: 06/40, To LMS: 01/44
7065	E	WD in Egypt, later sold to Egyptian Rly	-	Built: 1935, WD use: 01/40
7066	E	WD in Egypt	-	Built: 1935, WD use: 01/41
7067	E	WD in France, later sold to SNCB	-	Built: 1935, WD use: 07/41
7068	E	WD in Egypt, later sold to Egyptian Rly	-	WD use: 01/41
7069	P	Swanage Railway (at Poole)	-	WD use: 04/40
7070	E	WD in France	-	Built: 1935, WD use: 01/40
7071	E	WD in France	-	Built: 1935, WD use: 04/40
7072	E	WD in France	-	WD use: 04/40
7073	E	WD in France	-	WD use: 01/40
7074	C	BR Derby Works	06/62	-
7075	E	WD in France	-	WD use: 01/41
7076	C	BR Horwich Works	05/62	WD use: 11/39, To LMS: 01/40, WD use: 03/40, To LMS: 07/41
7077	E	WD in France	-	WD use: 12/39
7078	E	WD in France	-	WD use: 01/40
7079	C	BR Derby Works	09/56	Built as demonstrator locomotive
7080	C	Slag Reduction Co, Ickles	08/68	
7081	C	J Cashmore, Great Bridge	05/68	
7082	C	C F Booth, Rotherham	04/68	
7083	C	Slag Reduction Co, Ickles	06/68	
7084	C	Slag Reduction Co, Ickles	04/68	
7085	C	C F Booth, Rotherham	02/68	

LMS No.	Original LMS No.	Date Re No.	BR 1948 No.	Date Re No.	Built By	Works No. and Year of Build	Date Introduced	Depot of First Allocation	Date Withdrawn	Depot of Final Allocation
7086	-	-	12009	11/48	LMSR Derby	-	11/39	5B	09/67	8C
7087	-	-	12010	05/48	LMSR Derby	-	12/39	5B	09/67	8C
7088	-	-	12011	11/49	LMSR Derby	-	12/39	5B	03/66	5B
7089	-	-	12012	02/49	LMSR Derby	-	12/39	5B	12/67	8C
7090	-	-	12013	10/49	LMSR Derby	-	01/40	5B	11/67	8F
7091	-	-	12014	09/50	LMSR Derby	-	03/40	18A	10/67	8C
7092	-	-	12015	07/51	LMSR Derby	-	03/40	5B	10/67	8C
7093	-	-	12016	06/49	LMSR Derby	-	04/40	5B	09/67	8C
7094	-	-	12017	03/50	LMSR Derby	-	04/40	5B	10/67	8F
7095	-	-	12018	06/48	LMSR Derby	-	05/40	5B	10/67	8C
7096	-	-	12019	03/51	LMSR Derby	-	06/40	5B	10/67	8C
7097	-	-	12020	02/49	LMSR Derby	-	06/40	1A	11/67	8F
7098	-	-	12021	09/49	LMSR Derby	-	06/40	1A	10/67	8F
7099	-	-	12022	01/50	LMSR Derby	-	07/40	1A	11/66	H-Wks
7100	-	-	-	-	LMSR Derby	-	12/40	W Department	12/42	W Department
7101	-	-	-	-	LMSR Derby	-	01/41	W Department	12/42	W Department
7102	-	-	-	-	LMSR Derby	-	04/41	W Department	12/42	W Department
7103	-	-	-	-	LMSR Derby	-	04/41	W Department	12/42	W Department
7104	-	-	-	-	LMSR Derby	-	04/41	W Department	12/42	W Department
7105	-	-	-	-	LMSR Derby	-	07/41	W Department	12/42	W Department
7106	-	-	-	-	LMSR Derby	-	07/41	W Department	12/42	W Department
7107	-	-	-	-	LMSR Derby	-	07/41	W Department	12/42	W Department
7108	-	-	-	-	LMSR Derby	-	10/41	W Department	12/42	W Department
7109	-	-	-	-	LMSR Derby	-	10/41	W Department	12/42	W Department
7110	-	-	12023	05/48	LMSR Derby	-	01/42	68A	12/67	8F
7111	-	-	12024	06/50	LMSR Derby	-	01/42	68A	12/67	9D
7112	-	-	12025	10/52	LMSR Derby	-	02/42	68A	12/67	8C
7113	-	-	12026	10/48	LMSR Derby	-	02/42	68A	10/67	8C
7114	-	-	12027	10/51	LMSR Derby	-	03/42	68A	01/67	8C
7115	-	-	12028	12/48	LMSR Derby	-	03/42	68A	06/67	8C
7116	-	-	12029	09/51	LMSR Derby	-	03/42	5B	05/66	8C
7117	-	-	12030	12/48	LMSR Derby	-	03/42	1A	08/64	2F
7118	-	-	12031	01/49	LMSR Derby	-	07/42	1A	12/67	8F
7119	-	-	12032	04/49	LMSR Derby	-	06/42	18A	12/67	8F

7120-7132 See BR 1948 Nos 12033-12045
7400-7408 See LMS Nos 7050-7058

EXPERIMENTAL & TRIALS LOCOMOTIVES

ENGLISH ELECTRIC EXPERIMENTAL 0-6-0 SHUNTERS

Original No.	Revised No.	Date Re No.	Built By	Works No.	Date to Stock for Loan	Depot Allocation When Introduced for Loan	Date Withdrawn From Loan	Depot Allocation When Withdrawn From Loan
D226	D0226	08/59	EE.VF	2345/D226	07/57	30A	10/60	82B
D227	D0227	08/59	EE.VF	2346/D227	07/57	30A	09/59	30A

BRCW TYPE 4 Co-Co

Running No.	Name	Name Date	Built By	Works No.	Date to Stock for Loan	Depot Allocation When Introduced for Loan	Date Withdrawn From Loan	Depot Allocation When Withdrawn From Loan
D0260	Lion	04/62	BRCW	DEL 260	04/62	84A	10/63	34G

BRUSH TYPE 4 Co-Co

Original No.	BR 1957 No.	Date Re No.	Name	Name Date	Works No.	Date to Stock for Loan	Depot Allocation When Introduced for Loan	Date Sold to BR	Date Withdrawn	Depot of of Final Allocation
D0280	D1200	12/70	Falcon	10/61	280	10/61	34G	12/70	10/75	EJ

LMS No.	Disposal Code	Disposal Detail	Date Cut Up	Notes
7086	C	Slag Reduction Co, Ickles	09/68	
7087	C	Slag Reduction Co, Ickles	09/68	
7088	C	BR Derby Works	05/66	
7089	C	W Hatton, Bolton	07/68	
7090	C	Slag Reduction, Ickles	08/68	
7091	C	Slag Reduction, Ickles	04/68	
7092	C	Slag Reduction, Ickles	04/68	
7093	C	Slag Reduction, Ickles	10/68	
7094	C	Slag Reduction, Ickles	04/68	
7095	C	Slag Reduction, Ickles	04/68	
7096	C	Slag Reduction, Ickles	04/68	
7097	C	Slag Reduction, Ickles	08/68	
7098	C	Slag Reduction, Ickles	04/68	
7099	C	J Cashmore, Great Bridge	08/67	
7100	E	WD in Egypt	-	
7101	E	WD in Egypt, later sold to Egyptian Rly	-	
7102	E	WD in Egypt	-	
7103	F	WD in Egypt, later sold to Egyptian Rly	-	
7104	E	WD in Egypt, later sold to Egyptian Rly	-	
7105	E	WD in Egypt, later sold to Italian Rly	-	Returned to LMS: 01/42-03/42
7106	E	WD in N/Africa, later sold to Italian Rly	-	Returned to LMS: 01/42-03/42
7107	E	WD in Egypt, later sold to Egyptian Rly	-	
7108	E	WD in Egypt, later sold to Egyptian Rly	-	
7109	E	WD in N/Africa, later sold to Italian Rly	-	Returned to LMS: 10/42-12/42
7110	C	BR Bolton, by W Hatton	07/68	WD use: 01/42, To LMS: 05/42
7111	C	J Cashmore, Great Bridge	03/68	WD use: 01/42, To LMS: 03/42
7112	C	Slag Reduction, Ickles	10/68	WD use: 02/42, To LMS: 03/42
7113	C	Slag Reduction, Ickles	04/68	WD use: 02/42, To LMS: 05/42
7114	C	Slag Reduction, Ickles	04/68	WD use: 02/42, To LMS: 05/42
7115	C	C F Booth, Rotherham	04/68	WD use: 03/42, To LMS: 05/42
7116	C	BR Derby Works	07/66	
7117	C	BR Derby Works	09/64	Carried No. M7117 02/48-12/48
7118	C	J Cashmore, Great Bridge	07/68	
7119	C	J Cashmore, Great Bridge	07/68	

Original No.	Disposal Code	Disposal Detail	Date Cut Up	Notes
D226	P	Keighley & Worth Valley Railway	-	After loan, returned to EE.VF and stored
D227	C	RSH, Darlington	07/64	

Running No.	Disposal Code	Disposal Detail	Date Cut Up	Notes
D0260	C	BRCW, Smethwick	11/63	Built as Type 4 demonstrator

Original No.	Disposal Code	Disposal Detail	Date Cut Up	Notes
D0280	C	J Cashmore, Newport	03/76	Withdrawn: 05/74, R/I: 06/74, Stored: [U] 06/75

BRUSH DEMONSTRATOR 0-4-0

Allocated No.	Built By	Works No.	Date Introduced on Loan	Depot Allocation When Introduced for Loan	Date Returned to Builder	Depot Allocation at End of Loan	Disposal Code
D9998	Brush	-	09/61	Gloucester	12/62	Gloucester	S

EE TYPE 4 PROTOTYPE Co-Co

Running No.	Built By	Works No.	Date Introduced on Loan	Depot Allocation When Introduced for Loan	Date Withdrawn From Loan	Depot of Final Allocation	Disposal Code
DP2	EE.VF	3205/D722	05/62	1B	09/67	34G	C

GAS TURBINE PROTOTYPE 4-6-0

Running No.	Built By	Works No.	Date Introduced on Loan	Depot of First Allocation	Date Withdrawn	Depot of Final Allocation	Disposal Code
GT3	EE.VF	228/VF11657	05/61	-	10/62	-	C

BRUSH TYPE 5 PROTOTYPE Co-Co

Running No.	Name	Name Date	Built By	Works No.	Date Introduced on Loan	Depot of First Allocation	Date Withdrawn	Depot of Final Allocation	Disposal Code
HS4000	Kestrel	01/68	Brush	711	01/68	41A	01/71	41A	E

EE TYPE 5 PROTOTYPE Co-Co

Name	Name Date	Built By	Works No.	Date Introduced on Loan	Depot of First Allocation	Date Withdrawn	Depot of Final Allocation	Disposal Code
DELTIC	09/55	EE.DK	2007	10/55	8C	03/61	34B	P

YORKSHIRE ENGINE Co PROTOTYPE SHUNTERS

Name	Name Date	Built By	Works No.	Date Introduced on Loan	Depot of First Allocation	Date Withdrawn	Depot of Final Allocation	Disposal Code
TAURUS	03/61	YEC	2875	03/61	82B	03/64	30A	C
JANUS	06/56	YEC	2595	06/56	-	08/56	-	S

BRITISH RAILWAYS 1948 NUMBERS - DIESEL

LMS/BR PROTOTYPE Co-Co

BR 1948 No.	Built By	Works No.	Date Introduced	Depot of First Allocation	Date Withdrawn	Depot of Final Allocation	Disposal Code
10000	LMSR Derby	-	12/47	1B	12/63	1A	C
10001	BR Derby	-	07/48	1B	03/66	1A	C

LMR/FELL PROTOTYPE

BR 1948 No.	Built By	Works No.	Date Introduced	Depot of First Allocation	Date Withdrawn	Depot of Final Allocation	Disposal Code
10100	BR Derby	-	02/52	17A	11/58	17A	C

Allocated No.	Disposal Detail	Date Cut Up	Notes
D9998	Brush, Loughborough	-	Built as Brush demonstrator locomotive

Running No.	Disposal Detail	Date Cut Up	Notes
DP2	English Electric, Vulcan Foundry	10/68	Built as demonstrator, Stored: [U] 07/67

Running No.	Disposal Detail	Date Cut Up	Notes
GT3	T W Ward, Salford	02/66	

Running No.	Disposal Detail	Date Cut Up	Notes
HS4000	Sold to Soviet Railways	03/89	Used for research in USSR until 1988

Name	Disposal Detail	Date Cut Up	Notes
DELTIC	Science Museum, London	-	Built as demonstrator. Allocated No. DP1 (Not carried)

Name	Disposal Detail	Date Cut Up	Notes
TAURUS	Yorkshire Engine Co	1965	Built as demonstrator
JANUS	BSC Scunthorpe	-	Built as demonstrator

BR 1948 No.	Disposal Detail	Date Cut Up	Notes
10000	J Cashmore, Great Bridge	01/68	Stored: [U] 12/62
10001	Cox & Danks, Acton	02/68	

BR 1948 No.	Disposal Detail	Date Cut Up	Notes
10100	BR Derby Works	02/60	Completed 12/50 as private venture, sold to BR 05/55

SR/BR PROTOTYPE 1Co-Co1

BR 1948 No.	Built By	Works No.	Date Introduced	Depot of First Allocation	Date Withdrawn	Depot of Final Allocation	Disposal Code
10201	BR Ashford	-	11/50	73A	12/63	1A	C
10202	BR Ashford	-	08/51	17A	12/63	1A	C
10203	BR Brighton	-	03/54	70A	12/63	1A	C

NBL PROTOTYPE Bo-Bo

BR 1948 No.	Built By	Works No.	Date Introduced	Depot of First Allocation	Date Withdrawn	Depot of Final Allocation	Disposal Code
10800	NBL	26413	07/50	1A	08/59	1F	C

BR (BULLEID) 0-6-0 SHUNTER

BR 1948 No.	Built By	Works No.	Date Introduced	Depot of First Allocation	Date Withdrawn	Depot of Final Allocation	Disposal Code
11001	BR Ashford	-	05/50	75C	08/59	75C	C

HIBBERD 0-4-0 SHUNTER

BR 1948 No.	Built By	Works No.	Date Introduced	Depot of First Allocation	Date Withdrawn	Depot of Final Allocation	Disposal Code
11104	Hibberd	Planet 3466	05/50	W.Hartlepool	04/53	W.Hartlepool	D

11100-11115 See BR 1957 Nos D2200-D2214
11116-11120 See BR 1957 Nos D2500-D2504
11121-11135 See BR 1957 Nos D2215-D2229
11136-11143 See BR 1957 Nos D2550-D2557
11144-11148 See BR 1957 Nos D2505-D2509
11149-11160 See BR 1957 Nos D2230-D2241
11161-11176 See BR 1957 Nos D2558-D2573
11177-11186 See BR 1957 Nos D2400-D2409
11187-11211 See BR 1957 Nos D2000-D2024

11212-11229 See BR 1957 Nos D2242-D2259
11500-11502 See BR 1957 Nos D2950-D2952
11503-11506 See BR 1957 Nos D2953-D2956
11507-11508 See BR 1957 Nos D2957-D2958
11700-11707 See BR 1957 Nos D2700-D2707
11708-11719 See BR 1957 Nos D2708-D2719

12000-12002 See LMS Nos 7074, 7076, 7079
12003-12032 See LMS Nos 7080-7119

LMS/BR 0-6-0 SHUNTERS CLASS 11

BR 1948 No.	Date Re No.	LMS No.	Built By	Works No.	Date Introduced	Depot of First Allocation	Date Withdrawn	Depot of Final Allocation
12033	07/48	7120	LMSR Derby	-	04/45	18A	01/69	9A
12034	12/48	7121	LMSR Derby	-	05/45	18A	10/68	8J
12035	06/50	7122	LMSR Derby	-	06/45	5B	10/68	8J
12036	05/49	7123	LMSR Derby	-	08/45	5B	10/68	6A
12037	09/48	7124	LMSR Derby	-	10/45	5B	10/68	6A
12038	04/51	7125	LMSR Derby	-	12/45	5B	01/69	8F
12039	01/52	7126	LMSR Derby	-	08/47	21A	10/68	5A
12040	03/52	7127	LMSR Derby	-	09/47	21A	10/68	5A
12041	03/52	7128	LMSR Derby	-	10/47	21A	10/68	2E
12042	09/53	7129	LMSR Derby	-	11/47	21A	10/68	2E
12043	06/48	7130	BR Derby	-	02/48	21A	10/68	2E
12044	02/51	7131	BR Derby	-	03/48	21A	11/68	2E
12045	-	7132*	BR Derby	-	04/48	18A	01/69	1E
12046	-	7133*	BR Derby	-	05/48	18A	01/69	1E
12047	-	7134*	BR Derby	-	08/48	18A	01/69	5A
12048	-	7135*	BR Derby	-	12/48	18A	01/69	6A
12049	-	7136*	BR Derby	-	12/49	5B	10/71	1E
12050	-	7137*	BR Derby	-	01/49	5B	07/70	9A
12051	-	7138*	BR Derby	-	02/49	5B	10/71	8F
12052	-	7139*	BR Derby	-	03/49	5B	06/71	5A
12053	-	7140*	BR Derby	-	04/49	5B	04/71	1A
12054	-	7141*	BR Derby	-	05/49	5B	07/70	6A
12055	-	7142*	BR Derby	-	06/49	5B	06/71	5A
12056	-	7143*	BR Derby	-	07/49	18A	09/71	2F
12057	-	7144*	BR Derby	-	08/49	18A	01/69	8J
12058	-	7145*	BR Derby	-	10/49	18A	04/71	1A
12059	-	7146*	BR Derby	-	10/49	21A	01/69	2E
12060	-	7147*	BR Derby	-	11/49	21A	01/71	9A
12061	-	7148*	BR Derby	-	11/49	21A	10/71	8J

BR 1948 No.	Disposal Detail	Date Cut Up	Notes
10201	J Cashmore, Great Bridge	01/68	Stored: [U] 10/62
10202	J Cashmore, Great Bridge	01/68	Stored: [U] 01/63
10203	J Cashmore, Great Bridge	01/68	Stored: [U] 10/62

BR 1948 No.	Disposal Detail	Date Cut Up	Notes
10800	Brush, Loughborough	05/76	Locomotive used by Brush for development of ac traction equipment, named 'Hawk'

BR 1948 No.	Disposal Detail	Date Cut Up	Notes
11001	BR Ashford Works	12/59	Stored: [U] 01/59

BR 1948 No.	Disposal Detail	Date Cut Up	Notes
11104	To Departmental Stock - No. 52		Built for Departmental service

BR 1948 No.	Disposal Code	Disposal Detail	Date Cut Up	Notes
12033	C	J Cashmore, Great Bridge	08/70	
12034	C	J McWilliams, Shettleston	09/69	
12035	C	J McWilliams, Shettleston	09/69	
12036	C	J Cashmore, Great Bridge	02/70	
12037	C	J Cashmore, Great Bridge	07/70	
12038	C	G Cohen, Kettering	05/71	
12039	C	BR Bescot, by J Cashmore	12/69	
12040	C	J Cashmore, Great Bridge	02/70	
12041	C	BR Swindon Works	06/69	
12042	C	J Cashmore, Newport	11/69	
12043	C	J Cashmore, Great Bridge	11/69	Carried No. M7130 until 06/48
12044	C	J Cashmore, Great Bridge	11/69	
12045	C	J Cashmore, Great Bridge	07/69	Stored: [U] 02/69
12046	C	BR Bletchley, by G Cohen	01/70	
12047	C	J Cashmore, Great Bridge	12/69	Stored: [U] 12/68
12048	C	J Cashmore, Great Bridge	03/70	
12049	S	Days Roadstone, Brentford	-	
12050	C	NCB Philadelphia	11/71	PI
12051	C	J Cashmore, Great Bridge	07/73	
12052	P	Ayrshire Railway Project	-	PI
12053	C	Birds, Long Marston	11/71	
12054	C	A R Adams, Newport	04/84	PI
12055	C	C F Booth, Rotherham	05/72	
12056	C	G Cohen, Kettering	05/73	
12057	C	J Cashmore, Great Bridge	02/70	
12058	C	J Cashmore, Newport	02/73	
12059	C	J Cashmore, Great Bridge	01/70	
12060	C	NCB Philadelphia	11/85	
12061	P	Gwili Railway	-	PI

BR 1948 No.	Date Re No.	LMS No.	Built By	Works No.	Date Introduced	Depot of First Allocation	Date Withdrawn	Depot of Final Allocation
12062	-	7149*	BR Derby	-	12/49	21A	04/70	5A
12063	-	7150*	BR Derby	-	12/49	14A	01/72	8F
12064	-	7151*	BR Derby	-	12/49	14A	03/69	1A
12065	-	7152*	BR Derby	-	12/49	14A	05/71	8F
12066	-	7153*	BR Derby	-	12/49	14A	03/69	5A
12067	-	7154*	BR Derby	-	02/50	14A	01/69	1A
12068	-	7155*	BR Derby	-	03/50	14A	12/67	1A
12069	-	-	BR Derby	-	07/50	18A	03/71	5A
12070	-	-	BR Derby	-	08/50	18A	10/69	8F
12071	-	-	BR Derby	-	08/50	18A	10/71	8F
12072	-	-	BR Derby	-	08/50	18A	12/68	8F
12073	-	-	BR Derby	-	09/50	18A	11/71	8F
12074	-	-	BR Derby	-	09/50	21A	01/72	6A
12075	-	-	BR Derby	-	09/50	21A	11/71	8F
12076	-	-	BR Derby	-	10/50	21A	12/71	8F
12077	-	-	BR Derby	-	10/50	21A	10/71	8F
12078	-	-	BR Derby	-	10/50	5B	01/71	8F
12079	-	-	BR Derby	-	11/50	8C	08/71	12A
12080	-	-	BR Derby	-	11/50	68A	04/71	12A
12081	-	-	BR Derby	-	11/50	68A	06/70	9A
12082	-	-	BR Derby	-	11/50	68A	10/71	6G
12083	-	-	BR Derby	-	11/50	68A	10/71	12A
12084	-	-	BR Derby	-	12/50	12B	05/71	5A
12085	-	-	BR Derby	-	12/50	12B	05/71	12A
12086	-	-	BR Derby	-	12/50	12B	07/69	12A
12087	-	-	BR Derby	-	12/50	12B	06/71	2F
12088	-	-	BR Derby	-	06/51	3D	05/71	8J
12089	-	-	BR Derby	-	06/51	3D	09/70	1A
12090	-	-	BR Derby	-	06/51	3B	06/71	1E
12091	-	-	BR Derby	-	07/51	3A	06/70	5A
12092	-	-	BR Derby	-	08/51	3A	03/69	5A
12093	-	-	BR Derby	-	08/51	3B	05/71	5A
12094	-	-	BR Derby	-	09/51	3B	10/71	8J
12095	-	-	BR Derby	-	12/51	3D	03/69	9A
12096	-	-	BR Derby	-	12/51	6A	02/69	5A
12097	-	-	BR Derby	-	12/51	16A	05/71	8F
12098	-	-	BR Derby	-	02/52	16A	04/71	9A
12099	-	-	BR Derby	-	03/52	16A	05/71	1E
12100	-	-	BR Derby	-	04/52	16A	03/69	8F
12101	-	-	BR Derby	-	05/52	16A	08/70	1A
12102	-	-	BR Derby	-	06/52	16A	02/71	8F
12103	-	-	BR Darlington	-	03/52	30A	11/71	30A
12104	-	-	BR Darlington	-	04/52	30A	04/67	30A
12105	-	-	BR Darlington	-	04/52	30A	01/71	30A
12106	-	-	BR Darlington	-	04/52	30A	08/70	30A
12107	-	-	BR Darlington	-	05/52	30A	12/67	62A
12108	-	-	BR Darlington	-	05/52	30A	12/71	30A
12109	-	-	BR Darlington	-	06/52	30A	11/72	30A
12110	-	-	BR Darlington	-	06/52	30A	11/72	30A
12111	-	-	BR Darlington	-	07/52	30A	05/71	30A
12112	-	-	BR Darlington	-	07/52	34B	11/69	2E
12113	-	-	BR Darlington	-	07/52	53A	04/71	55D
12114	-	-	BR Darlington	-	07/52	53A	10/70	30A
12115	-	-	BR Darlington	-	08/52	53A	10/70	30A
12116	-	-	BR Darlington	-	08/52	53A	09/69	30A
12117	-	-	BR Darlington	-	08/52	53A	02/69	40B
12118	-	-	BR Darlington	-	09/52	53A	05/71	40B
12119	-	-	BR Darlington	-	09/52	53A	12/68	50B
12120	-	-	BR Darlington	-	09/52	53A	01/69	50B
12121	-	-	BR Darlington	-	09/52	53A	05/71	40B
12122	-	-	BR Darlington	-	09/52	53A	06/71	40B
12123	-	-	BR Darlington	-	10/52	34B	07/67	40B
12124	-	-	BR Darlington	-	10/52	34B	11/68	40B
12125	-	-	BR Darlington	-	10/52	34A	06/69	40B
12126	-	-	BR Darlington	-	10/52	34A	11/68	40B
12127	-	-	BR Darlington	-	10/52	31B	10/72	30A
12128	-	-	BR Darlington	-	10/52	31B	07/70	30A
12129	-	-	BR Darlington	-	11/52	31B	09/67	30A
12130	-	-	BR Darlington	-	11/52	31B	07/72	30A
12131	-	-	BR Darlington	-	11/52	31B	04/69	30A
12132	-	-	BR Darlington	-	11/52	31B	06/72	30A
12133	-	-	BR Darlington	-	12/52	31B	01/69	40B
12134	-	-	BR Darlington	-	12/52	30A	10/72	30A
12135	-	-	BR Darlington	-	12/52	30A	06/59	40B
12136	-	-	BR Darlington	-	12/52	30A	12/71	30A
12137	-	-	BR Darlington	-	12/52	31B	11/68	30A
12138	-	-	BR Darlington	-	12/52	31B	11/68	30A

13000-13366 See BR 1957 Nos D3000-D3366
15000-15003 See LNER Nos 8000-8003

BR 1948 No.	Disposal Code	Disposal Detail	Date Cut Up	Notes
12062	C	BREL Derby	12/71	
12063	C	NSF Nantgarw	11/87	PI
12064	C	J Cashmore, Great Bridge	01/70	
12065	C	C F Booth, Rotherham	08/72	
12066	C	J Cashmore, Great Bridge	02/70	
12067	C	G Cohen, Kettering	12/69	
12068	C	G Cohen, Kettering	06/68	
12069	C	Birds, Long Marston	03/72	
12070	C	C F Booth, Rotherham	08/70	
12071	P	South Yorkshire Railway	-	
12072	C	G Cohen, Kettering	12/69	
12073	C	C F Booth, Rotherham	06/72	
12074	P	South Yorkshire Railway	-	PI
12075	C	G Cohen, Kettering	05/73	
12076	C	G Cohen, Kettering	07/73	
12077	P	Midland Railway Centre, Butterley	-	
12078	C	G Cohen, Kettering	11/71	
12079	C	C F Booth, Rotherham	08/72	
12080	C	BREL Doncaster	03/72	
12081	C	C F Booth, Rotherham	05/72	
12082	P	South Yorkshire Railway	-	PI
12083	S	Tilcon, Grassington	-	
12084	C	NCB Philadelphia	11/85	PI
12085	C	T W Ward, Barrow	03/75	
12086	C	J Cashmore, Great Bridge	02/70	
12087	C	G Cohen, Kettering	05/73	
12088	P	South Yorkshire Railway	-	PI
12089	C	BREL Derby	12/71	
12090	C	C F Booth, Rotherham	06/72	
12091	C	BREL Derby	12/71	
12092	C	J Cashmore, Great Bridge	01/70	
12093	P	Ayrshire Railway Project	-	PI
12094	C	C F Booth, Rotherham	08/72	
12095	C	J Cashmore, Great Bridge	08/70	
12096	C	G Cohen, Kettering	12/69	
12097	C	BREL Doncaster	03/72	
12098	P	Stephenson Railway Museum	-	PI
12099	P	Severn Valley Railway	-	PI
12100	C	J Cashmore, Great Bridge	01/70	
12101	C	G Cohen, Kettering	08/71	
12102	C	G Cohen, Kettering	10/71	
12103	C	J Cashmore, Newport	03/73	Stored: [U] 05/69, R/I: 01/72
12104	C	BR Stratford, by G Cohen	10/67	
12105	C	BR Stratford Diesel Repair Shop	06/72	Stored: [U] 05/69
12106	C	C F Booth, Rotherham	01/71	
12107	C	J McWilliams, Shettleston	09/68	
12108	C	T W Ward, Beighton	12/72	
12109	C	Marple and Gillott, Attercliffe	08/73	
12110	C	Marple and Gillott, Attercliffe	08/73	Stored: [U] 11/68, R/I: 12/68
12111	C	BR Stratford Diesel Repair Shop	09/72	Stored: [U] 12/68, R/I: 01/69
12112	C	C F Booth, Rotherham	00/70	
12113	C	G Cohen, Kettering	08/71	Stored: [U] 11/68, R/I: 12/68
12114	C	BR Stratford Diesel Repair Shop	09/72	
12115	C	BR Stratford Diesel Repair Shop	07/72	Stored: [U] 05/69
12116	C	G Cohen, Kettering	12/69	
12117	C	Arnott Young, Parkgate	10/69	
12118	C	G Cohen, Kettering	09/71	
12119	C	NCB Philadelphia	11/85	PI
12120	C	NCB Philadelphia	03/80	PI
12121	C	C F Booth, Rotherham	04/72	
12122	C	NCB British Oak	07/85	
12123	C	T W Ward, Beighton	12/67	
12124	C	G Cohen, Kettering	07/69	
12125	C	J McWilliams, Shettleston	10/69	
12126	C	Arnott Young, Parkgate	07/69	
12127	C	BREL Doncaster	04/75	Stored: [U] 11/68, R/I: 12/68
12128	C	C F Booth, Rotherham	01/71	Stored: [U] 12/69
12129	C	Steelbreaking & Dismantling Co, Chesterfield	06/68	
12130	C	J Cashmore, Newport	02/73	
12131	P	North Norfolk Railway	-	PI
12132	C	J Cashmore, Newport	03/73	
12133	C	NCB Philadelphia	11/85	PI
12134	C	G Cohen, Kettering	03/73	
12135	C	J McWilliams, Shettleston	10/69	
12136	C	T W Ward, Beighton	12/72	
12137	C	BR Stratford, by J E McMurray	11/69	Stored: [U] 06/68
12138	C	G Cohen, Kettering	10/69	

LNER/BRUSH 0-6-0 SHUNTER

BR 1948 No.	Built By	Works No.	Date Introduced	Depot of First Allocation	Date Withdrawn	Depot of Final Allocation	Disposal Code
15004	Brush	-	04/49	31B	10/62	34E	C

LNER SIMPLEX 0-4-0 SHUNTERS

BR 1948 No.	Built By	Works No.	Date Introduced	Depot of First Allocation	Date Withdrawn	Depot of Final Allocation	Disposal Code
15097	Simplex	2126	1925	W Hartlepool	06/50	W Hartlepool	C
15098	Simplex	1931	1919	Lowerstoft	09/56	Ware	C
15099	Simplex	2037	1921	Kelso	11/56	Ware	C

15100 See GWR No 2

ENGLISH ELECTRIC 0-6-0 SHUNTERS

BR 1948 No.	Built By	Works No.	Date Introduced	Depot of First Allocation	Date Withdrawn	Depot of Final Allocation	Disposal Code
15101	BR Swindon	-	04/48	81A	08/67	2F	C
15102	BR Swindon	-	05/48	81A	07/67	2F	C
15103	BR Swindon	-	05/48	81A	08/67	2F	C
15104	BR Swindon	-	05/48	81A	07/67	2F	C
15105	BR Swindon	-	07/48	81A	08/67	2F	C
15106	BR Swindon	-	07/48	81A	08/67	2F	C
15107	BR Swindon	-	11/49	82B	06/58	82B	C

15201-15203 See SR Nos 1-3

ENGLISH ELECTRIC 0-6-0 SHUNTERS CLASS 12

BR 1948 No.	Built By	Works No.	Date Introduced	Depot of First Allocation	Date Withdrawn	Depot of Final Allocation	Disposal Code
15211	BR Ashford	-	04/49	75C	12/71	75C	C
15212	BR Ashford	-	04/49	75C	12/71	73C	C
15213	BR Ashford	-	04/49	75C	11/68	73C	C
15214	BR Ashford	-	05/49	75C	10/71	73F	C
15215	BR Ashford	-	05/49	75C	03/68	73C	C
15216	BR Ashford	-	06/49	75C	03/69	73F	C
15217	BR Ashford	-	06/49	75C	08/70	73C	C
15218	BR Ashford	-	06/49	73C	01/70	73C	C
15219	BR Ashford	-	07/49	73C	10/71	73F	C
15220	BR Ashford	-	07/49	73C	10/71	73C	C
15221	BR Ashford	-	08/49	73C	10/71	73C	C
15222	BR Ashford	-	09/49	73C	10/71	73C	C
15223	BR Ashford	-	10/49	73C	06/69	73F	C
15224	BR Ashford	-	10/49	73C	10/71	75C	P
15225	BR Ashford	-	11/49	73C	10/71	75C	C
15226	BR Ashford	-	12/50	73C	04/69	73F	C
15227	BR Ashford	-	02/51	73C	04/70	73F	C
15228	BR Ashford	-	04/51	73C	02/69	73F	C
15229	BR Ashford	-	07/51	73C	10/71	73C	C
15230	BR Ashford	-	09/51	73C	10/71	73F	C
15231	BR Ashford	-	10/51	73C	10/71	73F	C
15232	BR Ashford	-	11/51	73C	04/71	73F	C
15233	BR Ashford	-	11/51	73C	05/69	73F	C
15234	BR Ashford	-	12/51	73C	12/68	70D	C
15235	BR Ashford	-	12/51	73C	05/71	73C	C
15236	BR Ashford	-	01/52	73C	12/68	70D	C

BROWN-BOVERI GAS-TURBINE A1A-A1A

BR 1948 No.	Built By	Works No.	Date Introduced	Depot of First Allocation	Date Withdrawn	Depot of Final Allocation	Disposal Code
18000	Brown-Boveri	BB4559	05/50	81A	12/60	81A	E

BR 1948 No.	Disposal Detail	Date Cut Up	Notes
15004	BR Doncaster Works	05/63	Built as Brush Demonstrator 09/47, allocated No. 8004 never carried

BR 1948 No.	Disposal Detail	Date Cut Up	Notes
15097			
15098			Ex GER locomotive, carried No. 8430 from 06/25, 7591* from 03/42, 8188 from 06/46, 68188* from 01/48 and the BR No. 02/49
15099			Ex NBR locomotive, carried No. 8431 from 07/30, 7592* from 03/42, 8189 from 06/46, 68189 from 01/48 and the BR No. 02/49

BR 1948 No.	Disposal Detail	Date Cut Up	Notes
15101	G Cohen, Kettering	10/69	Allocated No. 502, never carried
15102	Steelbreaking & Dismantling Co, Chesterfield	09/68	
15103	Steelbreaking & Dismantling Co, Chesterfield	09/68	
15104	Steelbreaking & Dismantling Co, Chesterfield	09/68	
15105	G Cohen, Kettering	10/69	
15106	G Cohen, Kettering	09/69	
15107	BR Swindon Works	09/58	

BR 1948 No.	Disposal Detail	Date Cut Up	Notes
15211	J Cashmore, Newport	12/72	
15212	J Cashmore, Newport	12/72	
15213	BR Hither Green	03/70	
15214	J Cashmore, Newport	06/72	Stored: [U] 07/69, R/I: 01/71
15215	BR Swindon Works	08/68	
15216	BREL Swindon	02/70	
15217	BR Selhurst	07/71	
15218	C F Booth, Rotherham	08/70	
15219	J Cashmore, Newport	05/72	
15220	J Cashmore, Newport	11/72	
15221	J Cashmore, Newport	05/72	
15222	J Williams, Bloanyfan Quarry, Kidwelly	09/78	PI
15223	C F Booth, Rotherham	07/70	
15224	Lavender Line, Isfield	-	PI
15225	J Cashmore, Newport	04/72	
15226	J Cashmore, Newport	11/69	
15227	BREL Eastleigh	06/70	
15228	BR Swindon Works	07/69	
15229	J Cashmore, Newport	05/72	
15230	J Cashmore, Newport	11/72	Stored: [U] 11/68, R/I: 01/69
15231	Tilcon, Grassington	02/84	PI
15232	BREL Swindon	02/72	
15233	J Cashmore, Newport	11/69	
15234	BREL Swindon	02/70	
15235	J Cashmore, Newport	11/72	
15236	BREL Swindon	02/70	

BR 1948 No.	Disposal Detail	Date Cut Up	Notes
18000	UIC, Vienna Arsenal		

METROPOLITAN VICKERS GAS-TURBINE Co-Co

Original BR 1948 No.	First 1957 Electric No.	Date Re No.	Second 1957 Electric No.	Date Re No.	Built By	Works No.	Date Introduced	Depot of First Allocation	Date Withdrawn As G/Turbine	Date Introduced As Electric
18100	E1000	10/58	E2001	10/59	M.Vickers	-	12/51	81A	01/58	10/58

BRITISH RAILWAYS 1957 NUMBERS - DIESEL

BR TYPE 4 1Co-Co1 CLASS 44

BR 1957 No.	TOPS No.	Date Re No.	Name	Name Date	Built By	Works No.	Date Introduced	Depot of First Allocation	Date Withdrawn
D1	44001	02/74	Scafell Pike	07/59	BR Derby	-	04/59	17A	10/76
D2	44002	03/74	Helvellyn	09/59	BR Derby	-	09/59	1B	02/79
D3	44003	03/74	Skiddaw	09/59	BR Derby	-	09/59	1B	07/76
D4	44004	04/74	Great Gable	09/59	BR Derby	-	09/59	1B	11/80
D5	44005	11/73	Cross Fell	10/59	BR Derby	-	10/59	1B	04/78
D6	44006	10/73	Whernside	11/59	BR Derby	-	11/59	1B	01/77
D7	44007	02/74	Ingleborough	11/59	BR Derby	-	11/59	1B	11/80
D8	44008	03/74	Penyghent	12/59	BR Derby	-	12/59	1B	11/80
D9	44009	02/74	Snowdon	12/59	BR Derby	-	12/59	1B	03/79
D10	44010	02/74	Tryfan	02/60	BR Derby	-	02/60	1B	05/77

BR TYPE 4 1Co-Co1 CLASS 45

BR 1957 No.	TOPS No.	Date Re No.	Name	Name Date	Built By	Works No.	Date Introduced	Depot of First Allocation	Date Withdrawn
D11	45122	03/74	-	-	BR Derby	-	10/60	1B	04/87
D12	45011	02/74	-	-	BR Derby	-	11/60	1B	05/81
D13	45001	04/73	-	-	BR Derby	-	11/60	17A	01/86
D14	45015	03/74	-	-	BR Derby	-	11/60	17A	03/86
D15	45018	07/74	-	-	BR Derby	-	12/60	17A	01/81
D16	45016	04/74	-	-	BR Derby	-	12/60	17A	11/85
D17	45024	04/75	-	-	BR Derby	-	12/60	17A	10/80
D18	45121	02/74	-	-	BR Derby	-	12/60	17A	11/87
D19	45025	02/75	-	-	BR Derby	-	12/60	17A	05/81
D20	45013	03/74	-	-	BR Derby	-	02/61	17A	04/87
D21	45026	02/75	-	-	BR Derby	-	03/61	17A	04/86
D22	45132	07/74	-	-	BR Derby	-	04/61	17A	05/87
D23	45017	05/74	-	-	BR Derby	-	04/61	17A	08/85
D24	45027	02/75	-	-	BR Derby	-	04/61	17A	05/81
D25	45021	10/74	-	-	BR Derby	-	04/61	17A	12/80
D26	45020	08/74	-	-	BR Derby	-	04/61	17A	12/85
D27	45028	02/75	-	-	BR Derby	-	04/61	17A	01/81
D28	45124	04/74	-	-	BR Derby	-	05/61	17A	01/88
D29	45002	06/73	-	-	BR Derby	-	05/61	17A	09/84
D30	45029	04/75	-	-	BR Derby	-	05/61	17A	07/87
D31	45030	02/75	-	-	BR Derby	-	06/61	17A	11/80
D32	45126	05/74	-	-	BR Derby	-	06/61	17A	04/87
D33	45019	08/74	-	-	BR Derby	-	06/61	17A	09/85
D34	45119	01/74	-	-	BR Derby	-	07/61	17A	05/87
D35	45117	12/73	-	-	BR Derby	-	07/61	17A	05/86
D36	45031	05/75	-	-	BR Derby	-	07/61	17A	05/81
D37	45009	12/73	-	-	BR Derby	-	07/61	17A	09/86
D38	45032	03/75	-	-	BR Derby	-	07/61	17A	05/81
D39	45033	01/75	-	-	BR Derby	-	07/61	17A	02/88
D40	45133	03/74	-	-	BR Derby	-	07/61	17A	05/87
D41	45147	01/75	-	-	BR Derby	-	07/61	17A	01/85
D42	45034	02/75	-	-	BR Derby	-	08/61	17A	07/87
D43	45107	05/73	-	-	BR Derby	-	08/61	14A	07/88
D44	45035	02/75	-	-	BR Derby	-	09/61	14A	05/81
D45	45036	05/75	-	-	BR Derby	-	09/61	14A	05/86
D46	45037	01/75	-	-	BR Derby	-	10/61	14A	07/88
D47	45116	10/73	-	-	BR Derby	-	09/61	14A	12/86
D48	45038	03/75	-	-	BR Derby	-	10/61	14A	06/85
D49	45039	04/75	The Manchester Regiment	10/65	BR Derby	-	10/61	14A	12/80
D50	45040	01/75	The King's Shropshire Light Infantry	05/65	BR Crewe	-	05/62	17A	07/87
D51	45102	04/73	-	-	BR Crewe	-	06/62	17A	09/86
D52	45123	04/74	The Lancashire Fusilier	10/63	BR Crewe	-	06/62	17A	09/86

Original BR 1948 No.	Depot of Allocation As Electric	Date Withdrawn	Disposal Code	Disposal Detail	Date Cut Up	Notes
18100	ACL	04/68	C	J Cashmore, Great Bridge	11/72	Built as gas-turbine, rebuilt as prototype ac electric

BR 1957 No.	Depot of Final Allocation	Disposal Code	Disposal Detail	Date Cut Up	Notes
D1	TO	C	BREL Derby	02/77	Stored: [S] 01/69, R/I: 02/69
D2	TO	C	BREL Derby	10/79	Stored: [S] 01/69, R/I: 02/69
D3	TO	C	BREL Derby	07/76	Stored: [S] 01/69, R/I: 02/69, Stored: [U] 04/76
D4	TO	P	Midland Railway Centre, Butterley	-	Stored: [S] 01/69, R/I: 02/69
D5	TO	C	BREL Derby	12/78	Stored: [S] 01/69, R/I: 02/69
D6	TO	C	BREL Derby	02/78	Stored: [S] 01/69, R/I: 02/69, Stored: [U] 12/76
D7	TO	C	BREL Derby	11/81	Stored: [S] 01/69, R/I: 02/69
D8	TO	P	PeakRail, Matlock	-	Stored: [S] 01/69, R/I: 02/69
D9	TO	C	BREL Derby	07/80	Stored: [S] 01/69, R/I: 02/69
D10	TO	C	BREL Derby	07/78	Stored: [S] 01/69, R/I: 02/69, Stored: [U] 01/76, R/I: 04/76

BR 1957 No.	Depot of Final Allocation	Disposal Code	Disposal Detail	Date Cut Up	Notes
D11	TI	A	BR March		
D12	TI	C	BREL Derby	09/81	
D13	TO	C	M C Processors, Glasgow	11/88	
D14	TO	A	BR Toton [training compound]		
D15	TI	C	BREL Swindon	10/82	
D16	TO	C	V Berry, Leicester	10/86	
D17	TI	C	BREL Swindon	08/83	Stored: [U] 10/80
D18	TI	A	ABB-Transportation, Derby Litchurch		
D19	TI	C	BREL Derby	11/81	
D20	TI	A	BR March		
D21	TO	C	M C Processors, Glasgow	02/89	
D22	TI	P	Mid Hants Railway	-	
D23	TO	D	To Departmental Stock - ADB968024	-	
D24	TI	C	BREL Swindon	09/83	
D25	TI	C	BREL Swindon	04/83	Stored: [U] 10/80
D26	TO	C	V Berry, Leicester	07/88	
D27	TI	C	BREL Swindon	04/83	
D28	TI	C	M C Processors, Glasgow	02/92	
D29	TO	C	M C Processors, Glasgow	11/88	Stored: [U] 04/84
D30	TI	D	To Departmental Stock - 97410	-	Withdrawn: 01/81, R/I: 01/82, Loaned to CEGB Thorpe Marsh 05/82-07/82, R/I: 08/82
D31	TI	C	BREL Derby	03/81	
D32	TI	C	M C Processors, Glasgow	06/92	
D33	TO	C	V Berry, Leicester	10/86	
D34	TI	A	BR March		
D35	TO	C	V Berry, Leicester	01/87	
D36	TI	C	BREL Derby	10/81	
D37	TO	C	V Berry, Leicester	08/88	
D38	TI	C	BREL Swindon	09/83	
D39	TI	C	M C Processors, Glasgow	02/92	
D40	TI	P	Class 45/1 Group, MRC	-	
D41	TO	C	V Berry, at Patricroft	03/85	
D42	TI	D	To Departmental Stock - 97411	-	
D43	TI	C	M C Processors, Glasgow	03/90	
D44	TI	C	BREL Derby	11/81	
D45	TO	C	V Berry, Leicester	07/88	
D46	TI	C	M C Processors, Glasgow	11/92	Stored: [U] 09/82, R/I: 10/82
D47	TI	C	V Berry, Leicester	06/88	
D48	TO	C	V Berry, Leicester	10/86	
D49	TI	C	BREL Swindon	04/83	Stored: [U] 11/80
D50	TI	D	To Departmental Stock - 97412	-	
D51	TO	C	V Berry, Leicester	06/88	
D52	TO	C	V Berry, Leicester	07/88	

BR 1957 No.	TOPS No.	Date Re No.	Name	Name Date	Built By	Works No.	Date Introduced	Depot of First Allocation	Date Withdrawn
D53	45041	05/75	Royal Tank Regiment	09/64	BR Crewe	-	06/62	17A	05/88
D54	45023	01/75	The Royal Pioneer Corps	11/63	BR Crewe	-	08/62	17A	09/84
D55	45144	12/74	Royal Signals	06/65	BR Crewe	-	09/62	17A	12/87
D56	45137	09/74	The Bedfordshire and Hertfordshire Regiment (TA)	12/62	BR Crewe	-	10/62	17A	06/87
D57	45042	12/74	-	-	BR Crewe	-	06/63	17A	04/85
D58	45043	02/75	The King's Own Royal Border Regiment	05/63	BR Crewe	-	02/62	17A	09/84
D59	45104	04/73	The Royal Warwickshire Fusilier	05/64	BR Crewe	-	02/62	17A	04/88
D60	45022	11/74	Lytham St Annes	05/64	BR Crewe	-	02/62	17A	07/87
D61	45112	08/73	The Royal Army Ordnance Corps	09/65	BR Crewe	-	03/62	17A	05/87
D62	45143	11/74	5th Royal Inniskilling Dragoon Guards 1685-1985	11/64	BR Crewe	-	03/62	17A	05/87
D63	45044	03/75	Royal Inniskilling Fusilier	09/65	BR Crewe	-	03/62	17A	06/87
D64	45045	02/75	Coldstream Guardsman	04/65	BR Crewe	-	04/62	17A	05/83
D65	45111	08/73	Grenadier Guardsman	05/64	BR Crewe	-	04/62	17A	05/87
D66	45146	12/74	-	-	BR Crewe	-	04/62	17A	04/87
D67	45118	12/73	The Royal Artilleryman	09/65	BR Crewe	-	05/62	17A	05/87
D68	45046	02/75	Royal Fusilier	01/67	BR Crewe	-	10/60	5A	08/88
D69	45047	01/75	-	-	BR Crewe	-	10/60	5A	08/80
D70	45048	04/75	The Royal Marines	12/64	BR Crewe	-	11/60	5A	06/85
D71	45049	03/75	The Staffordshire Regiment (The Prince of Wales's)	05/66	BR Crewe	-	11/60	5A	02/88
D72	45050	02/75	-	-	BR Crewe	-	11/60	5A	09/84
D73	45110	07/73	-	-	BR Crewe	-	11/60	5A	07/88
D74	45051	01/75	-	-	BR Crewe	-	11/60	5A	04/87
D75	45052	01/75	-	-	BR Crewe	-	12/60	5A	08/88
D76	45053	02/75	-	-	BR Crewe	-	12/60	5A	11/83
D77	45004	09/73	Royal Irish Fusilier	09/65	BR Crewe	-	12/60	5A	12/85
D78	45150	07/75	-	-	BR Crewe	-	12/60	17A	02/88
D79	45005	11/73	-	-	BR Crewe	-	12/60	17A	03/86
D80	45113	10/73	-	-	BR Crewe	-	12/60	17A	08/88
D81	45115	10/73	-	-	BR Crewe	-	12/60	17A	05/88
D82	45141	11/74	-	-	BR Crewe	-	12/60	17A	08/88
D83	45142	11/74	-	-	BR Crewe	-	12/60	17A	06/87
D84	45055	01/75	Royal Corps of Transport	06/66	BR Crewe	-	12/60	17A	04/85
D85	45109	07/73	-	-	BR Crewe	-	02/61	17A	01/86
D86	45105	04/73	-	-	BR Crewe	-	03/61	17A	05/87
D87	45127	05/74	-	-	BR Crewe	-	02/61	17A	05/87
D88	45136	09/74	-	-	BR Crewe	-	03/61	17A	05/87
D89	45006	11/73	Honourable Artillery Company	06/65	BR Crewe	-	03/61	17A	09/86
D90	45008	11/73	-	-	BR Crewe	-	03/61	17A	12/80
D91	45056	03/75	-	-	BR Crewe	-	03/61	17A	12/85
D92	45138	10/74	-	-	BR Crewe	-	03/61	17A	05/87
D93	45057	01/75	-	-	BR Crewe	-	04/61	17A	01/85
D94	45114	09/73	-	-	BR Crewe	-	04/61	17A	02/87
D95	45054	05/75	-	-	BR Crewe	-	04/61	17A	01/85
D96	45101	03/73	-	-	BR Crewe	-	04/61	17A	11/86
D97	45058	01/75	-	-	BR Crewe	-	04/61	17A	09/87
D98	45059	01/75	Royal Engineer	12/66	BR Crewe	-	04/61	17A	03/86
D99	45135	09/74	3rd Carabinier	12/65	BR Crewe	-	05/61	17A	05/87
D100	45060	01/75	Sherwood Forester	09/61	BR Crewe	-	05/61	17A	12/85
D101	45061	07/75	-	-	BR Crewe	-	05/61	17A	08/81
D102	45140	10/74	-	-	BR Crewe	-	05/61	17A	03/88
D103	45062	01/75	-	-	BR Crewe	-	06/61	17A	07/87
D104	45063	01/75	-	-	BR Crewe	-	06/61	17A	05/86
D105	45064	01/75	-	-	BR Crewe	-	06/61	17A	01/85
D106	45106	05/73	-	-	BR Crewe	-	06/61	17A	02/89
D107	45120	02/74	-	-	BR Crewe	-	06/61	17A	03/87
D108	45012	03/74	-	-	BR Crewe	-	07/61	17A	08/88
D109	45139	10/74	-	-	BR Crewe	-	07/61	17A	04/87
D110	45065	01/75	-	-	BR Crewe	-	07/61	17A	03/85
D111	45129	06/74	-	-	BR Crewe	-	08/61	17A	06/87
D112	45010	12/73	-	-	BR Crewe	-	08/61	17A	03/85
D113	45128	06/74	-	-	BR Crewe	-	08/61	17A	04/89
D114	45066	02/75	-	-	BR Crewe	-	08/61	17A	07/87
D115	45067	01/75	-	-	BR Crewe	-	08/61	17A	07/77
D116	45103	04/73	-	-	BR Crewe	-	09/61	17A	08/88
D117	45130	06/74	-	-	BR Crewe	-	09/61	17A	05/87
D118	45068	06/75	-	-	BR Crewe	-	09/61	17A	01/86

BR 1957 No.	Depot of Final Allocation	Disposal Code	Disposal Detail	Date Cut Up	Notes
D53	TI	A	BR Thornaby		
D54	TO	C	V Berry, Leicester	10/86	Used as exciter at Willington CEGB 09/68-10/68
D55	TI	C	V Berry, Leicester	07/88	
D56	TI	A	BR March		
D57	TO	C	V Berry, Leicester	10/86	
D58	TO	C	V Berry, Leicester	10/86	
D59	TI	C	M C Processors, Glasgow	02/92	
D60	TI	D	To Departmental Stock - 97409	-	
D61	TI	P	Private on East Lancs Railway	-	
D62	TI	A	BR March		Named '5th Royal Inniskilling Dragoon Guards' until 06/85
D63	TI	C	M C Processors, Glasgow	02/89	
D64	TO	C	V Berry, Leicester	10/86	Stored: [U] 02/83
D65	TI	C	M C Processors, Glasgow	07/92	
D66	TI	C	M C Processors, Glasgow	04/92	
D67	TI	P	Northampton Steam Railway	-	
D68	TI	C	M C Processors, Glasgow	02/92	
D69	TI	C	BREL Derby	02/81	
D70	TO	C	M C Processors, Glasgow	01/89	
D71	TI	C	M C Processors, Glasgow	04/89	
D72	TO	C	V Berry, Leicester	01/87	
D73	TI	C	M C Processors, Glasgow	06/90	
D74	TI	C	M C Processors, Glasgow	01/89	
D75	TI	C	M C Processors, Glasgow	09/91	
D76	TO	C	BREL Crewe	11/88	Stored: [U] 11/81, R/I: 03/82
D77	TO	C	M C Processors, Glasgow	11/88	
D78	TI	C	M C Processors, Glasgow	12/91	Numbered 45054 between 01/75-07/75
D79	TO	C	V Berry, Leicester	07/88	
D80	TI	C	M C Processors, Glasgow	04/90	
D81	TI	C	M C Processors, Glasgow	06/90	
D82	TI	C	M C Processors, Glasgow	07/92	
D83	TI	A	BR March		
D84	TO	C	V Berry, Leicester	10/86	
D85	TO	C	V Berry, Leicester	10/86	
D86	TI	A	BR March (Sold to M. C. P)		
D87	TI	A	BR March		
D88	TI	C	M C Processors, Glasgow	04/92	
D89	TO	C	V Berry, Leicester	06/88	
D90	TO	C	BREL Swindon	09/83	Stored: [U] 08/80
D91	TO	C	V Berry, Leicester	10/86	
D92	TI	A	BR March		
D93	TO	C	V Berry, Leicester	01/87	
D94	TI	A	BR March		
D95	TO	C	BR Toton, by V Berry	11/85	
D96	TO	C	V Berry, Leicester	09/88	
D97	TI	A	BR March		
D98	TO	C	V Berry, Leicester	07/88	
D99	TI	P	PeakRail, Matlock	-	
D100	TO	P	PeakRail, Matlock	-	
D101	TO	C	BREL Swindon	04/82	
D102	TI	A	M C Processors, Glasgow		
D103	TI	A	BR March		
D104	TO	C	V Berry, Leicester	07/88	
D105	TO	C	V Berry, Leicester	07/88	
D106	TI	C	Booth-Roe, Rotherham	04/92	Withdrawn: 07/88, R/I: 08/88
D107	TI	A	M C Processors, Glasgow		
D108	TI	C	M C Processors, Glasgow	11/92	
D109	TI	A	BR March		
D110	TO	C	V Berry, Leicester	07/88	
D111	TI	C	V Berry, Leicester	08/88	
D112	TO	C	M C Processors, Glasgow	01/89	
D113	TI	C	M C Processors, Glasgow	07/92	Withdrawn: 08/88, R/I: 02/89
D114	TI	D	To Departmental Stock - 97413	-	
D115	TO	C	BREL Derby	06/80	
D116	TI	C	M C Processors, Glasgow	08/90	
D117	TI	C	M C Processors, Glasgow	06/92	
D118	TO	C	BR Allerton, by V Berry	04/86	

BR 1957 No.	TOPS No.	Date Re No.	Name	Name Date	Built By	Works No.	Date Introduced	Depot of First Allocation	Date Withdrawn
D119	45007	11/73	-	-	BR Crewe	-	09/61	17A	07/88
D120	45108	06/73	-	-	BR Crewe	-	09/61	17A	08/87
D121	45069	01/75	-	-	BR Crewe	-	10/61	17A	07/88
D122	45070	01/75	-	-	BR Crewe	-	10/61	17A	01/87
D123	45125	04/74	-	-	BR Crewe	-	10/61	17A	05/87
D124	45131	06/74	-	-	BR Crewe	-	10/61	17A	09/86
D125	45071	12/75	-	-	BR Crewe	-	11/61	17A	07/81
D126	45134	08/74	-	-	BR Crewe	-	11/61	17A	09/87
D127	45072	01/75	-	-	BR Crewe	-	11/61	17A	04/85
D128	45145	12/74	-	-	BR Crewe	-	11/61	17A	02/88
D129	45073	02/75	-	-	BR Crewe	-	11/61	17A	10/81
D130	45148	01/75	-	-	BR Crewe	-	12/61	17A	02/87
D131	45074	01/75	-	-	BR Crewe	-	12/61	17A	09/85
D132	45075	02/75	-	-	BR Crewe	-	12/61	17A	01/85
D133	45003	06/73	-	-	BR Crewe	-	12/61	17A	12/85
D134	45076	01/75	-	-	BR Crewe	-	12/61	17A	11/86
D135	45149	02/75	-	-	BR Crewe	-	12/61	17A	09/87
D136	45077	01/75	-	-	BR Crewe	-	12/61	17A	05/86
D137	45014	03/74	The Cheshire Regiment	06/66	BR Crewe	-	12/61	17A	03/86

BR TYPE 4 1Co-Co1 CLASS 46

BR 1957 No.	TOPS No.	Date Re No.	Name	Name Date	Built By	Works No.	Date Introduced	Depot of First Allocation	Date First Withdrawn	Date Reinstated
D138	46001	02/74	-	-	BR Derby	-	10/61	17A	12/80	09/81
D139	46002	02/74	-	-	BR Derby	-	11/61	17A	12/80	02/81
D140	46003	02/74	-	-	BR Derby	-	11/61	17A	10/78	-
D141	46004	01/74	-	-	BR Derby	-	11/61	17A	06/83	-
D142	46005	02/74	-	-	BR Derby	-	12/61	17A	12/77	-
D143	46006	02/74	-	-	BR Derby	-	12/61	17A	12/80	11/81
D144	46007	01/74	-	-	BR Derby	-	12/61	17A	12/80	11/81
D145	46008	09/73	-	-	BR Derby	-	12/61	17A	10/81	-
D146	46009	02/74	-	-	BR Derby	-	12/61	17A	10/83	-
D147	46010	04/74	-	-	BR Derby	-	12/61	17A	12/80	11/81
D148	46011	01/74	-	-	BR Derby	-	12/61	17A	11/84	-
D149	46012	01/74	-	-	BR Derby	-	12/61	17A	07/80	-
D150	46013	01/74	-	-	BR Derby	-	12/61	17A	08/80	-
D151	46014	12/73	-	-	BR Derby	-	01/62	17A	05/84	-
D152	46015	05/74	-	-	BR Derby	-	02/62	17A	12/80	-
D153	46016	04/74	-	-	BR Derby	-	01/62	17A	12/83	-
D154	46017	01/74	-	-	BR Derby	-	02/62	17A	12/80	11/81
D155	46018	02/74	-	-	BR Derby	-	02/62	17A	12/80	11/81
D156	46019	03/74	-	-	BR Derby	-	02/62	17A	12/80	-
D157	46020	02/74	-	-	BR Derby	-	03/62	17A	12/80	-
D158	46021	02/74	-	-	BR Derby	-	03/62	17A	12/80	11/81
D159	46022	04/74	-	-	BR Derby	-	03/62	17A	12/80	11/81
D160	46023	03/74	-	-	BR Derby	-	03/62	17A	11/83	-
D161	46024	01/74	-	-	BR Derby	-	03/62	17A	04/78	-
D162	46025	02/74	-	-	BR Derby	-	04/62	17A	12/80	11/81
D163	46026	02/74	Leicestershire & Derbyshire Yeomanry	04/62	BR Derby	-	04/62	17A	11/84	-
D164	46027	01/74	-	-	BR Derby	-	04/62	17A	11/84	-
D165	46028	04/74	-	-	BR Derby	-	05/62	17A	05/84	-
D166	46029	10/73	-	-	BR Derby	-	05/62	52A	01/83	-
D167	46030	02/74	-	-	BR Derby	-	05/62	52A	12/80	-
D168	46031	02/74	-	-	BR Derby	-	05/62	52A	04/83	-
D169	46032	02/74	-	-	BR Derby	-	06/62	52A	10/81	01/82
D170	46033	02/74	-	-	BR Derby	-	06/62	52A	05/81	01/82
D171	46034	02/74	-	-	BR Derby	-	07/62	52A	12/80	-
D172	46035	01/74	-	-	BR Derby	-	07/62	52A	11/84	-
D173	46036	11/73	-	-	BR Derby	-	07/62	52A	05/82	-
D174	46037	02/74	-	-	BR Derby	-	07/62	52A	12/80	12/81
D175	46038	02/74	-	-	BR Derby	-	08/62	52A	12/80	11/81
D176	46039	01/74	-	-	BR Derby	-	08/62	52A	10/83	-
D177	46040	02/74	-	-	BR Derby	-	08/62	52A	12/80	-
D178	46041	01/74	-	-	BR Derby	-	09/62	52A	12/80	-
D179	46042	02/74	-	-	BR Derby	-	09/62	52A	12/80	-
D180	46043	01/74	-	-	BR Derby	-	09/62	52A	12/80	-
D181	46044	01/74	-	-	BR Derby	-	09/62	52A	04/84	-
D182	46045	09/73	-	-	BR Derby	-	09/62	52A	11/84	-
D183	46046	02/74	-	-	BR Derby	-	10/62	52A	05/84	-
D184	46047	01/74	-	-	BR Derby	-	10/62	52A	09/84	-
D185	46048	03/74	-	-	BR Derby	-	10/62	52A	09/81	-
D186	46049	02/74	-	-	BR Derby	-	11/62	52A	12/82	-
D187	46050	02/74	-	-	BR Derby	-	11/62	52A	12/80	11/81
D188	46051	12/73	-	-	BR Derby	-	12/62	52A	12/83	-
D189	46052	02/74	-	-	BR Derby	-	01/63	52A	09/84	-

BR 1957 No.	Depot of Final Allocation	Disposal Code	Disposal Detail	Date Cut Up	Notes
D119	TI	C	M C Processors, Glasgow	11/92	
D120	TI	P	The Railway Age, Crewe	-	
D121	TO	C	V Berry, Leicester	07/88	
D122	TO	C	M C Processors, Glasgow	05/89	
D123	TI	P	HLPG Hull	-	
D124	TO	C	V Berry, Leicester	07/88	
D125	TO	C	BREL Swindon	07/83	
D126	TI	A	M C Processors, Glasgow		
D127	TO	C	V Berry, Leicester	10/86	
D128	TI	C	M C Processors, Glasgow	02/92	Withdrawn: 09/87, R/I: 10/87
D129	TO	C	BREL Derby	11/82	
D130	TI	C	M C Processors, Glasgow	06/92	
D131	TO	C	V Berry, Leicester	10/88	
D132	TO	C	V Berry, Leicester	10/86	
D133	TO	C	V Berry, Leicester	03/87	
D134	TI	A	BR March		
D135	TI	P	Lancastrian Carriage & Wagon Works	-	
D136	TO	C	V Berry, Leicester	09/88	
D137	TO	C	V Berry, at Ashburys	08/86	

BR 1957 No.	Date of Second Withdrawal	Depot of Final Allocation	Disposal Code	Disposal Detail	Date Cut Up	Notes
D138	12/81	GD	C	BREL Swindon	07/72	Stored: [S] 10/80
D139	09/81	GD	C	BREL Swindon	09/84	Stored: [S] 11/80
D140	-	LA	C	BREL Derby	03/80	
D141	-	GD	C	BREL Swindon	02/85	Stored: [S] 10/80, R/I: 10/80
D142	-	LA	C	BREL Derby	03/78	Stored: [U] 10/77
D143	01/82	GD	C	BREL Swindon	07/85	Stored: [S] 10/80, R/I: 10/80
D144	02/82	GD	C	BREL Swindon	06/85	Stored: [U] 11/80
D145	-	GD	C	BREL Swindon	12/82	
D146	-	GD	D	To Departmental Stock - 97401	-	
D147	11/84	GD	A	BR Doncaster		Stored: [U] 10/80
D148	-	GD	C	BREL Swindon	01/86	Stored: [S] 10/80, R/I: 10/80
D149	-	LA	C	BREL Swindon	10/80	
D150	-	LA	C	BREL Swindon	04/85	
D151	-	GD	C	BREL Swindon	02/86	
D152	-	LA	C	BREL Swindon	01/85	Stored: [S] 11/80
D153	-	GD	C	BREL Swindon	09/84	Stored: [S] 10/80, R/I: 10/80
D154	04/84	GD	C	BREL Swindon	06/86	Stored: [S] 10/80
D155	12/83	GD	C	BREL Swindon	04/85	Stored: [S] 10/80
D156	-	LA	C	BREL Swindon	08/83	Stored: [S] 10/80
D157	-	LA	C	BREL Swindon	12/84	Stored: [S] 10/80
D158	01/83	GD	C	BREL Swindon	06/85	Stored: [S] 10/80
D159	03/82	GD	C	BREL Swindon	10/83	Stored: [S] 10/80
D160	-	GD	D	To Departmental Stock - 97402	-	Stored: [S] 10/80, R/I: 10/80
D161	-	LA	C	BREL Derby	10/78	
D162	11/84	GD	C	BREL Doncaster	11/85	Stored: [S] 10/80, Withdrawn: 04/82, R/I: 05/82
D163	-	GD	C	BREL Doncaster	03/85	Stored: [S] 10/80, R/I: 11/80
D164	-	GD	C	V Berry, Leicester	10/86	Stored: [S] 10/80, R/I: 10/80
D165	-	GD	C	BREL Doncaster	01/06	Stored: [S] 10/80, R/I: 10/80
D166	-	GD	C	BREL Swindon	09/86	
D167	-	GD	C	BREL Swindon	11/82	
D168	-	GD	C	BREL Swindon	08/83	
D169	04/84	GD	C	BREL Doncaster	03/85	
D170	06/83	GD	C	BREL Swindon	10/84	
D171	-	GD	C	BREL Swindon	11/82	Stored: [U] 11/80
D172	-	GD	D	To Departmental Stock - 97403	-	
D173	-	GD	C	BREL Swindon	01/83	
D174	06/84	GD	C	BREL Doncaster	01/85	Stored: [U] 11/80
D175	03/82	GD	C	BREL Swindon	11/85	Stored: [U] 11/80
D176	-	GD	C	BREL Swindon	06/85	
D177	-	GD	C	BREL Derby	04/82	
D178	-	GD	C	BREL Swindon	08/83	
D179	-	GD	C	BREL Swindon	12/82	Stored: [U] 11/80
D180	-	GD	C	BREL Swindon	05/84	Stored: [U] 11/80
D181	-	GD	C	BREL Swindon	10/86	
D182	-	GD	D	To Departmental Stock - 97404	-	
D183	-	GD	C	BREL Doncaster	11/85	
D184	-	GD	C	BREL Swindon	01/86	
D185	-	GD	C	BREL Swindon	09/83	
D186	-	GD	C	BREL Swindon	10/85	
D187	10/82	GD	C	BREL Swindon	03/85	Stored: [U] 10/80
D188	-	GD	C	BREL Swindon	04/84	
D189	-	GD	C	BREL Doncaster	03/86	

BR 1957 No.	TOPS No.	Date Re No.	Name	Name Date	Built By	Works No.	Date Introduced	Depot of First Allocation	Date First Withdrawn	Date Reinstated
D190	46053	02/74	-	-	BR Derby	-	01/63	52A	02/81	-
D191	46054	02/74	-	-	BR Derby	-	01/63	52A	12/80	11/81
D192	46055	02/74	-	-	BR Derby	-	01/63	52A	05/81	01/82
D193	46056	11/73	-	-	BR Derby	-	01/63	52A	10/82	-

ENGLISH ELECTRIC TYPE 4 1Co-Co1 CLASS 40

BR 1957 No.	TOPS No.	Date Re No.	Name	Name Date	Built By	Works No.	Date Introduced	Depot of First Allocation	Date Withdrawn
D200	40122	04/74	-	-	EE.VF	2367/D395	03/58	30A	05/88
D201	40001	02/74	-	-	EE.VF	2368/D396	04/58	30A	04/84
D202	40002	02/74	-	-	EE.VF	2369/D397	04/58	30A	05/84
D203	40003	03/74	-	-	EE.VF	2370/D398	05/58	30A	09/82
D204	40004	03/74	-	-	EE.VF	2371/D399	05/58	30A	09/84
D205	40005	02/74	-	-	EE.VF	2372/D400	06/58	30A	01/76
D206	40006	02/74	-	-	EE.VF	2373/D401	07/58	34B	03/83
D207	40007	02/74	-	-	EE.VF	2374/D402	07/58	34B	02/83
D208	40008	02/74	-	-	EE.VF	2375/D403	08/58	34B	11/82
D209	40009	02/74	-	-	EE.VF	2376/D404	09/58	34B	11/84
D210	40010	02/74	Empress of Britain	05/60	EE.VF	2366/D427	05/59	1A	07/81
D211	40011	03/74	Mauretania	09/60	EE.VF	2667/D428	05/59	1A	10/80
D212	40012	02/74	Aureol	09/60	EE.VF	2668/D429	05/59	1A	02/85
D213	40013	03/74	Andania	06/62	EE.VF	2669/D430	06/59	1A	01/85
D214	40014	02/74	Antonia	05/61	EE.VF	2670/D431	06/59	1A	11/81
D215	40015	12/73	Aquitania	05/62	EE.VF	2671/D432	06/59	1A	11/84
D216	40016	11/73	Campania	05/62	EE.VF	2672/D433	06/59	5A	05/81
D217	40017	03/74	Carinthia	05/62	EE.VF	2673/D434	07/59	5A	02/81
D218	40018	03/74	Carmania	07/61	EE.VF	2674/D435	07/59	5A	09/81
D219	40019	01/74	Caronia	06/62	EE.VF	2675/D436	07/59	5A	12/81
D220	40020	02/74	Franconia	02/63	EE.VF	2676/D437	07/59	5A	08/82
D221	40021	02/74	Ivernia	03/61	EE.VF	2677/D438	07/59	5A	07/76
D222	40022	04/74	Laconia	10/62	EE.VF	2678/D439	08/59	5A	03/83
D223	40023	02/74	Lancastria	05/61	EE.VF	2679/D440	08/59	5A	05/81
D224	40024	03/74	Lucania	08/62	EE.VF	2680/D441	08/59	5A	06/84
D225	40025	03/74	Lusitania	03/62	EE.VF	2681/D442	08/59	5A	10/82
D226	40026	02/74	-	-	EE.VF	2682/D443	08/59	5A	08/80
D227	40027	12/73	Parthia	06/62	EE.VF	2683/D444	08/59	9A	04/83
D228	40028	02/74	Samaria	09/62	EE.VF	2684/D445	08/59	9A	10/84
D229	40029	03/74	Saxonia	03/63	EE.VF	2685/D446	09/59	12B	04/84
D230	40030	02/74	Scythia	04/61	EE.VF	2686/D447	09/59	12B	04/83
D231	40031	02/74	Sylvania	05/62	EE.VF	2687/D448	09/59	5A	05/81
D232	40032	03/74	Empress of Canada	03/61	EE.VF	2688/D449	09/59	5A	02/81
D233	40033	03/74	Empress of England	09/61	EE.VF	2689/D450	09/59	5A	08/84
D234	40034	02/74	Accra	05/62	EE.VF	2690/D451	09/59	5A	01/84
D235	40035	03/74	Apapa	05/62	EE.VF	2691/D452	10/59	5A	09/84
D236	40036	02/74	-	-	EE.VF	2692/D453	10/59	5A	01/82
D237	40037	02/74	-	-	EE.VF	2693/D454	10/59	52A	08/81
D238	40038	12/73	-	-	EE.VF	2694/D455	10/59	52A	12/80
D239	40039	02/74	-	-	EE.VF	2695/D456	10/59	52A	01/76
D240	40040	03/74	-	-	EE.VF	2715/D457	10/59	52A	07/80
D241	40041	02/74	-	-	EE.VF	2716/D458	10/59	52A	07/76
D242	40042	02/74	-	-	EE.VF	2717/D459	11/59	52A	12/80
D243	40043	01/74	-	-	EE.VF	2718/D460	11/59	52A	01/76
D244	40044	02/74	-	-	EE.VF	2719/D461	11/59	52A	01/85
D245	40045	02/74	-	-	EE.VF	2720/D462	11/59	52A	08/76
D246	40046	01/74	-	-	EE.VF	2721/D463	11/59	52A	02/83
D247	40047	02/74	-	-	EE.VF	2722/D464	11/59	52A	11/84
D248	40048	02/74	-	-	EE.VF	2723/D465	11/59	52A	10/77
D249	40049	02/74	-	-	EE.VF	2724/D466	11/59	52A	01/83
D250	40050	02/74	-	-	EE.VF	2772/D467	12/59	52A	08/83
D251	40051	10/73	-	-	EE.VF	2773/D468	12/59	50A	01/78
D252	40052	02/74	-	-	EE.VF	2774/D469	12/59	50A	06/83
D253	40053	02/74	-	-	EE.VF	2775/D470	01/60	50A	08/76
D254	40054	02/74	-	-	EE.VF	2776/D471	12/59	50A	12/77
D255	40055	04/74	-	-	EE.VF	2777/D472	01/60	17A	11/82
D256	40056	03/74	-	-	EE.VF	2778/D473	01/60	50A	09/84
D257	40057	02/74	-	-	EE.VF	2779/D474	02/60	50A	07/84
D258	40058	02/74	-	-	EE.VF	2780/D475	02/60	50A	09/84
D259	40059	03/74	-	-	EE.VF	2781/D476	02/60	50A	08/77
D260	40060	02/74	-	-	EE.VF	2782/D497	02/60	64B	01/85
D261	40061	03/74	-	-	EE.VF	2783/D498	02/60	64B	06/83
D262	40062	04/74	-	-	EE.VF	2784/D499	03/60	64B	11/81
D263	40063	04/74	-	-	EE.VF	2785/D500	03/60	64B	04/84
D264	40064	10/73	-	-	EE.VF	2786/D501	03/60	64B	04/82
D265	40065	09/74	-	-	EE.VF	2787/D502	03/60	64B	11/81
D266	40066	10/73	-	-	EE.VF	2788/D503	03/60	64B	04/81
D267	40067	02/74	-	-	EE.VF	2789/D504	03/60	1B	07/81

BR 1957 No.	Date of Second Withdrawal	Depot of Final Allocation	Disposal Code	Disposal Detail	Date Cut Up	Notes
D190	-	GD	C	BREL Derby	07/81	
D191	01/82	GD	C	BREL Swindon	08/83	Stored: [U] 10/80
D192	10/82	GD	C	BREL Swindon	11/84	
D193	-	GD	C	BREL Swindon	10/85	

BR 1957 No.	Depot of Final Allocation	Disposal Code	Disposal Detail	Date Cut Up	Notes
D200	SF	P	National Railway Museum, York	-	Withdrawn: 08/81, R/I: 04/83
D201	KD	C	BREL Swindon	02/87	
D202	KD	C	BREL Doncaster	03/85	
D203	IIM	C	BREL Doncaster	01/84	
D204	LO	C	BREL Crewe	09/86	
D205	HM	C	BREL Crewe	02/77	Stored: [U] 01/76
D206	HM	C	BREL Crewe	08/84	
D207	HM	C	BREL Doncaster	01/84	
D208	LO	C	BREL Crewe	04/88	
D209	LO	C	BREL Doncaster	03/85	
D210	KD	C	BREL Swindon	07/83	
D211	HM	C	BREL Swindon	11/80	
D212	KD	D	To Departmental Stock - 97407	-	
D213	LO	P	South Yorkshire Railway Society	-	Withdrawn: 10/84, R/I: 10/84
D214	LO	C	BREL Swindon	11/83	
D215	LO	C	BREL Swindon	11/86	
D216	KD	C	BREL Swindon	11/83	
D217	KD	C	BREL Swindon	08/81	
D218	SP	C	BREL Crewe	08/83	
D219	SP	C	BREL Doncaster	02/84	
D220	LO	C	BREL Crewe	03/87	
D221	LO	C	BREL Crewe	05/77	Stored: [U] 04/76
D222	KD	C	BREL Doncaster	10/84	
D223	KD	C	BREL Crewe	10/84	
D224	LO	C	BREL Crewe	09/85	
D225	LO	C	BREL Doncaster	05/85	
D226	LO	C	BREL Swindon	09/83	Name 'Media' allocated, never carried. Stored: [S] 01/76, R/I: 04/76
D227	LO	C	BREL Crewe	09/84	
D228	LO	C	BREL Crewe	02/88	
D229	LO	C	BREL Doncaster	11/84	
D230	LO	C	BREL Crewe	05/84	Stored: [U] 03/83
D231	LO	C	BREL Crewe	05/83	
D232	LO	C	BREL Swindon	04/83	
D233	LO	C	BREL Doncaster	03/85	Stored: [U] 07/84
D234	LO	C	BREL Doncaster	03/84	
D235	LO	C	BREL Crewe	06/85	Stored: [U] 06/65, R/I: 07/65
D236	HM	C	BREL Swindon	08/82	
D237	TF	C	BREL Swindon	11/83	
D238	HM	C	BREL Swindon	03/82	
D239	HM	C	BREL Crewe	08/76	Stored: [U] 01/76
D240	HM	C	BREL Doncaster	11/80	
D241	HA	C	BREL Crewe	11/78	Stored: [U] 02/76
D242	LO	C	BREL Derby	03/81	
D243	KD	C	BREL Crewe	04/77	
D244	LO	C	BREL Crewe	02/88	
D245	LO	C	BREL Derby	03/77	Stored: [S] 01/76, R/I: 06/76
D246	LO	C	V Berry, Leicester	06/87	Used after withdrawal as MOD training loco
D247	LO	C	BREL Doncaster	01/86	
D248	HA	C	BREL Doncaster	06/80	Stored: [U] 10/77
D249	LO	C	BREL Crewe	10/85	
D250	LO	C	BREL Doncaster	11/83	
D251	HM	C	BREL Doncaster	12/78	
D252	LO	C	BREL Crewe	10/83	
D253	GD	C	BREL Crewe	11/76	
D254	YK	C	BREL Crewe	09/78	
D255	LO	C	BREL Doncaster	05/83	
D256	KD	C	BREL Doncaster	04/85	
D257	LO	C	BREL Crewe	11/88	
D258	LO	C	BREL Crewe	01/88	Stored: [U] 08/84
D259	YK	C	BREL Doncaster	10/78	
D260	KD	D	To Departmental Stock - 97405	-	
D261	LO	C	BREL Crewe	03/84	
D262	KD	C	BREL Swindon	06/83	
D263	LO	C	V Berry, Leicester	06/87	
D264	LO	C	BREL Crewe	06/83	
D265	KD	C	BREL Crewe	02/85	
D266	HA	C	BREL Swindon	10/81	
D207	IIM	C	BREL Doncaster	04/82	

BR 1957 No.	TOPS No.	Date Re No.	Name	Name Date	Built By	Works Number	Date Introduced	Depot of First Allocation	Date Withdrawn
D268	40068	02/74	-	-	EE.VF	2790/D505	04/60	5A	07/83
D269	40069	02/74	-	-	EE.VF	2791/D506	04/60	1B	09/83
D270	40070	02/74	-	-	EE.VF	2792/D507	04/60	52A	06/81
D271	40071	03/74	-	-	EE.VF	2793/D508	04/60	52A	12/80
D272	40072	02/74	-	-	EE.VF	2794/D509	04/60	52A	08/77
D273	40073	03/74	-	-	EE.VF	2795/D510	05/60	52A	06/83
D274	40074	02/74	-	-	EE.VF	2796/D511	05/60	52A	03/84
D275	40075	02/74	-	-	EE.VF	2797/D512	05/60	50A	12/81
D276	40076	03/74	-	-	EE.VF	2798/D513	05/60	50A	04/83
D277	40077	03/74	-	-	EE.VF	2799/D514	05/60	50A	06/83
D278	40078	02/74	-	-	EE.VF	2800/D515	06/60	52A	08/81
D279	40079	02/74	-	-	EE.VF	2801/D516	06/60	52A	01/85
D280	40080	02/74	-	-	EE.VF	2802/D517	06/60	52A	09/83
D281	40081	02/74	-	-	EE.VF	2803/D518	06/60	50A	02/83
D282	40082	02/74	-	-	EE.VF	2804/D519	06/60	50A	11/84
D283	40083	02/74	-	-	EE.VF	2805/D520	07/60	50A	11/81
D284	40084	02/74	-	-	EE.VF	2806/D521	07/60	50A	05/83
D285	40085	02/74	-	-	EE.VF	2807/D522	07/60	50A	03/84
D286	40086	03/74	-	-	EE.VF	2808/D523	07/60	52A	01/85
D287	40087	04/74	-	-	EE.VF	2809/D524	08/60	12B	08/82
D288	40088	03/74	-	-	EE.VF	2811/D526	08/60	5A	02/82
D289	40089	02/74	-	-	EE.VF	2810/D525	08/60	5A	07/76
D290	40090	02/74	-	-	EE.VF	2812/D527	08/60	5A	11/83
D291	40091	02/74	-	-	EE.VF	2813/D528	09/60	5A	09/84
D292	40092	02/74	-	-	EE.VF	2814/D529	09/60	5A	11/82
D293	40093	11/73	-	-	EE.VF	2815/D530	09/60	5A	12/83
D294	40094	02/74	-	-	EE.VF	2816/D531	09/60	5A	10/82
D295	40095	02/74	-	-	EE.VF	2817/D532	10/60	5A	09/81
D296	40096	02/74	-	-	EE.VF	2818/D533	10/60	5A	12/83
D297	40097	03/74	-	-	EE.VF	2819/D534	10/60	5A	06/83
D298	40098	05/74	-	-	EE.VF	2820/D535	10/60	5A	04/81
D299	40099	05/74	-	-	EE.VF	2821/D536	10/60	5A	10/84
D300	40100	02/74	-	-	EE.VF	2822/D537	11/60	5A	12/80
D301	40101	02/74	-	-	EE.VF	2823/D538	11/60	5A	08/82
D302	40102	02/74	-	-	EE.VF	2824/D539	11/60	5A	01/76
D303	40103	02/74	-	-	EE.VF	2825/D540	11/60	5A	02/82
D304	40104	02/74	-	-	EE.VF	2826/D541	12/60	5A	01/84
D305	40105	03/74	-	-	EE.RSH	2725/8135	10/60	5A	12/80
D306	40106	03/74	-	-	EE.RSH	2726/8136	10/60	5A	04/83
D307	40107	10/73	-	-	EE.RSH	2727/8137	10/60	5A	12/81
D308	40108	04/74	-	-	EE.RSH	2728/8138	11/60	5A	08/80
D309	40109	03/74	-	-	EE.RSH	2729/8139	11/60	5A	12/80
D310	40110	04/74	-	-	EE.RSH	2730/8140	12/60	5A	12/80
D311	40111	05/74	-	-	EE.RSH	2731/8141	12/60	5A	05/81
D312	40112	02/74	-	-	EE.RSH	2732/8142	12/60	5A	12/80
D313	40113	06/74	-	-	EE.RSH	2733/8143	12/60	5A	10/81
D314	40114	04/74	-	-	EE.RSH	2734/8144	12/60	5A	12/80
D315	40115	02/74	-	-	EE.RSH	2850/8145	01/61	5A	03/82
D316	40116	05/74	-	-	EE.RSH	2851/8146	01/61	5A	02/81
D317	40117	11/73	-	-	EE.RSH	2852/8147	02/61	5A	09/81
D318	40118	12/73	-	-	EE.RSH	2853/8148	02/61	5A	02/85
D319	40119	11/73	-	-	EE.RSH	2854/8149	03/61	5A	12/80
D320	40120	02/74	-	-	EE.RSH	2855/8150	03/61	5A	05/81
D321	40121	02/74	-	-	EE.RSH	2856/8151	04/61	5A	03/83
D322	-	-	-	-	EE.RSH	2857/8152	04/61	5A	09/67
D323	40123	02/74	-	-	EE.RSH	2858/8153	05/61	5A	07/80
D324	40124	02/74	-	-	EE.RSH	2859/8154	06/61	5A	01/84
D325	40125	01/74	-	-	EE.VF	3071/D621	12/60	5A	05/81
D326	40126	03/74	-	-	EE.VF	3072/D622	12/60	5A	02/84
D327	40127	04/74	-	-	EE.VF	3073/D623	12/60	5A	02/82
D328	40128	04/74	-	-	EE.VF	3074/D624	01/61	5A	09/82
D329	40129	03/74	-	-	EE.VF	3075/D625	01/61	5A	05/84
D330	40130	02/74	-	-	EE.VF	3076/D626	02/61	5A	03/82
D331	40131	11/73	-	-	EE.VF	3077/D627	02/61	5A	10/83
D332	40132	03/74	-	-	EE.VF	3078/D628	02/61	5A	03/82
D333	40133	02/74	-	-	EE.VF	3079/D629	03/61	5A	01/84
D334	40134	05/74	-	-	EE.VF	3080/D630	03/61	5A	05/81
D335	40135	05/74	-	-	EE.VF	3081/D631	03/61	5A	01/85
D336	40136	02/74	-	-	EE.VF	3082/D632	03/61	5A	05/82
D337	40137	02/74	-	-	EE.VF	3083/D633	03/61	5A	01/81
D338	40138	02/74	-	-	EE.VF	3084/D634	04/61	5A	08/82
D339	40139	02/74	-	-	EE.VF	3085/D635	04/61	5A	02/82
D340	40140	02/74	-	-	EE.VF	3086/D636	04/61	5A	03/82
D341	40141	03/74	-	-	EE.VF	3087/D637	04/61	5A	10/83
D342	40142	05/74	-	-	EE.VF	3088/D638	05/61	5A	04/80
D343	40143	12/73	-	-	EE.VF	3089/D639	05/61	8A	01/85
D344	40144	05/74	-	-	EE.VF	3090/D640	05/61	8A	05/81
D345	40145	01/74	-	-	EE.VF	3091/D641	05/61	55H	06/83
D346	40146	02/74	-	-	EE.VF	3092/D642	05/61	55H	12/80
D347	40147	02/74	-	-	EE.VF	3093/D643	05/61	55H	12/80
D348	40148	02/74	-	-	EE.VF	3094/D644	06/61	55H	08/82
D349	40149	12/73	-	-	EE.VF	3095/D645	06/61	50A	08/81

BR 1957 No.	Depot of Final Allocation	Disposal Code	Disposal Detail	Date Cut Up	Notes
D268	HM	C	BREL Doncaster	10/83	
D269	LO	C	BREL Doncaster	09/84	
D270	GD	C	BREL Doncaster	03/82	
D271	HM	C	BREL Swindon	10/81	
D272	HA	C	BREL Glasgow	06/78	
D273	LO	C	BREL Crewe	09/84	
D274	KD	C	BREL Doncaster	12/84	
D275	GD	C	V Berry, Leicester	02/87	
D276	LO	C	BREL Doncaster	11/83	
D277	LO	C	BREL Doncaster	04/84	
D278	HM	C	BREL Swindon	10/83	
D279	LO	C	BREL Doncaster	02/85	
D280	LO	C	BREL Doncaster	08/84	
D281	LO	C	BREL Doncaster	09/83	
D282	LO	C	BREL Crewe	01/86	
D283	HM	C	BREL Swindon	09/85	
D284	HM	C	BREL Crewe	05/84	
D285	LO	C	BREL Doncaster	01/85	
D286	LO	C	BREL Doncaster	02/85	
D287	LO	C	BREL Doncaster	10/85	
D288	HM	C	BREL Crewe, by A Hampton	12/88	One cab preserved at The Railway Age, Crewe
D289	HA	C	BREL Crewe	10/78	Stored: [U] 02/76
D290	LO	C	BREL Doncaster	11/84	
D291	LO	C	BREL Crewe	12/88	Stored: [U] 08/84
D292	LO	C	BREL Swindon	03/86	
D293	LO	C	BREL Doncaster	05/84	
D294	LO	C	BREL Doncaster	08/85	
D295	LO	C	BREL Swindon	10/83	Stored: [U] 08/78, R/I: 11/78
D296	LO	C	BREL Doncaster	06/84	Frame used at Doncaster Works after body cut
D297	LO	C	BREL Doncaster	03/84	
D298	LO	C	BREL Swindon	04/83	
D299	LO	C	BREL Doncaster	04/85	Stored: [U] 10/80
D300	HM	C	BREL Swindon	06/81	
D301	KD	C	BREL Crewe	11/84	
D302	HM	C	BREL Crewe	01/77	Stored: [U] 01/76
D303	LO	C	BREL Crewe	04/83	
D304	LO	C	BREL Crewe	04/88	Stored: [U] 10/84
D305	KD	C	BREL Swindon	03/81	Stored: [U] 11/80
D306	LO	P	Nene Valley Railway, Peterborough	-	Stored: [S] 06/76, R/I: 08/76, Stored: [U] 03/83
D307	LO	C	BREL Crewe	09/84	Stored: [U] 08/80, R/I: 08/80
D308	LO	C	BREL Swindon	12/80	
D309	KD	C	BREL Swindon	03/84	Stored: [U] 11/80
D310	LO	C	BREL Swindon	09/83	
D311	LO	C	BREL Swindon	02/82	Stored: [U] 11/75, R/I: 01/76, Stored: [U] 04/81
D312	LO	C	BREL Swindon	10/85	
D313	LO	C	BREL Swindon	01/84	
D314	HM	C	BREL Swindon	08/82	Stored: [U] 10/80
D315	LO	C	BREL Crewe, by A Hampton	12/88	
D316	LO	C	BREL Swindon	07/81	
D317	LO	C	BREL Swindon	11/83	
D318	KD	D	To Departmental Stock - 97408	-	Stored: [U] 01/85, R/I: 02/85
D319	LO	C	BREL Swindon	10/82	
D320	LO	C	BREL Swindon	10/83	
D321	LO	C	BREL Crewe	12/83	
D322	W Lines	C	BR Crewe Workshops	09/67	
D323	HA	C	BREL Crewe	02/83	
D324	LO	C	BREL Doncaster	03/84	
D325	SP	C	BREL Swindon	12/83	
D326	KD	C	BREL Doncaster	04/84	
D327	SP	C	BREL Swindon	10/83	
D328	LO	C	BREL Doncaster	07/83	
D329	LO	C	BREL Doncaster	08/84	
D330	SP	C	BREL Swindon	05/83	Stored: [U] 06/65, R/I: 08/65, Stored: [U] 01/82
D331	CD	C	BREL Crewe	10/84	
D332	SP	C	V Berry, Leicester	01/87	
D333	LO	C	BREL Doncaster	03/84	
D334	LO	C	BREL Swindon	11/83	Stored: [U] 05/81
D335	LO	D	To Departmental Stock - 97406	-	
D336	SP	C	BREL Swindon	10/83	
D337	LO	C	BREL Swindon	12/81	Stored: [U] 06/65, R/I: 07/65
D338	KD	C	BREL Crewe	02/84	
D339	LO	C	BREL Crewe, by A Hampton	12/88	
D340	SP	C	BREL Crewe	08/83	
D341	CD	C	BREL Doncaster	08/84	
D342	HA	C	BREL Crewe	09/83	
D343	LO	C	BREL Crewe	09/86	
D344	KD	C	BREL Swindon	09/83	
D345	LO	P	Class 40 Preservation Society, Bury	-	Stored: [U] 05/83
D346	TE	C	BREL Swindon	12/83	Stored: [U] 10/80
D347	TE	C	BREL Swindon	08/83	Stored: [U] 09/80
D348	LO	C	BREL Doncaster	07/85	
D349	HM	C	BREL Swindon	05/86	

BR 1957 No.	TOPS No.	Date Re No.	Name	Name Date	Built By	Works No.	Date Introduced	Depot of First Allocation	Date Withdrawn
D350	40150	02/74	-	-	EE.VF	3096/D646	06/61	50A	01/85
D351	40151	01/74	-	-	EE.VF	3097/D647	06/61	50A	02/81
D352	40152	02/74	-	-	EE.VF	3098/D648	07/61	50A	01/85
D353	40153	02/74	-	-	EE.VF	3099/D649	07/61	50A	09/83
D354	40154	11/73	-	-	EE.VF	3100/D650	07/61	50A	01/82
D355	40155	02/74	-	-	EE.VF	3101/D651	08/61	50A	01/85
D356	40156	12/73	-	-	EE.VF	3102/D652	08/61	50A	07/80
D357	40157	04/74	-	-	EE.VF	3103/D653	08/61	64B	07/83
D358	40158	09/74	-	-	EE.VF	3104/D654	09/61	64B	12/83
D359	40159	10/73	-	-	EE.VF	3105/D655	09/61	64B	03/82
D360	40160	02/74	-	-	EE.VF	3106/D656	09/61	64B	11/84
D361	40161	03/74	-	-	EE.VF	3107/D657	10/61	64B	12/80
D362	40162	09/74	-	-	EE.VF	3108/D658	10/61	64B	12/82
D363	40163	04/74	-	-	EE.VF	3109/D659	10/61	64B	06/82
D364	40164	09/74	-	-	EE.VF	3110/D660	10/61	64B	07/83
D365	40165	04/74	-	-	EE.VF	3111/D661	11/61	64B	07/81
D366	40166	10/73	-	-	EE.VF	3112/D662	11/61	64B	02/82
D367	40167	04/74	-	-	EE.VF	3113/D663	11/61	64B	02/84
D368	40168	03/74	-	-	EE.VF	3114/D664	12/61	64B	11/84
D369	40169	02/74	-	-	EE.VF	3115/D665	12/61	5A	12/83
D370	40170	03/74	-	-	EE.VF	3116/D666	12/61	5A	12/83
D371	40171	02/74	-	-	EE.VF	3117/D667	12/61	5A	12/81
D372	40172	02/74	-	-	EE.VF	3118/D668	01/62	9A	09/83
D373	40173	03/74	-	-	EE.VF	3119/D669	01/62	1B	08/81
D374	40174	02/74	-	-	EE.VF	3120/D670	01/62	1B	05/84
D375	40175	09/73	-	-	EE.VF	3121/D671	02/62	1B	05/81
D376	40176	02/74	-	-	EE.VF	3122/D672	02/62	1B	05/81
D377	40177	03/74	-	-	EE.VF	3123/D673	02/62	1B	07/84
D378	40178	02/74	-	-	EE.VF	3124/D674	02/62	1B	06/81
D379	40179	02/74	-	-	EE.VF	3125/D675	02/62	1B	02/81
D380	40180	03/74	-	-	EE.VF	3126/D676	03/62	1B	06/83
D381	40181	02/74	-	-	EE.VF	3127/D677	03/62	1B	01/85
D382	40182	04/74	-	-	EE.VF	3128/D678	03/62	1B	06/82
D383	40183	03/74	-	-	EE.VF	3129/D679	03/62	1B	06/83
D384	40184	02/74	-	-	EE.VF	3130/D680	03/62	1B	12/82
D385	40185	02/74	-	-	EE.VF	3131/D681	03/62	50A	08/83
D386	40186	03/74	-	-	EE.VF	3132/D682	04/62	50A	12/82
D387	40187	03/74	-	-	EE.VF	3133/D683	04/62	50A	08/82
D388	40188	10/73	-	-	EE.VF	3134/D684	04/62	50A	08/83
D389	40189	05/74	-	-	EE.VF	3135/D685	05/62	50A	01/76
D390	40190	02/74	-	-	EE.VF	3136/D686	05/62	50A	01/76
D391	40191	10/73	-	-	EE.VF	3137/D687	05/62	50A	09/83
D392	40192	10/73	-	-	EE.VF	3138/D688	05/62	52A	01/85
D393	40193	04/74	-	-	EE.VF	3139/D689	06/62	52A	10/81
D394	40194	02/74	-	-	EE.VF	3140/D690	06/62	52A	01/85
D395	40195	03/74	-	-	EE.VF	3141/D691	06/62	52A	06/84
D396	40196	02/74	-	-	EE.VF	3142/D692	07/62	52A	05/84
D397	40197	02/74	-	-	EE.VF	3143/D693	07/62	52A	09/83
D398	40198	02/74	-	-	EE.VF	3144/D694	08/62	52A	01/83
D399	40199	02/74	-	-	EE.VF	3145/D695	09/62	52A	06/82

ENGLISH ELECTRIC TYPE 4 Co-Co CLASS 50

BR 1957 No.	TOPS No.	Date Re No.	Name	Name Date	Built By	Works No.	Date Introduced
D400	50050	02/74	[Fearless]	08/78-03/91	EE.VF	3770/D1141	10/67
D401	50001	02/74	Dreadnought	04/78	EE.VF	3772/D1143	12/67
D402	50002	04/74	Superb	03/78	EE.VF	3771/D1142	12/67
D403	50003	02/74	Temeraire	05/78	EE.VF	3773/D1144	01/68
D404	50004	02/74	St Vincent	05/78	EE.VF	3774/D1145	12/67
D405	50005	08/74	Collingwood	04/78	EE.VF	3775/D1146	01/68
D406	50006	03/74	Neptune	09/79	EE.VF	3776/D1147	04/68
D407	50007	04/74	Sir Edward Elgar	02/84	EE.VF	3777/D1148	03/68
D408	50008	02/74	Thunderer	09/78	EE.VF	3778/D1149	03/68
D409	50009	12/73	Conqueror	05/78	EE.VF	3779/D1150	03/68
D410	50010	03/74	Monarch	03/78	EE.VF	3780/D1151	03/68
D411	50011	02/74	Centurion	08/79	EE.VF	3781/D1152	04/68
D412	50012	02/74	Benbow	04/78	EE.VF	3782/D1153	04/68
D413	50013	05/74	Agincourt	04/78	EE.VF	3783/D1154	04/68
D414	50014	04/74	Warspite	05/78	EE.VF	3784/D1155	05/68
D415	50015	02/74	Valiant	04/78	EE.VF	3785/D1156	04/68
D416	50016	12/73	Barham	04/78	EE.VF	3786/D1157	05/68
D417	50017	02/74	Royal Oak	04/78	EE.VF	3787/D1158	04/68
D418	50018	02/74	Resolution	04/78	EE.VF	3788/D1159	04/68
D419	50019	12/73	Ramillies	04/78	EE.VF	3789/D1160	05/68

BR 1957 No.	Depot of Final Allocation	Disposal Code	Disposal Detail	Date Cut Up	Notes
D350	KD	C	BREL Crewe	03/87	
D351	HA	C	BREL Swindon	10/82	
D352	LO	C	BREL Doncaster	03/85	
D353	LO	C	BREL Crewe	05/84	Stored: [U] 08/83
D354	HM	C	BREL Swindon	11/85	
D355	KD	C	BREL Crewe	01/87	
D356	HM	C	BREL Swindon	12/80	
D357	CD	C	BREL Doncaster	10/83	
D358	KD	C	BREL Doncaster	09/84	Frame used at Doncaster Works after body cut
D359	LO	C	BREL Swindon	02/84	
D360	LO	C	BREL Crewe	02/87	
D361	HA	C	BREL Swindon	06/81	Stored: [U] 02/76, R/I: 02/76
D362	KD	C	BR Millerhill	04/86	
D363	LO	C	V Berry, Leicester	01/87	Stored: [U] 06/82
D364	CD	C	BREL Doncaster	12/83	
D365	HA	C	BREL Doncaster	06/83	
D366	LO	C	BREL Crewe	05/83	
D367	LO	C	BREL Doncaster	05/84	
D368	LO	C	BREL Crewe	07/86	
D369	LO	C	BREL Doncaster	10/84	
D370	LO	C	BREL Doncaster	10/84	
D371	LO	C	BREL Swindon	11/82	
D372	CD	C	BREL Doncaster	02/84	
D373	HA	C	Forth Ports Authority, Inverkeithing	04/85	
D374	LO	C	BREL Doncaster	11/84	
D375	KD	C	BREL Swindon	02/83	
D376	HM	C	BREL Swindon	12/85	
D377	LO	C	BREL Crewe	04/86	
D378	SP	C	BREL Swindon	09/83	
D379	SP	C	BREL Swindon	03/82	
D380	CD	C	BREL Crewe	01/84	
D381	LO	C	BREL Crewe	10/86	
D382	SP	C	BREL Crewe	02/84	
D383	LO	C	BREL Crewe	06/86	
D384	LO	C	BREL Doncaster	12/83	
D385	CD	C	BREL Doncaster	12/83	
D386	CD	C	BREL Doncaster	03/83	
D387	CD	C	BREL Doncaster	06/85	
D388	CD	C	BREL Crewe	07/84	
D389	SP	C	BREL Crewe	04/76	Stored: [U] 11/75
D390	SP	C	BREL Crewe	04/76	Stored: [U] 11/75
D391	CD	C	BREL Crewe	07/74	
D392	KD	C	BREL Doncaster	03/85	
D393	TE	C	BREL Swindon	07/86	
D394	LO	C	BREL Doncaster	02/85	
D395	LO	C	BREL Crewe	12/88	
D396	LO	C	BREL Doncaster	01/85	
D397	CD	C	BREL Doncaster	05/84	Frame used at Doncaster Works after body cut
D398	LO	C	BREL Doncaster	02/84	
D399	HM	C	BREL Doncaster	05/83	

BR 1957 No.	Depot of First Allocation	Date Withdrawn	Depot of Final Allocation	Disposal Code	Disposal Detail	Date Cut Up	Notes
D400	LMWL						Reverted to No. D400 03/91
D401	LMWL	04/91	LA	A	Booth-Roe, Rotherham		Stored: [U] 04/91
D402	LMWL	09/91	LA	P	Devon Diesel Group, Kingswear	-	
D403	LMWL	07/91	LA	C	M C Processors, Glasgow	05/92	
D404	LMWL	06/90	LA	C	Booth-Roe, Rotherham	04/92	
D405	LMWL	11/90	LA	C	BR Old Oak Common, by Coopers	03/91	
D406	LMWL	07/87	LA	C	V Berry, Leicester	03/88	
D407	LMWL						Named: Hercules 04/78-02/84 Withdrawn: 07/91, R/I: 03/92
D408	LMWL	05/92	LA	P	East Lancs Railway, Bury	-	Special duties only: 02/91 Stored: [S] 03/92
D409	LMWL	01/91	LA	C	BR Old Oak Common, by Coopers	02/91	
D410	LMWL	09/88	LA	C	BR Laira, by Coopers	05/92	
D411	LMWL	02/87	LA	C	BREL Crewe	09/92	
D412	LMWL	01/89	LA	C	V Berry, Leicester	05/89	
D413	LMWL	03/88	LA	C	BR Old Oak Common, By V Berry	06/89	
D414	LMWL	12/87	LA	C	V Berry, Leicester	05/89	
D415	LMWL	05/92	LA	P	East Lancs Railway, Bury	-	Special duties only: 02/91 Stored: [S] 03/92
D416	LMWL	08/90	LA	C	Booth-Roe, Rotherham	06/92	
D417	LMWL	09/91	LA	P	The Railway Age, Crewe	-	
D418	LMWL	07/91	LA	C	M C Processors, Glasgow	01/93	
D419	LMWL	09/90	LA	P	Class 50 Assoc, at Eridge	-	

BR 1957 No.	TOPS No.	Date Re No.	Name	Name Date	Built By	Works No.	Date Introduced
D420	50020	02/74	Revenge	07/78	EE.VF	3790/D1161	05/68
D421	50021	11/73	Rodney	07/78	EE.VF	3791/D1162	05/68
D422	50022	03/74	Anson	04/78	EE.VF	3792/D1163	05/68
D423	50023	12/73	Howe	05/78	EE.VF	3793/D1164	06/68
D424	50024	02/74	Vanguard	04/78	EE.VF	3794/D1165	06/68
D425	50025	02/74	Invincible	06/78	EE.VF	3795/D1166	07/68
D426	50026	02/74	Indomitable	03/78	EE.VF	3796/D1167	07/68
D427	50027	01/74	Lion	04/78	EE.VF	3797/D1168	06/68
D428	50028	02/74	Tiger	05/78	EE.VF	3798/D1169	07/68
D429	50029	03/74	Renown	11/78	EE.VF	3799/D1170	07/68
D430	50030	03/74	Repulse	04/78	EE.VF	3800/D1171	07/68
D431	50031	02/74	Hood	06/78	EE.VF	3801/D1172	07/68
D432	50032	02/74	Courageous	07/78	EE.VF	3802/D1173	07/68
D433	50033	04/74	Glorious	06/78	EE.VF	3803/D1174	08/68
D434	50034	03/74	Furious	04/78	EE.VF	3804/D1175	08/68
D435	50035	03/74	Ark Royal	01/78	EE.VF	3805/D1176	08/68
D436	50036	10/73	Victorious	05/78	EE.VF	3806/D1177	09/68
D437	50037	02/74	Illustrious	06/78	EE.VF	3807/D1178	09/68
D438	50038	02/74	Formidable	05/78	EE.VF	3808/D1179	10/68
D439	50039	02/74	Implacable	06/78	EE.VF	3809/D1180	10/68
D440	50040	02/74	Centurion	07/87	EE.VF	3810/D1181	10/68
D441	50041	03/74	Bulwark	05/78	EE.VF	3811/D1182	10/68
D442	50042	12/73	Triumph	10/78	EE.VF	3812/D1183	10/68
D443	50043	02/74	Eagle	06/78	EE.VF	3813/D1184	10/68
D444	50044	02/74	Exeter	04/78	EE.VF	3814/D1185	11/68
D445	50045	03/74	Achilles	04/78	EE.VF	3815/D1186	11/68
D446	50046	02/74	Ajax	10/78	EE.VF	3816/D1187	12/68
D447	50047	02/74	Swiftsure	05/78	EE.VF	3817/D1188	12/68
D448	50048	03/74	Dauntless	03/78	EE.VF	3818/D1189	12/68
D449	50049	02/74	Defiance	05/78	EE.VF	3819/D1190	12/68

NBL TYPE 4 A1A-A1A CLASS 41

BR 1957 No.	Name	Name Date	Built By	Works Number	Date Introduced	Depot of First Allocation	Date Withdrawn	Depot of Final Allocation
D600	Active	01/58	NBL	27660	01/58	82C	12/67	84A
D601	Ark Royal	03/58	NBL	27661	03/58	82C	12/67	84A
D602	Bulldog	11/58	NBL	27662	11/58	83D	12/67	84A
D603	Conquest	11/58	NBL	27663	11/58	83D	12/67	84A
D604	Cossack	01/59	NBL	27664	01/59	83D	12/67	84A

BR TYPE 4 B-B CLASS 42

BR 1957 No.	Name	Name Date	Built By	Works Number	Date Introduced	Depot of First Allocation	Date Withdrawn	Depot of Final Allocation
D800	Sir Brian Robertson	07/58	BR Swindon	-	08/58	82C	10/68	84A
D801	Vanguard	11/58	BR Swindon	-	11/58	83D	08/68	84A
D802	Formidable	12/58	BR Swindon	-	12/58	83D	10/68	84A
D803	Albion	03/59	BR Swindon	-	03/59	83D	01/72	83A
D804	Avenger	04/59	BR Swindon	-	04/59	83D	10/71	84A
D805	Benbow	05/59	BR Swindon	-	05/59	83D	10/72	84A
D806	Cambrian	06/59	BR Swindon	-	06/59	83D	11/72	84A
D807	Caradoc	06/59	BR Swindon	-	06/59	83D	09/72	84A
D808	Centaur	07/59	BR Swindon	-	07/59	83D	10/71	83A
D809	Champion	08/59	BR Swindon	-	08/59	83D	10/71	83A
D810	Cockade	09/59	BR Swindon	-	09/59	83D	12/72	84A
D811	Daring	10/59	BR Swindon	-	10/59	83D	01/72	83A
D812	Royal Naval Reserve 1859-1959	11/59	BR Swindon	-	11/59	83D	12/72	84A
D813	Diadem	12/59	BR Swindon	-	12/59	83D	01/72	83A
D814	Dragon	01/60	BR Swindon	-	01/60	83D	11/72	84A
D815	Druid	01/60	BR Swindon	-	01/60	83D	10/71	83A
D816	Eclipse	02/60	BR Swindon	-	02/60	83D	01/72	83A
D817	Foxhound	03/60	BR Swindon	-	03/60	83D	10/71	83A
D818	Glory	03/60	BR Swindon	-	03/60	83D	11/72	84A
D819	Goliath	04/60	BR Swindon	-	04/60	83D	10/71	83A
D820	Grenville	05/60	BR Swindon	-	05/60	83D	11/72	84A
D821	Greyhound	05/60	BR Swindon	-	05/60	83D	12/72	84A
D822	Hercules	06/60	BR Swindon	-	06/60	83D	10/71	84A
D823	Hermes	07/60	BR Swindon	-	07/60	83D	10/71	84A
D824	Highflyer	07/60	BR Swindon	-	07/60	83D	12/72	84A
D825	Intrepid	08/60	BR Swindon	-	08/60	83D	08/72	83A
D826	Jupiter	09/60	BR Swindon	-	09/60	83D	10/71	84A

BR 1957 No.	Depot of First Allocation	Date Withdrawn	Depot of Final Allocation	Disposal Code	Disposal Detail	Date Cut Up	Notes
D420	LMWL	07/90	LA	C	Booth-Roe, Rotherham	06/92	
D421	LMWL	04/90	LA	P	Gloucestershire Warwickshire Rly	-	
D422	LMWL	09/88	OC	C	V Berry, Leicester	05/89	
D423	LMWL	10/90	LA	A	Booth-Roe, Rotherham		
D424	LMWL	02/91	LA	C	BR Old Oak Common, By Coopers	07/91	
D425	LMWL	08/89	OC	C	BR Old Oak Common, By V Berry	10/89	
D426	D05	12/90	LA	P	Mid-Hants Railway	-	
D427	D05	07/91	LA	P	Mid-Hants Railway	-	
D428	D05	01/91	LA	C	BR Old Oak Common	07/91	
D429	D05	03/91	LA	A	BR Laira		Stored: [U] 03/92
D430	D05	04/92	LA	A	BR Laira		Stored: [U] 03/91
D431	D05	07/91	LA	P	Class 50 Fund, Hastings	-	
D432	D05	10/90	LA	C	BR Old Oak Common	03/91	
D433	D05						
D434	D05	06/90	OC	C	BR Old Oak Common	03/91	
D435	D05	08/90	LA	P	Class 50 Fund, Hastings	-	
D436	D05	04/91	LA	C	Booth-Roe, Rotherham	07/92	
D437	D05	09/91	LA	C	M C Processors, Glasgow	12/92	
D438	D05	09/88	OC	C	BR Old Oak Common, by V Berry	08/89	
D439	D05	06/89	OC	C	BR Old Oak Common, by Coopers	07/91	
D440	D05	08/90	LA	A	Booth-Roe, Rotherham		Named: Leviathan 09/78-07/87 Withdrawn: 11/89, R/I: [U] 11/89
D441	D05	04/90	LA	C	BR Old Oak Common	07/91	
D442	D05	10/90	LA	P	Bodmin & Wenford Railway	-	
D443	D05	02/91	LA	P	D318 Preservation Group, Tyseley	-	
D444	D05	01/91	LA	P	Class 50 Fund, Hastings	-	
D445	D05	12/90	LA	A	Booth-Roe, Rotherham		
D446	D05	03/92	LA	C	M C Processors, Glasgow	06/92	
D447	D05	04/88	LA	C	V Berry, Leicester	07/89	
D448	D05	07/91	LA	C	M C Processors, Glasgow	04/92	
D449	D05	08/91	LA	P	Class 50 Society, at Laira	-	Numbered: 50149 08/87-02/89

BR 1957 No.	Disposal Code	Disposal Detail	Date Cut Up	Notes
D600	C	D Woodham, Barry	03/70	
D601	C	D Woodham, Barry	06/80	
D602	C	J Cashmore, Newport	11/68	
D603	C	J Cashmore, Newport	11/68	
D604	C	J Cashmore, Newport	09/68	

BR 1957 No.	Disposal Code	Disposal Detail	Date Cut Up	Notes
D800	C	J Cashmore, Newport	07/69	Stored: [U] 09/68
D801	C	BREL Swindon	10/70	Stored: [U] 06/68
D802	C	BREL Swindon	11/70	
D803	C	BREL Swindon	10/72	
D804	C	BREL Swindon	03/72	
D805	C	BREL Swindon	05/73	
D806	C	BREL Swindon	04/75	
D807	C	BREL Swindon	11/72	
D808	C	BREL Swindon	02/72	
D809	C	BREL Swindon	10/72	
D810	C	BREL Swindon	09/73	
D811	C	BREL Swindon	10/72	Stored: [U] 11/68
D812	C	BREL Swindon	08/73	
D813	C	BREL Swindon	09/72	
D814	C	BREL Swindon	02/74	Withdrawn: 01/72, R/I: 05/72
D815	C	BREL Swindon	10/72	Stored: [U] 09/71
D816	C	BREL Swindon	09/72	
D817	C	BREL Swindon	03/72	
D818	C	BREL Swindon	11/85	
D819	C	BREL Swindon	03/72	
D820	C	BREL Swindon	08/73	
D821	P	Diesel Traction Group, on SVR	-	
D822	C	BREL Swindon	02/72	
D823	C	BREL Swindon	05/72	
D824	C	BREL Swindon	05/75	
D825	C	BREL Swindon	10/72	Withdrawn: 01/72, R/I: 05/72
D826	C	BREL Swindon	01/72	Used after withdrawal

BR 1957 No.	Name	Name Date	Built By	Works Number	Date Introduced	Depot of First Allocation	Date Withdrawn	Depot of Final Allocation
D827	Kelly	10/60	BR Swindon	-	10/60	83D	01/72	83A
D828	Magnificent	10/60	BR Swindon	-	10/60	83D	07/71	84A
D829	Magpie	11/60	BR Swindon	-	11/60	83D	08/72	83A
D830	Majestic	01/61	BR Swindon	-	01/61	83D	03/69	83A
D831	Monarch	01/61	BR Swindon	-	01/61	83D	10/71	84A
D832	Onslaught	02/61	BR Swindon	-	02/61	83D	12/72	84A

Class 42 continued at D866

NBL TYPE 4 B-B CLASS 43

BR 1957 No.	Name	Name Date	Built By	Works Number	Date Introduced	Depot of First Allocation	Date Withdrawn	Depot of Final Allocation
D833	Panther	07/60	NBL	27962	07/60	83D	10/71	83A
D834	Pathfinder	07/60	NBL	27963	07/60	83D	10/71	83A
D835	Pegasus	08/60	NBL	27964	08/60	83D	10/71	83A
D836	Powerful	09/60	NBL	27965	09/60	83D	05/71	83A
D837	Ramillies	11/60	NBL	27966	11/60	83D	05/71	83A
D838	Rapid	10/60	NBL	27967	09/60	83D	03/71	83A
D839	Relentless	11/60	NBL	27968	11/60	83D	10/71	83A
D840	Resistance	02/61	NBL	27969	02/61	83D	04/69	81A
D841	Roebuck	12/60	NBL	27970	12/60	83D	10/71	83A
D842	Royal Oak	12/60	NBL	27971	12/60	83D	10/71	83A
D843	Sharpshooter	01/61	NBL	27972	01/61	83D	05/71	83A
D844	Spartan	03/61	NBL	27973	03/61	83D	10/71	83A
D845	Sprightly	04/61	NBL	27974	04/61	83D	10/71	83A
D846	Steadfast	04/61	NBL	27975	04/61	83D	05/71	83A
D847	Strongbow	04/61	NBL	27976	04/61	83D	03/71	83A
D848	Sultan	04/61	NBL	27977	04/61	83D	03/69	81A
D849	Superb	05/61	NBL	27978	05/61	83D	05/71	83A
D850	Swift	06/61	NBL	27979	06/61	83D	05/71	83A
D851	Temeraire	07/61	NBL	27980	07/61	83D	05/71	83A
D852	Tenacious	07/61	NBL	27981	07/61	83D	10/71	83A
D853	Thruster	08/61	NBL	27982	08/61	83D	10/71	83A
D854	Tiger	09/61	NBL	27983	09/61	83D	10/71	83A
D855	Triumph	10/61	NBL	27984	10/61	83D	10/71	83A
D856	Trojan	11/61	NBL	27985	11/61	83D	05/71	83A
D857	Undaunted	12/61	NBL	27986	12/61	83D	10/71	83A
D858	Valorous	12/61	NBL	27987	12/61	83D	10/71	83A
D859	Vanquisher	01/62	NBL	27988	01/62	83D	03/71	83A
D860	Victorious	01/62	NBL	27989	01/62	83D	03/71	83A
D861	Vigilant	02/62	NBL	27990	02/62	83D	10/71	83A
D862	Viking	03/62	NBL	27991	02/62	83D	10/71	83A
D863	Warrior	04/62	NBL	27992	04/62	83D	03/69	83A
D864	Zambesi	05/62	NBL	27993	03/62	83D	03/71	83A
D865	Zealous	06/62	NBL	27994	06/62	83D	05/71	83A

BR TYPE 4 B-B CLASS 42

Class continued from D832

BR 1957 No.	Name	Name Date	Built By	Works Number	Date Introduced	Depot of First Allocation	Date Withdrawn	Depot of Final Allocation
D866	Zebra	03/61	BR Swindon	-	03/61	83D	01/72	83A
D867	Zenith	04/61	BR Swindon	-	04/61	83D	10/71	84A
D868	Zephyr	05/61	BR Swindon	-	05/61	83D	10/71	84A
D869	Zest	07/61	BR Swindon	-	07/61	83D	10/71	84A
D870	Zulu	10/61	BR Swindon	-	10/61	83D	08/71	84A

BR TYPE 4 C-C CLASS 52

BR 1957 No.	Name	Name Date	Built By	Works Number	Date Introduced	Depot of First Allocation	Date Withdrawn	Depot of Final Allocation
D1000	Western Enterprise	12/61	BR Swindon	-	12/61	83D	02/74	LA
D1001	Western Pathfinder	02/62	BR Swindon	-	02/62	83D	10/76	LA
D1002	Western Explorer	03/62	BR Swindon	-	03/62	83D	01/74	LA
D1003	Western Pioneer	04/62	BR Swindon	-	04/62	83D	01/75	LA
D1004	Western Crusader	05/62	BR Swindon	-	05/62	83D	08/73	LA
D1005	Western Venturer	06/62	BR Swindon	-	06/62	83D	11/76	LA
D1006	Western Stalwart	07/62	BR Swindon	-	07/62	83D	04/76	LA
D1007	Western Talisman	08/62	BR Swindon	-	08/62	83D	01/74	LA
D1008	Western Harrier	09/62	BR Swindon	-	09/62	83D	10/74	LA
D1009	Western Invader	09/62	BR Swindon	-	09/62	81A	11/76	LA

BR 1957 No.	Disposal Code	Disposal Detail	Date Cut Up	Notes
D827	C	BREL Swindon	10/72	
D828	C	BREL Swindon	03/72	Stored: [U] 07/71
D829	C	BREL Swindon	01/74	Withdrawn: 01/72, R/I: 05/72
D830	C	BREL Swindon	10/71	Stored: [U] 02/69
D831	C	BREL Swindon	06/72	
D832	P	East Lancs Railway	-	Stored: [U] 11/72, After withdrawal used by BR Research, Derby

BR 1957 No.	Disposal Code	Disposal Detail	Date Cut Up	Notes
D833	C	BREL Swindon	02/72	Stored: [U] 03/69, R/I: 09/69, Stored: [U] 10/71
D834	C	BREL Swindon	02/72	
D835	C	BREL Swindon	12/71	
D836	C	BREL Swindon	03/72	Stored: [U] 05/71
D837	C	BREL Swindon	06/72	Stored: [U] 04/71
D838	C	BREL Swindon	07/72	
D839	C	BREL Swindon	08/72	Stored: [S] 03/69, R/I: 04/69
D840	C	BREL Swindon	07/70	Stored: [U] 04/69
D841	C	BREL Swindon	02/72	Stored: [U] 03/69, R/I: 05/69, Stored: [U] 11/69, R/I: 06/70
D842	C	BREL Swindon	03/72	Stored: [U] 03/69, R/I: 07/69
D843	C	BREL Swindon	04/72	Stored: [U] 05/71
D844	C	BREL Swindon	05/72	
D845	C	BREL Swindon	05/72	Stored: [U] 03/69, R/I: 06/69, Stored: [U] 09/71
D846	C	BREL Swindon	12/71	Stored: [U] 05/71
D847	C	BREL Swindon	03/72	
D848	C	BREL Swindon	08/70	Stored: [U] 02/69
D849	C	BREL Swindon	07/72	Stored: [U] 05/71
D850	C	BREL Swindon	03/72	Stored: [U] 03/69, R/I: 05/69, Stored: [U] 05/71
D851	C	BREL Swindon	06/72	Stored: [U] 05/71
D852	C	BREL Swindon	06/72	Stored: [U] 03/69, R/I: 04/69, Stored: [U] 05/69, R/I: 11/69, Stored: [U] 09/71
D853	C	BREL Swindon	05/72	Stored: [U] 07/69, R/I: 06/70
D854	C	BREL Swindon	05/72	Stored: [U] 03/69, R/I: 04/69, Stored: [U] 05/69, R/I: 06/69, Stored: [U] 09/69, R/I: 12/69
D855	C	BREL Swindon	04/72	Stored: [U] 03/69, R/I: 04/69, Stored: [U] 05/69, R/I: 01/70
D856	C	BREL Swindon	01/72	Stored: [U] 05/71
D857	C	BREL Swindon	04/72	Stored: [U] 05/69, R/I: 09/69
D858	C	BREL Swindon	06/72	Stored: [U] 03/69, Withdrawn: 04/69, R/I: 10/69
D859	C	BREL Swindon	06/72	
D860	C	BREL Swindon	12/71	
D861	C	BREL Swindon	07/72	Stored: [U] 03/69, R/I: 07/69
D862	C	BREL Swindon	05/72	Stored: [U] 09/69, R/I: 10/69, Stored: [U] 01/70, R/I: 04/70
D863	C	J Cashmore, Newport	08/69	Stored: [U] 03/69
D864	C	BREL Swindon	11/71	
D865	C	BREL Swindon	06/72	Stored: [U] 04/69, R/I: 09/69, Stored: [U] 05/71

BR 1957 No.	Disposal Code	Disposal Detail	Date Cut Up	Notes
D866	C	BREL Swindon	10/72	
D867	C	BREL Swindon	09/72	Used after withdrawal
D868	C	BREL Swindon	04/72	
D869	C	BREL Swindon	06/72	
D870	C	BREL Swindon	05/72	Stored: [U] 07/71

BR 1957 No.	Disposal Code	Disposal Detail	Date Cut Up	Notes
D1000	C	BREL Swindon	07/74	
D1001	C	BREL Swindon	08/77	
D1002	C	BREL Swindon	06/74	
D1003	C	BREL Swindon	08/77	
D1004	C	BREL Swindon	09/74	
D1005	C	BREL Swindon	06/77	
D1006	C	BREL Swindon	03/77	
D1007	C	BREL Swindon	02/75	
D1008	C	BREL Swindon	10/75	
D1009	C	BREL Swindon	11/78	

BR 1957 No.	Name	Name Date	Built By	Works Number	Date Introduced	Depot of First Allocation	Date Withdrawn	Depot of Final Allocation
D1010	Western Campaigner	10/62	BR Swindon	-	10/62	81A	02/77	LA
D1011	Western Thunderer	10/62	BR Swindon	-	10/62	88A	10/75	LA
D1012	Western Firebrand	11/62	BR Swindon	-	11/62	88A	11/75	LA
D1013	Western Ranger	12/62	BR Swindon	-	12/62	88A	02/77	LA
D1014	Western Leviathan	12/62	BR Swindon	-	12/62	88A	08/74	LA
D1015	Western Champion	01/63	BR Swindon	-	01/63	88A	12/76	LA
D1016	Western Gladiator	02/63	BR Swindon	-	02/63	88A	12/75	LA
D1017	Western Warrior	03/63	BR Swindon	-	03/63	88A	08/73	LA
D1018	Western Buccaneer	04/63	BR Swindon	-	04/63	81A	06/73	LA
D1019	Western Challenger	05/63	BR Swindon	-	05/63	81A	05/73	LA
D1020	Western Hero	05/63	BR Swindon	-	05/63	81A	06/73	LA
D1021	Western Cavalier	06/63	BR Swindon	-	06/63	81A	08/76	LA
D1022	Western Sentinel	07/63	BR Swindon	-	07/63	81A	01/77	LA
D1023	Western Fusilier	09/63	BR Swindon	-	09/63	86A	02/77	LA
D1024	Western Huntsman	10/63	BR Swindon	-	10/63	86A	11/73	LA
D1025	Western Guardsman	11/63	BR Swindon	-	11/63	86A	10/75	LA
D1026	Western Centurion	12/63	BR Swindon	-	12/63	81A	10/75	LA
D1027	Western Lancer	01/64	BR Swindon	-	01/64	84A	12/75	LA
D1028	Western Hussar	02/64	BR Swindon	-	02/64	82A	10/76	LA
D1029	Western Legionnaire	07/64	BR Swindon	-	07/64	82A	11/74	LA
D1030	Western Musketeer	09/63	BR Crewe	-	12/63	81A	04/76	LA
D1031	Western Rifleman	12/63	BR Crewe	-	12/63	81A	02/75	LA
D1032	Western Marksman	12/63	BR Crewe	-	12/63	81A	05/73	LA
D1033	Western Trooper	01/64	BR Crewe	-	01/64	81A	09/76	LA
D1034	Western Dragoon	04/64	BR Crewe	-	04/64	82A	10/75	LA
D1035	Western Yeoman	07/62	BR Crewe	-	07/62	83D	01/75	LA
D1036	Western Emperor	08/62	BR Crewe	-	08/62	83D	11/76	LA
D1037	Western Empress	08/62	BR Crewe	-	08/62	83D	05/76	LA
D1038	Western Sovereign	09/62	BR Crewe	-	09/62	83D	10/73	LA
D1039	Western King	09/62	BR Crewe	-	09/62	83D	07/73	LA
D1040	Western Queen	09/62	BR Crewe	-	09/62	81A	02/76	LA
D1041	Western Prince	10/62	BR Crewe	-	10/62	81A	02/77	LA
D1042	Western Princess	10/62	BR Crewe	-	10/62	81A	07/73	LA
D1043	Western Duke	10/62	BR Crewe	-	10/62	81A	04/76	LA
D1044	Western Duchess	11/62	BR Crewe	-	11/62	88A	02/75	LA
D1045	Western Viscount	11/62	BR Crewe	-	11/62	88A	12/74	LA
D1046	Western Marquis	12/62	BR Crewe	-	12/62	82A	12/75	LA
D1047	Western Lord	02/63	BR Crewe	-	02/63	88A	02/76	LA
D1048	Western Lady	12/62	BR Crewe	-	12/62	88A	02/77	LA
D1049	Western Monarch	12/62	BR Crewe	-	12/62	88A	04/76	LA
D1050	Western Ruler	01/63	BR Crewe	-	01/63	88A	04/75	LA
D1051	Western Ambassador	01/63	BR Crewe	-	01/63	88A	09/76	LA
D1052	Western Viceroy	02/63	BR Crewe	-	02/63	88A	10/75	LA
D1053	Western Patriarch	02/63	BR Crewe	-	02/63	81A	12/76	LA
D1054	Western Governor	03/63	BR Crewe	-	03/63	88A	11/75	LA
D1055	Western Advocate	03/63	BR Crewe	-	03/63	88A	01/76	LA
D1056	Western Sultan	03/63	BR Crewe	-	03/63	88A	12/76	LA
D1057	Western Chieftain	04/63	BR Crewe	-	04/63	83A	05/76	LA
D1058	Western Nobleman	03/63	BR Crewe	-	03/63	82A	01/77	LA
D1059	Western Empire	04/63	BR Crewe	-	04/63	88A	10/75	LA
D1060	Western Dominion	04/63	BR Crewe	-	04/63	81A	11/73	LA
D1061	Western Envoy	04/63	BR Crewe	-	04/63	81A	10/74	LA
D1062	Western Courier	05/63	BR Crewe	-	05/63	81A	08/74	LA
D1063	Western Monitor	05/63	BR Crewe	-	05/63	81A	04/76	LA
D1064	Western Regent	05/63	BR Crewe	-	05/63	88A	12/75	LA
D1065	Western Consort	06/63	BR Crewe	-	06/63	81A	11/76	LA
D1066	Western Prefect	06/63	BR Crewe	-	06/63	88A	11/74	LA
D1067	Western Druid	07/63	BR Crewe	-	07/63	88A	01/76	LA
D1068	Western Reliance	07/63	BR Crewe	-	07/63	81A	10/76	LA
D1069	Western Vanguard	10/63	BR Crewe	-	10/63	86A	10/75	LA
D1070	Western Gauntlet	10/63	BR Crewe	-	10/63	86A	12/75	LA
D1071	Western Renown	11/63	BR Crewe	-	11/63	81A	12/76	LA
D1072	Western Glory	11/63	BR Crewe	-	11/63	81A	11/76	LA
D1073	Western Bulwark	12/63	BR Crewe	-	12/63	81A	08/74	LA

BR/BRUSH TYPE 4 Co-Co CLASS 47/48

BR 1957 No.	TOPS No.	Date Re No.	First TOPS Re No.	Date Re No.	Second TOPS Re No.	Date Re No.	Third TOPS Re No.	Date Re No.	Name	Name Date	Built By	Works No.	Date Introduced
D1100	47298	02/74							Pegasus	07/90	BR Crewe	-	07/66
D1101	47518	03/74									BR Crewe	-	08/66
D1102	47519	02/74									BR Crewe	-	09/66
D1103	47520	03/74									BR Crewe	-	10/66
D1104	47521	02/74									BR Crewe	-	10/66
D1105	47522	03/74							Doncaster Enterprise	10/87	BR Crewe	-	11/66
D1106	47523	02/74									BR Crewe	-	11/66
D1107	47524	02/74									BR Crewe	-	01/67
D1108	47525	02/74									BR Crewe	-	01/67

BR 1957 No.	Disposal Code	Disposal Detail	Date Cut Up	Notes
D1010	P	Foster Yeoman, with DEPG on WSR	-	Carries identification D1035 Western Yeoman
D1011	C	BREL Swindon	01/79	
D1012	C	BREL Swindon	04/79	
D1013	P	Western Loco Assoc, on SVR	-	
D1014	C	BREL Swindon	02/75	
D1015	P	Diesel Traction Group, at OC	-	
D1016	C	BREL Swindon	08/77	
D1017	C	BREL Swindon	03/75	
D1018	C	BREL Swindon	03/74	
D1019	C	BREL Swindon	10/74	
D1020	C	BREL Swindon	04/74	
D1021	C	BREL Swindon	03/79	
D1022	C	BREL Swindon	12/78	
D1023	P	National Railway Museum, York	-	
D1024	C	BREL Swindon	08/74	
D1025	C	BREL Swindon	01/79	
D1026	C	BREL Swindon	08/76	
D1027	C	BREL Swindon	06/76	
D1028	C	BREL Swindon	06/79	
D1029	C	BREL Swindon	05/75	Name changed to 'Western Legionaire' 09/67
D1030	C	BREL Swindon	09/76	
D1031	C	BREL Swindon	10/76	
D1032	C	BREL Swindon	12/74	
D1033	C	BREL Swindon	05/79	Used after withdrawal as steam heat unit
D1034	C	BREL Swindon	02/79	Used after withdrawal as steam heat unit
D1035	C	BREL Swindon	09/76	
D1036	C	BREL Swindon	02/77	
D1037	C	BREL Swindon	02/77	
D1038	C	BREL Swindon	11/74	
D1039	C	BREL Swindon	09/74	
D1040	C	BREL Swindon	08/76	Withdrawn: 10/75 R/I: 10/75
D1041	P	East Lancs Railway, Bury	-	
D1042	C	BREL Swindon	05/74	
D1043	C	BREL Swindon	02/77	
D1044	C	BREL Swindon	09/75	
D1045	C	BREL Swindon	08/75	
D1046	C	BREL Swindon	11/76	
D1047	C	BREL Swindon	09/76	
D1048	P	Bodmin & Wenford Railway	-	
D1049	C	BREL Swindon	02/77	
D1050	C	BREL Swindon	04/76	
D1051	C	BREL Swindon	08/77	
D1052	C	BREL Swindon	04/76	
D1053	C	BREL Swindon	06/77	
D1054	C	BREL Swindon	05/77	
D1055	C	BREL Swindon	06/76	
D1056	C	BREL Swindon	05/79	
D1057	C	BREL Swindon	06/77	
D1058	C	BREL Swindon	06/79	
D1059	C	BREL Swindon	07/76	
D1060	C	BREL Swindon	07/74	
D1061	C	BREL Swindon	08/75	
D1062	P	Western Loco Assoc, on SVR	-	
D1063	C	BREL Swindon	08/77	
D1064	C	BREL Swindon	07/77	
D1065	C	BREL Swindon	08/77	
D1066	C	BREL Swindon	05/75	
D1067	C	BREL Swindon	09/76	
D1068	C	BREL Swindon	08/77	
D1069	C	BREL Swindon	02/77	
D1070	C	BREL Swindon	05/79	
D1071	C	BREL Swindon	11/78	
D1072	C	BREL Swindon	04/77	
D1073	C	BREL Swindon	08/75	

BR 1957 No.	Depot of First Allocation	Date Withdrawn	Depot of Final Allocation	Disposal Code	Disposal Detail	Date Cut Up	Notes
D1100	50A						
D1101	50A	11/91	CD	A	BR Immingham		Withdrawn: 12/89, R/I: 12/89
D1102	50A						
D1103	50A						
D1104	50A						
D1105	50A						Stored: [U] 12/89
D1106	50A						
D1107	50A						
D1108	50A						

BR 1957 No.	TOPS No.	Date Re No.	First TOPS Re No.	Date Re No.	Second TOPS Re No.	Date Re No.	Third TOPS Re No.	Date Re No.	Name	Name Date	Built By	Works No.	Date Introduced
D1109	47526	03/74							[Northumbria]	07/86-05/91	BR Crewe	-	01/67
D1110	47527	02/74							[Kettering]	05/89-08/91	BR Crewe	-	01/67
D1111	47528	02/74							The Queens Own Mercian Yeomanry	11/89	BR Crewe	-	02/67

D1200 See Prototype No. D0280

BR 1957 No.	TOPS No.	Date Re No.	Name	Name Date	Built By	Works No.	Date Introduced
D1500	47401	11/73	[Star of the East]	05/91-06/92	Brush	342	09/62
D1501	47402	02/74	[Gateshead]	11/81-05/88	Brush	343	11/62
D1502	47403	02/74	[The Geordie]	02/82-04/87	Brush	344	11/62
D1503	47404	03/74	[Hadrian]	03/82-06/87	Brush	345	12/62
D1504	47405	02/74	[Northumbria]	10/82-05/86	Brush	346	01/63
D1505	47406	01/74	[Rail Riders]	12/81-05/88	Brush	347	01/63
D1506	47407	02/74	[Aycliffe]	11/84-05/88	Brush	348	01/63
D1507	47408	01/74	[Finsbury Park]	03/84-07/86	Brush	349	01/63
D1508	47409	02/74	[David Lloyd George]	09/85-04/87	Brush	350	01/63
D1509	47410	02/74			Brush	351	02/63
D1510	47411	01/74	[The Geordie]	04/87-05/88	Brush	352	02/63
D1511	47412	02/74			Brush	353	02/63
D1512	47413	02/74			Brush	354	03/63
D1513	47414	01/74			Brush	355	03/63
D1514	47415	02/74			Brush	357	03/63
D1515	47416	02/74			Brush	356	03/63
D1516	47417	02/74			Brush	358	04/63
D1517	47418	01/74			Brush	359	04/63
D1518	47419	02/74			Brush	413	04/63
D1519	47420	04/74			Brush	360	04/63
D1520	47421	01/74	[The Brontes of Haworth]	07/85-09/88	Brush	414	06/63
D1521	47001	02/74			Brush	415	06/63
D1522	47002	02/74			Brush	417	06/63
D1523	47003	01/74			Brush	416	06/63
D1524	47004	11/73			Brush	419	06/63
D1525	47422	02/74			Brush	418	06/63
D1526	47005	02/74			Brush	420	06/63
D1527	47423	02/74			Brush	421	06/63
D1528	47006	02/74			Brush	422	07/63
D1529	47007	02/74	[Stratford]	11/86-11/90	Brush	423	07/63
D1530	47008	11/74			Brush	424	07/63
D1531	47424	05/74	[The Brontes of Haworth]	10/88-11/90	Brush	425	07/63
D1532	47009	01/74			Brush	426	07/63
D1533	47425	03/74	[Holbeck]	04/86-02/91	Brush	427	08/63
D1534	47426	01/74			Brush	428	08/63
D1535	47427	03/74			Brush	429	08/63
D1536	47428	02/74			Brush	430	08/63
D1537	47010	02/74	[Xancidae]	05/89-10/89	Brush	431	08/63
D1538	47011	02/74			Brush	432	09/63
D1539	47012	05/74			Brush	433	09/63
D1540	47013	12/73			Brush	434	09/63
D1541	47429	02/74			Brush	436	09/63
D1542	47430	02/74			Brush	435	09/63
D1543	47014	02/74			Brush	438	09/63
D1544	47015	03/74			Brush	437	10/63
D1545	47431	02/74			Brush	439	10/63
D1546	47016	03/74	[The Toleman Group]	05/87-10/90	Brush	440	10/63
D1547	47432	07/74			Brush	441	10/63
D1548	47433	03/74			Brush	442	10/63
D1549	47434	06/74	[Pride of Huddersfield]	06/88-01/91	Brush	443	10/63
D1550	47435	02/75			BR Crewe	-	01/64
D1551	47529	03/74			BR Crewe	-	01/64
D1552	47436	11/73			BR Crewe	-	01/64
D1553	47437	12/73			BR Crewe	-	01/64
D1554	47438	10/73			BR Crewe	-	01/64
D1555	47439	12/73			BR Crewe	-	02/64
D1556	47440	03/74			BR Crewe	-	02/64
D1557	47441	02/74			BR Crewe	-	02/64
D1558	47442	03/74			BR Crewe	-	02/64
D1559	47443	03/74	North Eastern	05/88	BR Crewe	-	03/64
D1560	47444	02/74	[University of Nottingham]	05/81-06/90	BR Crewe	-	03/64
D1561	47445	11/73			BR Crewe	-	03/64
D1562	-	-			BR Crewe	-	03/64
D1563	47446	04/74			BR Crewe	-	04/64
D1564	47447	09/73			BR Crewe	-	03/64
D1565	47448	03/74	[Gateshead]	05/88-05/91	BR Crewe	-	03/64
D1566	47449	03/74			BR Crewe	-	03/64
D1567	47450	02/74			BR Crewe	-	03/64
D1568	47451	11/73			BR Crewe	-	03/64
D1569	47452	05/74	[Aycliffe]	07/88-08/91	BR Crewe	-	03/64

BR 1957 No.	Depot of First Allocation	Date Withdrawn	Depot of Final Allocation	Disposal Code	Disposal Detail	Date Cut Up	Notes
D1109	50A						
D1110	50A	11/92	BR	A	BR Bristol Bath Road		
D1111	50A						
D1500	34G	06/92	IM	A	BR Immingham		Named: North Eastern 12/81-05/88
D1501	34G	06/92	IM	A	BR Immingham		Stored: [U] 02/90, R/I: 03/90, Stored: [U] 12/90, R/I: 02/91, Withdrawn: 07/91, R/I: 10/91
D1502	34G	09/86	GD	A	BR Healey Mills		
D1503	34G	06/87	GD	C	V Berry, Leicester	01/90	
D1504	34G	03/86	GD	C	BREL Crewe, by A Hampton	12/88	
D1505	34G	08/90	IM	A	BR Scunthorpe		
D1506	34G	08/90	IM	A	BR Scunthorpe		
D1507	34G	07/86	GD	C	V Berry, Leicester	06/89	
D1508	34G	08/86	GD	C	V Berry, Leicester	05/89	
D1509	34G	05/87	GD	C	V Berry, Leicester	03/90	
D1510	34G	06/89	IM	A	BR Scunthorpe		
D1511	34G	05/87	GD	C	Booth-Roe, Rotherham	03/92	
D1512	34G	08/91	IM	A	BR Scunthorpe		Withdrawn: 04/91, R/I: 05/91 Stored: [U] 07/91
D1513	34G	03/86	GD	C	V Berry, Leicester	06/89	
D1514	34G	04/87	GD	C	V Berry, Leicester	06/90	
D1515	34G	03/86	GD	C	BREL Crewe	05/87	
D1516	34G	02/92	IM	A	BR Scunthorpe		Withdrawn: 04/91, R/I: 04/91, Stored: [U] 08/91
D1517	34G	02/91	IM	A	BR Scunthorpe		Stored: [U] 02/91
D1518	34G	02/87	GD	C	V Berry, Leicester	01/90	
D1519	34G	06/87	GD	C	V Berry, Leicester	06/89	
D1520	34G	09/91	CD	A	ABB Transportation, Crewe		
D1521	34G	11/86	BR	A	BR Crewe Basford Hall		
D1522	34G	06/91	TI	A	BR Doncaster, Belmont		Stored: [S] 07/90
D1523	34G	07/91	TI	C	Booth-Roe, Rotherham	07/92	Stored: [U] 04/91
D1524	34G						Withdrawn: 07/91, R/I: 11/91
D1525	34G	08/91	CD	A	Booth-Roe, Rotherham		Stored: [U] 07/91
D1526	34G	05/91	OC	A	Booth-Roe, Rotherham		
D1527	34G	07/92	OC	A	BR Old Oak Common		Withdrawn: 05/92, R/I: 05/92
D1528	34G	09/91	TI	A	Booth-Roe, Rotherham		Stored: [U] 06/91
D1529	34G	09/91	TI	A	BR Doncaster, Belmont		
D1530	34G	10/89	TI	A	BR Stratford		
D1531	34G	01/91	CD	A	ABB Transportation, Crewe		Stored: [U] 01/91
D1532	34G	05/90	TI	C	Booth-Roe, Rotherham	04/92	Stored: [U] 12/89
D1533	41A	12/92	OC	A	BR Old Oak Common		
D1534	41A	12/92	OC	A	BR Old Oak Common		
D1535	41A	02/90	CD	C	M C Processors, Glasgow	04/92	
D1536	41A	01/89	CD	C	V Berry, Leicester	06/90	
D1537	41A	12/92	IM	A	BR Immingham		
D1538	41A	04/87	BS	A	ABB Transportation, Crewe		
D1539	41A	12/89	TI	C	Booth-Roe, Rotherham	04/92	
D1540	41A	02/87	TI	C	BRML Doncaster, by C F Booth	10/88	
D1541	41A	01/87	IS	C	BREL Crewe, by A Hampton	01/89	
D1542	41A	02/92	OC	A	BR Old Oak Common		Stored: [S] 02/92
D1543	41A	07/91	TI	C	Booth-Roe, Rotherham	04/92	Stored: [U] 05/90
D1544	41A	05/87	BS	A	BR Crewe Basford Hall		
D1545	41A	07/92	OC	A	BR Old Oak Common		Stored: [S] 07/90, R/I 08/90
D1546	41A						
D1547	34G	04/92	BR	A	BR Bristol Bath Road		
D1548	41A						
D1549	41A	01/91	CD	A	BR Crewe, Basford Hall		Stored [U] 01/91
D1550	41A	03/90	CD	A	BR Crewe, Basford Hall		Stored: [U] 02/90
D1551	41A	02/87	CD	C	BREL Crewe	04/87	
D1552	41A	12/91	IS	A	BR Inverness		
D1553	41A	04/87	CD	C	Booth Roe, Rotherham	09/89	
D1554	41A	05/92	OC	A	BR Old Oak Common		
D1555	41A						
D1556	41A	08/91	OC	A	BR Old Oak Common		
D1557	41A	12/92	OC	A	BR Old Oak Common		
D1558	41A						Stored: [U] 04/91
D1559	41A						
D1560	41A	06/90	CD	A	BR Crewe, Basford Hall		Stored: [U] 06/90
D1561	41A	06/91	TI	A	BR Doncaster, Belmont		Stored: [U] 05/91
D1562	41A	06/71	SF	C	BREL Crewe	09/71	
D1563	41A	02/92	OC	A	BR Old Oak Common		
D1564	41A	03/91	TI	A	BR Doncaster, Belmont		Stored: [U] 03/91
D1565	41A	05/91	CD	A	BR Holbeck		Stored: [S] 08/90
D1566	41A						
D1567	41A	05/91	TI	A	Booth-Roe, Rotherham		Stored: [S] 07/90, R/I: 08/90
D1568	41A	06/91	TI	A	BR Doncaster, Belmont		Stored: [S] 07/90, R/I: 08/90, Stored: [S] 04/91
D1569	41A	08/91	OC	A	BR Old Oak Common		

BR 1957 No.	TOPS No.	Date Re No.	First TOPS Re No.	Date Re No.	Second TOPS Re No.	Date Re No.	Third TOPS Re No.	Date Re No.	Name	Name Date	Built By	Works No.	Date Introduced
D1570	47017	02/74									BR Crewe	-	03/64
D1571	47453	03/74									BR Crewe	-	04/64
D1572	47018	02/74									BR Crewe	-	04/64
D1573	47019	02/74									BR Crewe	-	04/64
D1574	47454	03/74									BR Crewe	-	04/64
D1575	47455	02/74									BR Crewe	-	04/64
D1576	47456	03/74									BR Crewe	-	04/64
D1577	47457	03/74							[Ben Line]	02/86-04/90	BR Crewe	-	04/64
D1578	47458	02/74							County of Cambridgeshire	01/91	BR Crewe	-	05/64
D1579	47459	02/74									BR Crewe	-	05/64
D1580	47460	02/74							[Great Eastern]	04/77-04/78	BR Crewe	-	05/64
D1581	47461	02/74							[Charles Rennie Mackintosh]	03/82-11/90	BR Crewe	-	05/64
D1582	47462	02/74							Cambridge Traction & Rolling Stock Maintenance Depot	09/90	BR Crewe	-	05/64
D1583	47020	03/74	47556	02/81	47844	02/90			Derby & Derbyshire Chamber of Commerce & Industry	01/91	BR Crewe	-	05/64
D1584	47021*	-	47531	05/74	47974	06/90	47531	06/92	[The Permanent Way Institution]	06/91-06/92	BR Crewe	-	05/64
D1585	47022*	-	47542	09/74							BR Crewe	-	05/64
D1586	47463	02/74									BR Crewe	-	05/64
D1587	47464	12/73									BR Crewe	-	05/64
D1588	47023*	-	47543	12/74							BR Crewe	-	05/64
D1589	47465	02/74									BR Crewe	-	05/64
D1590	47466	02/74									BR Crewe	-	05/64
D1591	47024	02/74	47557	12/79							BR Crewe	-	06/64
D1592	47025*	-	47544	02/75							BR Crewe	-	06/64
D1593	47467	11/73									BR Crewe	-	06/64
D1594	47468	06/74	47300	08/92							BR Crewe	-	06/64
D1595	47469	02/74							[Glasgow Chamber of Commerce]	03/83-03/89	BR Crewe	-	06/64
D1596	47470	01/74							[University of Edinburgh]	07/83-10/91	BR Crewe	-	06/64
D1597	47026	05/74	47597	12/83					Resilient	10/91	BR Crewe	-	06/64
D1598	47471	02/74							Norman Tunna GC	11/82	BR Crewe	-	06/64
D1599	47027	02/74	47558	11/80					Mayflower	10/82	BR Crewe	-	06/64
D1600	47472	01/74	97472	09/88	47472	06/89					BR Crewe	-	06/64
D1601	47473	01/74									BR Crewe	-	07/64
D1602	47474	02/74							Sir Rowland Hill	05/90	BR Crewe	-	07/64
D1603	47475	04/74							Restive	08/92	BR Crewe	-	07/64
D1604	47476	04/74							Night Mail	04/91	BR Crewe	-	07/64
D1605	47028	04/74	47559	11/80					[Sir Joshua Reynolds]	04/82-11/91	BR Crewe	-	07/64
D1606	47029	04/74	47635	01/86					[Jimmy Milne]	04/87-05/91	BR Crewe	-	07/64
D1607	47477	03/74									BR Crewe	-	07/64
D1608	47478	01/74									BR Crewe	-	07/64
D1609	47030	12/73	47618	08/84	47836	08/89			[Fair Rosamund]	10/84-09/89	BR Crewe	-	08/64
D1610	47031	04/74	47560	07/80	47832	06/89			Tamar	04/82	BR Crewe	-	08/64
D1611	47032	09/73	47662	12/86	47817	07/89					BR Crewe	-	08/64
D1612	47479	10/73							[Track 29]	07/90-09/92	BR Crewe	-	08/64
D1613	47033	03/74									BR Crewe	-	08/64
D1614	47034	03/74	47561	10/80	97561	09/88	47973	07/89	Derby Evening Telegraph	09/90	BR Crewe	-	08/64
D1615	47035	02/74	47594	09/83					Resourceful	10/91	BR Crewe	-	08/64
D1616	47480	11/73	97480	09/88	47971	07/89			Robin Hood	11/79	BR Crewe	-	09/64
D1617	47036	03/74	47562	11/79	47672	07/91	47562	11/92	[Sir William Burrell]	09/83-03/92	BR Crewe	-	09/64
D1618	47037	03/74	47563	01/80	47831	06/89			Bolton Wanderer	06/89	BR Crewe	-	09/64
D1619	47038	02/74	47564	07/80					[Colossus]	10/86-05/91	BR Crewe	-	09/64
D1620	47039	02/74	47565	11/80							BR Crewe	-	09/64
D1621	47040	03/74	47642	03/86					Resolute	08/92	BR Crewe	-	09/64
D1622	47041	02/74	47630	11/85							BR Crewe	-	09/64
D1623	47042	02/74	47586	03/83	47676	07/91			Northamptonshire	09/89	BR Crewe	-	10/64
D1624	47043	02/74	47566	05/80							BR Crewe	-	10/64
D1625	47044	10/73	47567	10/80					Red Star	07/85	BR Crewe	-	10/64
D1626	47045	04/74	47568	12/80					Royal Engineers Postal & Courier Services	03/90	BR Crewe	-	10/64
D1627	47481	05/74									BR Crewe	-	10/64
D1628	47046	11/73	47601	12/75	47901	11/79					BR Crewe	-	10/64
D1629	47047	02/74	47569	01/81					The Gloucestershire Regiment	07/90	BR Crewe	-	10/64
D1630	47048	03/74	47570	10/80	47849	01/90					BR Crewe	-	10/64
D1631	47049	02/74									BR Crewe	-	10/64
D1632	47050	02/74									BR Crewe	-	11/64
D1633	47051	02/74									BR Crewe	-	11/64
D1634	47052	03/74									BR Crewe	-	11/64
D1635	47053	03/74							Cory Brothers 1842-1992	09/92	BR Crewe	-	11/64
D1636	47482	06/74									BR Crewe	-	12/64
D1637	47483	03/74									BR Crewe	-	12/64
D1638	47054	01/74							[Xancidae]	03/91-07/92	BR Crewe	-	12/64
D1639	47055	05/74	47652	07/86	47807	07/89					BR Crewe	-	12/64
D1640	47056	02/74	47654	08/86	47809	07/89			Finsbury Park	08/86	BR Crewe	-	12/64
D1641	47057*	-	47532	06/74							BR Crewe	-	12/64
D1642	47058*	-	47547	09/74					[University of Oxford]	10/90-05/92	BR Crewe	-	01/65
D1643	47059	02/74	47631	11/85							BR Crewe	-	01/65

BR 1957 No.	Depot of First Allocation	Date Withdrawn	Depot of Final Allocation	Disposal Code	Disposal Detail	Date Cut Up	Notes
D1570	55A	03/91	TI	C	Booth-Roe, Rotherham	04/92	Stored: [U] 02/87, R/I: 05/87, Stored: [U] 09/89
D1571	55A	12/92	OC	A	BR Old Oak Common		
D1572	55A	11/91	TI	A	BR Doncaster, Belmont		
D1573	55A						
D1574	52A	01/91	TI	A	BR Doncaster, Belmont		
D1575	52A	03/90	CD	A	ABB Transportation, Crewe		
D1576	52A	10/91	CD	A	Booth-Roe, Rotherham		Stored: [U] 09/91
D1577	52A	02/92	OC	A	BR Old Oak Common		Stored: [S] 01/92
D1578	52A						
D1579	52A	03/92	CD	A	Booth-Roe, Rotherham		
D1580	52A	02/92	CD	A	BR Crewe		Unofficial Name. Stored [U] 01/91, R/I: 01/91, Stored: [U] 12/91, Withdrawn: 01/92, R/I: 02/92
D1581	52A	04/91	CD	A	BR Crewe, Basford Hall		Stored: [U] 01/91
D1582	52A						
D1583	86A						
D1584	87E						
D1585	87E	03/89	BR	A	BR Stratford		
D1586	86A						Stored: [U] 02/90
D1587	86A	10/86	IS	C	BREL Crewe	11/88	
D1588	86A						
D1589	86A	11/91	OC	A	BR Old Oak Common		
D1590	86A	08/91	CD	A	BR Holbeck		
D1591	86A						Stored: [U] 11/89
D1592	86A	02/90	CD	C	M C Processors, Glasgow	07/91	
D1593	86A						
D1594	86A						
D1595	86A	03/89	IS	C	M C Processors, Glasgow	12/89	
D1596	86A	11/91	CD	A	ABB Transportation, Crewe		Stored: [U] 09/91
D1597	86A						
D1598	86A						
D1599	87E						
D1600	87E	09/91	OC	A	BR Old Oak Common		Withdrawn: 06/91, R/I: 06/91
D1601	87E						
D1602	87E						
D1603	87E						
D1604	87E						
D1605	87E						Stored: [U] 10/90
D1606	87E						
D1607	87E	11/92	CD	A	Booth-Roe, Rotherham		Stored: [U] 01/90, R/I: 02/91
D1608	87E						Stored: [U] 02/87, R/I: 03/87
D1609	87E						
D1610	87E						
D1611	87E						
D1612	87E	09/92	CD	A	Booth-Roe, Rotherham		
D1613	85A						
D1614	85A						Named: Midland Counties 150, 1839-1989 05/89-03/90
D1615	85A						
D1616	16A						
D1617	16A						
D1618	16A						Named: Womens Guild 04/87-05/89
D1619	16A						
D1620	16A						
D1621	16A						Named: Strathisla 09/86-12/91
D1622	16A						
D1623	16A						
D1624	16A						
D1625	16A						
D1626	16A						
D1627	16A						
D1628	16A	03/90	CF	C	M C Processors, Glasgow	03/92	
D1629	16A						
D1630	16A						
D1631	5A						
D1632	5A						
D1633	5A						
D1634	5A						
D1635	5A						
D1636	86A						
D1637	86A						
D1638	86A	07/92	IM	A	BR Immingham		
D1639	86A						
D1640	86A						
D1641	86A						Stored: [U] 07/91, R/I: 08/91
D1642	86A						
D1643	86A						

BR 1957 No.	TOPS No.	Date Re No.	First TOPS Re No.	Date Re No.	Second TOPS Re No.	Date Re No.	Third TOPS Re No.	Date Re No.	Name	Name Date	Built By	Works No.	Date Introduced
D1644	47060	03/74							Halewood Silver Jubilee 1988	10/88	BR Crewe	-	01/65
D1645	47061	04/74	47649	05/86	47830	07/89					BR Crewe	-	01/65
D1646	47062*	-	47545	11/74	97545	09/88	47972	07/89	The Royal Army Ordnance Corps	02/93	BR Crewe	-	01/65
D1647	47063	02/74									BR Crewe	-	01/65
D1648	47064	02/74	47639	02/86	47851	02/90			[Industry Year 1986]	06/86-10/88	BR Crewe	-	01/65
D1649	47065*	-	47535	04/74					University of Leicester	05/82	BR Crewe	-	01/65
D1650	47066	01/74	47661	12/86	47816	02/89					BR Crewe	-	01/65
D1651	47067*	-	47533	09/74							BR Crewe	-	01/65
D1652	47068	03/74	47632	12/85	47848	01/90					BR Crewe	-	02/65
D1653	47069	01/74	47638	02/86	47845	11/89			County of Kent	06/86	BR Crewe	-	02/65
D1654	47070	02/74	47620	09/84	47835	08/89			Windsor Castle	07/85	BR Crewe	-	02/65
D1655	47071*	-	47536	10/74							BR Crewe	-	02/65
D1656	47072	01/74	47609	04/84	47834	08/89			Fire Fly	08/85	BR Crewe	-	02/65
D1657	47073*	-	47537	08/74					[Sir Gwynedd/County of Gwynedd]	11/82-10/92	BR Crewe	-	02/65
D1658	47074	02/74	47646	04/86	47852	02/90					BR Crewe	-	02/65
D1659	47075	03/74	47645	03/86					[Robert F Fairlie]	05/86-02/90	BR Crewe	-	02/65
D1660	47076	02/74	47625	11/84					Resplendent	10/91	BR Crewe	-	03/65
D1661	47077	02/74	47613	05/84	47840	11/89			North Star	03/65	BR Crewe	-	03/65
D1662	47484	12/73							Isambard Kingdom Brunel	03/65	BR Crewe	-	03/65
D1663	47078	02/74	47628	01/85					[Sir Daniel Gooch]	05/65-12/89	BR Crewe	-	03/65
D1664	47079	02/74							[George Jackson Churchward]	05/65-08/87	BR Crewe	-	03/65
D1665	47080	03/74	47612	06/84	47838	09/89	47612	02/93	[Titan]	04/66-10/89	BR Crewe	-	03/65
D1666	47081	02/74	47606	02/84	47842	01/90	47606	02/93	[Odin]	03/65-07/90	BR Crewe	-	03/65
D1667	47082	02/74	47626	12/84					Atlas	06/66	BR Crewe	-	03/65
D1668	47083	03/74	47633	12/85					[Orion]	10/65-12/85	BR Crewe	-	03/65
D1669	47084*	-	47538	11/74					[Python]	03/66-01/73	BR Crewe	-	03/65
D1670	47085	02/74							[Conidae]	07/88-02/91	BR Crewe	-	04/65
D1671	-	-							[Thor]	09/65-12/65	BR Crewe	-	04/65
D1672	47086	01/74	47641	03/86					[Fife Region]	10/86-08/91	BR Crewe	-	04/65
D1673	47087	02/74	47624	10/84					Cyclops	06/66	BR Crewe	-	05/65
D1674	47088	02/74	47653	08/86	47808	02/89			Samson	09/65	BR Crewe	-	05/65
D1675	47089	01/74							[Amazon]	11/65-06/87	BR Crewe	-	05/65
D1676	47090	02/74	47623	11/84	47843	02/90			[Vulcan]	10/65-06/89	BR Crewe	-	05/65
D1677	47091	03/74	47647	04/86	47846	01/90			Thor	08/66	BR Crewe	-	05/65
D1678	47092*	-	47534	05/74							BR Crewe	-	05/65
D1679	47093	02/74									BR Crewe	-	05/65
D1680	47094	04/74									BR Crewe	-	05/65
D1681	47095	03/74									BR Crewe	-	05/65
D1682	47096	05/74									Brush	444	10/63
D1683	47485	12/73									Brush	445	10/63
D1684	47097	02/74									Brush	446	10/63
D1685	47098	03/74									Brush	447	11/63
D1686	47099	03/74									Brush	448	11/63
D1687	47100	03/74									Brush	449	11/63
D1688	47101	03/74									Brush	450	11/63
D1689	47486	02/74									Brush	451	11/63
D1690	47102	02/74									Brush	452	11/63
D1691	47103	02/74									Brush	453	11/63
D1692	47104	03/74									Brush	454	12/63
D1693	47105	01/74									Brush	455	12/63
D1694	47106	03/74									Brush	456	12/63
D1695	47107	02/74									Brush	457	12/63
D1696	47108	03/74									Brush	458	12/63
D1697	47109	03/74									Brush	460	12/63
D1698	47110	02/74									Brush	459	12/63
D1699	47111	04/74									Brush	461	12/63
D1700	47112	02/74									Brush	462	01/64
D1701	47113	02/74									Brush	463	01/64
D1702	47114	03/74									Brush	464	12/65
D1703	47115	10/73									Brush	465	09/65
D1704	47116	01/74									Brush	466	07/66
D1705	47117	03/74									Brush	467	11/65
D1706	47118	10/73									Brush	468	12/65
D1707	47487	03/74									Brush	469	01/64
D1708	47119	05/74							[Arcidae]	09/88-10/89	Brush	470	01/64
D1709	47120	03/74							[RAF Kinloss]	06/85-10/86	Brush	471	01/64
D1710	47121	02/74									Brush	472	01/64
D1711	47122	04/74									Brush	473	01/64
D1712	47123	03/74									Brush	474	01/64
D1713	47488	02/74							Rail Riders	05/88	Brush	475	01/64
D1714	47124	04/74									Brush	476	02/64
D1715	47125	05/74	47548*	-					[Tonnidae]	09/88-08/91	Brush	477	02/64
D1716	47489	12/73							Crewe Diesel Depot	06/90	Brush	487	02/64
D1717	47126	01/74	47555	09/74					[The Commonwealth Spirit]	04/79-01/92	Brush	488	02/64
D1718	47127*	-	47539	07/74					Rochdale Pioneers	11/82	Brush	489	02/64
D1719	47128	04/74	47656	09/86	47811	08/89					Brush	490	02/64
D1720	47129	02/74	47658	10/86	47813	07/89					Brush	491	03/64
D1721	47130	02/74									Brush	492	03/64
D1722	47131	05/74									Brush	493	03/64

BR 1957 No.	Depot of First Allocation	Date Withdrawn	Depot of Final Allocation	Disposal Code	Disposal Detail	Date Cut Up	Notes
D1644	86A						
D1645	86A						
D1646	86A						
D1647	86A						
D1648	86A						
D1649	86A						
D1650	86A						
D1651	86A	06/91	CD	A	BR Old Oak Common		
D1652	86A						
D1653	86A						
D1654	86A						
D1655	86A						
D1656	87E						
D1657	87E						
D1658	87E	03/92	CD	A	Booth-Roe, Rotherham		Withdrawn: 04/91, R/I: 01/92
D1659	87E	02/90	CD	C	M C Processors, Glasgow	04/92	Stored: [U] 08/89
D1660	87E						Named: City of Truro 06/65 - 10/88
D1661	87E						Plates off: 11/89-11/90
D1662	87E						
D1663	87E						
D1664	87E						Nameplate altered to G J Churchward From 03/79
D1665	87E						Stored: [U] 06/92
D1666	87E						Stored: [U] 06/92
D1667	87E						
D1668	86A	04/91	IS	A	ABB Transportation, Crewe		Stored: [U]11/90
D1669	87E	08/90	BR	S	DML Devonport Dockyard		
D1670	86A						Named: Mammoth 08/65-08/88
D1671	86A	04/66	86A	C	R S Hayes, Bridgend	06/66	Stored: [U] 12/65
D1672	86A						Named: Colossus 08/65-10/86
D1673	86A						
D1674	86A						Named: Samson 04/86-10/88
D1675	86A	06/87	BR	C	Coopers Metals, Brightside	04/89	
D1676	86A						
D1677	86A						
D1678	86A	06/91	CD	A	ABB Transportation, Crewe		Stored: [U] 06/91
D1679	86A	01/88	TI	C	M C Processors, Glasgow	11/90	Stored: [U] 11/87
D1680	86A	02/92	IM	A	BR Scunthorpe		
D1681	86A						Stored: [U] 12/92
D1682	81A	07/91	TI	A	BR Tinsley		Stored: [S] 07/90, R/I: 08/90
D1683	81A						
D1684	81A	04/90	TI	C	Booth-Roe, Rotherham	04/92	Withdrawn: 01/89, R/I: 01/89
D1685	2B	06/91	TI	C	BR Eastleigh	09/92	Withdrawn: 03/89, R/I: 06/89
							Stored: [U] 02/90, R/I: 09/90
D1686	2B	11/91	TI	A	BR Doncaster, Belmont		
D1687	2B	07/91	TI	A	BR Doncaster, Belmont		Stored: [S] 07/90, R/I: 09/90
D1688	2B	03/89	CD	A	BR Crewe Basford Hall		
D1689	82A	07/87	BS	C	V Berry, Leicester	06/89	
D1690	86A	07/91	TI	A	BR Tinsley		Stored: [S] 07/90, R/I: 09/90
D1691	86A	06/87	CD	C	V Berry, Leicester	06/89	
D1692	82A	06/88	TI	C	M C Processors Glasgow	11/90	Withdrawn: 04/88, R/I: 05/88
D1693	82A						
D1694	81A	02/88	TI	C	V Berry, Leicester	06/89	
D1695	2B	06/91	TI	A	BR Doncaster, Belmont		
D1696	16C						
D1697	86A	08/87	ED	C	M C Processors, Glasgow	12/89	
D1698	81A	05/89	TI	A	BR Thornaby		
D1699	81A	02/86	CD	C	BR Canton	03/87	Stored: [U] 02/86
D1700	81A	11/91	OC	A	BR Old Oak Common		
D1701	81A	01/88	CD	C	V Berry, Leicester	02/90	
D1702	41A						
D1703	41A	04/91	IM	A	BR Scunthorpe		
D1704	41A	07/90	CD	A	ABB Transportation, Crewe		
D1705	41A	07/91	TI	A	BR Doncaster, Belmont		
D1706	41A	03/91	TI	A	BR Immingham		Stored: [S] 07/90, R/I: 09/90 Stored: [U] 11/90
D1707	81A	12/88	CF	C	M C Processors, Glasgow	06/89	
D1708	81A	05/92	IM	A	BR Immingham		
D1709	2B	03/91	TI	A	BR Doncaster, Belmont		Stored: [U] 07/90
D1710	81A						Stored: [U] 02/90, R/I: 04/90, Stored: [U] 07/91
D1711	81A	09/87	TI	C	M C Processors, Glasgow	10/89	
D1712	2B	07/91	TI	A	BR Doncaster Belmont		Stored: [S] 07/90
D1713	2B						
D1714	81A	06/89	TI	C	M C Processors, Glasgow	11/90	
D1715	81A						
D1716	2B						
D1717	81A						
D1718	86A						
D1719	86A						
D1720	82A						
D1721	86A	03/88	TI	C	M C Prossors, Glasgow	11/90	
D1722	82A	02/87	BR	C	V Berry, Leicester	07/88	

BR 1957 No.	TOPS No.	Date Re No.	First TOPS Re No.	Date Re No.	Second TOPS Re No.	Date Re No.	Third TOPS Re No.	Date Re No.	Name	Name Date	Built By	Works No.	Date Introduced
D1723	47132*	-	47540	10/74	47975	08/90			The Institution of Civil Engineers	09/91	Brush	494	03/64
D1724	47133*	-	47549	02/75					[Royal Mail]	09/86-04/91	Brush	495	03/64
D1725	47490	02/74							[Bristol Bath Road]	09/88-05/91	Brush	496	03/64
D1726	47134	02/74	47622	10/84	47841	11/89			The Institution of Mechanical Engineers	05/87	Brush	497	03/64
D1727	47135	02/74	47664	02/87	47819	07/89					Brush	498	03/64
D1728	47136	05/74	47621	09/84	47839	10/89			[Royal County of Berkshire]	06/85-12/89	Brush	499	04/64
D1729	47137	02/74									Brush	500	04/64
D1730	47138	02/74	47607	03/84	47821	04/89			Royal Worcester	09/85	Brush	501	04/64
D1731	47139*	-	47550	02/75					University of Dundee	05/82	Brush	502	06/64
D1732	47140	02/74									Brush	503	04/64
D1733	47141	02/74	47614	07/84	47853	02/90					Brush	504	06/64
D1734	-	-									Brush	505	07/64
D1735	47142	04/74							The Sapper	10/87	Brush	506	05/64
D1736	47143	02/74									Brush	508	05/64
D1737	47144	02/74									Brush	507	05/64
D1738	47145	03/74									Brush	509	05/64
D1739	47146	03/74									Brush	510	06/64
D1740	47147	03/74									Brush	511	06/64
D1741	47148	02/74									Brush	512	06/64
D1742	47149	02/74	47617	07/84	47677	07/91			University of Stirling	09/84	Brush	513	05/64
D1743	47150	02/74									Brush	514	06/64
D1744	47151	03/74	47648	05/86	47850	01/90					Brush	515	06/64
D1745	47152	02/74									Brush	516	07/6
D1746	47153	05/74	47551	02/75	47801	09/89	47551	05/92			Brush	517	07/64
D1747	47154*	-	47546	11/74	47976	06/90			Aviemore Centre	05/85	Brush	518	07/64
D1748	47155	02/74	47660	12/86	47815	08/89					Brush	519	07/64
D1749	47156	02/74									Brush	486	07/64
D1750	47157	03/74									Brush	478	07/64
D1751	47158	02/74	47634	12/85					Holbeck	05/91	Brush	479	07/64
D1752	47159	03/74									Brush	480	08/64
D1753	47491	04/74							[Horwich Enterprise]	06/85-10/92	Brush	481	08/64
D1754	47160	02/74	47605	02/84							Brush	482	08/64
D1755	47161*	-	47541	09/74					[The Queen Mother]	10/82-02/91	Brush	483	08/64
D1756	47162	02/74									Brush	484	08/64
D1757	47163	02/74	47610	04/84	47823	04/89			SS Great Britain	10/92	Brush	485	09/64
D1758	47164	01/74	47571	12/79	47822	04/89					Brush	520	05/64
D1759	47165	02/74	47590	06/83	47825	05/89			Thomas Telford	10/83	Brush	521	08/64
D1760	47492	02/74							[The Enterprising Scot]	09/86-01/92	Brush	522	09/64
D1761	47166	10/73	47611	05/84	47837	09/89			[Thames]	09/84-10/89	Brush	523	09/64
D1762	47167	12/73	47580	05/80					County of Essex	08/79	Brush	524	09/64
D1763	47168	09/73	47572	01/81					Ely Cathedral	05/86	Brush	525	09/64
D1764	47169	02/74	47581	12/79					[Great Eastern]	03/79-07/90	Brush	526	09/64
D1765	47170	02/74	47582	02/81					County of Norfolk	08/79	Brush	527	10/64
D1766	47171	03/74	47592	08/83					County of Avon	09/83	Brush	528	09/64
D1767	47172	03/74	47583	11/80					County of Hertfordshire	07/79	Brush	529	10/64
D1768	47173	01/74	47573	02/81					[The London Standard]	06/86-01/92	Brush	530	10/64
D1769	47174	02/74	47574	01/81					Benjamin Gimbert GC	07/87	Brush	531	10/64
D1770	47175	02/74	47575	03/81					City of Hereford	06/85	Brush	532	10/64
D1771	47176	08/73	47576	06/80					Kings Lynn	11/86	Brush	533	10/64
D1772	47177	02/74	47599	10/83							Brush	534	10/64
D1773	47178	03/74	47588	06/83					Resurgent	04/92	Brush	535	09/64
D1774	47179	07/74	47577	03/81	47847	01/90			[Benjamin Gimbert GC]	09/81-07/87	Brush	536	10/64
D1775	47180	05/74	47584	12/80					County of Suffolk	05/79	Brush	537	10/64
D1776	47181	09/73	47578	01/81					[The Royal Society of Edinburgh]	01/85-07/91	Brush	538	10/64
D1777	47182	04/74	47598	12/83							Brush	539	11/64
D1778	47183	02/74	47579	03/81					James Nightall GC	09/81	Brush	540	10/64
D1779	47184	02/74	47585	01/81					[County of Cambridgeshire]	05/79-11/90	Brush	541	10/64
D1780	47185	11/73	47602	11/83	47824	04/89			[Glorious Devon]	08/85-02/93	Brush	542	10/64
D1781	47186	02/74									Brush	543	11/64
D1782	47301	01/74									Brush	544	11/64
D1783	47302	11/73									Brush	545	11/64
D1784	47303	02/74									Brush	546	11/64
D1785	47304	03/74									Brush	547	12/64
D1786	47305	02/74									Brush	548	11/64
D1787	47306	04/74									Brush	549	11/64
D1788	47307	02/74									Brush	550	12/64
D1789	47308	11/73									Brush	551	11/64
D1790	47309	12/73							The Halewood Transmission	09/90	Brush	552	12/64
D1791	47310	02/74							Henry Ford	09/91	Brush	553	12/64
D1792	47311	02/74							[Warrington Yard]	06/88-11/91	Brush	554	01/65
D1793	47312	04/74									Brush	555	12/64
D1794	47313	04/74									Brush	556	12/64
D1795	47314	01/74							Transmark	12/90	Brush	557	01/65
D1796	47315	04/74									Brush	558	01/65
D1797	47316	05/74									Brush	559	01/65
D1798	47317	03/74							Willesden Yard	07/88	Brush	560	01/65
D1799	47318	02/74									Brush	561	01/65
D1800	47319	02/74							Norsk Hydro	03/88	Brush	562	01/65

BR 1957 No.	Depot of First Allocation	Date Withdrawn	Depot of Final Allocation	Disposal Code	Disposal Detail	Date Cut Up	Notes
D1723	82A						
D1724	82A	08/91	BR	A	BR Crewe		Stored: [U] 06/91
D1725	82A						
D1726	86A						
D1727	81A						
D1728	86A						
D1729	87E	08/87	ED	C	M C Processors, Glasgow	03/92	Stored: [U] 07/87
D1730	86A						
D1731	81A						
D1732	86A	02/88	TI	C	V Berry, Leicester	07/89	
D1733	86A						
D1734	82A	03/65	82A	C	BR Crewe Works	04/65	
D1735	86A						
D1736	86A	12/89	TI	A	BR Doncaster		Stored: 11/89
D1737	86A						
D1738	86A						Withdrawn: 06/88, R/I: 11/88
D1739	82A						Stored: [S] 07/90, R/I: 08/90
D1740	81A						
D1741	86A	07/87	BR	C	M C Processors, Glasgow	10/89	
D1742	86A						
D1743	86A						
D1744	86A						
D1745	81A						
D1746	81A						Withdrawn: 08/91, R/I: 12/91, Withdrawn: 06/92. R/I: 06/92
D1747	81A						
D1748	87E						
D1749	82A						
D1750	82A						
D1751	87E						Named: Henry Ford 07/81-03/91
D1752	87E	05/88	TI	A	BR Thornaby		
D1753	2B						
D1754	87E						
D1755	87E						Stored: [U] 02/91,
D1756	87E	01/87	BR	C	BREL Crewe	05/87	
D1757	86A						
D1758	41A						
D1759	41A						
D1760	41A						
D1761	41A						Withdrawn: 10/91, R/I: 11/91
D1762	41A						
D1763	41A						
D1764	41A						
D1765	41A						
D1766	41A						
D1767	41A						
D1768	41A						
D1769	41A						Named: Lloyds List 12/85-04/87
D1770	41A						
D1771	41A						
D1772	41A						
D1773	41A						Stored: [U] 06/91, R/I: 09/91 Named: Carlisle Currock 07/88-06/91
D1774	41A						
D1775	41A						
D1776	41A						Stored: [U] 03/91, Withdrawn: 08/91, R/I: 09/92
D1777	41A						
D1778	41A						
D1779	41A	02/91	CD	A	BR Holbeck		Stored: [U] 11/90
D1780	41A						Stored: [U] 11/92
D1781	41A						
D1782	41A						
D1783	41A						
D1784	41A						
D1785	41A						
D1786	41A						
D1787	41A						
D1788	41A						
D1789	41A						
D1790	41A						
D1791	41A						
D1792	41A	11/91	TI	A	Booth-Roe, Doncaster		
D1793	41A						
D1794	41A						
D1795	41A						
D1796	41A						
D1797	41A						
D1798	41A						
D1799	41A						
D1800	41A						

BR 1957 No.	TOPS No.	Date Re No.	First TOPS Re No.	Date Re No.	Second TOPS Re No.	Date Re No.	Third TOPS Re No.	Date Re No.	Name	Name Date	Built By	Works No.	Date Introduced
D1801	47320	03/74									Brush	563	01/65
D1802	47321	04/74									Brush	564	01/65
D1803	47322	03/74									Brush	565	01/65
D1804	47323	12/73									Brush	566	02/65
D1805	47324	04/74							[Glossidae]	09/88-07/92	Brush	567	01/65
D1806	47325	03/74									Brush	568	01/65
D1807	47326	03/74									Brush	569	01/65
D1808	47327	03/74									Brush	570	01/65
D1809	47328	04/74									Brush	571	01/65
D1810	47329	05/74									Brush	572	02/65
D1811	47330	04/74							Tren Nwyddau Amlwch/ Amlwch Freighter	06/90	Brush	573	02/65
D1812	47331	05/74									Brush	574	02/65
D1813	47332	03/74									Brush	575	02/65
D1814	47333	03/74							Civil Link	05/90	Brush	576	02/65
D1815	47334	03/74									Brush	577	02/65
D1816	47335	02/74									Brush	578	02/65
D1817	47336	05/74									Brush	579	03/65
D1818	47337	09/73							[Herbert Austin]	04/86-07/90	Brush	580	02/65
D1819	47338	03/74							Warrington Yard	01/92	Brush	581	02/65
D1820	47339	04/74									Brush	582	02/65
D1821	47340	03/74									Brush	583	03/65
D1822	47341	03/74									Brush	584	03/65
D1823	47342	03/74									Brush	585	03/65
D1824	47343	11/73									Brush	586	03/65
D1825	47344	03/74									Brush	587	03/65
D1826	47345	09/73									Brush	588	03/65
D1827	47346	12/73									Brush	589	03/65
D1828	47347	04/74									Brush	590	03/65
D1829	47348	11/73							St Christophers Railway Home	05/87	Brush	591	03/65
D1830	47349	10/73									Brush	592	03/65
D1831	47350	10/73							[British Petroleum]	04/87-03/90	Brush	593	05/65
D1832	47351	01/74									Brush	594	04/65
D1833	47352	03/74									Brush	595	05/65
D1834	47353	05/74									Brush	596	04/65
D1835	47354	04/74									Brush	597	05/65
D1836	47355	03/74									Brush	598	04/65
D1837	47187	04/74									Brush	599	05/65
D1838	47188	05/74									Brush	600	05/65
D1839	47189	04/74									Brush	601	04/65
D1840	47190	02/74							[Pectinidae]	09/88-02/91	Brush	602	05/65
D1841	47191	03/74									Brush	603	05/65
D1842	47192	04/74									BR Crewe	-	05/65
D1843	47193	04/74							Lucinidae	08/88	BR Crewe	-	05/65
D1844	47194	04/74							Carlisle Currock	01/93	BR Crewe	-	05/65
D1845	47195	03/74							[Muricidae]	09/88-06/91	BR Crewe	-	05/65
D1846	47196	02/74							[Haliotidae]	10/88-06/91	BR Crewe	-	05/65
D1847	47197	02/74									BR Crewe	-	06/65
D1848	47198	02/74									BR Crewe	-	06/65
D1849	47199	03/74									BR Crewe	-	06/65
D1850	47200	01/74									BR Crewe	-	06/65
D1851	47201	04/74									BR Crewe	-	06/65
D1852	47202	12/73									BR Crewe	-	06/65
D1853	47203	02/74									BR Crewe	-	06/65
D1854	47204	03/74									BR Crewe	-	07/65
D1855	47205	05/74									BR Crewe	-	07/65
D1856	47206	05/74									BR Crewe	-	08/65
D1857	47207	03/74							Bulmers of Hereford	12/87	BR Crewe	-	08/65
D1858	47208	02/74									BR Crewe	-	08/65
D1859	47209	02/74							Herbert Austin	08/90	BR Crewe	-	08/65
D1860	47210	02/74							Blue Circle Cement	01/90	BR Crewe	-	08/65
D1861	47211	01/74									BR Crewe	-	08/65
D1862	47212	02/74									Brush	624	05/65
D1863	47213	02/74									Brush	625	05/65
D1864	47214	02/74							Distillers MG	11/90	Brush	626	05/65
D1865	47215	02/74									Brush	627	05/65
D1866	47216	02/74	47299	12/81							Brush	628	05/65
D1867	47217	01/74									Brush	629	07/65
D1868	47218	02/74							United Transport Europe	07/90	Brush	630	05/65
D1869	47219	02/74							Arnold Kunzler	10/91	Brush	631	06/65
D1870	47220	02/74									Brush	632	06/65
D1871	47221	01/74									Brush	633	06/65
D1872	47222	01/74							[Appleby-Frodingham]	06/82-04/91	Brush	634	06/65
D1873	47223	02/74							British Petroleum	08/90	Brush	635	05/65
D1874	47224	02/74							Arcidae	06/91	Brush	636	06/65
D1875	47356	03/74									Brush	637	07/65
D1876	47357	05/74							The Permanent Way Institution	06/92	Brush	638	06/65
D1877	47358	03/74									Brush	639	06/65
D1878	47359	02/74									Brush	640	06/65
D1879	47360	02/74									Brush	641	07/65
D1880	47361	02/74							Wilton Endeavour	07/83	Brush	642	02/66

BR 1957 No.	Depot of First Allocation	Date Withdrawn	Depot of Final Allocation	Disposal Code	Disposal Detail	Date Cut Up	Notes
D1801	41A						
D1802	41A						
D1803	41A						
D1804	41A						
D1805	41A	07/92	IM	A	BR Immingham		Stored: [S] 04/92
D1806	41A						Stored: [U] 02/87 R/I: 09/87
D1807	D16						
D1808	D16						
D1809	D16						
D1810	D16						
D1811	D16						
D1812	D16						
D1813	D16						
D1814	D16						
D1815	D16						
D1816	D16						
D1817	D16						Stored: [U] 12/92
D1818	D16						
D1819	D16						Stored: [S] 07/90, R/I: 08/90
D1820	D16						Withdrawn: 08/91, R/I: 08/91
D1821	D16						
D1822	D16						
D1823	D16	11/88	CF	C	M C Processors, Glasgow	03/92	
D1824	D16	07/92	CD	C	M C Processors, Glasgow	11/92	
D1825	D16						
D1826	D16						
D1827	D16						
D1828	D16						
D1829	D16						Stored: [U] 02/91. R/I: 06/92
D1830	D16						
D1831	D16						
D1832	D16						Stored: [S] 08/90
D1833	D16						
D1834	D16						
D1835	D16						
D1836	D16						
D1837	5A						
D1838	5A						Stored: [U] 01/89
D1839	5A	03/89	TI	C	M C Processors, Glasgow	03/92	Withdrawn: 07/87, R/I: [S] 01/88
D1840	5A						Stored: [U] 02/91, R/I: 06/91
D1841	5A	07/87	CD	A	BR Springs Branch		
D1842	5A	05/88	CD	P	The Railway Age, Crewe	-	
D1843	5A						
D1844	5A						Named: Bullidae 08/88-03/91
D1845	5A	11/91	TI	A	BR Tinsley		Stored: [U] 10/91
D1846	5A						
D1847	5A						
D1848	5A	04/89	CF	A	BR Cardiff		
D1849	5A	08/87	CD	A	BR Carlisle		
D1850	5A						Stored: [S] 07/90, R/I: 10/90
D1851	5A						
D1852	5A	03/87	BR	C	BR Bristol Bath Road	09/91	
D1853	5A	03/89	TI	C	V Berry, Leicester	06/90	
D1854	5A						
D1855	5A						
D1856	5A						
D1857	5A						
D1858	5A	01/80	HA	C	BR Dundee	03/80	Stored: [U] 10/79
D1859	5A						
D1860	5A						
D1861	5A						
D1862	41A						
D1863	41A						
D1864	41A						Stored: [U] 02/90, Named: Tinsley Traction Depot 09/87-01/90
D1865	41A	03/91	TI	A	BR Eastleigh West		Stored: [U] 09/89
D1866	41A						
D1867	41A						
D1868	41A						
D1869	41A						Named: W A Camwell 19/10/91 only
D1870	41A						Stored: [U] 08/90, R/I: 09/90 Stored: [U] 06/92
D1871	41A						
D1872	41A						Stored: [S] 06/91, R/I: 07/91
D1873	41A						
D1874	41A						
D1875	40B						
D1876	41A						
D1877	41A						Stored: 08/91
D1878	41A						
D1879	40B						
D1880	41A						

BR 1957 No.	TOPS No.	Date Re No.	First TOPS Re No.	Date Re No.	Second TOPS Re No.	Date Re No.	Third TOPS Re No.	Date Re No.	Name	Name Date	Built By	Works No.	Date Introduced
D1881	47362	02/74									Brush	643	06/65
D1882	47363	02/74							[Billingham Enterprise]	12/85-11/91	Brush	644	07/65
D1883	47364	04/74									Brush	645	07/65
D1884	47365	03/74							ICI Diamond Jubilee	09/86	Brush	646	07/65
D1885	47366	05/74							[The Institution of Civil Engineers]	05/86-09/91	Brush	647	07/65
D1886	47367	05/74									Brush	648	08/65
D1887	47368	02/74							Neritidae	10/88	Brush	649	08/65
D1888	47369	02/74									Brush	650	07/65
D1889	47370	02/74									Brush	651	07/65
D1890	47371	02/74									Brush	652	07/65
D1891	47372	02/74									Brush	653	08/65
D1892	47373	02/74									Brush	654	08/65
D1893	47374	02/74							[Petrolea]	06/86-03/88	Brush	655	08/65
D1894	47375	02/74							Tinsley Traction Depot Quality Approved	02/90	Brush	656	12/65
D1895	47376	02/74									Brush	657	09/65
D1896	47377	02/74									Brush	658	09/65
D1897	47378	03/74									Brush	659	09/65
D1898	47379	02/74							[Total Energy]	04/86-07/91	Brush	660	08/65
D1899	47380	02/74							[Immingham]	09/87-07/92	Brush	661	08/65
D1900	47381	02/74									Brush	662	09/65
D1901	47225	03/74									Brush	663	10/65
D1902	47226	03/74									Brush	664	10/65
D1903	47227	04/74									Brush	665	09/65
D1904	47228	05/74									Brush	666	09/65
D1905	47229	10/73									Brush	667	09/65
D1906	47230	01/74									Brush	668	10/65
D1907	47231	10/73							The Silcock Express	07/88	Brush	669	09/65
D1908	-	-									Brush	670	10/65
D1909	47232	03/74	47665	03/87	47820	09/89					Brush	671	11/65
D1910	47233	02/74							[Strombidae]	09/88-01/91	Brush	672	11/65
D1911	47234	09/73									Brush	673	11/65
D1912	47235	05/74									Brush	674	10/65
D1913	47236	03/74									Brush	675	10/65
D1914	47237	03/74									Brush	676	11/65
D1915	47238	12/73							Bescot Yard	10/88	Brush	677	12/65
D1916	47239	04/74	47657	10/86	47812	07/89					Brush	678	11/65
D1917	47240	11/73	47663	01/87	47818	02/89					Brush	679	12/65
D1918	47241	05/74									Brush	680	11/65
D1919	47242	03/74	47659	11/86	47814	07/89					Brush	681	01/66
D1920	47243	12/73	47636	01/86					Sir John De Graeme	04/86	Brush	682	01/66
D1921	47244	05/74	47640	02/86					University of Strathclyde	04/86	Brush	683	01/66
D1922	47245	10/73									Brush	684	01/66
D1923	47246	02/74	47644	03/86					[The Permanent Way Institution]	06/86-05/91	Brush	685	01/66
D1924	47247	12/73	47655	09/86	47810	08/89					Brush	686	01/66
D1925	47248	11/73	47616	06/84	47671	07/91			[Y Ddraig Goch/The Red Dragon]	07/85-02/93	Brush	687	01/66
D1926	47249	01/74									Brush	688	01/66
D1927	47250	05/74	47600	11/83					[Dewi Sant/Saint David]	07/85-05/91	Brush	689	02/66
D1928	47251	03/74	47589	05/83	47827	05/89					Brush	690	02/66
D1929	47252	03/74	47615	06/84					[Castell Caerffilli/Caerphilly Castle]	04/85-04/92	Brush	691	02/66
D1930	47253*	-	47530	11/73							Brush	692	02/66
D1931	47254	03/74	47651	07/86	47806	07/89					Brush	693	03/66
D1932	47493	05/74	47701	01/79					Old Oak Common Traction & Rolling Stock Depot	08/91	Brush	694	03/66
D1933	47255	02/74	47596	09/83					Aldeburgh Festival	06/84	Brush	695	03/66
D1934	47256	02/74									Brush	696	03/66
D1935	47257	06/74	47650	07/86	47805	08/89			Bristol Bath Road	06/91	Brush	697	03/66
D1936	47494	02/74	47706	06/79					[Strathclyde]	05/79-04/86	Brush	698	03/66
D1937	47495	02/74	47704	02/79					[Dunedin]	02/79-10/91	Brush	699	03/66
D1938	47258	12/73									Brush	700	06/66
D1939	47496	03/74	47710	10/79					Capital Radio - Help a London Child	08/91	Brush	701	04/66
D1940	47497	06/74	47717	09/88					[Tayside Region]	09/88-12/90	Brush	702	06/66
D1941	47498	02/74	47711	09/79					[Greyfriars Bobby]	04/81-12/90	Brush	703	07/66
D1942	47499	02/74	47709	09/79					[The Lord Provost]	09/79-12/90	Brush	704	06/66
D1943	47500	03/74							Restormel	11/92	Brush	705	06/66
D1944	47501	02/74							Craftsman	10/87	Brush	706	07/66
D1945	47502	02/74	47715	03/85					Haymarket	08/85	Brush	707	07/66
D1946	47503	05/74							[The Geordie]	05/88-04/91	Brush	708	07/66
D1947	47504	03/74	47702	03/79					Saint Cuthbert	03/79	Brush	709	07/66
D1948	47505	02/74	47712	09/79					Lady Diana Spencer	04/81	Brush	610	08/66
D1949	47506	01/74	47707	05/79					Holyrood	05/79	Brush	611	08/66
D1950	47259	03/74	47552	02/75	47802	07/89					Brush	612	09/66
D1951	47507	02/74	47716	03/85					Duke of Edinburgh's Award	07/85	Brush	613	10/66
D1952	47508	02/74							[SS Great Britain]	07/85-10/92	Brush	614	10/66
D1953	47509	02/74							[Albion]	03/79-11/92	Brush	615	11/66
D1954	47510	02/74	47713	01/85					[Tayside Region]	05/85-07/88	Brush	616	11/66
D1955	47511	03/74	47714	03/85					[Grampian Region]	05/85-02/89	Brush	617	12/66
D1956	47260	02/74	47553	09/74	47803	03/89			[Womans Guild]	06/89-07/92	Brush	618	12/66
D1957	47261*	-	47554	10/74	47705	03/79			[Lothian]	04/79-05/89	Brush	619	01/67

BR 1957 No.	Depot of First Allocation	Date Withdrawn	Depot of Final Allocation	Disposal Code	Disposal Detail	Date Cut Up	Notes
D1881	41A						
D1882	40B						
D1883	40B						
D1884	40B						
D1885	40B						
D1886	40B						
D1887	40B						
D1888	41A						
D1889	41A						
D1890	41A						
D1891	41A						
D1892	41A	06/92	IM	A	BR Scunthorpe		Stored: [U] 09/91
D1893	40B	07/92	IM	A	BR Immingham		
D1894	40B						
D1895	41A						
D1896	41A						
D1897	41A						
D1898	40B						
D1899	40B	07/92	IM	A	BR Scunthorpe		
D1900	40B	06/92	IM	A	BR Scunthorpe		
D1901	86A						
D1902	86A						
D1903	86A						Stored; [S] 04/92
D1904	86A						
D1905	86A						
D1906	86A	01/87	CF	C	Booth-Roe, Rotherham	04/89	
D1907	86A						
D1908	86A	08/69	86A	C	BR Crewe Works	10/69	
D1909	86A						Stored: [U] 01/92
D1910	86A	02/91	IM	A	BR Frodingham		Plates originally read 'Stombidae' Stored: [U] 02/91
D1911	86						
D1912	86A	11/88	TI	C	V Derry, Leicester	06/90	
D1913	86A						
D1914	86A						
D1915	86A						
D1916	86A						
D1917	86A						
D1918	86A						
D1919	86A						
D1920	86A						
D1921	86A						
D1922	86A						
D1923	86A						Withdrawn: 07/91, R/I: 02/92
D1924	86A						
D1925	82A						
D1926	82A						
D1927	82A						
D1928	82A						
D1929	82A						
D1930	82A						
D1931	82A						
D1932	82A						Named: Saint Andrew 01/79-10/90
D1933	82A						
D1934	82A						
D1935	82A						
D1936	86A						
D1937	86A						
D1938	82A						Named: Sir Walter Scott 10/79 - 09/90
D1939	LMWL						
D1940	LMWL						
D1941	LMWL						
D1942	LMWL						Named: Great Western 02/79-09/91
D1943	LMWL						
D1944	LMWL						
D1945	D16						Stored: [U] 09/91, R/I: 09/91
D1946	D16						
D1947	D16						
D1948	LMWL						
D1949	LMWL						
D1950	LMWL						
D1951	LMWL						Named: Great Britain 03/79-07/85
D1952	LMWL						
D1953	LMWL	11/92	BR	A	BR Bristol Bath Road		Named: Fair Rosamund 03/79-10/84
D1954	LMWL	07/88	ED	C	V Berry, Leicester	04/90	Named: Thames 03/79-09/84
D1955	LMWL						
D1956	LMWL						
D1957	LMWL						

BR 1957 No.	TOPS No.	Date Re No.	First TOPS Re No.	Date Re No.	Second TOPS Re No.	Date Re No.	Third TOPS Re No.	Date Re No.	Name	Name Date	Built By	Works No.	Date Introduced
D1958	47512	03/74									Brush	620	02/67
D1959	47513	02/74							Severn	03/79	Brush	621	02/67
D1960	47514	03/74	47703	03/79					The Queen Mother	03/91	Brush	622	07/67
D1961	47515	03/74							[Night Mail]	09/86-02/91	Brush	623	01/68
D1962	47262	04/74	47608	03/84	47833	07/89					BR Crewe	-	09/65
D1963	47263	02/74	47587	03/83					[Ruskin College Oxford]	10/90-05/92	BR Crewe	-	09/65
D1964	47264	07/74	47619	08/84	47829	06/89					BR Crewe	-	09/65
D1965	47265	02/74	47591	07/83	47804	07/89			Kettering	08/92	BR Crewe	-	10/65
D1966	47266	02/74	47629	11/85	47828	06/89					BR Crewe	-	10/65
D1967	47267	03/74	47603	12/83					[County of Somerset]	08/85-09/92	BR Crewe	-	11/65
D1968	47516	10/73	47708	06/79					Templecombe	06/91	BR Crewe	-	10/65
D1969	47268	04/74	47595	09/83	47675	08/91			Confederation of British Industry	11/83	BR Crewe	-	10/65
D1970	47269	02/74	47643	03/86							BR Crewe	-	10/65
D1971	47270	03/74									BR Crewe	-	10/65
D1972	47271	09/74	47604	12/83	47674	07/91			Women's Royal Voluntary Service	08/88	BR Crewe	-	11/65
D1973	47272	09/74	47593	08/83	47673	07/91			Galloway Princess	09/83	BR Crewe	-	11/65
D1974	47273	04/74	47627	12/84					[City of Oxford]	05/85-10/92	BR Crewe	-	11/65
D1975	47517	04/74							[Andrew Carnegie]	08/85-11/92	BR Crewe	-	11/65
D1976	47274	09/74	47637	01/86	47826	07/89			[Springburn]	06/87-11/89	BR Crewe	-	11/65
D1977	47275	03/74									BR Crewe	-	12/65
D1978	47276	02/74									BR Crewe	-	12/65
D1979	47277	10/73									BR Crewe	-	12/65
D1980	47278	03/74							Vasidae	11/88	BR Crewe	-	12/65
D1981	47279	02/74									BR Crewe	-	12/65
D1982	47280	03/74							Pedigree	01/86	BR Crewe	-	12/65
D1983	47281	03/74									BR Crewe	-	12/65
D1984	47282	03/74									BR Crewe	-	01/66
D1985	47283	04/74							Johnnie Walker	09/88	BR Crewe	-	01/66
D1986	47284	03/74									BR Crewe	-	01/66
D1987	47285	12/73									BR Crewe	-	01/66
D1988	47286	02/74									BR Crewe	-	02/66
D1989	47287	02/74									BR Crewe	-	02/66
D1990	47288	02/74									BR Crewe	-	02/66
D1991	47289	11/73									BR Crewe	-	03/66
D1992	47290	09/73									BR Crewe	-	03/66
D1993	47291	02/74							The Port of Felixstowe	03/87	BR Crewe	-	03/66
D1994	47292	12/73									BR Crewe	-	04/66
D1995	47293	03/74									BR Crewe	-	04/66
D1996	47294	02/74									BR Crewe	-	06/66
D1997	47295	03/74									BR Crewe	-	06/66
D1998	47296	02/74									BR Crewe	-	06/66
D1999	47297	02/74									BR Crewe	-	06/66

BR DIESEL MECHANICAL 0-6-0 CLASS 03

BR 1957 No.	BR 1948 No.	TOPS No.	Date Re No.	Built By	Works No.	Date Introduced	Depot of First Allocation	Date Withdrawn	Depot of Final Allocation
D2000	11187*	-	-	BR Swindon	-	12/57	34A	05/69	75C
D2001	11188*	-	-	BR Swindon	-	12/57	34A	06/69	73F
D2002	11189*	-	-	BR Swindon	-	12/57	34A	06/69	75C
D2003	11190*	-	-	BR Swindon	-	12/57	34A	05/69	73C
D2004	11191*	03004	02/74	BR Swindon	-	01/58	31A	05/76	CA
D2005	11192*	03005	02/74	BR Swindon	-	01/58	31A	11/76	CA
D2006	11193*	-	-	BR Swindon	-	01/58	31A	10/72	31A
D2007	11194*	03007	02/74	BR Swindon	-	01/58	31A	05/76	CA
D2008	11195*	03008	03/74	BR Swindon	-	02/58	31A	12/78	CA
D2009	11196*	03009	06/74	BR Swindon	-	02/58	31A	07/76	CR
D2010	11197*	03010	02/74	BR Swindon	-	02/58	31A	11/74	TE
D2011	11198*	-	-	BR Swindon	-	02/58	31A	10/72	31B
D2012	11199*	03012	02/74	BR Swindon	-	03/58	31A	12/75	MR
D2013	11200*	03013	02/74	BR Swindon	-	03/58	31A	07/76	YK
D2014	11201*	03014	03/74	BR Swindon	-	03/58	31A	06/74	SF
D2015	11202*	-	-	BR Swindon	-	03/58	31A	07/71	31B
D2016	11203*	03016	02/74	BR Swindon	-	04/58	31A	12/78	CA
D2017	11204*	03017	03/74	BR Swindon	-	04/58	31A	02/82	MR
D2018	11205*	03018	03/74	BR Swindon	-	04/58	34D	11/75	NR
D2019	11206*	-	-	BR Swindon	-	05/58	34D	07/71	32A
D2020	11207*	03020	05/74	BR Swindon	-	06/58	40B	12/75	NR
D2021	11208*	03021	05/74	BR Swindon	-	07/58	40B	11/82	GD
D2022	11209*	03022	05/74	BR Swindon	-	07/58	40B	11/82	GD
D2023	11210*	-	-	BR Swindon	-	08/58	40A	07/71	40A
D2024	11211*	-	-	BR Swindon	-	08/58	40A	07/71	40A
D2025	-	03025	05/74	BR Swindon	-	09/58	40A	09/77	LN
D2026	-	03026	05/74	BR Swindon	-	09/58	40A	02/83	LN
D2027	-	03027	05/74	BR Swindon	-	10/58	40A	01/76	CR

56

BR 1957 No.	Depot of First Allocation	Date Withdrawn	Depot of Final Allocation	Disposal Code	Disposal Detail	Date Cut Up	Notes
D1958	LMWL	10/91	CD	A	BR Crewe		Stored: [U] 02/91
D1959	LMWL						
D1960	LMWL						Named: St Mungo 03/79-12/90
D1961	LMWL	04/91	BR	A	BR Holbeck		
D1962	86A						
D1963	86A						
D1964	86A						
D1965	86A						
D1966	86A						Stored: [U] 09/91
D1967	86A						
D1968	64B						Named Waverley: 06/79-10/90
D1969	64B						
D1970	64B	04/91	IS	A	BR Inverness		
D1971	64B						
D1972	64B						Stored: [S] 06/92
D1973	64B						
D1974	64B						
D1975	64B						
D1976	64B						
D1977	52A	09/86	SF	C	C F Booth, Rotherham	03/89	Stored: [U] 08/86
D1978	52A						
D1979	52A						
D1980	52A						
D1981	52A						
D1982	52A						
D1983	52A						
D1984	52A	09/86	BR	C	C F Booth, Rotherham	03/87	
D1985	52A						
D1986	52A						
D1987	52A						
D1988	52A						
D1989	52A						
D1990	50A						
D1991	50A						
D1992	50A						
D1993	50A						
D1994	50A						
D1995	50A						
D1996	50A						
D1997	50A						
D1998	50A						
D1999	50A						

BR 1957 No.	Disposal Code	Disposal Detail	Date Cut Up	Notes
D2000	C	Steelbreaking and Dismantling Co, Chesterfield	08/69	Carried No. 11187 when built.
D2001	C	C F Booth, Rotherham	10/70	
D2002	C	BR Kentish Town, by Ingot Metals	09/69	
D2003	C	BR Kentish Town, by Ingot Metals	09/69	
D2004	C	G Cohen, Kettering	11/76	
D2005	C	BREL Doncaster	03/77	
D2006	C	BREL Swindon	06/73	
D2007	C	G Cohen, Kettering	11/76	
D2008	C	BREL Swindon	03/79	
D2009	C	G Cohen, Kettering	11/76	
D2010	E	Exported to Trieste, Italy [05/76]	-	After withdrawal to P Wood, Queenborough
D2011	C	BREL Swindon	08/73	
D2012	C	Meyer Newman, Snailwell	01/91	PI
D2013	C	BREL Doncaster	03/77	
D2014	C	BREL Doncaster	04/76	
D2015	C	G Cohen, Kettering	07/72	
D2016	C	BR Swindon Works	05/79	
D2017	C	BREL Swindon	00/82	
D2018	S	600 Fragmentisers, Willesden [No.2/600]		
D2019	E	Exported to Brescia, Italy [09/72]	-	After withdrawal to P Wood, Queenborough
D2020	S	Meyer Newman, Snailwell [F134L]		
D2021	C	BREL Swindon	03/83	
D2022	P	Swindon and Cricklade Railway	-	
D2023	P	Kent and East Sussex Railway [3]	-	PI
D2024	P	Kent and East Sussex Railway [4]	-	PI
D2025	C	BREL Swindon	03/78	Stored: [U] 07/77
D2026	C	C F Booth, Rotherham	04/84	Withdrawn: 11/82, R/I: 11/82
D2027	P	South Yorkshire Railway	-	PI

57

BR 1957 No.	BR 1948 No.	TOPS No.	Date Re No.	Built By	Works No.	Date Introduced	Depot of First Allocation	Date Withdrawn	Depot of Final Allocation
D2028	-	-	-	BR Swindon	-	10/58	31A	12/69	70D
D2029	-	03029	04/74	BR Swindon	-	10/57	31A	09/79	NR
D2030	-	-	-	BR Swindon	-	10/58	31A	08/69	75A
D2031	-	-	-	BR Swindon	-	11/58	31A	05/69	75A
D2032	-	-	-	BR Swindon	-	11/58	32A	07/71	32A
D2033	-	-	-	BR Swindon	-	12/58	32A	12/71	32A
D2034	-	03034	04/74	BR Swindon	-	01/59	32A	02/83	LN
D2035	-	03035	04/74	BR Swindon	-	02/59	32A	07/76	NR
D2036	-	-	-	BR Swindon	-	02/59	32A	12/71	32A
D2037	-	03037	04/74	BR Swindon	-	02/59	32A	09/76	NR
D2038	-	-	-	BR Swindon	-	02/59	32A	03/72	32A
D2039	-	-	-	BR Swindon	-	03/59	32A	02/72	32A
D2040	-	-	-	BR Swindon	-	03/59	32B	06/69	73F
D2041	-	-	-	BR Swindon	-	04/59	32B	02/70	75C
D2042	-	-	-	BR Swindon	-	04/59	32B	06/69	75C
D2043	-	-	-	BR Swindon	-	05/59	32B	09/71	73F
D2044	-	03044	02/74	BR Doncaster	-	11/58	52F	01/76	DN
D2045	-	03045	12/73	BR Doncaster	-	12/58	52F	02/79	41E
D2046	-	-	-	BR Doncaster	-	12/58	52F	10/71	51L
D2047	-	03047	03/74	BR Doncaster	-	12/58	52B	07/79	HS
D2048	-	-	-	BR Doncaster	-	12/58	52A	10/72	32A
D2049	-	-	-	BR Doncaster	-	01/59	52C	08/71	GO
D2050	-	03050	04/74	BR Doncaster	-	01/59	52B	08/78	CA
D2051	-	-	-	BR Doncaster	-	01/59	53A	12/72	30E
D2052	-	-	-	BR Doncaster	-	02/59	53A	05/72	52A
D2053	-	-	-	BR Doncaster	-	04/59	53A	05/72	52A
D2054	-	-	-	BR Doncaster	-	04/59	53A	11/72	55B
D2055	-	03055	01/74	BR Doncaster	-	04/59	52A	06/74	SF
D2056	-	03056	02/74	BR Doncaster	-	05/59	52A	06/80	GD
D2057	-	-	-	BR Doncaster	-	05/59	52A	10/71	51L
D2058	-	03058	02/74	BR Doncaster	-	05/59	52A	06/75	GD
D2059	-	03059	01/74	BR Doncaster	-	06/59	52A	07/87	NC
D2060	-	03060	06/74	BR Doncaster	-	06/59	52A	12/82	HS
D2061	-	03061	02/74	BR Doncaster	-	06/59	52A	10/80	GD
D2062	-	03062	01/74	BR Doncaster	-	07/59	50C	12/80	NR
D2063	-	03063	02/74	BR Doncaster	-	07/59	50A	11/87	GD
D2064	-	03064	02/74	BR Doncaster	-	07/59	50A	06/81	GD
D2065	-	-	-	BR Doncaster	-	08/59	50A	12/72	TE
D2066	-	03066	02/74	BR Doncaster	-	08/59	50A	01/88	GD
D2067	-	03067	02/74	BR Doncaster	-	09/59	51L	08/81	GD
D2068	-	03068	03/74	BR Doncaster	-	09/59	51C	04/76	TE
D2069	-	03069	10/73	BR Doncaster	-	09/59	51C	12/83	GD
D2070	-	-	-	BR Doncaster	-	10/59	51C	11/71	51L
D2071	-	-	-	BR Doncaster	-	10/59	56G	05/72	52A
D2072	-	03072	02/74	BR Doncaster	-	10/59	56G	03/81	DN
D2073	-	03073	03/74	BR Doncaster	-	11/59	56G	03/89	BD
D2074	-	-	-	BR Doncaster	-	11/59	56G	05/72	52A
D2075	-	03075	02/74	BR Doncaster	-	11/59	56G	07/76	TE
D2076	-	03076	11/73	BR Doncaster	-	12/59	51A	03/76	TE
D2077	-	-	-	BR Doncaster	-	12/59	51C	10/72	51L
D2078	-	03078	02/74	BR Doncaster	-	12/59	51L	01/88	GD
D2079	-	03079	01/74	BR Doncaster	-	01/60	51L		
D2080	-	03080	02/74	BR Doncaster	-	01/60	51A	12/80	DN
D2081	-	03081	05/74	BR Doncaster	-	01/60	56G	12/80	MR
D2082	-	-	-	BR Doncaster	-	02/59	75A	12/69	70D
D2083	-	-	-	BR Doncaster	-	03/59	73C	06/69	70D
D2084	-	03084	12/73	BR Doncaster	-	03/59	73F	07/87	NC
D2085	-	-	-	BR Doncaster	-	04/59	71A	12/69	70D
D2086	-	03086	12/73	BR Swindon	-	05/59	82C	11/83	BC
D2087	-	-	-	BR Swindon	-	06/59	82C	06/71	LE
D2088	-	-	-	BR Swindon	-	07/59	82C	06/72	30E
D2089	-	03089	02/74	BR Doncaster	-	05/60	56G	11/87	NC
D2090	-	03090	02/74	BR Doncaster	-	05/60	56G	07/76	YK
D2091	-	03091	02/74	BR Doncaster	-	06/60	56G	03/74	HS
D2092	-	03092	05/74	BR Doncaster	-	06/60	52E	08/77	SF
D2093	-	-	-	BR Doncaster	-	06/60	52E	10/71	51L
D2094	-	03094	02/74	BR Doncaster	-	07/60	50A	01/88	GD
D2095	-	03095	02/74	BR Doncaster	-	07/60	50A	12/75	DN
D2096	-	03096	02/74	BR Doncaster	-	07/60	50A	12/76	TE
D2097	-	03097	02/74	BR Doncaster	-	07/60	50A	06/76	HM
D2098	-	03098	02/74	BR Doncaster	-	07/60	50A	11/75	DN
D2099	-	03099	02/74	BR Doncaster	-	08/60	50A	02/76	TE
D2100	-	-	-	BR Doncaster	-	08/60	50A	11/71	50C
D2101	-	-	-	BR Doncaster	-	08/60	50A	11/71	55B
D2102	-	03102	02/74	BR Doncaster	-	09/60	50A	02/76	GD
D2103	-	03103	03/74	BR Doncaster	-	09/60	50A	02/79	NR
D2104	-	03104	02/74	BR Doncaster	-	09/60	50A	06/75	GD
D2105	-	03105	02/74	BR Doncaster	-	09/60	50A	02/76	GD
D2106	-	03106	02/74	BR Doncaster	-	09/60	50A	09/75	GD
D2107	-	03107	02/74	BR Doncaster	-	10/60	50A	07/81	GD
D2108	-	03108	02/74	BR Doncaster	-	10/60	41A	11/76	GD
D2109	-	03109	05/74	BR Doncaster	-	10/60	51A	07/75	CR

BR 1957 No.	Disposal Code	Disposal Detail	Date Cut Up	Notes
D2028	C	BREL Doncaster	02/72	
D2029	C	BREL Doncaster	10/79	Stored: [S] 02/77, R/I: 04/77
D2030	C	C F Booth, Rotherham	11/70	
D2031	C	Pollock and Brown, Southampton	09/69	
D2032	E	Exported to Brescia, Italy [10/83]	-	After withdrawal to P Wood, Queenborough
D2033	E	Exported to Brescia, Italy [10/83]	-	After withdrawal to P Wood, Queenborough
D2034	C	C F Booth, Rotherham	08/83	Withdrawn: 12/76, R/I: 07/77
D2035	C	G Cohen, Kettering	09/77	
D2036	E	Exported to Brescia, Italy [10/83]	-	After withdrawal to P Wood Queenborough
D2037	S	Oxcroft DP, Clowne	-	
D2038	C	T W Ward, Beighton	05/73	
D2039	C	T W Ward, Beighton	05/77	
D2040	C	C F Booth, Rotherham	10/70	
D2041	P	Colne Valley Railway [No. 1]	-	PI
D2042	C	BR Kentish Town, by Ingot Metals	09/69	
D2043	C	P Wood, Queenborough	06/72	
D2044	C	G Cohen, Kettering	09/76	
D2045	C	BREL Doncaster	09/79	Stored: [U] 01/69, R/I: 03/69, Stored: [U] 04/77
D2046	S	Gulf Oil, Waterston	-	Rebuilt by Hunslet as 6644 of 1967
D2047	C	BREL Doncaster	10/79	Withdrawn: 06/76, R/I: 11/76
D2048	C	BREL Swindon	05/73	
D2049	C	NCB British Oak by Wath Skip Hire	11/85	
D2050	C	C F Booth, Rotherham	05/79	Withdrawn: 06/76, R/I: 11/76
D2051	S	Ford Motors, Dagenham [No. 4]	-	
D2052	C	A Draper, Hull	12/73	
D2053	C	A Draper, Hull	11/73	
D2054	C	C F Booth, Rotherham	09/82	PI
D2055	C	BREL Doncaster	11/74	
D2056	C	BREL Doncaster	03/81	
D2057	C	C F Booth, Rotherham	05/86	Rebuilt by Hunslet as 6645 of 1967, PI
D2058	C	BREL Doncaster	02/77	
D2059	P	Isle of Wight Steam Railway	-	
D2060	C	BREL Doncaster	07/83	
D2061	C	BREL Swindon	02/82	
D2062	P	Dean Forest Railway	-	Withdrawn: 06/76, R/I: 09/76, Stored: [U] 10/80
D2063	P	Colne Valley Railway	-	
D2064	C	BREL Doncaster	08/81	
D2065	C	C F Booth, Rotherham	07/73	
D2066	P	South Yorkshire Railway	-	PI
D2067	C	BREL Doncaster	08/82	
D2068	C	G Cohen, Kettering	09/76	
D2069	P	Gloucester & Warwickshire Railway	-	PI
D2070	P	South Yorkshire Railway	-	PI
D2071	C	A Draper, Hull	11/73	
D2072	P	Lakeside Railway	-	
D2073	P	The Railway Age, Crewe	-	
D2074	C	A Draper, Hull	11/73	
D2075	C	BREL Doncaster	01/79	
D2076	C	G Cohen, Kettering	10/76	
D2077	C	BREL Swindon	05/73	
D2078	P	Stephenson Railway Museum	-	
D2079				To Departmental Stock as No. 97805 04/84-02/89
D2080	C	BREL Swindon	03/81	Stored: [U] 11/00
D2081	E	Exported to Maldegem, Belgium [11/81]	-	Stored: [U] 11/80
D2082	C	BREL Doncaster	09/71	
D2083	C	Pollock and Brown, Southampton	11/69	
D2084	P	Private in Derbyshire	-	
D2085	C	BREL Doncaster	03/72	
D2086	C	BREL Doncaster	03/84	
D2087	C	Pounds Shipbreakers, Fratton	08/73	Stored: [U] 04/71
D2088	C	C F Booth, Rotherham	03/73	
D2089	P	Mangapps Farm, Burnham on Crouch	-	
D2090	P	National Railway Museum, York	-	
D2091	C	BREL Doncaster	03/77	
D2092	C	BREL Doncaster	01/78	
D2093	C	C F Booth, Rotherham	04/86	Rebuilt by Hunslet as 6643 of 1967, PI
D2094	P	South Yorkshire Railway	-	
D2095	C	G Cohen, Kettering	10/76	
D2096	C	BREL Doncaster	03/77	
D2097	C	BREL Doncaster	03/79	
D2098	E	Exported to Italy [05/76]	-	After withdrawal to P Wood Queenborough
D2099	P	South Yorkshire Railway	-	
D2100	C	G Cohen, Kettering	07/72	
D2101	C	T W Ward, Beighton	06/72	
D2102	C	G Cohen, Kettering	09/76	
D2103	C	BREL Doncaster	05/79	
D2104	C	BREL Doncaster	02/76	
D2105	C	G Cohen, Kettering	09/76	
D2106	C	W Heselwood, Attercliffe	05/76	
D2107	C	BREL Doncaster	10/82	
D2108	C	BREL Doncaster	03/77	
D2109	C	BREL Doncaster	04/76	

BR 1957 No.	BR 1948 No.	TOPS No.	Date Re No.	Built By	Works No.	Date Introduced	Depot of First Allocation	Date Withdrawn	Depot of Final Allocation
D2110	-	03110	02/74	BR Doncaster	-	11/60	50A	02/76	GD
D2111	-	03111	11/73	BR Doncaster	-	11/60	50A	07/80	GD
D2112	-	03112	03/74	BR Doncaster	-	11/60	50A	07/87	GD
D2113	-	03113	04/74	BR Doncaster	-	12/60	50A	08/75	YK
D2114	-	-	-	BR Swindon	-	07/59	87C	05/68	82C
D2115	-	-	-	BR Swindon	-	08/59	87C	05/68	85A
D2116	-	-	-	BR Swindon	-	08/59	87C	10/71	12C
D2117	-	-	-	BR Swindon	-	09/59	87C	10/71	8F
D2118	-	-	-	BR Swindon	-	09/59	87C	06/72	12C
D2119	-	03119	03/74	BR Swindon	-	09/59	87C	02/86	LE
D2120	-	03120	03/74	BR Swindon	-	10/59	87C	02/86	LE
D2121	-	03121	01/74	BR Swindon	-	10/59	87C	05/81	BR
D2122	-	-	-	BR Swindon	-	10/59	87C	11/72	82A
D2123	-	-	-	BR Swindon	-	11/59	87C	11/68	82C
D2124	-	-	-	BR Swindon	-	12/59	87C	02/70	12C
D2125	-	-	-	BR Swindon	-	12/59	87C	11/68	82C
D2126	-	-	-	BR Swindon	-	12/59	87C	10/71	8F
D2127	-	-	-	BR Swindon	-	12/59	87C	05/68	85A
D2128	-	03128	02/74	BR Swindon	-	01/60	83D	07/76	BR
D2129	-	03129	02/74	BR Swindon	-	01/60	83A	12/81	BH
D2130	-	-	-	BR Swindon	-	01/60	83A	08/72	16C
D2131	-	-	-	BR Swindon	-	02/60	83B	06/68	82C
D2132	-	-	-	BR Swindon	-	02/60	83B	05/69	82C
D2133	-	-	-	BR Swindon	-	02/60	83B	07/69	82A
D2134	-	03134	02/74	BR Swindon	-	02/60	83D	07/76	BR
D2135	-	03135	03/75	BR Swindon	-	03/60	82E	01/76	CR
D2136	-	-	-	BR Swindon	-	03/60	85B	01/72	82A
D2137	-	03137	03/74	BR Swindon	-	04/60	85B	07/76	BG
D2138	-	-	-	BR Swindon	-	04/60	85B	05/69	82C
D2139	-	-	-	BR Swindon	-	04/60	85B	05/68	85A
D2140	-	-	-	BR Swindon	-	04/60	83B	04/70	84A
D2141	-	03141	03/74	BR Swindon	-	05/60	83B	07/85	LE
D2142	-	03142	02/74	BR Swindon	-	05/60	83B	10/83	LE
D2143	-	-	-	BR Swindon	-	01/61	82C	06/68	87E
D2144	-	03144	02/74	BR Swindon	-	01/61	82C	02/86	LE
D2145	-	03145	02/74	BR Swindon	-	02/61	82C	07/85	LE
D2146	-	-	-	BR Swindon	-	02/61	82C	09/68	82C
D2147	-	03147	02/74	BR Swindon	-	06/60	50A	09/75	GD
D2148	-	-	-	BR Swindon	-	06/60	50A	11/71	55C
D2149	-	03149	04/74	BR Swindon	-	06/60	52E	11/82	LN
D2150	-	-	-	BR Swindon	-	06/60	50A	11/72	55B
D2151	-	03151	05/74	BR Swindon	-	07/60	50A	07/85	LE
D2152	-	03152	05/74	BR Swindon	-	07/60	50A	10/83	LE
D2153	-	03153	03/74	BR Swindon	-	07/60	50A	11/75	TE
D2154	-	03154	02/74	BR Swindon	-	07/60	50A	09/83	MR
D2155	-	03155	03/74	BR Swindon	-	08/60	50A	06/75	NR
D2156	-	03156	01/74	BR Swindon	-	08/60	50A	11/75	GD
D2157	-	03157	06/74	BR Swindon	-	08/60	50A	12/75	BG
D2158	-	03158	03/74	BR Swindon	-	09/60	50A	07/87	NC
D2159	-	03159	05/74	BR Swindon	-	09/60	50A	10/77	TE
D2160	-	03160	03/75	BR Swindon	-	09/60	50A	12/81	SF
D2161	-	03161	03/75	BR Swindon	-	09/60	50A	12/81	SF
D2162	-	03162	02/74	BR Swindon	-	10/60	50A	03/89	BD
D2163	-	03163	02/74	BR Swindon	-	10/60	50A	01/76	GD
D2164	-	03164	01/74	BR Swindon	-	10/60	50A	01/76	SF
D2165	-	03165	02/74	BR Swindon	-	10/60	50A	08/75	GD
D2166	-	03166	02/74	BR Swindon	-	10/60	51A	11/75	HM
D2167	-	03167	02/74	BR Swindon	-	11/60	50A	07/75	YK
D2168	-	03168	04/74	BR Swindon	-	11/60	50B	08/81	SF
D2169	-	03169	05/74	BR Swindon	-	11/60	50B	11/75	BG
D2170	-	03170	03/74	BR Swindon	-	11/60	50B	03/89	BD
D2171	-	03171	05/74	BR Swindon	-	11/60	50B	10/77	SF
D2172	-	03172	02/74	BR Swindon	-	11/60	50B	05/76	YK
D2173	-	-	-	BR Swindon	-	11/60	50B	11/73	BG
D2174	-	03174	03/74	BR Swindon	-	12/60	50B	11/75	HM
D2175	-	03175	04/74	BR Swindon	-	12/60	73F	05/83	MR
D2176	-	-	-	BR Swindon	-	12/61	73F	05/68	C-WKS
D2177	-	-	-	BR Swindon	-	12/61	73F	09/68	85A
D2178	-	-	-	BR Swindon	-	01/62	73F	09/69	84A
D2179	-	03179	12/73	BR Swindon	-	01/62	73C		
D2180	-	03180	01/74	BR Swindon	-	02/62	73C	03/84	NC
D2181	-	-	-	BR Swindon	-	02/62	82C	05/68	87E
D2182	-	-	-	BR Swindon	-	03/62	83B	05/68	87E
D2183	-	-	-	BR Swindon	-	03/62	83B	09/68	87E
D2184	-	-	-	BR Swindon	-	04/62	87F	12/68	82A
D2185	-	-	-	BR Swindon	-	05/62	87C	12/68	85A
D2186	-	-	-	BR Swindon	-	05/62	82C	09/69	82C
D2187	-	-	-	BR Swindon	-	03/61	82C	05/68	82C
D2188	-	-	-	BR Swindon	-	03/61	82C	05/68	83B
D2189	-	03189	02/74	BR Swindon	-	03/61	82C	03/86	CH

BR 1957 No.	Disposal Code	Disposal Detail	Date Cut Up	Notes
D2110	C	G Cohen, Kettering	09/76	
D2111	C	BREL Swindon	02/81	Withdrawn: 06/76, R/I: 11/76
D2112	P	Nene Valley Railway	-	PI
D2113	P	Heritage & Maritime Museum, Milford Haven	-	
D2114	C	Birds, Long Marston	01/75	Stored: [U] 07/67
D2115	C	G Cohen, Kingsbury	09/68	Stored: [U] 09/67
D2116	C	Marple and Gillott, Sheffield	08/73	
D2117	P	Lakeside Railway [No. 8]	-	
D2118	S	Anglia Building Products, Lenwade	-	
D2119	P	Dean Forest Railway	-	Stored: [U] 07/69, Withdrawn: 01/70, R/I: 12/70, PI
D2120	P	W H McAlpine, Fawley Hill	-	Stored: [S] 08/78, R/I: 11/78
D2121	C	BREL Swindon	09/85	
D2122	C	J Cashmore at Duport Steel, Briton Ferry	08/76	
D2123	C	Birds, Bristol	11/78	PI, Stored: [U] 08/68
D2124	C	Slag Reduction, Barrow	09/70	
D2125	C	Birds, Cardiff	06/76	PI, Stored: [U] 10/68
D2126	C	P Wood, Queenborough	09/73	
D2127	C	G Cohen, Kingsbury	10/68	Stored: [U] 01/67
D2128	E	Exported to Maldegem, Belgium [12/76]	-	Withdrawn: 01/70, R/I: 04/70
D2129	C	C F Booth, Rotherham	03/83	In use after withdrawal
D2130	C	C F Booth, Rotherham	03/73	Withdrawn: 06/76, R/I: 07/72
D2131	C	G Cohen, Kettering	11/68	Stored: [U] 06/86
D2132	C	C F Booth, Rotherham	11/84	Stored: [U] 11/68, R/I: 01/69, Stored: [U] 05/69, PI
D2133	S	British Cellophane, Bridgwater	-	
D2134	P	South Yorkshire Railway	-	PI
D2135	C	P Wood, Queenborough	07/76	Stored: [U] 08/67, R/I: 11/67
D2136	C	Robinson and Hanon, Blaydon	06/72	
D2137	C	BREL Doncaster	12/77	Stored: [U] 08/67, R/I: 11/67
D2138	P	Midland Railway Centre, Butterley	-	Stored: [U] 05/69, PI
D2139	P	South Yorkshire Railway	-	Stored: [U] 08/67, PI
D2140	C	BREL Swindon	02/72	
D2141	S	J M DeMulder, Shilton	-	
D2142	C	BREL Swindon	09/85	Stored: [U] 05/83
D2143	C	J Cashmores, Newport	12/68	Stored: [U] 06/68
D2144	S	MOD Long Marston [Western Waggoner]	-	
D2145	S	J M DeMulder, Shilton	-	
D2146	C	Birds, Long Marston	07/78	Stored: [U] 08/68
D2147	C	C F Booth, Doncaster	04/76	
D2148	P	Steamport, Southport	-	PI
D2149	C	BREL Doncaster	06/83	Stored: [U] 01/77
D2150	S	British Salt, Middlewich	-	
D2151	C	C F Booth, Rotherham	10/85	Stored: [U] 05/85
D2152	P	Swindon Railway Workshops Ltd	-	
D2153	E	Exported to Trieste, Italy [05/76]	-	After withdrawal to P Wood, Queenborough
D2154	C	BREL Doncaster	11/83	Stored: [S] 07/77
D2155	C	BREL Doncaster	04/76	
D2156	E	Exported to Trieste, Italy [05/76]	-	After withdrawal to P Wood, Queenborough
D2157	E	Exported to Trieste, Italy [02/77]	-	After withdrawal to P Wood, Queenborough
D2158	P	Private in Derbyshire	-	
D2159	C	BREL Swindon	10/78	Stored: [U] 07/77
D2160	C	C F Booth, Rotherham	04/83	Stored: [U] 04/81
D2161	C	C F Booth, Rotherham	04/83	Withdrawn: 09/76, R/I: 02/77, Stored: [U] 04/81
D2162	P	Llangollen Railway, by Wirral Metropolitan Council	-	
D2163	C	G Cohen, Kettering	09/76	
D2164	E	Exported to Trieste, Italy [02/77]	-	After withdrawal to P Wood, Queenborough
D2165	C	C F Booth, Doncaster	02/76	
D2166	C	C F Booth, Rotherham	06/76	
D2167	C	BREL Doncaster	06/76	
D2168	C	BREL Doncaster	04/82	
D2169	C	BR Botanic Gardens, by A Draper	06/76	
D2170	S	Otis Euro-Trancrail, Manchester	-	
D2171	C	G Cohen, Kettering	04/78	Stored: [S] 07/77, R/I: 09/77
D2172	C	G Cohen, Kettering	08/77	
D2173	C	BREL Doncaster	03/77	
D2174	C	C F Booth, Doncaster	05/76	
D2175	C	BREL Doncaster	11/83	Stored: [S] 07/67, R/I: 11/67, Stored: [S] 06/83
D2176	C	G Cohen, Kettering	11/71	
D2177	C	Birds, Long Marston	06/70	Stored: [S] 12/67, R/I: 02/68, Stored: [U] 03/68, R/I: 03/68, Stored: [U] 07/68
D2178	P	Caerphilly Railway Centre	-	PI
D2179				To Departmental Stock as No. 97807 07/87-02/89
D2180	P	South Yorkshire Railway	-	PI
D2181	C	Marple & Gillott, Sheffield	01/87	Stored: [S] 07/67, PI
D2182	P	Gloucestershire Warwickshire Railway	-	Stored: [S] 07/67, PI
D2183	C	Birds, Long Marston	03/70	Stored: [S] 07/67, R/I: 06/68, Stored: [U] 08/68
D2184	P	Colne Valley Railway	-	Stored: [S] 08/67, PI
D2185	C	Birds, Long Marston	06/78	Stored: [S] 08/68, R/I: 11/68, PI
D2186	C	A R Adams, Newport	01/81	
D2187	C	Birds, Long Marston	06/78	Stored: [S] 08/67, PI
D2188	C	Birds, Long Marston	02/78	Stored: [U] 08/67, R/I: 11/67, Stored: [S] 01/68, PI
D2189	P	Steamport, Southport	-	Stored: [U] 10/67, R/I: 11/67, Withdrawn: 03/86,

BR 1957 No.	BR 1948 No.	TOPS No.	Date Re No.	Built By	Works No.	Date Introduced	Depot of First Allocation	Date Withdrawn	Depot of Final Allocation
D2190	-	-	-	BR Swindon	-	04/61	82C	12/68	82A
D2191	-	-	-	BR Swindon	-	04/61	82C	05/68	87E
D2192	-	-	-	BR Swindon	-	05/61	82C	01/69	82C
D2193	-	-	-	BR Swindon	-	05/61	82C	01/69	82A
D2194	-	-	-	BR Swindon	-	05/61	82C	09/68	85A
D2195	-	-	-	BR Swindon	-	06/61	82C	09/68	82A
D2196	-	03196	01/74	BR Swindon	-	06/61	82C	06/83	BC
D2197	-	03197	01/74	BR Swindon	-	06/61	83D	07/87	NC
D2198	-	-	-	BR Swindon	-	06/61	8A	11/70	8F
D2199	-	-	-	BR Swindon	-	06/61	8C	06/72	12C

Class 03 continued from D2370

DREWRY DIESEL MECHANICAL 0-6-0 CLASS 04

BR 1957 No.	Date Re No.	BR 1948 No.	Built By	Works No.	Date Introduced	Depot of First Allocation	Date Withdrawn	Depot of Final Allocation
D2200	09/60	11100	DC.VF	2397/D142	05/52	32B	04/68	C-WKS
D2201	12/61	11101	DC.VF	2398/D143	05/52	32E	04/68	C-WKS
D2202	01/58	11102	DC.VF	2399/D144	05/52	31B	02/68	C-WKS
D2203	10/58	11103	DC.VF	2400/D145	06/52	31B	12/67	C-WKS
D2204	05/58	11105	DC.VF	2485/D211	03/53	51C	10/69	55F
D2205	01/59	11106	DC.VF	2486/D212	03/53	51C	07/69	51L
D2206	07/58	11107	DC.VF	2487/D213	03/53	51C	07/69	51A
D2207	10/58	11108	DC.VF	2482/D208	02/53	30F	12/67	C-WKS
D2208	09/58	11109	DC.VF	2483/D209	02/53	30F	07/68	5A
D2209	09/58	11110	DC.VF	2484/D210	03/53	30F	07/68	8J
D2210	06/58	11111	DC.VF	2508/D242	09/54	30F	05/70	32A
D2211	07/59	11112	DC.VF	2509/D243	09/54	32B	07/70	16C
D2212	12/58	11113	DC.VF	2510/D244	09/54	30F	11/70	32A
D2213	07/61	11114	DC.VF	2529/D257	10/54	34A	08/68	8H
D2214	02/58	11115	DC.VF	2530/D258	10/54	30F	07/68	8F
D2215	03/60	11121	DC.VF	2538/D264	07/55	30A	02/69	30A
D2216	01/59	11122	DC.VF	2539/D265	07/55	30A	05/71	30A
D2217	11/60	11123	DC.VF	2540/D266	08/55	30A	05/72	30A
D2218	02/61	11124	DC.VF	2541/D267	08/55	32A	08/68	5A
D2219	12/59	11125	DC.VF	2542/D268	08/55	32A	04/68	8H
D2220	11/59	11126	DC.VF	2543/D269	09/55	32A	02/68	2F
D2221	04/59	11127	DC.VF	2544/D270	09/55	30A	07/68	8J
D2222	01/60	11128	DC.VF	2545/D271	09/55	30A	11/68	52A
D2223	09/58	11129	DC.VF	2546/D272	10/55	30A	05/71	30E
D2224	03/60	11130	DC.VF	2547/D273	10/55	30A	04/68	9D
D2225	09/59	11131	DC.VF	2548/D274	10/55	30A	03/69	8F
D2226	02/62	11132	DC.VF	2549/D275	11/55	30A	04/68	9D
D2227	10/59	11133	DC.VF	2550/D276	11/55	30A	04/68	9D
D2228	10/59	11134	DC.VF	2551/D277	11/55	30A	07/68	8F
D2229	10/60	11135	DC.VF	2552/D278	11/55	30A	12/69	52A
D2230	02/59	11149	DC.VF	2554/D280	01/56	51C	03/68	52A
D2231	02/59	11150	DC.VF	2555/D281	01/56	51C	06/69	16C
D2232	03/59	11151	DC.VF	2556/D282	01/56	51C	03/68	52A
D2233	12/59	11152	DC.VF	2557/D283	02/56	40B	09/68	5A
D2234	02/58	11153	DC.VF	2558/D284	02/56	40B	04/68	9D
D2235	08/59	11154	DC.VF	2559/D285	03/56	40B	04/68	8H
D2236	02/62	11155	DC.VF	2560/D286	03/56	34A	02/68	W-WKS
D2237	11/60	11156	DC.VF	2561/D287	04/56	35A	10/69	52A
D2238	04/61	11157	DC.VF	2562/D288	05/56	38A	07/68	8H
D2239	02/59	11158	DC.VF	2563/D289	05/56	35A	09/71	75C
D2240	09/60	11159	DC.VF	2564/D290	06/56	35A	04/68	8H
D2241	08/59	11160	DC.VF	2565/D291	06/56	35A	05/71	30E
D2242	01/59	11212	DC.RSH	2572/7858	10/56	50B	10/69	55H
D2243	01/59	11213	DC.RSH	2575/7862	10/56	50B	07/69	51L
D2244	01/59	11214	DC.RSH	2576/7863	11/56	50B	06/70	55H
D2245	02/59	11215	DC.RSH	2577/7864	11/56	50B	12/68	50D
D2246	03/59	11216	DC.RSH	2578/7865	12/56	50B	07/68	55G
D2247	03/59	11217	DC.RSH	2579/7866	12/56	52E	11/69	55B
D2248	03/59	11218	DC.RSH	2580/7867	01/57	52E	06/70	55F
D2249	11/58	11219	DC.RSH	2581/7868	01/57	52E	12/70	30E
D2250	02/62	11220	DC.RSH	2594/7871	05/57	73C	06/68	73F
D2251	08/63	11221	DC.RSH	2595/7872	05/57	73C	12/68	70D
D2252	01/62	11222	DC.RSH	2596/7873	05/57	75A	10/68	73C
D2253	07/61	11223	DC.RSH	2597/7874	05/57	73C	03/69	8J
D2254	11/61	11224	DC.RSH	2598/7875	06/57	73C	06/67	75C
D2255	10/62	11225	DC.RSH	2599/7876	06/57	73C	03/68	70D
D2256	02/63	11226	DC.RSH	2600/7877	10/57	73C	09/68	73F
D2257	11/63	11227	DC.RSH	2601/7878	10/57	72D	01/68	73F
D2258	05/63	11228	DC.RSH	2602/7879	10/57	72D	09/70	16C
D2259	06/63	11229	DC.RSH	2603/7889	10/57	72D	12/68	73F
D2260	-	-	DC.RSH	2604/7890	12/57	56G	10/70	55F
D2261	-	-	DC.RSH	2605/7891	12/57	56G	03/70	55F

BR 1957 No.	Disposal Code	Disposal Detail	Date Cut Up	Notes
D2190	C	Birds, Long Marston	06/70	Stored: [U] 11/68
D2191	C	G Cohen, Kingsbury	11/68	Stored: [S] 07/67
D2192	P	Paignton & Dartmouth Railway	-	Stored: [S] 10/68, R/I: 11/68, Stored: [U] 01/69
D2193	C	A R Adams, Newport	01/81	Stored: [U] 01/69
D2194	C	Birds, Long Marston	07/78	Stored: [U] 09/68, PI
D2195	C	Duport Steel, Llanelly	09/81	Stored: [U] 08/68, PI
D2196	P	Steamtown, Carnforth	-	Stored: [U] 08/67, R/I: 12/67, PI
D2197	P	South Yorkshire Railway	-	Stored: [U] 07/67, R/I: 11/67, Stored: [U] 01/68, R/I: 01/68, Withdrawn: 12/85, R/I: 03/86
D2198	C	BREL, Doncaster	03/72	Stored: [S] 03/70
D2199	P	South Yorkshire Railway	-	PI

BR 1957 No.	Disposal Code	Disposal Detail	Date Cut Up	Notes
D2200	C	A Finlay, Birdwell	07/68	
D2201	C	A Finlay, Birdwell	07/68	
D2202	C	G Cohen, Kettering	06/68	
D2203	P	Yorkshire Dales Railway	-	PI
D2204	C	Duport Steel, Briton Ferry	09/79	
D2205	P	West Somerset Railway	-	PI
D2206	C	Hughes Bolckow, Blyth	12/69	
D2207	P	North Yorkshire Moors Railway	-	PI
D2208	C	NCB Silverwood	05/79	Stored: [U] 10/66, R/I: 05/67, PI
D2209	C	NCB Kiveton Park	11/85	PI
D2210	C	A King, Norwich	09/70	
D2211	C	Rees Industries, Saron Works, Llanelly	11/80	Stored: [U] 10/66, R/I 05/67, PI
D2212	C	C F Booth, Rotherham	07/72	
D2213	C	NCB Manvers Main	02/78	PI
D2214	C	C F Booth, Rotherham	08/69	
D2215	C	Birds, Long Marston	02/79	
D2216	E	Exported to Brescia, Italy [09/72]	-	Stored: [U] 12/68
D2217	C	C F Booth, Rotherham	10/73	Stored: [U] 05/69, Withdrawan: 12/71, R/I: 01/72
D2218	C	Steelbreaking and Dismantling, Chesterfield	08/69	
D2219	C	Barrow Coking Plane, Barnsley	05/77	PI
D2220	C	G Cohen, Kettering	05/68	
D2221	C	C F Booth, Rotherham	08/69	
D2222	C	C F Booth, Rotherham	06/69	
D2223	C	G Cohen, Kettering	11/71	Stored:[U] 01/68, R/I: 05/69
D2224	C	A Draper, Hull	08/68	
D2225	C	NCB Wath	07/85	PI
D2226	C	A Draper, Hull	08/68	
D2227	C	A Draper, Hull	08/68	
D2228	C	Bowaters, Sittingbourne	03/79	PI
D2229	P	South Yorkshire Railway	-	PI
D2230	C	Steelbreaking and Dismantling, Chesterfield	01/69	
D2231	C	Steelbreaking and Dismantling, Chesterfield	05/70	
D2232	E	Exported to Rome, Italy [00/72]	-	
D2233	C	Steelbreaking and Dismantling, Chesterfield	08/68	
D2234	C	A Draper, Hull	08/68	
D2235	C	A Finlay, Birdwell	07/68	
D2236	C	G Cohen, Kettering	05/68	
D2237	C	Hughes Bolckow, Blyth	03/70	
D2238	C	NCB Manvers Main	11/82	PI
D2239	C	C F Booth, Rotherham	04/86	PI
D2240	C	A King, Norwich	10/70	
D2241	C	G Cohen, Kettering	11/76	PI
D2242	C	C F Booth, Rotherham	06/70	Stored: [U] 10/69
D2243	C	Middlesbrough and Hartlepool Docks	03/73	PI
D2244	C	A R Adams, Newport	01/81	
D2245	P	Battlefield Line [No. 2]	-	
D2246	S	British Coal, West Drayton	-	
D2247	C	Duport Steel, Briton Ferry	09/79	
D2248	C	Carol & Good, Rotherham	04/87	Stored: [U] 09/69, R/I: 10/69, PI
D2249	C	P Wood, Queenborough	10/72	Stored: [U] 10/69
D2250	C	Pounds Shipbreakers, Fratton	06/69	
D2251	C	Pollock and Brown, Southampton	09/69	
D2252	C	Pollock and Brown, Southampton	11/69	
D2253	C	G Cohen, Kettering	12/69	
D2254	C	BR Selhurst	07/67	
D2555	C	Pollock and Brown, Southampton	08/68	
D2256	C	Pollock and Brown, Southampton	08/69	
D2257	C	Pollock and Brown, Southampton	07/68	
D2258	C	C F Booth, Rotherham	01/87	PI
D2259	C	Bowaters, Sittingbourne	01/78	PI
D2260	C	P D Fuels, Coed Bach	06/83	PI
D2261	C	C F Booth, Rotherham	08/70	

BR 1957 No.	Date Re No.	BR 1948 No.	Built By	Works No.	Date Introduced	Depot of First Allocation	Date Withdrawn	Depot of Final Allocation
D2262	-	-	DC.RSH	2606/7892	12/57	55D	09/68	51A
D2263	-	-	DC.RSH	2607/7893	12/57	55D	11/67	55H
D2264	-	-	DC.RSH	2608/7894	12/57	56G	10/69	55H
D2265	-	-	DC.RSH	2609/7895	12/57	56G	03/70	55F
D2266	-	-	DC.RSH	2610/7896	01/58	55D	11/67	56A
D2267	-	-	DC.RSH	2611/7897	01/58	55D	12/69	50D
D2268	-	-	DC.RSH	2612/7898	02/58	55D	06/68	50D
D2269	-	-	DC.RSH	2613/7911	02/58	55D	08/68	73C
D2270	-	-	DC.RSH	2614/7912	03/58	55D	02/68	55B
D2271	-	-	DC.RSH	2615/7913	03/58	55B	10/79	55F
D2272	-	-	DC.RSH	2616/7914	03/58	55A	10/70	55F
D2273	-	-	DC.RSH	2617/7915	04/58	55A	10/67	56G
D2274	-	-	DC.RSH	2620/7918	07/59	71B	05/69	8J
D2275	-	-	DC.RSH	2619/7917	07/59	71B	10/67	30A
D2276	-	-	DC.RSH	2622/7920	08/59	73F	08/69	30A
D2277	-	-	DC.RSH	2621/7919	09/59	73F	05/69	30E
D2278	-	-	DC.RSH	2618/7916	11/59	73C	04/70	30E
D2279	-	-	DC.RSH	2656/8097	02/60	75C	05/71	30E
D2280	-	-	DC.RSH	2657/8098	02/60	75C	03/71	30E
D2281	-	-	DC.RSH	2658/8099	02/60	75A	10/68	30E
D2282	-	-	DC.RSH	2659/8100	02/60	75A	12/70	30E
D2283	-	-	DC.RSH	2660/8101	03/60	73F	11/69	30E
D2284	-	-	DC.RSH	2661/8102	03/60	73C	04/71	30E
D2285	-	-	DC.RSH	2662/8103	03/60	73F	08/69	30E
D2286	-	-	DC.RSH	2663/8104	03/60	75A	05/68	73C
D2287	-	-	DC.RSH	2667/8120	04/60	73H	06/68	73F
D2288	-	-	DC.RSH	2668/8121	04/60	71A	12/67	70F
D2289	-	-	DC.RSH	2669/8122	04/60	71A	09/71	70D
D2290	-	-	DC.RSH	2670/8123	04/60	70C	11/67	70D
D2291	-	-	DC.RSH	2671/8124	06/60	71A	12/67	70D
D2292	-	-	DC.RSH	2672/8125	06/60	73C	12/67	70D
D2293	-	-	DC.RSH	2673/8126	06/60	73F	04/71	73F
D2294	-	-	DC.RSH	2674/8127	06/60	75C	02/71	70D
D2295	-	-	DC.RSH	2675/8128	06/60	70C	04/71	70D
D2296	-	-	DC.RSH	2677/8155	09/60	40F	11/69	30E
D2297	-	-	DC.RSH	2678/8156	09/60	40A	07/70	30E
D2298	-	-	DC.RSH	2679/8157	09/60	40A	12/68	52A
D2299	-	-	DC.RSH	2680/8158	09/60	40A	01/70	52A
D2300	-	-	DC.RSH	2681/8159	10/60	40A	05/69	8J
D2301	-	-	DC.RSH	2682/8160	10/60	41J	09/68	1E
D2302	-	-	DC.RSH	2683/8161	10/60	40A	06/69	16C
D2303	-	-	DC.RSH	2684/8162	11/60	40A	11/67	32A
D2304	-	-	DC.RSH	2685/8163	11/60	51A	02/68	51A
D2305	-	-	DC.RSH	2686/8164	11/60	51A	02/68	51A
D2306	-	-	DC.RSH	2687/8165	11/60	51A	02/68	51L
D2307	-	-	DC.RSH	2688/8166	11/60	51A	02/68	51L
D2308	-	-	DC.RSH	2689/8167	11/60	51A	02/68	51A
D2309	-	-	DC.RSH	2690/8168	11/60	51A	07/68	55G
D2310	-	-	DC.RSH	2691/8169	12/60	52A	01/69	52A
D2311	-	-	DC.RSH	2692/8170	12/60	52A	02/68	52A
D2312	-	-	DC.RSH	2693/8171	01/61	52A	02/68	52A
D2313	-	-	DC.RSH	2694/8172	01/61	52A	02/68	55G
D2314	-	-	DC.RSH	2695/8173	01/61	52A	02/68	52A
D2315	-	-	DC.RSH	2696/8174	02/61	52H	02/68	51L
D2316	-	-	DC.RSH	2697/8175	02/61	51A	02/68	51L
D2317	-	-	DC.RSH	2698/8176	02/61	51A	08/69	52A
D2318	-	-	DC.RSH	2699/8177	02/61	51A	02/68	51A
D2319	-	-	DC.RSH	2700/8178	02/61	51A	02/68	51A
D2320	-	-	DC.RSH	2701/8179	03/61	51A	02/68	51L
D2321	-	-	DC.RSH	2702/8180	03/61	52E	07/68	52A
D2322	-	-	DC.RSH	2703/8181	03/61	52B	08/68	52A
D2323	-	-	DC.RSH	2704/8182	03/61	55E	07/68	55G
D2324	-	-	DC.RSH	2705/8183	04/61	55E	07/68	55B
D2325	-	-	DC.RSH	2706/8184	04/61	52A	07/68	50D
D2326	-	-	DC.RSH	2707/8185	05/61	52H	08/68	52A
D2327	-	-	DC.RSH	2708/8186	05/61	52H	08/68	52A
D2328	-	-	DC.RSH	2709/8187	05/61	52H	09/68	52A
D2329	-	-	DC.RSH	2710/8188	06/61	52B	07/68	52A
D2330	-	-	DC.RSH	2711/8189	06/61	52A	07/69	52A
D2331	-	-	DC.RSH	2712/8190	06/61	51A	07/68	51L
D2332	-	-	DC.RSH	2713/8191	06/61	52B	06/69	52A
D2333	-	-	DC.RSH	2714/8192	07/61	52A	06/69	52A
D2334	-	-	DC.RSH	2715/8193	07/61	52B	07/68	51A
D2335	-	-	DC.RSH	2716/8194	08/61	52C	07/68	51A
D2336	-	-	DC.RSH	2717/8195	08/61	51A	07/68	51A
D2337	-	-	DC.RSH	2718/8196	08/61	51A	07/68	51A
D2338	-	-	DC.RSH	2719/8197	09/61	51A	07/68	51L
D2339	-	-	DC.RSH	2720/8198	10/61	52B	10/67	55A
D2340	-	-	Drewry Car Co	2593/7870	03/62	51A	10/68	55F
D2341	DS 1173	03/67	DC.VF	2217/D46	03/67	73C	12/68	73F

BR 1957 No.	Disposal Code	Disposal Detail	Date Cut Up	Notes
D2262	C	Ford Motors, Dagenham	07/78	PI
D2263	C	Pollock and Brown, Southampton	12/68	
D2264	C	C F Booth, Rotherham	06/70	Stored: [U] 08/69
D2265	C	C F Booth, Rotherham	09/70	
D2266	C	C F Booth, Rotherham	04/68	
D2267	S	Ford Motors, Dagenham (1)	-	
D2268	C	T W Ward, Beighton	11/68	
D2269	C	Pounds Shipbreakers, Fratton	06/69	
D2270	C	Duport Steel, Briton Ferry	09/79	
D2271	P	West Somerset Railway	-	Stored: [U] 09/69, PI
D2272	S	British Fuels, Blackburn	-	
D2273	C	C F Booth, Rotherham	11/68	
D2274	C	NCB Maltby Main	09/80	PI
D2275	C	BR Swindon Works	02/68	
D2276	C	A R Adams, Newport	05/77	PI
D2277	C	H Brahams, Bury St Edmunds	12/69	
D2278	C	BR Stratford, by Pounds of Fratton	02/71	
D2279	P	East Anglian Railway Museum	-	Stored: [U] 08/70, PI
D2280	S	Ford Motors, Dagenham [P1381C]	-	
D2281	C	Duport Steel, Briton Ferry	08/71	
D2282	C	P Wood, Queenborough	08/72	
D2283	C	BR Stratford, by Hartwood	08/70	
D2284	P	South Yorkshire Railway	-	PI
D2285	C	H Brahams, Bury St Edmunds	03/70	
D2286	C	Pounds Shipbreakers, Fratton	06/69	
D2287	C	Pounds Shipbreakers, Fratton	08/69	
D2288	C	Pollock and Brown, Southampton	08/68	
D2289	E	Exported to Brescia, Italy [04/72]	-	
D2290	C	Pollock and Brown, Southampton	08/68	
D2291	C	Pollock and Brown, Southampton	07/68	
D2292	C	Pollock and Brown, Southampton	09/68	
D2293	C	P Wood, Queenborough	10/72	
D2294	C	P Wood, Queenborough	10/85	PI
D2295	E	Exported to Brescia, Italy [09/72]	-	
D2296	C	H Brahams, Bury St Edmunds	04/70	
D2297	C	H Brahams, Bury St Edmunds	09/70	
D2298	P	Buckinghamshire Railway Centre, Aylesbury	-	
D2299	C	C F Booth, Rotherham	02/84	PI
D2300	C	NCB Manton	08/86	PI
D2301	C	Steelbreaking and Dismantling, Chesterfield	01/70	
D2302	S	Potter Group, Ely	-	
D2303	C	BR Doncaster Works	01/69	
D2304	C	Duport Steel, Llanelly	05/77	PI
D2305	C	Duport Steel, Llanelly	09/81	PI
D2306	C	Duport Steel, Llanelly	09/81	PI
D2307	C	Duport Steel, Llanelly	10/79	PI
D2308	C	Duport Steel, Llanelly	05/80	PI
D2309	C	C F Booth, Rotherham	05/69	
D2310	S	Coal Mechanisation, Tolworth	-	
D2311	C	Slag Reduction, Ickles	09/68	
D2312	C	C F Booth, Rotherham	12/68	
D2313	C	Slag Reduction, Ickles	09/68	
D2314	C	Slag Reduction, Ickles	09/68	
D2315	C	Hughes Bolckow, Blyth	08/68	
D2316	C	C F Booth, Rotherham	01/69	
D2317	C	NCB Cortonwood	09/86	PI
D2318	C	Hughes Bolckow, Blyth	07/68	
D2319	C	Arnott Young, Dinsdale	12/68	
D2320	C	C F Booth, Rotherham	03/69	
D2321	C	G Cohen, Middlesbrough	12/68	
D2322	C	NCB Kiveton	11/85	PI
D2323	C	C F Booth, Rotherham	04/69	
D2324	S	Redland, Mountsorral	-	
D2325	P	Mangapps Farm, Burnham on Crouch	-	PI
D2326	C	NCB Manvers Main	02/76	
D2327	C	Coopers Metals, Attercliffe	02/84	PI
D2328	C	NCB Cortonwood	07/86	PI
D2329	C	BR York Layerthorpe	04/70	
D2330	C	BR Thornaby, by M Turnbull	08/70	
D2331	C	T W Ward, Beighton	11/68	
D2332	C	NCB Dinnington	07/86	PI
D2333	S	Ford Motors, Dagenham [P1062C]	-	
D2334	P	South Yorkshire Railway	-	PI
D2335	C	NCB Maltby	02/80	PI
D2336	C	NCB Manvers Main	02/78	
D2337	P	South Yorkshire Railway	-	PI
D2338	C	T W Ward, Beighton	11/68	
D2339	C	Hughes Bolckow, Blyth	06/68	
D2340	C	Duport Steel, Briton Ferry	09/79	Built 1956, loaned to BR at 51C
D2341	C	Pollock and Brown, Southampton	08/69	Pre 1967 in Departmental service

BR DIESEL MECHANICAL 0-6-0 CLASS 03

BR 1957 No.	BR 1948 No.	TOPS No.	Date Re No.	Built By	Works No.	Date Introduced	Depot of First Allocation	Date Withdrawn	Depot of Final Allocation
continued from D2199									
D2370	-	03370	03/74	BR Swindon	-	05/67	31A	12/82	NR
D2371	-	03371	02/74	BR Swindon	-	05/67	31A	11/87	GD
D2372	-	-	-	BR Swindon	-	07/61	8D	11/70	8F
D2373	-	-	-	BR Swindon	-	08/61	8A	05/68	9D
D2374	-	-	-	BR Swindon	-	08/61	8G	05/68	8H
D2375	-	-	-	BR Swindon	-	08/61	8G	05/68	8H
D2376	-	-	-	BR Swindon	-	08/61	8G	05/68	8F
D2377	-	-	-	BR Swindon	-	09/61	17A	05/68	8F
D2378	-	-	-	BR Swindon	-	09/61	17A	06/71	87A
D2379	-	-	-	BR Swindon	-	10/61	17A	05/68	8F
D2380	-	-	-	BR Swindon	-	10/61	17A	05/68	1F
D2381	-	-	-	BR Swindon	-	10/61	17A	06/72	16C
D2382	-	03382	09/74	BR Swindon	-	11/61	17A	10/83	LE
D2383	-	-	-	BR Swindon	-	11/61	17A	04/71	16C
D2384	-	-	-	BR Swindon	-	11/61	17A	05/68	5A
D2385	-	-	-	BR Doncaster	-	03/61	6C	02/70	8H
D2386	-	03386	07/74	BR Doncaster	-	03/61	6F	03/76	16C
D2387	-	-	-	BR Doncaster	-	04/61	8G	12/72	16C
D2388	-	-	-	BR Doncaster	-	04/61	6C	07/72	16C
D2389	-	03389	02/74	BR Doncaster	-	05/61	9G	02/83	LN
D2390	-	-	-	BR Doncaster	-	06/61	9G	05/68	10A
D2391	-	-	-	BR Doncaster	-	06/61	9G	11/70	8F
D2392	-	-	-	BR Doncaster	-	06/61	9G	06/71	8F
D2393	-	-	-	BR Doncaster	-	07/61	8A	12/69	8M
D2394	-	-	-	BR Doncaster	-	08/61	8D	11/68	12C
D2395	-	-	-	BR Doncaster	-	08/61	21C	05/68	5B
D2396	-	-	-	BR Doncaster	-	08/61	21E	05/68	8J
D2397	-	03397	03/77	BR Doncaster	-	09/61	75C	07/87	NC
D2398	-	-	-	BR Doncaster	-	09/61	73C	10/71	70F
D2399	-	03399	03/77	BR Doncaster	-	09/61	73F	07/87	NC

ANDREW BARCLAY DIESEL MECHANICAL 0-6-0

BR 1957 No.	Date Re No.	BR 1948 No.	Built By	Works No.	Date Introduced	Depot of First Allocation	Date Withdrawn	Depot of Final Allocation
D2400	11/60	11177	A.Barclay	402	07/56	34A	10/67	41E
D2401	11/60	11178	A.Barclay	403	07/56	34A	12/68	41E
D2402	02/61	11179	A.Barclay	404	09/56	40A	09/67	41E
D2403	11/60	11180	A.Barclay	405	10/56	40A	01/69	41E
D2404	01/61	11181	A.Barclay	406	11/56	40A	01/69	41E
D2405	04/59	11182	A.Barclay	407	12/56	40A	12/68	41E
D2406	12/60	11183	A.Barclay	408	12/56	40A	05/67	40A
D2407	11/60	11184	A.Barclay	409	01/57	40A	01/69	41E
D2408	04/59	11185	A.Barclay	410	02/57	40A	05/67	40B
D2409	01/61	11186	A.Barclay	411	03/57	40A	12/68	41E

ANDREW BARCLAY DIESEL MECHANICAL 0-4-0 CLASS 06

BR 1957 No.	TOPS No.	Date Re No.	Built By	Works No.	Date Introduced	Depot of First Allocation	Date Withdrawn	Depot of Final Allocation
D2410	-	-	A.Barclay	425	06/58	60A	01/69	65A
D2411	-	-	A.Barclay	426	07/58	60A	06/68	67C
D2412	-	-	A.Barclay	427	09/58	60A	06/68	62B
D2413	06001	04/74	A.Barclay	428	10/58	60A	09/76	ED
D2414	06002	04/75	A.Barclay	429	10/58	61C	09/81	DE
D2415	-	-	A.Barclay	430	11/58	61C	06/68	64H
D2416	-	-	A.Barclay	431	11/58	61A	11/72	62B
D2417	-	-	A.Barclay	432	12/58	61A	06/68	64H
D2418	-	-	A.Barclay	433	01/59	61A	12/68	64H
D2419	-	-	A.Barclay	434	01/59	61A	01/69	64H
D2420	06003	04/74	A.Barclay	435	02/59	61A	02/81	DE
D2421	06004	04/74	A.Barclay	436	02/59	61A	03/79	DT
D2422	06005	12/74	A.Barclay	437	03/59	61A	10/80	DE
D2423	06006	11/74	A.Barclay	438	04/59	61A	06/80	DE
D2424	-	-	A.Barclay	439	04/59	61A	11/72	65A

BR 1957 No.	Disposal Code	Disposal Detail	Date Cut Up	Notes
D2370	C	BREL Doncaster	07/83	Departmental No. 91 until 07/67
D2371	P	Privately at Rowden Mill Station	-	Departmental No. 92 until 07/67
D2372	C	G Cohen, Kettering	10/71	Stored: [S] 04/70
D2373	C	NCB Manvers Main	03/82	PI
D2374	C	G Cohen, Kettering	12/68	
D2375	C	G Cohen, Kettering	11/68	
D2376	C	G Cohen, Kettering	11/68	
D2377	C	BREL Swindon	02/70	
D2378	C	BREL Swindon	10/72	Stored: [U] 04/71
D2379	C	BREL Swindon	02/70	
D2380	C	G Cohen, Kettering	12/68	
D2381	P	Steamtown, Carnforth	-	
D2382	C	BREL Swindon	02/86	Withdrawn: 06/72, R/I: 08/72, Stored: [U] 05/83
D2383	C	T W Ward, Beighton	02/72	
D2384	C	G Cohen, Kettering	11/68	
D2385	C	C F Booth, Rotherham	10/70	Stored: [U] 01/70
D2386	C	G Cohen, Kettering	12/76	Stored: [U] 12/75
D2387	C	C F Booth, Rotherham	04/73	Stored: [U] 11/72
D2388	C	C F Booth, Rotherham	02/73	
D2389	C	C F Booth, Rotherham	07/83	
D2390	C	G Cohen, Kettering	11/68	
D2391	C	G Cohen, Kettering	12/71	Stored: [U] 03/70
D2392	C	C F Booth, Rotherham	03/72	
D2393	C	C F Booth, Rotherham	06/71	
D2394	C	C F Booth, Rotherham	09/69	
D2395	C	G Cohen, Kettering	12/68	
D2396	C	G Cohen, Kettering	11/68	
D2397	C	V Berry, Leicester	01/91	Withdrawn: 05/71, R/I: 10/71, Stored: [U] 10/72, R/I: 11/72
D2398	C	Pounds Shipbreakers, Fratton	09/72	
D2399	P	Mangapps Farm, Burnham on Crouch	-	Stored: [U] 10/72, R/I: 11/72

BR 1957 No.	Disposal Code	Disposal Detail	Date Cut Up	Notes
D2400	C	Slag Reduction, Ickles	04/68	
D2401	C	C F Booth, Rotherham	04/69	
D2402	C	C F Booth, Rotherham	04/68	
D2403	C	C F Booth, Rotherham	05/69	
D2404	C	C F Booth, Rotherham	05/69	Stored: [U] 01/69
D2405	C	C F Booth, Rotherham	03/70	
D2406	C	T W Ward, Beighton	04/68	
D2407	C	C F Booth, Rotherham	05/69	
D2408	C	T W Ward, Beighton	04/68	
D2409	C	C F Booth, Rotherham	03/70	

BR 1957 No.	Disposal Code	Disposal Detail	Date Cut Up	Notes
D2410	C	G H Campbell, Airdrie	06/69	
D2411	C	G H Campbell, Airdrie	01/69	
D2412	C	G H Campbell, Airdrie	01/69	
D2413	C	G H Campbell, Airdrie	04/77	Stored: [U] 08/76
D2414	C	BREL Swindon	04/82	After withdrawal to Reading as spares to No. 97804
D2415	C	G H Campbell, Airdrie	10/68	
D2416	C	BREL Glasgow	11/73	
D2417	C	G H Campbell, Airdrie	01/69	
D2418	C	G H Campbell, Airdrie	12/68	
D2419	C	G H Campbell, Airdrie	02/69	
D2420	D	To Departmental Stock - 97804	-	
D2421	C	BREL Glasgow	02/80	
D2422	C	BR Dundee	07/83	
D2423	C	BR Dundee	07/83	
D2424	C	BREL Glasgow	01/74	

BR 1957 No.	TOPS No.	Date Re No.	Built By	Works No.	Date Introduced	Depot of First Allocation	Date Withdrawn	Depot of Final Allocation
D2425	-	-	A.Barclay	452	11/59	66D	06/68	65A
D2426	06007	04/75	A.Barclay	453	11/59	66B	09/77	DT
D2427	-	-	A.Barclay	454	12/59	66B	09/69	65A
D2428	-	-	A.Barclay	455	01/60	66B	06/68	66A
D2429	-	-	A.Barclay	456	01/60	66B	04/69	62B
D2430	-	-	A.Barclay	457	01/60	66C	06/68	62A
D2431	-	-	A.Barclay	458	02/60	66C	11/71	66A
D2432	-	-	A.Barclay	459	02/60	66C	12/68	65A
D2433	-	-	A.Barclay	460	02/60	66C	06/72	65A
D2434	-	-	A.Barclay	461	03/60	67C	07/69	65A
D2435	-	-	A.Barclay	462	03/60	67C	11/71	64H
D2436	-	-	A.Barclay	463	04/60	67C	11/71	65A
D2437	06008	04/75	A.Barclay	464	04/60	67C	10/80	DT
D2438	-	-	A.Barclay	465	05/60	67C	11/72	62C
D2439	-	-	A.Barclay	466	07/60	67A	11/71	62C
D2440	06009	04/75	A.Barclay	467	08/60	67A	08/75	DT
D2441	-	-	A.Barclay	468	08/60	67A	03/67	66C
D2442	-	-	A.Barclay	469	09/60	67A	11/72	62C
D2443	-	-	A.Barclay	470	10/60	67A	06/72	62B
D2444	06010	12/74	A.Barclay	471	10/60	67A	06/75	DE

HUDSWELL CLARKE DIESEL MECHANICAL 0-6-0

BR 1957 No.	Date Re No.	BR 1948 No.	Built By	Works No.	Date Introduced	Depot of First Allocation	Date Withdrawn	Depot of Final Allocation
D2500	06/61	11116	H.Clarke	D898	12/55	6C	05/67	12C
D2501	05/61	11117	H.Clarke	D899	01/56	6C	02/67	12C
D2502	01/62	11118	H.Clarke	D900	01/56	6C	10/67	12C
D2503	09/61	11119	H.Clarke	D901	02/56	6C	08/67	12C
D2504	05/58	11120	H.Clarke	D902	02/56	6C	03/67	12C
D2505	08/60	11144	H.Clarke	D938	04/56	6C	08/67	12C
D2506	12/61	11145	H.Clarke	D939	04/56	6C	12/67	12C
D2507	10/61	11146	H.Clarke	D940	05/56	6C	03/67	12C
D2508	05/57	11147	H.Clarke	D941	06/56	6C	05/67	12C
D2509	03/61	11148	H.Clarke	D942	07/56	6C	08/67	12C

HUDSWELL CLARKE DIESEL MECHANICAL 0-6-0

BR 1957 No.	Built By	Works No.	Date Introduced	Depot of First Allocation	Date Withdrawn	Depot of Final Allocation	Disposal Code
D2510	H.Clarke	D1201	08/61	6C	08/67	1F	C
D2511	H.Clarke	D1202	08/61	12E	12/67	12C	P
D2512	H.Clarke	D1203	09/61	12E	05/67	12C	C
D2513	H.Clarke	D1204	10/61	12E	08/67	12C	C
D2514	H.Clarke	D1205	10/61	12E	08/67	12C	C
D2515	H.Clarke	D1206	10/61	12E	08/67	12C	C
D2516	H.Clarke	D1207	11/61	12E	08/67	12C	C
D2517	H.Clarke	D1208	11/61	12E	02/67	12A	C
D2518	H.Clarke	D1209	11/61	1C	02/67	5A	C
D2519	H.Clarke	D1210	11/61	5B	07/67	5A	C

HUNSLET DIESEL MECHANICAL 0-6-0 CLASS 05

BR 1957 No.	Date Re No.	BR 1948 No.	TOPS Re No.	Date	Built By	Works	Date Introduced	Depot of First Allocation	Date Withdrawn	Depot of Final Allocation
D2550	08/58	11136	-	-	Hunslet	4866	10/55	32B	10/66	32B
D2551	10/59	11137	-	-	Hunslet	4867	10/55	32B	01/68	8C
D2552	02/60	11138	-	-	Hunslet	4868	02/56	32B	06/67	8C
D2553	02/60	11139	-	-	Hunslet	4869	02/56	32B	01/68	8C
D2554	01/59	11140	05001	08/74	Hunslet	4870	03/56	30F	01/81	RY
D2555	08/58	11141	-	-	Hunslet	4871	06/56	32F	01/68	8C
D2556	05/58	11142	-	-	Hunslet	4872	07/56	32B	08/67	8C
D2557	04/59	11143	-	-	Hunslet	4873	08/56	32B	04/67	8E
D2558	03/59	11161	-	-	Hunslet	4896	10/56	32A	07/67	8F
D2559	05/59	11162	-	-	Hunslet	4897	10/56	32A	08/67	8C
D2560	06/58	11163	-	-	Hunslet	4898	02/57	32A	11/67	8C
D2561	10/58	11164	-	-	Hunslet	4899	03/57	41A	08/67	8F
D2562	12/60	11165	-	-	Hunslet	5000	03/57	41A	01/68	8C

BR 1957 No.	Disposal Code	Disposal Detail	Date Cut Up	Notes
D2425	C	G H Campbell, Airdrie	11/68	
D2426	C	BREL Glasgow	08/79	Withdrawn: 01/69, R/I: 03/69
D2427	C	J McWilliam, Shettleston	11/71	
D2428	C	G H Campbell, Airdrie	02/69	
D2429	C	J McWilliam, Shettleston	10/71	
D2430	C	G H Campbell, Airdrie	01/69	
D2431	C	BREL Glasgow	03/72	
D2432	E	Exported to Trieste, Italy [03/77]	-	
D2433	C	BREL Glasgow	05/73	
D2434	C	J McWilliam, Shettleston	10/71	
D2435	C	G H Campbell, Airdrie	10/74	
D2436	C	BREL Glasgow	12/73	
D2437	C	BR Polmadie, by G H Campbell	08/83	
D2438	C	G H Campbell, Airdrie	12/74	
D2439	C	BREL Glasgow	03/72	
D2440	C	G H Campbell, Airdrie	07/77	
D2441	C	Slag Reduction, Ickles	08/67	
D2442	C	G H Campbell, Airdrie	09/74	
D2443	C	BREL Glasgow	05/73	
D2444	C	BREL Glasgow	04/79	

BR 1957 No.	Disposal Code	Disposal Detail	Date Cut Up	Notes
D2500	C	C F Booth, Rotherham	04/68	
D2501	C	Slag Reduction, Ickles	08/67	
D2502	C	C F Booth, Rotherham	03/68	
D2503	C	C F Booth, Rotherham	04/68	
D2504	C	C F Booth, Rotherham	03/68	
D2505	C	C F Booth, Rotherham	03/68	
D2506	C	Steelbreaking and Dismantling, Chesterfield	05/70	
D2507	C	Slag Reduction, Ickles	08/67	
D2508	C	C F Booth, Rotherham	04/68	
D2509	C	C F Booth, Rotherham	04/68	

BR 1957 No.	Disposal Detail	Date Cut Up	Notes
D2510	C F Booth, Rotherham	04/68	
D2511	Keighley & Worth Valley Railway	-	
D2512	C F Booth, Rotherham	02/68	
D2513	NCB Cadeby	10/75	PI
D2514	C F Booth, Rotherham	04/68	
D2515	BR Bolton, by W Hatton	05/68	
D2516	C F Booth, Rotherham	05/68	
D2517	Slag Reduction, Ickles	08/67	
D2518	NCB Hatfield	06/73	PI
D2519	Marple & Gillott at K&WVR	04/85	PI, To K&WVR from 04/82

BR 1957 No.	Disposal Code	Disposal Detail	Date Cut Up	Notes
D2550	C	BR Doncaster	11/66	
D2551	C	C F Booth, Rotherham	06/68	
D2552	C	C F Booth, Rotherham	04/68	
D2553	C	C F Booth, Rotherham	06/68	
D2554	D	To Departmental Stock - 97803	-	
D2555	C	C F Booth, Rotherham	07/68	
D2556	C	G H Campbell, Airdrie	03/68	
D2557	C	C F Booth, Rotherham	04/68	
D2558	C	C F Booth, Rotherham	03/68	
D2559	C	G H Campbell, Airdrie	03/68	
D2560	C	Slag Reduction, Ickles	07/68	
D2561	C	Duport Steel, Llanelly	10/72	PI
D2562	C	Slag Reduction, Ickles	04/68	

BR 1957 No.	Date Re No.	BR 1948 No.	TOPS Re No.	Date	Built By	Works No.	Date Introduced	Depot of First Allocation	Date Withdrawn	Depot of Final Allocation
D2563	10/64	11166	-	-	Hunslet	5001	06/57	32A	08/67	8C
D2564	07/58	11167	-	-	Hunslet	5002	06/57	32A	08/67	8C
D2565	03/63	11168	-	-	Hunslet	5003	06/57	32A	03/67	8F
D2566	01/61	11169	-	-	Hunslet	5004	08/57	32A	01/68	8C
D2567	10/58	11170	-	-	Hunslet	5005	08/57	32A	11/67	8C
D2568	11/58	11171	-	-	Hunslet	5006	09/57	32A	08/67	8F
D2569	07/65	11172	-	-	Hunslet	5007	09/57	32A	08/67	8F
D2570	11/58	11173	-	-	Hunslet	5008	10/57	32A	07/67	8F
D2571	10/58	11174	-	-	Hunslet	5009	10/57	32A	01/68	8C
D2572	09/59	11175	-	-	Hunslet	5010	12/57	32A	02/67	30A
D2573	05/58	11176	-	-	Hunslet	5011	01/58	32A	01/68	8C
D2574	-	-	-	-	Hunslet	5456	07/58	68C	06/68	62B
D2575	-	-	-	-	Hunslet	5457	07/58	68C	06/68	62B
D2576	-	-	-	-	Hunslet	5458	07/58	62A	06/68	62A
D2577	-	-	-	-	Hunslet	5459	07/58	62A	06/67	62A
D2578	-	-	-	-	Hunslet	5460	11/58	62A	07/67	62A
D2579	-	-	-	-	Hunslet	5461	12/58	62A	06/68	62A
D2580	-	-	-	-	Hunslet	5462	12/58	62A	06/68	62A
D2581	-	-	-	-	Hunslet	5463	12/58	62A	06/68	62A
D2582	-	-	-	-	Hunslet	5464	01/59	62A	06/68	62A
D2583	-	-	-	-	Hunslet	5465	01/59	62A	06/68	62A
D2584	-	-	-	-	Hunslet	5466	02/59	62A	07/67	62A
D2585	-	-	-	-	Hunslet	5467	04/59	62A	06/68	62A
D2586	-	-	-	-	Hunslet	5635	11/59	56B	03/67	51C
D2587	-	-	-	-	Hunslet	5636	11/59	56A	12/67	62C
D2588	-	-	-	-	Hunslet	5637	11/59	56A	03/67	51C
D2589	-	-	-	-	Hunslet	5638	11/59	56A	12/67	62A
D2590	-	-	-	-	Hunslet	5639	12/59	56A	06/68	64H
D2591	-	-	-	-	Hunslet	5640	12/59	56A	03/67	51C
D2592	-	-	-	-	Hunslet	5641	12/59	56A	03/68	64B
D2593	-	-	-	-	Hunslet	5642	12/59	56A	12/67	62C
D2594	-	-	-	-	Hunslet	5643	01/60	52E	03/67	51C
D2595	-	-	-	-	Hunslet	5644	01/60	50A	06/68	62A
D2596	-	-	-	-	Hunslet	5645	02/60	55A	06/68	64H
D2597	-	-	-	-	Hunslet	5646	06/60	55A	12/67	64H
D2598	-	-	-	-	Hunslet	5647	06/60	50A	12/67	50D
D2599	-	-	-	-	Hunslet	5648	06/60	50A	12/67	50D
D2600	-	-	-	-	Hunslet	5649	06/60	50A	12/67	50D
D2601	-	-	-	-	Hunslet	5650	08/60	50A	12/67	50D
D2602	-	-	-	-	Hunslet	5651	08/60	50A	07/67	50D
D2603	-	-	-	-	Hunslet	5652	09/60	56B	07/67	6G
D2604	-	-	-	-	Hunslet	5653	09/60	56B	12/67	6G
D2605	-	-	-	-	Hunslet	5654	09/60	56B	12/67	6G
D2606	-	-	-	-	Hunslet	5655	09/60	56B	02/67	6G
D2607	-	-	-	-	Hunslet	5656	10/60	56B	12/67	6G
D2608	-	-	-	-	Hunslet	5657	11/60	56B	12/67	62C
D2609	-	-	-	-	Hunslet	5658	11/60	50D	12/67	50D
D2610	-	-	-	-	Hunslet	5659	12/60	50D	12/67	50D
D2611	-	-	-	-	Hunslet	5660	01/61	50D	12/67	50D
D2612	-	-	-	-	Hunslet	5661	01/61	51A	02/61	51A
D2613	-	-	-	-	Hunslet	5662	02/61	50D	12/67	50D
D2614	-	-	-	-	Hunslet	5663	02/61	50D	03/67	50D
D2615	-	-	-	-	Hunslet	5664	02/61	52E	12/67	52A
D2616	-	-	-	-	Hunslet	5665	03/61	52E	12/67	50D
D2617	-	-	-	-	Hunslet	5666	03/61	51A	12/67	62C
D2618	-	-	-	-	Hunslet	5667	03/61	51A	06/68	64H

NORTH BRITISH DIESEL HYDRAULIC 0-4-0

BR 1957 No.	Date Re No.	BR 1948 No.	Built By	Works No.	Date Introduced	Depot of First Allocation	Date Withdrawn	Depot of Final Allocation
D2700	02/58	11700	NBL	27100	07/53	51C	11/63	50D
D2701	06/59	11701	NBL	27101	11/53	51C	03/67	50D
D2702	06/59	11702	NBL	27102	08/54	51C	03/67	50D
D2703	02/60	11703	NBL	27431	08/55	62B	02/68	65F
D2704	12/59	11704	NBL	27432	09/55	64A	06/67	62C
D2705	08/59	11705	NBL	27433	09/55	64A	08/67	64H
D2706	07/60	11706	NBL	27434	12/55	64A	03/67	64H
D2707	02/59	11707	NBL	27435	02/56	64A	03/67	62C

BR 1957 No.	Disposal Code	Disposal Detail	Date Cut Up	Notes
D2563	C	C F Booth, Rotherham	04/68	
D2564	C	C F Booth, Rotherham	04/68	
D2565	C	C F Booth, Rotherham	03/68	
D2566	C	C F Booth, Rotherham	06/68	
D2567	C	Slag Reduction, Ickles	09/68	
D2568	C	Duport Steel, Briton Ferry	05/69	PI
D2569	C	Duport Steel, Briton Ferry	06/69	PI
D2570	C	Duport Steel, Briton Ferry	06/71	PI
D2571	C	BR Glasgow Works	10/68	
D2572	C	Slag Reduction, Ickles	08/67	
D2573	C	J McWilliam, Shettleston	09/69	
D2574	C	G H Campbell, Airdrie	01/69	
D2575	C	G H Campbell, Airdrie	01/69	
D2576	C	Machinery and Scrap Ltd, Motherwell	08/68	
D2577	C	Machinery and Scrap Ltd, Motherwell	09/67	
D2578	P	Bulmers Railway Centre	-	Rebuilt by Hunslet in 1968 as works No. 6999
D2579	C	G H Campbell, Airdrie	01/69	
D2580	C	G H Campbell, Airdrie	12/68	
D2581	C	G H Campbell, Airdrie	11/68	
D2582	C	G H Campbell, Airdrie	01/69	
D2583	C	G H Campbell, Airdrie	01/69	
D2584	C	Hunslet, Leeds	08/68	Used to provide spares
D2585	C	G H Campbell, Airdrie	01/69	
D2586	C	Slag Reduction, Ickles	10/67	
D2587	P	South Yorkshire Railway	-	Rebuilt by Hunslet in 1968 as works No. 7180, PI
D2588	C	Slag Reduction, Ickles	07/67	
D2589	C	G H Campbell, Airdrie	12/65	
D2590	C	G H Campbell, Airdrie	01/69	
D2591	C	Slag Reduction, Ickles	07/67	
D2592	C	G H Campbell, Airdrie	02/69	
D2593	C	Hunslet, Leeds	10/68	
D2594	C	Slag Reduction, Ickles	07/67	
D2595	P	Steamport, Southport	-	Rebuilt by Hunslet in 1969 as works No. 7179
D2596	C	G H Campbell, Airdrie	01/69	
D2597	C	G H Campbell, Airdrie	01/69	
D2598	C	NCB Philadelphia	05/75	PI
D2599	C	NCB Askern	05/81	PI
D2600	C	Duport Steel, Briton Ferry	06/71	PI
D2601	C	Duport Steel, Llanelly	07/79	PI
D2602	C	Slag Reduction, Ickles	12/67	
D2603	C	C F Booth, Rotherham	04/68	Used after withdrawal at Menai Bridge
D2604	C	G Cohen, Morriston	06/68	
D2605	C	G Cohen, Morriston	06/68	
D2606	C	Slag Reduction, Ickles	12/67	
D2607	C	Coopers Metals, Sheffield	06/84	
D2608	C	G H Campbell, Airdrie	01/69	
D2609	C	C F Booth, Rotherham	05/68	
D2610	C	C F Booth, Rotherham	08/68	
D2611	C	NCB Yorkshire Main	12/76	PI
D2612	D	To Departmental Stock - No. 88	-	
D2613	C	NCB Bentley	06/77	PI
D2614	C	A Draper, Hull	12/67	
D2615	C	C F Booth, Rotherham	08/68	As Departmental No. 89 from 01/64 11/67
D2616	C	NCB Hatfield Main	06/73	PI
D2617	C	Hunslet, Leeds	04/76	Used to provide spares
D2618	C	G H Campbell, Airdrie	02/69	

BR 1957 No.	Disposal Code	Disposal Detail	Date Cut Up	Notes
D2700	C	BR Darlington Works	11/64	Carried No. D2600 04/02/58 - 16/02/58. Stored: [U] 08/55, R/I: 06/57
D2701	C	A Draper, Hull	09/67	
D2702	C	A Draper, Hull	09/67	
D2703	C	Shipbreaking Industries, Faslane	05/68	
D2704	C	Arnott Young, Carmyle	10/67	
D2705	C	J N Connel, Coatbridge	11/67	
D2706	C	Slag Reduction, Ickles	08/67	
D2707	C	Slag Reduction, Ickles	08/67	

NORTH BRITISH DIESEL HYDRAULIC 0-4-0

BR 1957 No.	Date Re No.	BR 1948 No.	Built By	Works No.	Date Introduced	Depot of First Allocation	Date Withdrawn	Depot of Final Allocation
D2708	10/61	11708	NBL	27703	08/57	62B	02/67	C-WKS
D2709	05/62	11709	NBL	27704	09/57	62B	02/67	W-WKS
D2710	12/61	11710	NBL	27705	09/57	62B	03/67	64H
D2711	04/61	11711	NBL	27706	09/57	62B	02/67	C-WKS
D2712	06/62	11712	NBL	27707	09/57	62B	03/67	64H
D2713	08/61	11713	NBL	27708	09/57	62B	03/67	62B
D2714	06/61	11714	NBL	27709	09/57	62B	03/67	64H
D2715	01/63	11715	NBL	27710	10/57	62B	03/67	64H
D2716	11/62	11716	NBL	27711	10/57	62B	03/67	62C
D2717	11/62	11717	NBL	27712	10/57	62B	07/67	64H
D2718	02/63	11718	NBL	27713	10/57	62C	07/67	62C
D2719	04/61	11719	NBL	27714	11/57	64A	02/67	C-WKS
D2720	-	-	NBL	27815	06/58	64A	07/67	64H
D2721	-	-	NBL	27816	06/58	64A	07/67	64H
D2722	-	-	NBL	27817	07/58	64A	02/67	C-WKS
D2723	-	-	NBL	27818	06/58	64A	07/67	64H
D2724	-	-	NBL	27819	07/58	64A	03/67	64H
D2725	-	-	NBL	27820	07/58	64A	06/67	64H
D2726	-	-	NBL	27821	07/58	64A	02/67	W-WKS
D2727	-	-	NBL	27822	08/58	64A	03/67	64H
D2728	-	-	NBL	27823	08/58	64A	07/67	64H
D2729	-	-	NBL	27824	09/58	64A	03/67	64H
D2730	-	-	NBL	27825	09/58	64A	10/67	65A
D2731	-	-	NBL	27826	09/58	64A	10/67	65A
D2732	-	-	NBL	27827	10/58	64A	02/67	W-WKS
D2733	-	-	NBL	27828	10/58	65G	02/67	W-WKS
D2734	-	-	NBL	27829	10/58	65G	09/67	65A
D2735	-	-	NBL	27830	11/58	65G	03/67	65A
D2736	-	-	NBL	27831	11/58	65G	03/67	65A
D2737	-	-	NBL	27832	11/58	65G	03/67	65A
D2738	-	-	NBL	27833	11/58	65G	06/67	65A
D2739	-	-	NBL	27834	12/58	63B	03/67	65A
D2740	-	-	NBL	27835	12/58	63B	03/67	65A
D2741	-	-	NBL	27836	01/59	63B	02/67	C-WKS
D2742	-	-	NBL	27837	01/59	63B	02/67	C-WKS
D2743	-	-	NBL	27838	03/59	62C	02/67	C-WKS
D2744	-	-	NBL	27839	04/59	62C	07/67	64H
D2745	-	-	NBL	27998	01/60	64A	04/67	64H
D2746	-	-	NBL	27999	02/60	64A	03/67	64H
D2747	-	-	NBL	28000	02/60	64A	07/67	64H
D2748	-	-	NBL	28001	02/60	64A	07/67	64H
D2749	-	-	NBL	28002	02/60	64A	07/67	64H
D2750	-	-	NBL	28003	02/60	64A	07/67	64H
D2751	-	-	NBL	28004	03/60	64A	07/67	64H
D2752	-	-	NBL	28005	03/60	64B	03/67	64B
D2753	-	-	NBL	28006	04/60	64B	06/67	64B
D2754	-	-	NBL	28007	04/60	64C	07/67	64H
D2755	-	-	NBL	28008	05/60	64C	07/67	64H
D2756	-	-	NBL	28009	05/60	65A	02/68	65A
D2757	-	-	NBL	28010	05/60	65A	06/67	65A
D2758	-	-	NBL	28011	05/60	65A	02/68	65A
D2759	-	-	NBL	28012	06/60	65A	10/67	65A
D2760	-	-	NBL	28013	06/60	65A	02/68	65A
D2761	-	-	NBL	28014	06/60	65A	10/67	65A
D2762	-	-	NBL	28015	06/60	65A	03/67	65A
D2763	-	-	NBL	28016	07/60	65A	06/67	65A
D2764	-	-	NBL	28017	07/60	65A	02/68	65A
D2765	-	-	NBL	28018	07/60	65A	03/67	65A
D2766	-	-	NBL	28019	07/60	65A	03/67	65A
D2767	-	-	NBL	28020	08/60	65A	06/67	65A
D2768	-	-	NBL	28021	08/60	65C	02/68	65A
D2769	-	-	NBL	28022	09/60	65C	02/68	65A
D2770	-	-	NBL	28023	09/60	65C	02/68	65A
D2771	-	-	NBL	28024	09/60	65D	03/67	65A
D2772	-	-	NBL	28025	09/60	65D	03/67	65A
D2773	-	-	NBL	28026	10/60	65F	02/68	65A
D2774	-	-	NBL	28027	10/60	65F	06/67	65A
D2775	-	-	NBL	28028	10/60	65F	02/68	65A
D2776	-	-	NBL	28029	10/60	65K	10/67	65F
D2777	-	-	NBL	28030	10/60	65K	03/67	65A
D2778	-	-	NBL	28031	11/60	65K	03/67	65F
D2779	-	-	NBL	28032	02/61	64F	02/68	65A
D2780	-	-	NBL	28033	03/61	65A	02/68	65A

BR 1957 No.	Disposal Code	Disposal Detail	Date Cut Up	Notes
D2708	C	Slag Reduction, Ickles	09/67	
D2709	C	Slag Reduction, Ickles	09/67	
D2710	C	Slag Reduction, Ickles	08/67	
D2711	C	Slag Reduction, Ickles	09/67	
D2712	C	Slag Reduction, Ickles	08/67	
D2713	C	Slag Reduction, Ickles	08/67	
D2714	C	Slag Reduction, Ickles	09/67	
D2715	C	Slag Reduction, Ickles	08/67	
D2716	C	Slag Reduction, Ickles	08/67	
D2717	C	Argosy Salvage, Shettleston	12/67	
D2718	C	Argosy Salvage, Shettleston	12/67	
D2719	C	Slag Reduction, Ickles	11/67	
D2720	C	J N Connel, Coatbridge	07/71	PI
D2721	C	Argosy Salvage, Shettleston	12/67	
D2722	C	Slag Reduction, Ickles	11/67	
D2723	C	J N Connel, Coatbridge	11/67	
D2724	C	Slag Reduction, Ickles	08/67	
D2725	C	Motherwell Machinery and Scrap Co	09/67	
D2726	C	P Wood, Queenborough	10/71	PI
D2727	C	Slag Reduction, Ickles	08/67	
D2728	C	Argosy Salvage, Shettleston	12/67	
D2729	C	Slag Reduction, Ickles	08/67	
D2730	C	Argosy Salvage, Shettleston	06/68	
D2731	C	Argosy Salvage, Shettleston	06/68	
D2732	C	Slag Reduction, Ickles	10/67	
D2733	C	Slag Reduction, Ickles	09/67	
D2734	C	Argosy Salvage, Shettleston	12/67	
D2735	C	Slag Reduction, Ickles	08/67	
D2736	C	Birds, Cardiff	07/69	PI
D2737	C	Slag Reduction, Ickles	07/67	
D2738	C	NCB Killoch Colliery	08/79	PI
D2739	C	Birds, Long Marston	09/69	
D2740	C	Slag Reduction, Ickles	08/67	
D2741	C	Slag Reduction, Ickles	09/67	
D2742	C	Slag Reduction, Ickles	09/67	
D2743	C	Slag Reduction, Ickles	11/67	
D2744	C	Argosy Salvage, Shettleston	11/67	
D2745	C	Slag Reduction, Ickles	08/67	
D2746	C	Slag Reduction, Ickles	08/67	
D2747	C	Argosy Salvage, Shettleston	12/67	
D2748	C	Argosy Salvage, Shettleston	12/67	
D2749	C	Argosy Salvage, Shettleston	12/67	
D2750	C	J N Connel, Coatbridge	11/67	
D2751	C	J N Connel, Coatbridge	11/67	
D2752	C	Slag Reduction, Ickles	07/67	
D2753	C	Motherwell Machinery and Scrap Co	10/67	
D2754	C	J N Connel, Coatbridge	06/68	
D2755	C	Argosy Salvage, Shettleston	01/68	
D2756	C	G H Campbell, Airdrie	11/68	
D2757	C	Birds, Cardiff	10/70	PI
D2758	C	G H Campbell, Airdrie	11/68	
D2759	C	Argosy Salvage, Shettleston	05/68	
D2760	C	G H Campbell, Airdrie	11/68	
D2761	C	Argosy Salvage, Shettleston	05/68	
D2762	C	Slag Reduction, Ickles	08/67	
D2763	C	BSC Landore	04/77	PI
D2764	C	Barnes and Bell, Coatbridge	05/68	
D2765	C	Slag Reduction, Ickles	07/67	
D2766	C	Slag Reduction, Ickles	08/67	
D2767	P	East Lancs Railway	-	PI
D2768	C	Shipbreaking Industries, Faslane	05/68	Stored: [U] 01/68
D2769	C	G H Campbell, Airdrie	10/68	
D2770	C	G H Campbell, Airdrie	10/68	
D2771	C	Slag Reduction, Ickles	08/67	
D2772	C	Slag Reduction, Ickles	08/67	
D2773	C	Barnes and Bell, Coatbridge	05/68	
D2774	P	East Lancs Railway	-	Stored: [U] 01/68, PI
D2775	C	G H Campbell, Airdrie	10/68	
D2776	C	Argosy Salvage, Shettleston	05/68	
D2777	C	Birds, Cardiff	05/68	
D2778	C	Slag Reduction, Ickles	09/67	
D2779	C	G H Campbell, Airdrie	11/08	
D2780	C	Barnes and Bell, Coatbridge	05/68	

73

YORKSHIRE ENGINE DIESEL HYDRAULIC 0-4-0 CLASS 02

BR 1957 No.	TOPS No.	Date Re No.	Built By	Works No.	Date Introduced	Depot of First Allocation	Date Withdrawn	Depot of Final Allocation
D2850	-	-	YEC	2809	09/60	27A	07/60	8J
D2851	02001	10/74	YEC	2810	10/60	27A	06/75	AN
D2852	02002*	-	YEC	2811	10/60	27A	10/73	AN
D2853	02003	05/74	YEC	2812	10/60	9A	06/75	AN
D2854	-	-	YEC	2813	11/60	27A	02/70	8J
D2855	-	-	YEC	2814	11/60	27A	10/70	8J
D2856	02004	02/74	YEC	2815	11/60	27A	06/75	AN
D2857	-	-	YEC	2816	12/60	27A	04/71	8J
D2858	-	-	YEC	2817	12/60	26A	02/70	9A
D2859	-	-	YEC	2818	12/60	17B	03/70	50D
D2860	-	-	YEC	2843	09/61	24F	12/70	8J
D2861	-	-	YEC	2844	10/61	24F	12/69	10D
D2862	-	-	YEC	2845	10/61	24F	12/69	10D
D2863	-	-	YEC	2846	10/61	24F	12/69	10D
D2864	-	-	YEC	2847	10/61	8B	02/70	9A
D2865	-	-	YEC	2848	11/61	8B	03/70	50D
D2866	-	-	YEC	2849	11/61	27A	02/70	9A
D2867	-	-	YEC	2850	11/61	27A	09/70	6A
D2868	-	-	YEC	2851	11/61	27A	12/69	10D
D2869	-	-	YEC	2852	12/61	27A	12/69	9A

NORTH BRITISH DIESEL HYDRAULIC 0-4-0

BR 1957 No.	Built By	Works No.	Date Introduced	Depot of First Allocation	Date Withdrawn	Depot of Final Allocation	Disposal Code
D2900	NBL	27751	04/58	1D	02/67	C-WKS	C
D2901	NBL	27752	04/58	1D	02/67	C-WKS	C
D2902	NBL	27753	05/58	1D	02/67	C-WKS	C
D2903	NBL	27754	05/58	1D	02/67	W-WKS	C
D2904	NBL	27755	06/58	1D	02/67	W-WKS	C
D2905	NBL	27756	06/58	1D	02/67	5A	C
D2906	NBL	27757	07/58	1D	02/67	5A	C
D2907	NBL	27758	09/58	1D	02/67	5A	C
D2908	NBL	27759	09/58	2A	02/67	1F	C
D2909	NBL	27760	10/58	2A	02/67	5B	C
D2910	NBL	27761	10/58	2A	02/67	5B	C
D2911	NBL	27995	12/59	2B	02/67	5B	C
D2912	NBL	27996	12/59	8A	02/67	1F	C
D2913	NBL	27997	12/59	8A	02/67	5A	C

HUNSLET DIESEL MECHANICAL 0-4-0

BR 1957 No.	Date Re No.	BR 1948 No.	Built By	Works No.	Date Introduced	Depot of First Allocation	Date Withdrawn	Depot of Final Allocation
D2950	04/58	11500	Hunslet	4625	11/54	32B	12/67	50D
D2951	11/57	11501	Hunslet	4626	11/54	32B	12/67	50D
D2952	05/58	11502	Hunslet	4627	01/55	32B	12/66	32B

ANDREW BARCLAY DIESEL MECHANICAL 0-4-0 CLASS 01

BR 1957 No.	Date Re No.	BR 1948 No.	TOPS No.	Date Re No.	Built By	Works No.	Date Introduced	Depot of First Allocation	Date Withdrawn	Depot of Final Allocation
D2953	10/60	11503	-	-	A.Barclay	395	01/56	30A	06/66	30A
D2954	11/60	11504	01001	06/74	A.Barclay	396	01/56	30A	09/79	HD
D2955	07/61	11505	01002	06/74	A.Barclay	397	02/56	30A	03/81	HD
D2956	03/59	11506	-	-	A.Barclay	398	03/56	30A	05/66	36A
D2956	07/67	No 81	-	-	A.Barclay	424	07/67	34E	11/67	36A

RUSTON & HORNSBY DIESEL MECHANICAL 0-4-0

BR 1957 No.	Date Re No.	BR 1948 No.	Built By	Works No.	Date Introduced	Depot of First Allocation	Date Withdrawn	Depot of Final Allocation
D2957	04/58	11507	R.Hornsby	390774	03/56	40B	03/67	50D
D2958	03/58	11508	R.Hornsby	390777	05/56	40B	01/68	30A

BR 1957 No.	Disposal Code	Disposal Detail	Date Cut Up	Notes
D2850	C	W Heselwood, Attercliffe	06/71	Stored: [S] 04/70, R/I: 06/70
D2851	C	Arnott Young, Dudley Hill	01/76	Stored: [U] 07/71, R/I: 10/71, Stored: 11/71
D2852	C	BR Allerton	03/76	Stored: [U] 10/71
D2853	S	Lunt, Comley and Pitt, Shutt End	-	
D2854	P	South Yorkshire Railway	-	
D2855	C	W Heselwood, Attercliffe	06/71	Stored: [U] 05/70
D2856	C	Redland Roadstone, Mountsorrel	10/86	PI
D2857	S	Birds, Long Marston	-	Stored: [U] 06/69, R/I: 12/69, Stored: [U] 01/70
D2858	S	Adamson, Butterley	-	
D2859	C	Birds, Long Marston	03/71	
D2860	P	National Railway Museum, York	-	Stored: [U] 04/70, R/I: 06/70, Rebuilt 1978 by Thomas Hill
D2861	C	C F Booth, Rotherham	05/71	Stored: [U] 07/69
D2862	C	NCB Norton Colliery	04/79	Stored: [U] 07/69, PI
D2863	C	T W Ward, Beighton	05/71	Stored: [U] 05/69
D2864	C	W Heselwood, Attercliffe	08/70	
D2865	C	V Berry, Leicester	05/85	P
D2866	P	Brechin Railway	-	
D2867	S	Redland Roadstone, Mountsorrel	-	
D2868	S	Lunt, Comley and Pitt, Shutt End	-	Stored: [U] 05/69, R/I: 09/69, Stored: [U] 09/69
D2869	C	T W Ward, Beighton	08/71	Stored: [U] 05/69

BR 1957 No.	Disposal Detail	Date Cut Up	Notes
D2900	Slag Reduction, Ickles	10/67	
D2901	Slag Reduction, Ickles	10/67	
D2902	Slag Reduction, Ickles	10/67	
D2903	Slag Reduction, Ickles	09/67	
D2904	Slag Reduction, Ickles	11/67	
D2905	Slag Reduction, Ickles	11/67	
D2906	Slag Reduction, Ickles	11/67	
D2907	Slag Reduction, Ickles	09/67	
D2908	Slag Reduction, Ickles	12/67	
D2909	Slag Reduction, Ickles	10/67	
D2910	Slag Reduction, Ickles	11/67	
D2911	Slag Reduction, Ickles	11/67	
D2912	Slag Reduction, Ickles	09/67	
D2913	Slag Reduction, Ickles	09/67	

BR 1957 No.	Disposal Code	Disposal Detail	Date Cut Up	Notes
D2950	C	Thyssen, Llanelly	01/83	PI
D2951	C	C F Booth, Rotherham	04/68	
D2952	C	Slag Reduction, Ickles	08/67	

BR 1957 No.	Disposal Code	Disposal Detail	Date Cut Up	Notes
D2953	P	South Yorkshire Railway	-	PI
D2954	C	BR Holyhead Breakwater	02/82	
D2955	C	BR Holyhead Breakwater	02/82	
D2956	P	East Lancashire Railway	-	
D2956	C	BSC Briton Ferry	08/69	Second locomotive to carry No. D2956, PI

BR 1957 No.	Disposal Code	Disposal Detail	Date Cut Up	Notes
D2957	C	Slag Reduction, Ickles	08/67	
D2958	C	C F Booth, Rotherham	10/84	PI

RUSTON & HORNSBY DIESEL ELECTRIC 0-6-0 CLASS 07

BR 1957 No.	TOPS No.	Date Re No.	Built By	Works No.	Date Introduced	Depot of First Allocation	Date Withdrawn	Depot of Final Allocation
D2985	07001	01/74	R.Hornsby	480686	06/62	71I	07/77	EH
D2986	07002	12/73	R.Hornsby	480687	06/62	71I	07/77	EH
D2987	07003	10/73	R.Hornsby	480688	06/62	71I	10/76	EH
D2988	07004*	-	R.Hornsby	480689	06/62	71I	05/73	EH
D2989	07005	02/74	R.Hornsby	480690	06/62	71I	07/77	EH
D2990	07006	01/74	R.Hornsby	480691	07/62	71I	07/77	EH
D2991	07007*	-	R.Hornsby	480692	07/62	71I	05/73	EH
D2992	07008*	-	R.Hornsby	480693	09/62	71I	05/73	EH
D2993	07009	01/74	R.Hornsby	480694	09/62	71I	10/76	EH
D2994	07010	04/74	R.Hornsby	480695	09/62	71I	10/76	EH
D2995	07011	01/74	R.Hornsby	480696	10/62	71I	07/77	EH
D2996	07012	04/74	R.Hornsby	480697	10/62	71I	07/77	EH
D2997	07013	12/73	R.Hornsby	480698	11/62	71I	07/77	EH
D2998	07014*	-	R.Hornsby	480699	11/62	71I	05/73	EH

BEYER PEACOCK/BRUSH DEMONSTRATOR 0-4-0

BR 1957 No.	Built By	Works No.	Date Introduced	Depot of First Allocation	Date Withdrawn	Depot of Final Allocation	Disposal Code
D2999	B.Peacock	100/7861	09/60	30A	10/67	30A	C

BR DIESEL ELECTRIC 0-6-0 CLASSES 08, 09 & 10

BR 1957 No.	Date Re No.	BR 1948 No.	First TOPS No.	Date Re No.	Second TOPS No.	Date Re No.	Name	Name Date	Built By	Works No.	Date Introduced	Depot of First Allocation	Date Withdrawn	Depot of Final Allocation
D3000	05/61	13000							BR Derby		10/52	84E	11/72	82A
D3001	05/61	13001							BR Derby		10/52	84E	11/72	30A
D3002	02/60	13002							BR Derby		10/52	84E	07/72	82A
D3003	07/60	13003							BR Derby		10/52	84E	07/72	82A
D3004	11/57	13004	08001	02/74					BR Derby		11/52	84E	06/78	GD
D3005	09/58	13005	08002	03/74					BR Derby		11/52	67C	09/77	DR
D3006	07/57	13006							BR Derby		12/52	67B	11/72	40B
D3007	11/60	13007	08003	02/74					BR Derby		12/52	67B	12/77	DN
D3008	06/58	13008	08004	02/74					BR Derby		12/52	67B	08/83	TE
D3009	07/57	13009	08005	03/74					BR Derby		12/52	67B	10/78	BS
D3010	06/58	13010	08006	03/74					BR Derby		12/52	73C	02/80	DN
D3011	05/58	13011							BR Derby		12/52	73C	10/72	70D
D3012	02/60	13012	08007*						BR Derby		12/52	73C	01/73	70D
D3013	06/58	13013							BR Derby		12/52	73C	10/72	70D
D3014	12/60	13014							BR Derby		12/52	73C	10/72	70D
D3015	09/60	13015	08008	03/74					BR Derby		01/53	1A	11/83	GD
D3016	11/57	13016	08009	05/74					BR Derby		01/53	1A	11/75	AF
D3017	05/57	13017	08010	05/74					BR Derby		01/53	1A	12/77	WN
D3018	07/58	13018	08011	06/74					BR Derby		02/53	1A	12/91	RG
D3019	06/57	13019	08012*						BR Derby		02/53	1A	06/73	AN
D3020	02/58	13020	08013*						BR Derby		04/53	3D	12/73	SY
D3021	02/58	13021	08014	03/74					BR Derby		04/53	3C	05/60	BU
D3022	02/61	13022	08015	01/74					BR Derby		05/53	14A	09/80	TI
D3023	11/57	13023	08016	02/74					BR Derby		05/53	14A	05/80	NH
D3024	11/60	13024	08017*						BR Derby		06/53	14A	08/73	NH
D3025	05/60	13025	08018	05/74					BR Derby		07/53	82C	07/83	AN
D3026	01/59	13026							BR Derby		10/53	84E	11/72	41A
D3027	03/58	13027	08019	06/74					BR Derby		10/53	84E	09/83	BY
D3028	03/59	13028	08020*						BR Derby		10/53	84E	08/73	CD
D3029	12/57	13029	08021	04/74					BR Derby		10/53	84E	04/86	TI
D3030	10/58	13030	08022	02/74					BR Derby		10/53	81A	03/85	TI
D3031	10/58	13031	08023	09/74					BR Derby		11/53	81A	09/83	CH
D3032	01/58	13032	08024	03/74					BR Derby		11/53	81A	12/82	TI
D3033	05/58	13033	08025	03/74					BR Derby		11/53	81A	12/77	BS
D3034	09/59	13034							BR Derby		12/53	84B	11/72	41A
D3035	05/59	13035							BR Derby		12/53	84B	12/72	41A
D3036	11/57	13036	08026	04/74					BR Derby		12/53	84B	07/82	CA
D3037	01/59	13037							BR Derby		12/53	84B	12/72	41A
D3038	10/59	13038							BR Derby		12/53	84B	12/72	GA
D3039	01/59	13039	08027	03/74					BR Derby		12/53	84B	11/80	TO
D3040	11/59	13040	08028	09/74					BR Derby		01/54	70B	04/81	BU
D3041	12/57	13041	08029	05/74					BR Derby		01/54	70B	03/78	EJ
D3042	06/57	13042	08030	01/74					BR Derby		02/54	70B	07/82	EH
D3043	08/58	13043	08031	02/74					BR Derby		02/54	73C	07/82	DR
D3044	03/58	13044	08032	03/74					BR Derby		03/54	75C	08/74	TO
D3045	09/58	13045							BR Derby		03/54	73C	11/72	70D
D3046	11/57	13046	08033	03/74					BR Derby		04/54	75C	01/85	TI

BR 1957 No.	Disposal Code	Disposal Detail	Date Cut Up	Notes
D2985	P	South Yorkshire Railway	-	PI
D2986	C	P D Fuels, Kidwelly	11/86	PI
D2987	C	British Industrial Sand, Oakamoor	06/85	PI
D2988	C	BREL Eastleigh by M Claydon	11/73	
D2989	S	ICI Wilton	-	Rebuilt by Resco
D2990	C	P D Fuels, Kidwelly	10/84	PI
D2991	D	To Departmental Stock	-	
D2992	C	BREL Eastleigh	07/76	
D2993	E	Exported to Trieste, Italy [03/77]	-	
D2994	P	West Somerset Railway	-	
D2995	S	ICI Wilton	-	Rebuilt by Resco
D2996	S	P D Fuels, Coed Bach	-	
D2997	S	Dow Chemicals, Kings Lynn	-	Rebuilt by Resco
D2998	C	BREL Eastleigh	08/76	

BR 1957 No.	Disposal Code	Disposal Detail	Date Cut Up	Notes
D2999		C F Booth, Rotherham	10/70	Built as Brush demonstrator

BR 1957 No.	Disposal Code	Disposal Detail	Date Cut Up	Notes
D3000	P	Brighton Railway Museum	-	PI
D3001	C	BREL Doncaster	11/75	
D3002	P	Plym Valley Railway	-	PI
D3003	C	Wanstrow, Childrens Playground	12/91	PI
D3004	C	BREL Swindon	03/79	
D3005	C	BREL Swindon	01/78	
D3006	D	To Departmental Stock - 966507	-	
D3007	C	BREL Glasgow	07/79	Stored: [U] 03/77, R/I: 09/77
D3008	C	BREL Swindon	09/86	
D3009	C	Marple & Gillott, Sheffield	12/85	
D3010	C	BREL Swindon	04/80	Stored: [U] 04/69, R/I: 05/69
D3011	C	Marple & Gillott, Sheffield	12/85	PI
D3012	C	BREL Swindon	10/73	
D3013	C	J Cashmore, Newport	10/73	
D3014	P	Paignton & Dartmouth Railway	-	PI
D3015	C	BREL Swindon	06/86	
D3016	C	BREL Swindon	06/76	
D3017	C	BREL Eastleigh	02/78	
D3018	P	Chinnor & Princes Risborough Railway	-	
D3019	P	South Yorkshire Railway	-	PI
D3020	C	BREL Derby	04/75	Stored: [U] 09/73
D3021	C	C F Booth, Rotherham	11/81	
D3022	P	Severn Valley Railway	-	
D3023	P	South Yorkshire Railway	-	PI
D3024	C	BREL Doncaster	10/74	
D3025	C	BREL Swindon	09/85	
D3026	C	BREL Swindon	04/73	
D3027	C	BREL Swindon	07/86	
D3028	C	BREL Doncaster	11/75	
D3029	P	Birmingham Railway Museum	-	
D3030	S	Arthur Guinness, Park Royal	-	Withdrawn: 07/82, R/I: 07/82
D3031	C	BREL Swindon	03/87	
D3032	C	BREL Doncaster	09/83	
D3033	C	BREL Swindon	03/78	
D3034	C	BREL Derby	11/73	
D3035	D	To Departmental Stock - 966508	-	
D3036	C	BREL Swindon	04/86	
D3037	D	To Departmental Stock - 966510	-	
D3038	C	Bates Colliery, Blyth	06/80	PI
D3039	C	BREL Swindon	04/82	
D3040	C	BREL Swindon	04/82	
D3041	C	BREL Swindon	05/78	Stored: [U] 10/68, R/I: 11/68
D3042	C	BREL Swindon	10/84	
D3043	C	BR Immingham	03/88	Withdrawn: 12/81, R/I: 07/82
D3044	S	Foster Yeoman, Merehead	-	
D3045	C	BREL Glasgow	04/76	
D3046	C	BREL Swindon	05/86	

BR 1957 No.	Date Re No.	BR 1948 No.	First TOPS No.	Date Re No.	Second TOPS No.	Date Re No.	Name Date	Built By	Works No.	Date Introduced	Depot of First Allocation	Date Withdrawn	Depot of Final Allocation
D3047	01/58	13047	08034*					BR Derby		04/54	73C	07/73	EH
D3048	11/57	13048	08035	03/74				BR Derby		04/54	75C	09/79	LE
D3049	11/57	13049	08036	06/74				BR Derby		04/54	75C	09/81	WN
D3050	12/57	13050	08037	04/74				BR Derby		05/54	5B	02/80	BU
D3051	01/58	13051	08038*					BR Derby		05/54	5B	06/73	WN
D3052	05/58	13052	08039*					BR Derby		05/54	5B	12/73	WN
D3053	12/59	13053	08040*					BR Derby		05/54	5B	06/73	BY
D3054	01/58	13054	08041	05/74				BR Derby		06/54	5B	08/78	BS
D3055	02/58	13055	08042	04/74				BR Derby		06/54	5B	03/79	YK
D3056	10/58	13056	08043	02/74				BR Derby		07/54	18A	12/77	BU
D3057	12/58	13057	08044	04/74				BR Derby		07/54	15A	01/78	GD
D3058	11/57	13058	08045	07/74				BR Derby		08/54	15A	07/82	TO
D3059	05/58	13059	08046	07/74				BR Derby		08/54	15A	05/80	BU
D3060	09/59	13060	08047	02/74				BR Darlington		08/53	36B	10/79	WH
D3061	05/60	13061	08048	02/74				BR Darlington		08/53	36B	12/77	WH
D3062	06/59	13062	08049	02/74				BR Darlington		08/53	36B	05/81	WH
D3063	06/58	13063	08050	02/74				BR Darlington		09/53	36B	10/81	WH
D3064	12/60	13064	08051	02/74				BR Darlington		09/53	36B	07/82	WH
D3065	11/60	13065	08052	02/74				BR Darlington		10/53	38E	01/81	CA
D3066	05/58	13066	08053	08/74				BR Darlington		10/53	38E	03/81	TE
D3067	11/58	13067	08054	04/74				BR Darlington		10/53	38E	02/80	GD
D3068	05/60	13068	08055	02/74				BR Darlington		10/53	38E	11/80	CD
D3069	05/58	13069						BR Darlington		11/53	38E	01/73	SF
D3070	05/57	13070	08056	02/74				BR Darlington		11/53	53C	08/86	KY
D3071	07/57	13071	08057	03/74				BR Darlington		11/53	53A	07/76	HM
D3072	01/58	13072	08058	02/74				BR Darlington		12/53	53A	04/82	TE
D3073	06/57	13073	08059	02/74				BR Darlington		12/53	53A	09/80	TE
D3074	08/57	13074	08060	05/74				BR Darlington		12/53	53A	06/84	LN
D3075	10/57	13075	08061	03/74				BR Darlington		12/53	53A	05/84	BG
D3076	02/58	13076	08062	02/74				BR Darlington		12/53	53A	09/84	YC
D3077	03/58	13077	08063	03/74				BR Darlington		01/54	53A	07/84	TE
D3078	10/57	13078						BR Darlington		01/54	53A	11/72	51L
D3079	08/57	13079	08064	03/74				BR Darlington		01/54	53A	12/84	YC
D3080	04/58	13080	08065	02/74				BR Darlington		01/54	53A	09/77	IM
D3081	09/57	13081	08066	01/74				BR Darlington		02/54	53A	09/77	IM
D3082	03/58	13082	08067	03/74				BR Derby		10/54	21A	07/83	TS
D3083	02/59	13083	08068	03/74				BR Derby		10/54	16A	07/83	TS
D3084	08/59	13084	08069	06/74				BR Derby		10/54	16A	02/83	KD
D3085	05/59	13085	08070	09/74				BR Derby		11/54	16A	12/77	CH
D3086	01/60	13086	08071	03/74				BR Derby		11/54	16A	02/78	DN
D3087	03/59	13087	08072*					BR Derby		10/54	8C	06/73	SP
D3088	05/61	13088	08073*					BR Derby		10/54	8C	12/73	BS
D3089	01/59	13089	08074	08/74				BR Derby		10/54	5B	12/73	SP
D3090	02/59	13090	08075	02/74				BR Derby		10/54	5B	12/81	SB
D3091	10/59	13091	08076	03/74				BR Derby		11/54	5B	09/80	IM
D3092	01/58	13092						BR Derby		11/54	75C	10/72	73C
D3093	12/58	13093						BR Derby		12/54	75C	05/72	73F
D3094	10/57	13094						BR Derby		12/54	75C	10/72	73F
D3095	03/59	13095						BR Derby		12/54	75C	05/72	73F
D3096	02/58	13096						BR Derby		12/54	75C	05/72	75C
D3097	09/58	13097						BR Derby		01/55	75C	05/72	73C
D3098	07/57	13098						BR Derby		01/55	75C	10/72	73F
D3099	06/59	13099						BR Derby		01/55	75C	10/72	73F
D3100	02/59	13100						BR Derby		01/55	75C	10/72	75C
D3101	12/58	13101						BR Derby		02/55	75C	05/72	73F
D3102	03/57	13102	08077	02/74				BR Derby		02/55	86F	11/77	CA
D3103	12/58	13103	08078	05/74				BR Derby		02/55	86E	08/83	KD
D3104	02/59	13104	08079	05/74				BR Derby		02/55	86E	12/83	CH
D3105	10/59	13105	08080	06/74				BR Derby		02/55	84C	11/80	CD
D3106	06/59	13106	08081	05/74				BR Derby		03/55	84C	04/80	GD
D3107	09/59	13107	08082	04/74				BR Derby		03/55	84C	11/80	BY
D3108	08/59	13108	08083	05/74				BR Derby		03/55	84C	07/84	CA
D3109	10/58	13109	08084	04/74				BR Derby		03/55	84C	11/80	NH
D3110	03/59	13110	08085	06/74				BR Derby		03/55	84C	03/86	GD
D3111	11/60	13111	08086	09/74				BR Derby		03/55	84G	08/80	HM
D3112	01/61	13112	08087	05/74				BR Derby		04/55	84F	11/79	GD
D3113	03/62	13113	08088	03/74				BR Derby		04/55	84F	09/83	CD
D3114	08/60	13114	08089	02/74				BR Derby		04/55	84G	09/80	CA
D3115	11/61	13115	08090	03/74				BR Derby		04/55	84G	11/77	CA
D3116	09/60	13116	08091	04/74				BR Derby		04/55	84G	11/82	HM
D3117	05/62	13117						BR Derby		06/55	18A	07/67	16A
D3118	09/60	13118						BR Derby		06/55	18A	07/67	16A
D3119	08/61	13119						BR Derby		06/55	18A	07/67	16A
D3120	06/62	13120						BR Derby		07/55	18A	04/67	16A
D3121	09/62	13121						BR Derby		08/55	18A	07/67	16A
D3122	12/61	13122						BR Derby		09/55	18A	12/66	16A
D3123	11/61	13123						BR Derby		09/55	18A	12/66	16A
D3124	11/58	13124						BR Derby		09/55	18A	12/66	16A
D3125	08/60	13125						BR Derby		07/57	18A	07/67	16A
D3126	04/61	13126						BR Derby		07/57	18A	12/66	16A
D3127	09/62	13127	08092	03/74				BR Darlington		07/54	35A	11/78	MR

BR 1957 No.	Disposal Code	Disposal Detail	Date Cut Up	Notes
D3047	E	Lamco Mining Co, Liberia [02/75]	-	
D3048	C	BREL Swindon	01/80	Stored: [U] 03/78, R/I: 01/79
D3049	C	BREL Swindon	03/83	
D3050	C	BREL Swindon	10/80	
D3051	C	BREL Derby	08/73	
D3052	C	J Cashmore, Newport	08/74	
D3053	C	BREL Doncaster	10/75	
D3054	C	BREL Swindon	04/79	
D3055	C	BREL Doncaster	05/79	
D3056	C	BREL Swindon	04/78	
D3057	C	BREL Doncaster	01/79	
D3058	C	BREL Swindon	12/84	
D3059	P	Brechin Railway	-	PI
D3060	C	BREL Swindon	06/80	
D3061	C	BREL Doncaster	05/78	
D3062	C	BREL Swindon	05/83	
D3063	C	BREL Swindon	01/83	
D3064	C	BREL Swindon	05/86	
D3065	C	BR March, by C F Booth	01/82	
D3066	C	BREL Swindon	06/81	Stored: [U] 03/77, R/I: 12/77
D3067	S	Tilcon Ltd, Grassington	-	
D3068	C	BREL Swindon	04/83	
D3069	D	To Departmental Stock - 966509	-	
D3070	C	V Berry, Leicester	05/89	
D3071	C	BREL Doncaster	05/77	
D3072	C	BREL Doncaster	05/83	
D3073	C	BREL Swindon	04/85	Stored: [U] 05/77, R/I: 09/77
D3074	S	Arthur Guinness, Park Royal	-	
D3075	C	BREL Doncaster	04/85	
D3076	C	BREL Doncaster	03/85	
D3077	C	BREL Doncaster	04/85	
D3078	D	To Departmental Stock - 966506	-	
D3079	P	National Railway Museum, York	-	
D3080	C	BREL Doncaster	11/78	
D3081	C	BREL Doncaster	11/78	
D3082	C	BREL Swindon	11/86	
D3083	C	BREL Swindon	10/86	
D3084	C	BREL Doncaster	03/84	
D3085	D	To Departmental Stock - 97802	-	
D3086	C	BREL Doncaster	07/78	
D3087	C	Birchills Power Station, by T Ward	05/83	PI
D3088	C	NCB Bates Colliery, Blyth	10/85	PI
D3089	C	BREL Derby	06/77	
D3090	C	BREL Swindon	04/82	
D3091	C	BREL Swindon	04/82	
D3092	E	Lamco Mining Co, Liberia [05/74]	-	
D3093	C	G Cohen, Kettering	01/74	
D3094	E	Lamco Mining Co, Liberia [05/74]	-	
D3095	C	BREL Swindon	12/73	
D3096	C	BR Selhurst	09/72	
D3097	C	BREL Swindon	12/73	
D3098	E	Lamco Mining Co, Liberia [05/74]	-	
D3099	C	P Wood, Queenborough	07/77	PI
D3100	E	Lamco Mining Co, Liberia [05/74]	-	
D3101	P	Great Central Railway, Loughborough	-	PI
D3102	S	Wiggins Teape, Fort William	-	Stored: [U] 08/76, R/I: 12/76, Stored: [U] 10/77
D3103	C	BREL Swindon	04/87	
D3104	C	BREL Swindon	05/86	
D3105	C	BREL Swindon	05/81	
D3106	C	BREL Swindon	07/80	
D3107	C	BREL Swindon	02/82	
D3108	C	BREL Doncaster	02/85	
D3109	C	BREL Swindon	02/82	
D3110	A	RFS Locomotives, Doncaster	-	
D3111	C	BREL Swindon	01/81	
D3112	C	BREL Swindon	05/80	
D3113	C	BREL Swindon	11/86	
D3114	C	BREL Swindon	12/80	
D3115	C	BREL Swindon	01/79	Stored: [U] 10/77
D3116	C	BREL Doncaster	01/84	
D3117	C	J Cashmore, Great Bridge	11/67	
D3118	C	J Cashmore, Great Bridge	07/67	
D3119	C	J Cashmore, Great Bridge	08/67	
D3120	C	G Cohen, Kettering	06/68	
D3121	C	BR Workshops, Derby	05/68	
D3122	C	Steelbreaking and Dismantling Co, Chesterfield	07/67	
D3123	C	J Cashmore, Great Bridge	06/67	
D3124	C	J Cashmore, Great Bridge	06/67	
D3125	C	G Cohen, Kettering	03/68	
D3126	C	J Cashmore, Great Bridge	06/67	
D3127	C	BREL Doncaster	02/79	

BR 1957 No.	Date Re No.	BR 1948 No.	First TOPS No.	Date Re No.	Second TOPS No.	Date Re No.	Name	Name Date	Built By	Works No.	Date Introduced	Depot of First Allocation	Date Withdrawn	Depot of Final Allocation
D3128	08/57	13128	08093	02/74					BR Darlington		08/54	35A	07/81	YK
D3129	11/57	13129	08094	02/74					BR Darlington		08/54	35A	06/83	CA
D3130	11/57	13130	08095	02/74					BR Darlington		08/54	35A	02/83	MR
D3131	12/57	13131	08096	02/74					BR Darlington		08/54	35A	09/84	YC
D3132	06/57	13132	08097	02/74					BR Darlington		10/54	65F	03/81	HM
D3133	08/57	13133	08098	02/74					BR Darlington		10/54	65F	11/80	HM
D3134	05/57	13134	08099	03/74					BR Darlington		10/54	65F	09/82	BG
D3135	12/57	13135	08100	03/74					BR Darlington		11/54	65F	12/82	CA
D3136	11/57	13136	08101	05/74					BR Darlington		12/54	65F	07/83	LN
D3137	05/58	13137							BR Darlington		03/55	53A	07/70	31B
D3138	10/58	13138							BR Darlington		03/55	53A	04/72	31B
D3139	09/58	13139							BR Darlington		03/55	53A	06/68	51L
D3140	09/58	13140							BR Darlington		04/55	53C	06/68	51L
D3141	06/59	13141							BR Darlington		05/55	53C	07/71	51L
D3142	07/60	13142							BR Darlington		06/55	53A	06/68	51L
D3143	11/59	13143							BR Darlington		06/55	53A	08/69	51L
D3144	02/60	13144							BR Darlington		06/55	51B	07/69	51L
D3145	05/57	13145							BR Darlington		06/55	51B	04/72	51L
D3146	01/58	13146							BR Darlington		07/55	51B	02/68	51L
D3147	07/59	13147							BR Darlington		07/55	51B	05/68	51L
D3148	11/58	13148							BR Darlington		07/55	51B	03/68	51L
D3149	09/59	13149							BR Darlington		07/55	51B	07/70	51L
D3150	12/60	13150							BR Darlington		07/55	51B	06/68	51L
D3151	02/59	13151							BR Darlington		08/55	51B	12/67	51L
D3152	02/60	13152							BR Darlington		02/55	40B	10/67	16B
D3153	04/58	13153							BR Darlington		02/55	40B	07/67	16B
D3154	12/58	13154							BR Darlington		02/55	40B	07/67	16B
D3155	02/60	13155							BR Darlington		02/55	40B	09/67	40B
D3156	08/58	13156							BR Darlington		03/55	40B	07/67	16B
D3157	08/59	13157							BR Darlington		03/55	40B	10/67	40B
D3158	12/59	13158							BR Darlington		03/55	40B	02/67	40B
D3159	01/60	13159							BR Darlington		08/55	40B	09/67	40B
D3160	04/60	13160							BR Darlington		08/55	40B	11/67	16B
D3161	11/58	13161							BR Darlington		09/55	40B	09/67	40B
D3162	06/60	13162							BR Darlington		09/55	40B	12/67	16B
D3163	07/60	13163							BR Darlington		09/55	30A	09/67	40B
D3164	03/61	13164							BR Darlington		09/55	30A	12/67	16B
D3165	04/58	13165							BR Darlington		10/55	34A	07/67	16B
D3166	11/61	13166							BR Darlington		10/55	34A	07/67	16B
D3167	07/62	13167	08102	04/74					BR Derby		08/55	21A	03/88	DR
D3168	04/62	13168	08103	04/74					BR Derby		08/55	21A	09/83	DY
D3169	02/59	13169	08104	04/74					BR Derby		09/55	2A	07/82	KD
D3170	06/57	13170	08105	05/74					BR Derby		09/55	12A	07/83	KD
D3171	02/61	13171	08106	03/74					BR Derby		09/55	12A	01/82	KD
D3172	09/59	13172							BR Derby		09/55	5B	05/72	12A
D3173	02/60	13173	08107	06/74					BR Derby		09/55	5B	12/82	KD
D3174	02/62	13174	08108	02/74					BR Derby		09/55	5B	07/84	CA
D3175	02/61	13175	08109	02/74					BR Derby		10/55	5B	04/81	DN
D3176	09/61	13176	08110	05/74					BR Derby		10/55	5B	10/79	LE
D3177	08/58	13177	08111	03/74					BR Derby		10/55	6A	02/77	GL
D3178	05/62	13178	08112	09/74					BR Derby		10/55	6A	07/82	KD
D3179	11/60	13179	08113	04/74					BR Derby		10/55	17A	03/84	CF
D3180	12/59	13180	08114	03/74					BR Derby		11/55	17A	11/83	DR
D3181	02/58	13181	08115	02/74					BR Derby		11/55	17A	07/84	DR
D3182	05/61	13182	08116	03/74					BR Derby		10/55	82B	08/82	GD
D3183	08/59	13183							BR Derby		10/55	82B	12/72	SW
D3184	10/61	13184	08117	02/74					BR Derby		10/55	82B	02/77	GL
D3185	02/60	13185	08118	03/74					BR Derby		11/55	82B	03/80	LE
D3186	05/59	13186	08119	02/74					BR Derby		11/55	82B	02/77	GL
D3187	06/61	13187	08120	02/74					BR Derby		11/55	82B	10/81	DN
D3188	10/59	13188	08121	10/74					BR Derby		11/55	86E	06/84	AN
D3189	07/59	13189	08122	04/74					BR Derby		11/55	86E	06/77	EJ
D3190	02/58	13190	08123	04/74					BR Derby		11/55	86E	03/84	CD
D3191	01/61	13191	08124	04/74					BR Derby		11/55	84B	03/81	NR
D3192	04/59	13192	08125	02/74					BR Derby		11/55	84G	12/81	CD
D3193	02/61	13193							BR Derby		11/55	84G	09/67	6D
D3194	10/59	13194	08126	03/74					BR Derby		11/55	84G	11/80	SP
D3195	11/59	13195	08127	02/74					BR Derby		11/55	81D	08/80	GD
D3196	12/51	13196	08128	02/74					BR Derby		12/55	81D	08/80	DR
D3197	02/60	13197	08129	04/74					BR Derby		12/55	66A	07/83	SP
D3198	09/60	13198	08130	03/74					BR Derby		12/55	66A	12/82	AN
D3199	02/59	13199	08131	05/74					BR Derby		12/55	66A	03/81	DR
D3200	12/59	13200	08132	04/74					BR Derby		12/55	66A	01/86	CD
D3201	03/60	13201	08133	10/73					BR Derby		12/55	66A	09/80	LN
D3202	07/60	13202	08134	05/74					BR Derby		12/55	66B	09/81	SY
D3203	12/58	13203	08135	03/74					BR Derby		12/55	66B	02/77	GL
D3204	04/60	13204	08136	03/74					BR Derby		12/55	66B	08/83	YK
D3205	02/60	13205	08137	05/74					BR Derby		12/55	66B	11/82	LN
D3206	08/58	13206	08138	03/74					BR Derby		12/55	66C	01/78	NR
D3207	02/58	13207	08139	03/74					BR Derby		01/56	65A	09/80	MR
D3208	07/58	13208	08140	04/74					BR Derby		02/56	65A	06/77	SW
D3209	12/59	13209	08141	02/74					BR Derby		02/56	65A	01/88	TI

BR 1957 No.	Disposal Code	Disposal Detail	Date Cut Up	Notes
D3128	C	BREL Swindon	03/82	
D3129	C	BREL Doncaster	02/85	
D3130	C	BREL Swindon	09/86	
D3131	C	BREL Doncaster	04/85	
D3132	C	BREL Swindon	06/81	
D3133	C	BREL Swindon	01/81	
D3134	C	BREL Doncaster	10/83	
D3135	C	BREL Doncaster	02/85	
D3136	C	BREL Doncaster	11/84	
D3137	C	BR Doncaster, by C F Booth	11/70	
D3138	C	C F Booth, Rotherham	08/73	
D3139	C	C F Booth, Rotherham	05/69	
D3140	C	C F Booth, Rotherham	04/69	
D3141	C	C F Booth, Rotherham	05/72	
D3142	C	C F Booth, Rotherham	03/69	
D3143	C	G Cohen, Kettering	02/70	
D3144	C	Hughes, Bolckow	11/69	
D3145	C	C F Booth, Rotherham	12/72	
D3146	C	Hughes, Bolckow	05/68	
D3147	C	C F Booth, Rotherham	02/69	
D3148	C	C F Booth, Rotherham	05/69	
D3149	C	C F Booth, Rotherham	11/70	
D3150	C	C F Booth, Rotherham	04/69	
D3151	C	Hughes, Bolckow	04/68	
D3152	C	Slag Reduction, Ickles	03/68	
D3153	C	C F Booth, Rotherham	03/68	
D3154	C	C F Booth, Rotherham	04/68	
D3155	C	C F Booth, Rotherham	01/68	
D3156	C	C F Booth, Rotherham	04/68	
D3157	C	Slag Reduction, Ickles	07/68	
D3158	C	Slag Reduction, Ickles	07/68	
D3159	C	C F Booth, Rotherham	01/68	
D3160	C	Slag Reduction, Ickles	04/68	
D3161	C	C F Booth, Rotherham	01/68	
D3162	C	Slag Reduction, Ickles	07/68	
D3163	C	C F Booth, Rotherham	01/68	
D3164	C	Slag Reduction, Ickles	09/68	
D3165	C	C F Booth, Rotherham	04/68	
D3166	C	C F Booth, Rotherham,	02/68	
D3167	P	Lincoln City Council	-	Withdrawn: 06/84, R/I: 06/84, Withdrawn: 03/85, R/I: 05/85
D3168	C	BREL Swindon	04/86	
D3169	C	BREL Doncaster	03/64	
D3170	C	BREL Doncaster	08/84	
D3171	C	BREL Swindon	01/83	
D3172	C	BREL Derby	07/72	
D3173	C	BREL Doncaster	03/84	
D3174	P	East Kent Railway	-	
D3175	C	BREL Swindon	10/81	
D3176	C	BREL Swindon	04/80	
D3177	D	To Departmental Stock - 968012	-	Stored: [U] 12/76
D3178	C	BREL Swindon	04/84	
D3179	S	BCOE Gwaun-Cae-Gurwen	-	
D3180	P	Great Central Railway	-	
D3181	C	BREL Doncaster	11/85	
D3182	C	BREL Doncaster	07/83	
D3183	C	BC Merthyr Vale	12/87	
D3184	D	To Departmental Stock - 968010	-	Stored: [U] 12/76
D3185	C	BREL Swindon	08/80	
D3186	D	To Departmental Stock - 968011	-	Stored: [U] 12/76
D3187	C	BREL Swindon	08/82	
D3188	C	BREL Swindon	11/85	
D3189	C	BREL Doncaster	12/77	Stored: [U] 05/77
D3190	P	Cholsey & Wallingford Railway	-	
D3191	C	BREL Swindon	02/82	
D3192	C	BREL Swindon	02/84	
D3193	C	BR Workshops Derby	09/67	
D3194	C	BREL Swindon	12/81	
D3195	C	BREL Swindon	12/80	
D3196	C	BREL Swindon	02/84	
D3197	C	BREL Doncaster	05/84	
D3198	C	BREL Swindon	04/85	
D3199	C	BREL Swindon	08/81	
D3200	C	Birds, Long Marston	07/88	
D3201	S	Sheerness Iron & Steel Co	-	
D3202	C	BREL Swindon	12/86	
D3203	C	BREL Swindon	05/77	Stored: [U] 12/76
D3204	C	BREL Doncaster	12/84	
D3205	C	BREL Doncaster	09/85	
D3206	C	BREL Swindon	05/78	
D3207	C	BREL Swindon	04/82	
D3208	C	BREL Swindon	09/77	Stored: [U] 05/77
D3209	A	Booth-Roe, Rotherham		

BR 1957 No.	Date Re No.	BR 1948 No.	First TOPS No.	Date Re No.	Second TOPS No.	Date Re No.	Name	Name Date	Built By	Works No.	Date Introduced	Depot of First Allocation	Date Withdrawn	Depot of Final Allocation
D3210	01/60	13210	08142	04/74					BR Derby		02/56	62A	11/83	YK
D3211	06/59	13211	08143	10/73					BR Derby		02/56	62A	03/76	ED
D3212	03/60	13212	08144	09/74					BR Derby		02/56	65C	12/77	AY
D3213	04/60	13213	08145	09/74					BR Derby		03/56	65C	12/77	DT
D3214	12/60	13214	08146	03/74					BR Derby		03/56	65C	12/80	DR
D3215	02/61	13215	08147	03/74					BR Derby		03/56	65B	07/83	GD
D3216	03/60	13216	08148	02/74					BR Derby		03/56	65B	03/84	GD
D3217	09/57	13217	08149	01/74					BR Darlington		04/55	75C	04/81	DN
D3218	05/57	13218	08150	12/73					BR Darlington		04/55	75C	10/84	TE
D3219	02/59	13219	08151	12/73					BR Darlington		04/55	75C	01/79	EH
D3220	10/58	13220	08152	12/73					BR Darlington		04/55	75C	07/80	SU
D3221	11/58	13221	08153	09/74					BR Darlington		04/55	75C	12/82	CH
D3222	04/58	13222	08154	12/73					BR Darlington		06/55	75C	10/79	SU
D3223	09/59	13223	08155	01/74					BR Darlington		07/55	75C	01/79	AF
D3224	03/58	13224	08156	01/74					BR Darlington		06/55	75C	01/79	AF
D3225	11/59	13225	08157	01/74					BR Darlington		06/55	75C	04/77	AF
D3226	08/57	13226	08158	12/73					BR Darlington		07/55	75C	03/79	SU
D3227	05/62	13227	08159	02/74					BR Darlington		12/55	51B	11/85	GD
D3228	11/61	13228	08160	05/74					BR Darlington		12/55	51D	12/85	BY
D3229	01/60	13229	08161	02/74					BR Darlington		12/55	51D	11/85	TE
D3230	10/59	13230	08162	03/74					BR Darlington		12/55	53A	07/80	HM
D3231	03/60	13231	08163	03/74					BR Darlington		12/55	53A	12/82	HM
D3232	10/59	13232	08164	03/74					BR Darlington		01/56	53A	03/86	GD
D3233	08/59	13233	08165	04/74					BR Darlington		01/56	53A	07/80	BG
D3234	02/60	13234	08166	03/74					BR Darlington		01/56	53A	07/82	BG
D3235	01/60	13235	08167	02/74					BR Darlington		02/56	53A	09/77	DN
D3236	11/59	13236	08168	05/74					BR Darlington		02/56	53A	03/88	YK
D3237	08/62	13237	08169	02/74					BR Darlington		02/56	50A	03/81	YK
D3238	08/63	13238	08170	02/74					BR Darlington		02/56	50A	03/86	GD
D3239	04/61	13239	08171	02/74					BR Darlington		02/56	50A	04/83	YK
D3240	01/58	13240	08172	03/74					BR Darlington		02/56	50A	08/85	BH
D3241	06/57	13241	08173	04/74					BR Darlington		03/56	50A	12/78	ED
D3242	05/61	13242	08174	03/74					BR Darlington		03/56	50A	08/81	TE
D3243	10/60	13243	08175	12/74					BR Darlington		03/56	50A	12/78	GM
D3244	02/61	13244	08176	02/74					BR Darlington		03/56	50A	11/85	GD
D3245	05/61	13245	08177	02/74					BR Derby		05/56	5B	10/88	YK
D3246	12/60	13246	08178	05/74					BR Derby		05/56	16A	05/82	BY
D3247	03/61	13247	08179	08/74					BR Derby		05/56	16A	06/75	CD
D3248	10/61	13248	08180	04/74					BR Derby		05/56	17A	05/81	SB
D3249	02/61	13249	08181	09/74					BR Derby		05/56	17A	07/82	DY
D3250	09/61	13250	08182	02/74					BR Derby		06/56	17A	03/81	SY
D3251	10/61	13251	08183	03/74					BR Derby		06/56	19A	03/84	LN
D3252	02/60	13252	08184	03/74					BR Derby		06/56	19A	06/81	DR
D3253	08/61	13253	08185	02/74					BR Derby		06/56	19A	02/82	TE
D3254	09/59	13254	08186	05/74					BR Derby		06/56	19A	03/85	BH
D3255	03/59	13255							BR Derby		06/56	82B	12/72	85B
D3256	08/61	13256	08187	04/74					BR Derby		06/56	82B	07/83	CF
D3257	02/58	13257	08188	03/74					BR Derby		06/56	82B	04/83	CF
D3258	05/60	13258	08189	03/74					BR Derby		06/56	87F	12/81	CF
D3259	11/61	13259	08190	03/74					BR Derby		06/56	87F	10/80	CF
D3260	03/61	13260	08191	04/74					BR Derby		06/56	87F	01/88	LE
D3261	09/60	13261							BR Derby		06/56	87F	12/72	86A
D3262	06/61	13262	08192	05/74					BR Derby		07/56	87F	07/82	ML
D3263	05/61	13263	08193	05/74					BR Derby		08/56	87F	07/83	CF
D3264	09/61	13264	08194	04/74					BR Derby		08/56	87B	11/80	CF
D3265	02/59	13265	08195	02/74					BR Derby		08/56	87B	09/83	CF
D3266	10/59	13266	08196	03/74					BR Derby		08/56	87B	10/83	GM
D3267	06/61	13267	08197	02/74					BR Derby		08/56	87B	04/82	YK
D3268	12/60	13268	08198	03/74					BR Derby		09/56	81D	12/80	SW
D3269	12/60	13269	08199	04/74					BR Derby		09/56	81D	01/83	DY
D3270	12/59	13270	08200	12/73					BR Derby		09/56	70C	04/86	TE
D3271	02/62	13271	08201	12/73					BR Derby		09/56	70C	12/82	WN
D3272	08/61	13272	08202	01/74					BR Derby		10/56	70C	05/89	CF
D3273	06/61	13273	08203	03/74	08991	01/86	[Kidwelly]	01/86-06/87	BR Derby		10/56	70C	06/87	LE
D3274	09/59	13274	08204	05/74					BR Derby		10/56	70C	09/83	EH
D3275	10/58	13275	08205	04/74					BR Derby		10/56	65A	11/83	NR
D3276	10/60	13276	08206	02/74					BR Derby		10/56	65A	02/88	HO
D3277	02/61	13277	08207	12/73					BR Derby		10/56	65C	09/80	CA
D3278	03/59	13278	08208	02/74					BR Derby		11/56	65B	10/84	TI
D3279	02/59	13279	08209	02/74					BR Derby		11/56	65B	02/83	TI
D3280	11/61	13280	08210	03/74					BR Derby		11/56	65B	09/88	TI
D3281	02/60	13281	08211	03/74					BR Derby		11/56	65F	03/85	TE
D3282	12/59	13282	08212	09/74					BR Derby		11/56	66B	09/81	TE
D3283	02/61	13283	08213	03/74					BR Derby		11/56	66B	11/80	AN
D3284	04/62	13284	08214	02/74					BR Derby		11/56	66B	02/84	GD
D3285	12/60	13285	08215	02/74					BR Derby		11/56	66B	09/82	TE
D3286	09/60	13286	08216	05/74					BR Derby		11/56	66B	11/80	DY
D3287	02/58	13287	08217	09/74					BR Derby		11/56	66B	04/82	GD
D3288	02/62	13288	08218	03/74					BR Derby		12/56	19A	07/80	BR
D3289	01/62	13289	08219	02/74					BR Derby		12/56	19A	03/83	TI
D3290	09/62	13290	08220	08/74					BR Derby		12/56	16A	03/86	CD
D3291	07/62	13291	08221	03/74					BR Derby		12/56	19A	03/81	BY

BR 1957 No.	Disposal Code	Disposal Detail	Date Cut Up	Notes
D3210	C	BREL Swindon	03/87	
D3211	C	BREL Swindon	02/77	Stored: [U] 05/76
D3212	C	BREL Glasgow	02/80	
D3213	C	BREL Glasgow	07/78	
D3214	C	BREL Swindon	05/81	Stored: [U] 10/80
D3215	C	BREL Doncaster	12/84	
D3216	C	BREL Doncaster	09/84	
D3217	C	BREL Swindon	02/84	Stored: [U] 07/80
D3218	C	BREL Swindon	09/86	
D3219	C	BREL Eastleigh	05/80	
D3220	C	BREL Swindon	11/80	
D3221	C	BREL Swindon	12/86	
D3222	C	BREL Swindon	03/80	
D3223	C	BREL Swindon	03/80	
D3224	C	BREL Swindon	02/80	
D3225	S	Independent Sea Terminals, Ridham Docks	-	
D3226	C	BREL Swindon	10/79	
D3227	C	BREL Doncaster	06/86	
D3228	C	BREL Doncaster	09/86	
D3229	C	BREL Doncaster	04/86	
D3230	C	BREL Swindon	01/81	
D3231	C	BREL Doncaster	02/84	
D3232	S	RFS Industries, Doncaster [002]	-	
D3233	C	BREL Swindon	10/80	
D3234	C	BREL Doncaster	10/83	
D3235	C	BREL Glasgow	07/79	
D3236	S	ABB Transportation, York [002]	-	
D3237	C	BREL Swindon	10/81	
D3238	A	RFS Locomotives, Kilnhurst		
D3239	C	BREL Doncaster	09/83	
D3240	C	BREL Doncaster	11/85	
D3241	D	To Departmental Stock - P01	-	
D3242	C	BREL Swindon	03/82	
D3243	C	J R Adams, Glasgow	04/86	
D3244	C	BREL Doncaster	05/86	
D3245	S	ABB Transportation, Crewe	-	
D3246	C	BREL Swindon	03/84	
D3247	C	BREL Swindon	08/76	
D3248	C	BREL Swindon	08/81	Stored: [U] 08/76, R/I: 12/76, Withdrawn: 01/79, R/I: 02/79
D3249	C	BREL Swindon	11/84	
D3250	C	BREL Swindon	03/82	
D3251	C	BREL Doncaster	10/84	
D3252	C	BREL Swindon	04/82	
D3253	C	BREL Swindon	10/82	
D3254	C	BREL Swindon	08/86	
D3255	P	Brighton Railway Museum	-	PI
D3256	C	BREL Swindon	07/86	
D3257	C	BREL Swindon	09/86	
D3258	C	BREL Swindon	03/84	Stored: [U] 11/81
D3259	C	BREL Swindon	12/80	
D3260	C	V Berry, Leicester	10/89	
D3261	P	Brighton Railway Museum	-	PI
D3262	C	BREL Swindon	02/85	
D3263	C	BREL Swindon	05/86	
D3264	C	BREL Swindon	11/83	
D3265	P	Llangollen Railway	-	
D3266	C	J R Adams, Glasgow	01/86	Stored: [U] 12/80, R/I: 02/81, Stored: [U] 06/83
D3267	C	BREL Doncaster	02/84	
D3268	C	BREL Swindon	11/81	Stored: [U] 07/80
D3269	C	BREL Doncaster	07/84	
D3270	C	Booth-Roe, Rotherham	09/92	
D3271	C	BREL Doncaster	05/84	
D3272	S	Potter Group, Ely	-	
D3273	C	V Berry, Leicester	-	Withdrawn: 05/85, R/I: 08/85
D3274	C	BREL Swindon	09/84	
D3275	C	BREL Doncaster	08/84	
D3276	C	V Berry, Leicester	07/89	
D3277	C	BREL Swindon	04/83	
D3278	C	BREL Doncaster	11/85	
D3279	C	BREL Doncaster	08/83	
D3280	C	Booth-Roe, Rotherham	12/89	
D3281	C	BREL Doncaster	03/86	Stored: [U] 03/77, R/I: 04/77
D3282	C	BREL Swindon	02/82	
D3283	C	BREL Swindon	08/81	
D3284	C	BREL Doncaster	01/86	
D3285	C	BREL Doncaster	05/83	
D3286	S	Sheerness Iron & Steel Co	-	
D3287	C	BREL Doncaster	05/83	
D3288	C	BREL Swindon	08/81	
D3289	C	BREL Doncaster	08/83	
D3290	P	Steamtown, Carnforth	-	
D3291	C	BREL Swindon	08/81	

BR 1957 No.	Date Re No.	BR 1948 No.	First TOPS No.	Date Re No.	Second TOPS No.	Date Re No.	Name	Name Date	Built By	Works No.	Date Introduced	Depot of First Allocation	Date Withdrawn	Depot of Final Allocation
D3292	02/62	13292	08222	03/74					BR Derby		12/56	19A	12/84	BN
D3293	11/62	13293	08223	02/74					BR Derby		12/56	19A	06/79	TI
D3294	03/63	13294	08224	02/74					BR Derby		12/56	20B	07/88	DR
D3295	12/62	13295	08225	01/74					BR Derby		01/57	20B	10/85	TE
D3296	06/62	13296	08226	02/74					BR Derby		01/57	20B	11/85	DR
D3297	10/61	13297	08227	04/74					BR Derby		02/57	55B	10/83	DT
D3298	11/59	13298	08228	03/74					BR Darlington		03/56	34A	04/85	NC
D3299	09/58	13299	08229	03/74					BR Darlington		04/56	30A	12/77	FH
D3300	12/58	13300	08230	10/73					BR Darlington		04/56	30A	09/80	MR
D3301	09/57	13301	08231	05/74					BR Darlington		04/56	30A	06/81	TE
D3302	01/59	13302	08232	03/74					BR Darlington		04/56	30A	09/82	TE
D3303	09/58	13303	08233	11/73					BR Darlington		04/56	30A	05/81	SF
D3304	11/62	13304	08234	04/74					BR Darlington		04/56	34E	09/82	CW
D3305	05/61	13305	08235	05/74					BR Darlington		05/56	34E	07/83	BY
D3306	06/62	13306	08236	05/74					BR Darlington		05/56	34E	09/75	WN
D3307	05/60	13307	08237	04/74					BR Darlington		05/56	34A	10/84	BN
D3308	10/58	13308	08238	02/74					BR Darlington		05/56	34A	03/84	GL
D3309	11/58	13309	08239	03/74					BR Darlington		05/56	34A	07/84	NL
D3310	06/59	13310	08240	03/74					BR Darlington		06/56	34A	07/82	CA
D3311	04/59	13311	08241	02/74					BR Darlington		06/56	34A	10/81	SF
D3312	06/60	13312	08242	02/74					BR Darlington		06/56	34A	12/86	LN
D3313	06/59	13313	08243	03/74					BR Darlington		06/56	50A	02/86	KY
D3314	02/58	13314	08244	02/74					BR Darlington		07/56	50A	05/86	TI
D3315	06/63	13315	08245	02/74					BR Darlington		07/56	50A	05/84	NL
D3316	01/62	13316	08246	09/74					BR Darlington		09/56	52E	08/82	GM
D3317	12/61	13317	08247	04/74					BR Darlington		09/56	52E	03/81	DR
D3318	09/58	13318	08248	03/74					BR Darlington		10/56	50A	09/84	BG
D3319	09/58	13319	08249	02/74					BR Darlington		10/56	50A	05/85	LN
D3320	02/60	13320	08250	04/74					BR Darlington		10/56	50A	12/88	NC
D3321	01/59	13321	08251	02/74					BR Darlington		10/56	52E	09/80	TE
D3322	12/59	13322	08252	02/74					BR Darlington		11/56	52E	08/81	YK
D3323	09/60	13323	08253	03/74					BR Darlington		11/56	53A	02/87	YC
D3324	03/60	13324	08254	02/74					BR Darlington		11/56	52E	11/86	MR
D3325	01/62	13325	08255	02/74					BR Darlington		11/56	34A	10/84	GD
D3326	05/61	13326	08256	05/74					BR Darlington		11/56	35A	02/84	CR
D3327	02/61	13327	08257	02/74					BR Darlington		12/56	31B	10/85	NC
D3328	03/60	13328	08258	02/74					BR Darlington		12/56	36B	12/88	CA
D3329	01/61	13329	08259	02/74	08992	01/86	[Gwendraeth]	01/86-06/87	BR Darlington		12/56	36B	06/87	LE
D3330	11/62	13330	08260	03/74					BR Darlington		12/56	36B	08/83	SB
D3331	08/62	13331	08261	05/74					BR Darlington		12/56	41A	12/83	CR
D3332	08/60	13332	08262	01/74					BR Darlington		01/57	41A	01/84	OC
D3333	10/61	13333	08263	02/74					BR Darlington		01/57	41A	03/84	GD
D3334	01/59	13334	08264	03/74					BR Darlington		01/57	41A	07/84	HM
D3335	02/59	13335	08265	03/74					BR Darlington		01/57	41A	12/81	SP
D3336	02/62	13336	08266	02/74					BR Darlington		02/57	41A	03/85	SB
D3337	06/60	13337	08267	02/74					BR Derby		03/57	62A	12/77	BH
D3338	02/60	13338	08268	03/74					BR Derby		03/57	62A	07/85	GD
D3339	10/60	13339	08269	04/74					BR Derby		03/57	62A	07/84	HM
D3340	05/61	13340	08270	09/74					BR Derby		03/57	62A	12/82	AN
D3341	07/61	13341	08271	04/75					BR Derby		04/57	62A	09/81	DT
D3342	08/59	13342	08272	04/74					BR Derby		04/57	62C	07/87	TI
D3343	12/61	13343	08273	03/74					BR Derby		04/57	62C	12/82	AN
D3344	02/62	13344	08274	04/75					BR Derby		04/57	62C	12/84	GD
D3345	04/62	13345	08275	03/74					BR Derby		04/57	62C	02/83	TO
D3346	07/62	13346	08276	12/74					BR Derby		04/57	62C	06/77	DE
D3347	07/61	13347	08277	12/73					BR Derby		05/57	62B	11/81	LN
D3348	09/62	13348	08278	10/74					BR Derby		05/57	66B	06/77	HN
D3349	11/61	13349	08279	09/74					BR Derby		05/57	66B	04/81	DT
D3350	02/60	13350	08280	09/74					BR Derby		05/57	66B	10/80	ML
D3351	02/62	13351	08281	02/74					BR Derby		05/57	66B	12/80	NA
D3352	08/62	13352	08282	02/74					BR Derby		06/57	87F	11/80	AN
D3353	11/62	13353	08283	02/74					BR Derby		06/57	87F	02/86	LO
D3354	03/62	13354	08284	02/74					BR Derby		06/57	87F	05/85	SP
D3355	12/61	13355	08285	02/74					BR Derby		06/57	87F	09/88	DR
D3356	04/62	13356	08286	03/74					BR Derby		06/56	87F	04/82	LE
D3357	07/62	13357	08287	02/74					BR Derby		06/57	87F	02/83	TI
D3358			08288	03/74					BR Derby		06/57	87F	01/83	SW
D3359			08289	02/74					BR Derby		06/57	87F	12/85	DY
D3360			08290	02/74					BR Derby		08/57	87F	04/82	AN
D3361			08291	02/74					BR Derby		07/57	87F	01/83	AN
D3362			08292	05/74					BR Derby		05/57	82C	05/84	ED
D3363	01/62	13363	08293	06/74					BR Derby		05/57	88B	12/85	SP
D3364	02/62	13364	08294	06/74					BR Derby		05/57	88B	11/80	BS
D3365	04/63	13365	08295	02/74					BR Derby		05/57	88B	05/88	TE
D3366	03/62	13366	08296	02/74					BR Derby		05/57	88B	10/88	YK
D3367			08297	03/74					BR Derby		07/57	10A	08/88	DR
D3368			08298	04/74					BR Derby		08/57	10A	03/81	LO
D3369			08299	02/74					BR Derby		07/57	18A	07/83	SP
D3370			08300	02/74					BR Derby		08/57	10A	07/83	AN
D3371			08301	02/74					BR Derby		08/57	10A	12/81	SP
D3372			08302	03/74					BR Derby		08/57	26A	06/81	SP

BR 1957 No.	Disposal Code	Disposal Detail	Date Cut Up	Notes
D3292	A	BR Bounds Green		Stored: [U] 09/83
D3293	C	BREL Swindon	02/80	
D3294	A	BR Doncaster		
D3295	C	BREL Doncaster	02/86	
D3296	C	BREL Doncaster	05/86	
D3297	C	J R Adams, Glasgow	04/86	Stored: [U] 07/83
D3298	C	BREL Doncaster	01/86	
D3299	C	BREL Doncaster	12/78	
D3300	C	BREL Swindon	05/83	Stored: [U] 09/80
D3301	C	BREL Swindon	10/81	
D3302	C	BREL Doncaster	06/83	
D3303	C	BR Stratford, by G Morris	02/82	
D3304	C	BREL Swindon	12/82	
D3305	C	BREL Doncaster	09/84	
D3306	C	BREL Swindon	09/76	
D3307	C	BREL Doncaster	10/86	
D3308	P	Dean Forest Railway	-	
D3309	A	BR Neville Hill		
D3310	C	BREL Swindon	02/85	Stored: [U] 08/77, R/I: 09/77
D3311	C	BREL Swindon	05/82	
D3312	C	BR Lincoln	04/90	
D3313	C	Birds, Long Marston	06/92	
D3314	C	Booth-Roe, Rotherham	12/89	
D3315	C	BREL Doncaster	11/85	
D3316	C	D Christie, Camlackie	04/87	
D3317	C	BREL Swindon	11/81	As PO1: 08/77-05/78
D3318	C	BREL Doncaster	07/85	
D3319	C	BREL Swindon	03/87	
D3320	C	M C Processors, Glasgow	08/91	
D3321	C	BREL Swindon	01/81	Stored: [U] 07/78, R/I: 08/78
D3322	C	BREL Swindon	03/82	
D3323	C	V Berry, Leicester	07/88	Stored: [U] 01/87
D3324	A	BR Gateshead		Withdrawn: 03/86, R/I: 04/86
D3325	C	BREL Doncaster	01/86	
D3326	C	BREL Doncaster	01/85	
D3327	C	V Berry, Leicester	09/89	
D3328	C	V Berry, Leicester	10/89	
D3329	C	BREL Crewe	10/89	Stored: [U] 07/80, R/I: 12/80, Withdrawn: 03/84, R/I: 09/84
D3330	C	BREL Doncaster	10/83	
D3331	C	BREL Doncaster	01/85	
D3332	C	BREL Swindon	11/84	Withdrawn: 10/83, R/I: 10/83
D3333	C	BREL Doncaster	08/84	
D3334	C	BREL Doncaster	11/84	
D3335	C	BREL Swindon	03/87	
D3336	P	Keighley & Worth Valley Railway	-	
D3337	D	To Departmental Stock - 97801	-	
D3338	C	BREL Doncaster	02/86	
D3339	C	BREL Doncaster	06/85	
D3340	C	BREL Doncaster	01/84	
D3341	C	BREL Swindon	04/83	
D3342	C	RFS Industries, Doncaster	01/91	
D3343	C	BREL Doncaster	03/84	
D3344	C	BREL Doncaster	07/85	
D3345	C	BREL Doncaster	07/84	
D3346	C	BREL Doncaster	11/77	
D3347	C	BREL Doncaster	07/83	
D3348	C	BREL Glasgow	10/77	
D3349	C	BREL Swindon	04/83	
D3350	C	BREL Swindon	03/82	
D3351	C	BREL Swindon	01/83	
D3352	C	BREL Swindon	08/81	
D3353	C	C F Booth, Rotherham	11/87	
D3354	C	C F Booth, Rotherham	10/87	
D3355	A	BR Doncaster		
D3356	C	BR Polmadie	01/86	
D3357	C	BREL Doncaster	09/83	
D3358	P	Mid-Hants Railway	-	
D3359	C	Birds, Long Marston	06/92	
D3360	C	BREL Swindon	03/84	
D3361	C	BREL Doncaster	01/84	
D3362	S	Deanside Transit, Glasgow	-	Stored: [U] 10/83, R/I: 12/83
D3363	C	C F Booth, Rotherham	10/87	
D3364	C	BREL Swindon	12/81	
D3365	A	BR Thornaby		
D3366	S	ABB Transportation, Crewe	-	Stored: [U] 09/83, R/I: 12/83
D3367	C	V Berry, Leicester	12/89	Stored: [S] 03/88
D3368	C	BREL Swindon	07/81	
D3369	C	BREL Swindon	10/82	
D3370	C	BREL Doncaster	04/84	
D3371	C	BREL Swindon	05/85	
D3372	C	BREL Swindon	03/82	

BR 1957 No.	Date Re No.	BR 1948 No.	First TOPS No.	Date Re No.	Second TOPS No.	Date Re No.	Name	Name Date	Built By	Works No.	Date Introduced	Depot of First Allocation	Date Withdrawn	Depot of Final Allocation
D3373			08303	02/74					BR Derby		08/57	26A	12/81	DY
D3374			08304	03/74					BR Derby		08/57	26A	05/85	LN
D3375			08305	02/74					BR Derby		08/57	55D	10/88	HO
D3376			08306	02/74					BR Derby		08/57	55D	12/77	HM
D3377			08307	03/74					BR Derby		08/57	55D	12/77	HM
D3378			08308	04/74					BR Derby		08/57	55D	02/92	DY
D3379			08309	02/74					BR Derby		09/57	55D	07/92	KY
D3380			08310	02/74					BR Derby		09/57	56D	09/77	TE
D3381			08311	04/74					BR Derby		09/57	55E	12/82	HM
D3382			08312	09/74					BR Derby		09/57	66B	05/83	ML
D3383			08313	07/74					BR Derby		09/57	66B	05/82	GD
D3384			08314	09/74					BR Derby		09/57	66B	02/81	ML
D3385			08315	02/74					BR Derby		09/57	66B	02/79	MR
D3386			08316	09/74					BR Derby		09/57	65A	03/76	HN
D3387			08317	05/74					BR Derby		10/57	65A	10/82	BG
D3388			08318	10/74					BR Derby		10/57	65A	10/76	HN
D3389			08319	09/74					BR Derby		10/57	65A	04/83	ML
D3390			08320	09/74					BR Derby		10/57	65A	12/82	TO
D3391			08321	04/74					BR Derby		10/57	65A	06/84	ML
D3392			08322	04/74					BR Derby		10/57	65A	09/83	CF
D3393			08323	09/74					BR Derby		10/57	65A	07/81	SU
D3394			08324	02/74					BR Derby		10/57	65A	12/84	MR
D3395			08325	04/74					BR Derby		11/57	65A	07/84	GD
D3396			08326	09/74					BR Derby		11/57	65A	10/83	ML
D3397			08327	03/74					BR Derby		11/57	88B	02/83	BS
D3398			08328	02/74					BR Derby		11/57	88B	12/81	AN
D3399			08329	02/74					BR Derby		11/57	88B	07/84	LR
D3400			08330	07/74					BR Derby		11/57	88B	12/84	HM
D3401			08331	02/74					BR Derby		11/57	88B	03/88	DR
D3402			08332	07/74					BR Derby		11/57	88B	09/84	HM
D3403			08333	05/74					BR Derby		11/57	88B	12/82	DY
D3404			08334	11/74					BR Derby		12/57	88B	07/86	CF
D3405			08335	04/74					BR Derby		11/57	88B	01/87	TI
D3406			08336	03/74					BR Derby		12/57	88B	04/81	YK
D3407			08337	03/74					BR Derby		12/57	88B	02/87	DR
D3408			08338	06/74					BR Derby		01/58	65E	07/85	MG
D3409			08339	04/74					BR Derby		01/58	65E	10/84	YC
D3410			08340	02/74					BR Derby		01/58	65C	08/81	SP
D3411			08341	04/75					BR Derby		01/58	65C	10/83	DT
D3412			08342	03/74					BR Derby		01/58	65G	07/82	SP
D3413			08343	09/74					BR Derby		01/58	65G	10/83	IS
D3414			08344	06/74					BR Derby		02/58	65G	05/84	AY
D3415			08345	09/74					BR Derby		02/58	65G	10/83	AY
D3416			08346	04/74					BR Derby		02/58	65B	03/84	DT
D3417			08347	09/74					BR Derby		02/58	65B	10/83	GM
D3418			08348	12/73					BR Derby		02/58	65B	07/82	ED
D3419			08349	03/74					BR Crewe		11/57	88B	07/83	CF
D3420			08350	03/74					BR Crewe		12/57	88B	01/84	CF
D3421			08351	04/74					BR Crewe		12/57	88B	01/84	CF
D3422			08352	03/74					BR Crewe		12/57	88B	07/83	CF
D3423			08353	03/74					BR Crewe		01/58	88B	07/81	CF
D3424			08354	02/74					BR Crewe		02/58	88B	07/86	CF
D3425			08355	03/74					BR Crewe		02/58	88B	04/83	SP
D3426			08356	04/74					BR Crewe		02/58	88B	11/82	SP
D3427			08357	02/74					BR Crewe		03/58	88B	09/77	IM
D3428			08358	02/74					BR Crewe		03/58	88B	10/77	IM
D3429			08359	03/74					BR Crewe		03/58	82B	01/84	CF
D3430			08360	04/74					BR Crewe		04/58	82B	01/83	MG
D3431			08361	03/74					BR Crewe		04/58	84B	02/88	AF
D3432			08362	03/74					BR Crewe		04/58	87B	03/82	MG
D3433			08363	04/74					BR Crewe		04/58	87B	08/82	RG
D3434			08364	03/74					BR Crewe		05/58	87B	01/83	LE
D3435			08365	03/74					BR Crewe		05/58	87B	09/80	MG
D3436			08366	03/74					BR Crewe		05/58	87B	12/80	MG
D3437			08367	04/74					BR Crewe		06/58	87B	09/88	DR
D3438			08368	04/74					BR Crewe		06/58	87B	12/80	MG
D3439									BR Darlington		09/57	34A	09/68	41E
D3440									BR Darlington		09/57	34A	12/68	41E
D3441									BR Darlington		10/57	34A	09/68	41E
D3442									BR Darlington		10/57	34A	07/68	16A
D3443									BR Darlington		10/57	34A	12/68	30A
D3444									BR Darlington		10/57	34A	12/68	31B
D3445									BR Darlington		10/57	35A	11/68	30A
D3446									BR Darlington		10/57	35A	06/68	16A
D3447									BR Darlington		10/57	35A	03/68	16A
D3448									BR Darlington		11/57	35A	06/68	16A
D3449									BR Darlington		11/57	35A	12/67	16A
D3450									BR Darlington		11/57	35A	07/68	16A
D3451									BR Darlington		11/57	35A	07/68	34E

BR 1957 No.	Disposal Code	Disposal Detail	Date Cut Up	Notes
D3373	C	BREL Swindon	03/86	
D3374	C	BREL Swindon	12/86	Stored: [U] 09/83, R/I: 10/83
D3375	A	BR Knottingley		
D3376	C	BREL Swindon	12/77	
D3377	C	BREL Swindon	06/78	
D3378	A	BR Hull		
D3379	A	BR Knottingley		Withdrawn: 10/88, R/I: 12/88
D3380	C	BREL Doncaster	10/78	
D3381	C	BREL Doncaster	02/84	
D3382	C	J R Adams, Glasgow	04/86	
D3383	C	BREL Doncaster	02/83	
D3384	C	BREL Glasgow	09/81	Stored: [U] 10/80
D3385	C	BREL Doncaster	12/80	
D3386	C	BREL Swindon	03/77	Stored: [U] 02/76
D3387	C	BREL Doncaster	10/83	
D3388	C	BREL Eastleigh	05/77	Stored: [U] 08/76
D3389	C	J R Adams, Glasgow	01/86	
D3390	S	ECC, Burngullow	-	
D3391	C	J R Adams, Glasgow	01/86	
D3392	C	BREL Swindon	11/86	
D3393	C	BREL Swindon	02/82	
D3394	C	BREL Doncaster	10/85	
D3395	C	BREL Swindon	09/85	
D3396	C	J R Adams, Glasgow	04/86	Stored: [U] 07/83
D3397	C	BREL Swindon	02/87	
D3398	C	BREL Swindon	12/86	
D3399	C	BREL Swindon	03/86	
D3400	C	BREL Doncaster	09/85	
D3401	S	RFS Industries, Doncaster [001]	-	OOS
D3402	C	BREL Doncaster	05/85	
D3403	C	BREL Swindon	04/86	
D3404	A	Booth-Roe, Rotherham		
D3405	C	Thomas Hill, Rotherham	09/89	
D3406	C	BREL Swindon	03/82	
D3407	C	RFS Industries, Doncaster	01/89	Stored: [U] 09/83, R/I: 10/83
D3408	C	V Berry, Leicester	10/89	
D3409	C	BREL Doncaster	04/85	
D3410	C	BREL Swindon	03/82	
D3411	C	J R Adams, Glasgow	04/86	Stored: [U] 07/83
D3412	C	BREL Swindon	11/83	
D3413	C	BR Eastleigh	01/86	Stored: [U] 07/83
D3414	C	J R Adams, Glasgow	03/86	
D3415	S	Deanside Transport, Glasgow	-	Stored: [U] 07/83
D3416	C	J R Adams, Glasgow	04/86	
D3417	C	J R Adams, Glasgow	04/86	Stored: [U] 07/83
D3418	C	BR Eastfield	01/86	
D3419	C	BREL Swindon	06/86	
D3420	P	North Staffordshire Railway	-	
D3421	C	BREL Swindon	04/86	
D3422	C	BREL Swindon	09/86	
D3423	C	BREL Swindon	03/82	
D3424	C	M C Processors, Glasgow	04/91	Stored: [U] 05/78, R/I: 03/79
D3425	C	BREL Doncaster	02/84	
D3426	C	BREL Swindon	12/83	
D3427	C	BREL Swindon	04/79	
D3428	C	BREL Doncaster	11/77	
D3429	P	Peak Railway, Darley Dale	-	
D3430	C	BREL Swindon	04/87	
D3431	C	M C Processors, Glasgow	04/91	Stored: [U] 05/78, R/I: 01/79
D3432	C	BREL Swindon	04/83	
D3433	C	BREL Swindon	04/83	
D3434	C	BREL Swindon	11/83	Withdrawn: 12/81, R/I: 01/82
D3435	C	BREL Swindon	04/81	
D3436	C	BREL Swindon	02/82	
D3437	A	BR Doncaster		
D3438	C	BREL Swindon	09/84	
D3439	C	Steelbreaking & Dismantling Co Chesterfield	01/69	
D3440	C	G Cohen, Kettering	06/69	
D3441	C	G Cohen, Kettering	01/69	
D3442	C	C F Booth, Rotherham	01/69	
D3443	C	C F Booth, Rotherham	06/69	
D3444	C	G Cohen, Kettering	02/69	
D3445	C	Cox & Danks, Park Royal	04/69	
D3446	C	Steelbreaking & Dismantling Co Chesterfield	08/69	
D3447	C	C F Booth, Rotherham	02/69	
D3448	C	Steelbreaking & Dismantling Co Chesterfield	09/69	
D3449	C	C F Booth, Rotherham	09/68	
D3450	C	C F Booth, Rotherham	01/69	
D3451	C	C F Booth, Rotherham	06/69	

BR 1957 No.	Date Re No.	BR 1948 No.	First TOPS No.	Date Re No.	Second TOPS No.	Date Re No.	Name Date	Name Date	Built By	Works No.	Date Introduced	Depot of First Allocation	Date Withdrawn	Depot of Final Allocation
D3452									BR Darlington		11/57	35A	07/68	16A
D3453									BR Darlington		11/57	35A	07/68	34E
D3454			08369	02/74					BR Darlington		05/57	56G	03/85	YC
D3455			08370	02/74					BR Darlington		05/57	56G	03/84	GD
D3456			08371	03/74					BR Darlington		05/57	65G	10/83	HM
D3457			08372	02/74					BR Darlington		06/57	55B	10/84	HM
D3458			08373	03/74					BR Darlington		06/57	55B	05/85	GD
D3459			08374	02/74					BR Darlington		06/57	73C	12/82	BY
D3460			08375	12/73					BR Darlington		06/57	73C	11/91	CF
D3461			08376	01/74					BR Darlington		06/57	73C	07/83	SP
D3462			08377	02/74					BR Darlington		06/57	73C	06/83	LA
D3463			08378	12/73					BR Darlington		07/57	73C	07/81	SU
D3464			08379	01/74					BR Darlington		07/57	73C	12/80	SU
D3465			08380	12/73					BR Darlington		07/57	73C	08/82	ED
D3466			08381	01/74					BR Darlington		07/57	73C	12/82	WN
D3467			08382	03/74					BR Darlington		08/57	73C	03/85	WN
D3468			08383	01/74					BR Darlington		08/57	73C	03/85	TE
D3469			08384	01/74					BR Darlington		08/57	74A	01/86	GD
D3470			08385	02/74					BR Darlington		08/57	74A	09/88	DR
D3471			08386	02/74					BR Darlington		08/57	73C	01/87	LN
D3472			08387	04/74					BR Darlington		08/57	73C	05/82	SU
D3473									BR Darlington		12/57	35A	08/68	16A
D3474									BR Darlington		12/57	35A	06/68	36A
D3475									BR Darlington		12/57	35A	07/68	16A
D3476									BR Darlington		12/57	34A	06/68	16A
D3477									BR Darlington		12/57	34A	07/68	30A
D3478									BR Darlington		01/58	34C	06/68	40A
D3479									BR Darlington		01/58	36A	04/69	36A
D3480									BR Darlington		01/58	36A	08/68	36A
D3481									BR Darlington		02/58	36A	12/68	36A
D3482									BR Darlington		02/58	36A	07/68	16A
D3483									BR Darlington		03/58	36A	03/69	36A
D3484									BR Darlington		03/58	36A	08/68	36A
D3485									BR Darlington		03/58	35A	12/68	31B
D3486									BR Darlington		03/58	35A	11/70	31B
D3487									BR Darlington		03/58	35A	07/68	34E
D3488									BR Darlington		03/58	35A	12/68	31B
D3489									BR Darlington		04/58	35A	04/68	16A
D3490									BR Darlington		04/58	34A	07/68	16A
D3491									BR Darlington		04/58	31B	12/68	31B
D3492									BR Darlington		04/58	31B	06/69	31B
D3493									BR Darlington		04/58	31B	07/68	16A
D3494									BR Darlington		05/58	31B	05/68	30E
D3495									BR Darlington		05/58	30A	07/68	16A
D3496									BR Darlington		06/58	30A	11/67	30E
D3497									BR Darlington		12/57	30A	04/68	16B
D3498									BR Darlington		12/57	30A	06/68	16A
D3499									BR Doncaster		12/57	30A	01/68	16A
D3500									BR Doncaster		01/58	30A	01/68	16A
D3501									BR Doncaster		01/58	30A	07/68	16A
D3502									BR Doncaster		01/58	30A	02/68	16A
D3503			08388	02/74					BR Derby		04/58	82B		
D3504			08389	02/74					BR Derby		04/58	82B		
D3505			08390	06/74					BR Derby		04/58	82B		
D3506			08391	02/74					BR Derby		04/58	82B	08/85	BG
D3507			08392	03/74					BR Derby		04/58	82B	02/84	BG
D3508			08393	02/74					BR Derby		04/58	82B		
D3509			08394	03/74					BR Derby		05/58	83D	11/87	LE
D3510			08395	03/74					BR Derby		05/58	83D	07/88	DR
D3511			08396	01/74					BR Derby		05/58	83D	03/86	SP
D3512			08397	03/74					BR Derby		05/58	83D		
D3513			08398	02/74					BR Derby		05/58	83D	07/85	BR
D3514			08399	06/74					BR Derby		05/58	83D	11/92	CD
D3515			08400	03/74					BR Derby		05/58	83D	12/87	LE
D3516			08401	03/74					BR Derby		05/58	83D		
D3517			08402	04/74					BR Derby		05/58	83D		
D3518			08403	05/74					BR Derby		05/58	83D	09/85	BY
D3519			08404	03/74					BR Derby		05/58	83A	06/78	FH
D3520			08405	02/74					BR Derby		06/58	83A		
D3521			08406	03/74					BR Derby		06/58	83A	04/88	CA
D3522			08407	04/74					BR Derby		06/58	83A	02/93	SF
D3523			08408	05/74					BR Derby		06/58	83A	04/88	SF
D3524			08409	03/74					BR Derby		06/58	83A	06/86	SF
D3525			08410	03/74					BR Derby		06/58	83A		
D3526			08411	04/74					BR Derby		06/58	83F		
D3527			08412	02/74					BR Derby		06/58	83F	09/85	MR
D3528			08413	02/74					BR Derby		06/58	66B		
D3529			08414	12/74					BR Derby		06/58	66C		
D3530			08415	05/74					BR Derby		07/58	65G		
D3531			08416	03/74					BR Derby		07/58	65G	02/92	TO

BR 1957 No.	Disposal Code	Disposal Detail	Date Cut Up	Notes
D3452	P	Bodmin & Wenford Railway	-	PI
D3453	C	C F Booth, Rotherham	05/69	
D3454	C	BREL Doncaster	06/85	
D3455	C	BREL Doncaster	10/85	
D3456	C	BREL Doncaster	05/85	
D3457	C	BREL Doncaster	05/85	
D3458	C	BREL Doncaster	03/86	
D3459	C	BREL Doncaster	09/84	
D3460	A	BR Cardiff		Stored: [U] 04/83, R/I: 10/83, Withdrawn: 12/85, R/I: 01/86, Stored: [S] 08/91
D3461	C	BREL Doncaster	04/84	
D3462	P	Dean Forest Railway	-	
D3463	C	BREL Swindon	03/82	
D3464	C	BREL Swindon	03/81	
D3465	C	BREL Swindon	05/83	
D3466	C	BREL Doncaster	05/84	
D3467	C	BREL Doncaster	08/86	
D3468	C	BREL Doncaster	05/86	
D3469	C	BREL Doncaster	04/86	Stored: [U] 09/83, R/I: 11/83
D3470	A	BR Doncaster		Stored: [U] 09/83, R/I: 10/83, Stored: [U] 02/86, R/I: 05/86
D3471	C	BR Lincoln	04/90	Withdrawn: 03/85, R/I: 05/85
D3472	C	BREL Swindon	03/86	
D3473	C	C F Booth, Rotherham	01/69	
D3474	C	C F Booth, Rotherham	01/69	
D3475	C	C F Booth, Rotherham	01/69	
D3476	P	South Yorkshire Railway	-	PI
D3477	C	G Cohen, Kettering	08/69	
D3478	C	C F Booth, Rotherham	07/69	
D3479	C	G Cohen, Kettering	07/69	
D3480	C	Steelbreaking & Dismantling Co Chesterfield	01/69	
D3481	C	C F Booth, Rotherham	05/69	
D3482	C	C F Booth, Rotherham	01/69	
D3483	C	G Cohen, Kettering	10/69	
D3484	C	Steelbreaking & Dismantling Co Chesterfield	01/69	
D3485	C	G Cohen, Kettering	07/69	
D3486	C	C F Booth, Rotherham	04/71	
D3487	C	C F Booth, Rotherham	06/69	
D3488	C	G Cohen, Kettering	11/69	
D3489	S	Felixstowe Dock & Railway Co	-	
D3490	C	C F Booth, Rotherham	01/69	
D3491	C	G Cohen, Kettering	06/69	
D3492	C	G Cohen, Kettering	10/69	
D3493	C	C F Booth, Rotherham	01/69	
D3494	C	G Cohen, Kettering	01/69	
D3495	C	C F Booth, Rotherham	01/69	
D3496	C	BR Workshops, Doncaster	09/68	
D3497	C	ECC, Fowey	02/90	
D3498	C	C F Booth, Rotherham	01/69	
D3499	C	C F Booth, Rotherham	09/68	
D3500	C	J Cashmore, Great Bridge	09/68	
D3501	C	C F Booth, Rotherham	12/68	
D3502	C	J Cashmore, Great Bridge	07/68	
D3503				
D3504				
D3505				
D3506	C	BREL Doncaster	01/86	
D3507	C	BREL Doncaster	09/85	Withdrawn: 10/80, R/I: 10/80
D3508				
D3509	C	V Berry, Leicester	09/89	
D3510	C	Booth-Roe, Rotherham	04/89	Withdrawn: 02/86, R/I: 04/86
D3511	C	C F Booth, Rotherham	11/87	
D3512				
D3513	S	ECC Marsh Mills	-	
D3514	A	BRML Springburn		
D3515	C	V Berry, Leicester	10/89	
D3516				
D3517				
D3518	C	BREL Swindon	04/86	Withdrawn: 10/81, R/I: [S] 01/83
D3519	C	BREL Doncaster	01/79	
D3520				
D3521	C	Booth-Roe, Rotherham	10/92	
D3522	S	BR Cheriton		
D3523	C	M C Processors, Glasgow	10/91	Withdrawn: 12/84, R/I: 02/85, Stored: [S] 03/88
D3524	C	BREL Doncaster	09/86	
D3525				
D3526				
D3527	C	Booth-Roe, Rotherham	10/92	Stored: [U] 02/85
D3528				
D3529				
D3530				Withdrawn: 07/81, R/I: [S] 01/83, R/I: 01/85
D3531	S	RFS Locomotives, Kilnhurst	-	Stored: [U] 07/82, R/I: 08/82

89

BR 1957 No.	Date Re No.	BR 1948 No.	First TOPS No.	Date Re No.	Second TOPS No.	Date Re No.	Name	Name Date	Built By	Works No.	Date Introduced	Depot of First Allocation	Date Withdrawn	Depot of Final Allocation
D3532			08417	11/73					BR Derby		07/58	65D		
D3533			08418	05/74					BR Derby		07/58	65I		
D3534			08419	07/74					BR Derby		07/58	65I		
D3535			08420	03/74					BR Derby		07/58	63B	09/88	DR
D3536			08421	03/74	09201	10/92			BR Derby		08/58	63B		
D3537			08422	03/74					BR Derby		08/58	63B	12/85	BY
D3538			08423	03/74					BR Derby		08/58	63B	11/88	SP
D3539			08424	12/74					BR Derby		08/58	63B	06/85	TJ
D3540			08425	04/75					BR Derby		08/58	63B	09/85	DT
D3541			08426	12/74					BR Derby		08/58	63A	07/76	DE
D3542			08427	04/74					BR Derby		08/58	63A	11/86	MR
D3543			08428	12/74					BR Derby		08/58	63A		
D3544			08429	02/74					BR Derby		09/58	63A	09/83	SB
D3545			08430	09/74					BR Derby		09/58	63A	12/85	ED
D3546			08431	12/74					BR Derby		09/58	61B	07/85	AN
D3547			08432	02/74					BR Derby		09/58	61B	10/83	HS
D3548			08433	04/74					BR Derby		09/58	61B	03/88	TJ
D3549			08434	02/74					BR Derby		09/58	61B	02/92	DY
D3550			08435	03/74					BR Derby		10/58	61B	11/80	BG
D3551			08436	03/74					BR Derby		10/58	61B	01/92	DR
D3552			08437	10/74					BR Derby		10/58	61A	12/84	ML
D3553			08438	04/74					BR Derby		10/58	61A	03/86	MR
D3554			08439	02/74					BR Derby		10/58	64F	10/88	IM
D3555			08440	03/74					BR Derby		10/58	64F	05/90	SF
D3556			08441	12/74					BR Derby		10/58	64F		
D3557			08442	12/74					BR Derby		11/58	64E		
D3558			08443	12/74					BR Derby		11/58	64E	07/85	TJ
D3559			08444	03/74					BR Derby		11/58	64F	10/86	CF
D3560			08445	04/75					BR Derby		11/58	64B		
D3561			08446	09/74					BR Derby		11/58	64B	12/85	TJ
D3562			08447	09/74					BR Derby		11/58	65C		
D3563			08448	09/74					BR Derby		11/58	67C		
D3564			08449	06/74					BR Derby		11/58	67C		
D3565			08450	03/74					BR Derby		12/58	12A	05/86	CD
D3566			08451	06/74					BR Derby		12/58	12A		
D3567			08452	01/74					BR Derby		12/58	12A	06/85	TJ
D3568			08453	09/74					BR Derby		12/58	17B	10/80	YK
D3569			08454	02/74					BR Derby		12/58	17B		
D3570			08455	02/74					BR Derby		12/58	17B	09/85	DY
D3571			08456	02/74					BR Derby		01/59	17B	04/88	DR
D3572			08457	05/74					BR Derby		01/59	17B	09/85	BY
D3573			08458	04/74					BR Crewe		07/58	14D	08/88	SF
D3574			08459	04/74					BR Crewe		08/58	18A	10/88	DR
D3575			08460	03/74					BR Crewe		08/58	18A		
D3576			08461	04/74					BR Crewe		08/58	6C	02/88	SF
D3577			08462	03/74	08994	09/87	Gwendraeth	09/87	BR Crewe		08/58	6C		
D3578			08463	02/74					BR Crewe		09/58	8A	03/89	DY
D3579			08464	02/74					BR Crewe		09/58	8A	09/85	AN
D3580			08465	02/74					BR Crewe		10/58	24K	09/85	DY
D3581			08466	10/73					BR Crewe		10/58	11B		
D3582			08467	02/74					BR Crewe		10/58	21A	11/81	SY
D3583			08468	03/74					BR Crewe		11/58	21A	11/88	SP
D3584			08469	05/74					BR Crewe		11/58	21A	12/81	CD
D3585			08470	03/74					BR Crewe		11/58	17B	03/86	CD
D3586			08471	05/74					BR Crewe		11/58	17B	09/85	LR
D3587			08472	02/74					BR Crewe		11/58	17B		
D3588			08473	04/74					BR Crewe		11/58	26A	02/86	CF
D3589			08474	03/74					BR Crewe		12/58	26A	08/86	LR
D3590			08475	03/74					BR Crewe		12/59	26A	03/87	SP
D3591			08476	06/74					BR Crewe		12/58	26A	09/85	DT
D3592			08477	04/74					BR Crewe		01/59	26A	12/86	SP
D3593			08478	03/74					BR Horwich		08/58	83E	04/88	IM
D3594			08479	03/74					BR Horwich		09/58	83E	11/91	CF
D3595			08480	03/74					BR Horwich		09/58	83F		
D3596			08481	03/74					BR Horwich		09/58	83G		
D3597			08482	03/74					BR Horwich		09/58	83G		
D3598			08483	01/74					BR Horwich		10/58	83C		
D3599			08484	02/74					BR Horwich		10/58	81A		
D3600			08485	04/75					BR Horwich		10/58	81A		
D3601			08486	03/74					BR Horwich		10/58	81A	11/86	CF
D3602			08487	03/74					BR Horwich		10/58	81A	02/88	SF
D3603			08488	03/74					BR Horwich		10/58	81A	02/86	CF
D3604			08489	03/74					BR Horwich		11/58	81A		
D3605			08490	05/74					BR Horwich		11/58	81A	11/85	TJ
D3606			08491	03/74					BR Horwich		11/58	81A	09/85	CF
D3607			08492	03/74					BR Horwich		11/58	81A		
D3608			08493	02/74					BR Horwich		02/58	30A		
D3609			08494	05/74					BR Horwich		03/58	30A	08/88	BY
D3610			08495	03/74					BR Horwich		03/58	31A		
D3611			08496	03/74					BR Horwich		03/58	31A	06/90	CA

BR 1957 No.	Disposal Code	Disposal Detail	Date Cut Up	Notes
D3532				Withdrawn: 03/82, R/I: 06/82
D3533				Stored: [U] 02/85, R/I: 12/85
D3534				Withdrawn: 12/81, R/I: 04/83
D3535	A	BR Doncaster		Worked after withdrawal
D3536				
D3537	C	Birds, Long Marston	09/88	
D3538	S	Trafford Park Estates	-	Stored: [U] 07/82, R/I: 02/83
D3539	C	J R Adams, Glasgow	04/86	Stored: [U] 05/85
D3540	C	BREL Swindon	08/86	Stored: [U] 01/84
D3541	C	BREL Doncaster	05/77	
D3542	A	BR March		
D3543				
D3544	C	BREL Doncaster	11/83	
D3545	C	J McWilliam, Shettleston	05/86	Stored: [U] 07/84, R/I: 12/84
D3546	S	Texas Metals, Hyde	-	Stored: [U] 11/84
D3547	C	BREL Doncaster	02/84	
D3548	C	J McWilliam, Shettleston	05/86	
D3549	A	BR Derby		Stored: [U] 01/87, R/I: 02/87
D3550	C	BREL Swindon	05/83	
D3551	A	BR Doncaster		
D3552	C	J R Adams, Glasgow	04/86	
D3553	C	Booth-Roe, Rotherham	10/92	
D3554	A	BR Immingham		
D3555	C	BR Stratford, by M C Processors	06/91	
D3556				
D3557				
D3558	S	Scottish Grain Distillers, Cambus	-	
D3559	P	Bodmin and Wenford Railway	-	
D3560				
D3561	C	J McWilliam, Shettleston	05/86	
D3562				
D3563				Stored: [U] 03/82, R/I: 09/83
D3564				Stored: [U] 05/87
D3565	C	Birds, Long Marston	06/92	
D3566				Withdrawn: 09/81, R/I: [S] 01/83, R/I: 10/84
D3567	C	BREL Swindon	03/87	Stored: [U] 12/82, R/I: 02/83, Stored: [U] 03/85
D3568	C	BREL Swindon	08/82	
D3569				Withdrawn: 10/81, R/I: [S] 01/83, R/I: 01/85
D3570	C	BREL Swindon	03/87	Withdrawn: 12/81, R/I: [S] 01/83, R/I: 12/84
D3571	C	C F Booth, Rotherham	05/89	Stored: [U] 03/88
D3572	C	BREL Swindon	05/86	Stored: [U] 08/81, Withdrawn: 12/81, R/I: [S] 01/83, R/I: 11/84
D3573	C	M C Processors, Glasgow	08/91	Stored: [U] 08/77, R/I: 09/77
D3574	C	V Berry, Leicester	10/89	Stored: [U] 00/77, R/I: 09/77
D3575				Stored: [U] 11/86, R/I: 12/86
D3576	C	M C Processors, Glasgow	08/91	
D3577				Stored: [U] 05/82, R/I: 02/83, Stored: [U] 11/86, R/I: 09/87
D3578	C	Booth-Roe, Rotherham	05/92	Stored: [U] 04/82, R/I: 02/83
D3579	C	BREL Swindon	05/87	Withdrawn: 12/81, R/I: [S] 01/83, R/I: 09/83
D3580	C	BREL Doncaster	07/86	Stored: 02/85
D3581				Withdrawn: 12/81, R/I: [S] 01/83, R/I: 04/83
D3582	C	BREL Swindon	11/82	
D3583	A	BR Springs Branch		Stored: [U] 03/88
D3584	C	BREL Swindon	07/86	
D3585	S	ABB Transportation, Crewe	-	
D3586	P	Severn Valley Railway	-	Stored: [U] 05/83
D3587				Withdrawn: 04/82, R/I: 02/83
D3588	A	BR Leicester		
D3589	C	V Berry, Leicester	07/88	
D3590	C	C F Booth, Rotherham	11/87	
D3591	P	Swanage Railway	-	Stored: [U] 10/83
D3592	C	BR Allerton, by V Berry	03/90	
D3593	A	BR Immingham		
D3594	A	BR Cardiff		Stored: [U] 08/91
D3595				
D3596				
D3597				
D3598				
D3599				
D3600				Stored: [U] 04/88, R/I: 05/88
D3601	C	M C Processors, Glasgow	04/91	
D3602	C	M C Processors, Glasgow	06/91	Withdrawn: 05/80, R/I: 05/86
D3603	C	M C Processors, Glasgow	04/91	Withdrawn: 10/86, R/I: 11/86
D3604				Stored: [U] 11/86
D3605	P	Strathspey Railway	-	
D3606	C	M C Processors, Glasgow	04/91	
D3607				
D3608				
D3609	C	M C Processors, Glasgow	08/91	Stored: [U] 03/88
D3610				Stored: [U] 08/76, R/I: 01/78, Withdrawn: 11/81, R/I: [S] 01/83, R/I: 06/85
D3611	A	BR Cambridge		Stored: [U] 10/86

BR 1957 No.	Date Re No.	BR 1948 No.	First TOPS No.	Date Re No.	Second TOPS No.	Date Re No.	Name	Name Date	Built By	Works No.	Date Introduced	Depot of First Allocation	Date Withdrawn	Depot of Final Allocation
D3612									BR Darlington		06/58	36E	03/69	40A
D3613									BR Darlington		06/58	36E	02/69	40A
D3614									BR Darlington		07/58	36E	02/69	40A
D3615									BR Darlington		07/58	36E	12/68	31B
D3616									BR Darlington		07/58	36E	07/69	31B
D3617									BR Darlington		08/58	36E	07/69	31B
D3618									BR Darlington		08/58	36E	04/69	40A
D3619									BR Darlington		08/58	36E	02/69	40A
D3620									BR Darlington		08/58	36E	02/67	30A
D3621									BR Darlington		09/58	36E	07/69	36A
D3622									BR Darlington		09/58	36E	02/69	36A
D3623									BR Darlington		09/58	36E	02/69	36A
D3624									BR Darlington		09/58	40E	11/68	40A
D3625									BR Darlington		10/58	40E	08/68	16B
D3626									BR Darlington		10/58	40E	08/68	16B
D3627									BR Darlington		10/58	40E	08/68	16B
D3628									BR Darlington		10/58	40E	12/67	16B
D3629									BR Darlington		10/58	40E	05/69	30A
D3630									BR Darlington		10/58	40E	12/68	30A
D3631									BR Darlington		10/58	40E	06/67	30A
D3632									BR Darlington		10/58	30A	08/68	16A
D3633									BR Darlington		10/58	30A	12/68	30A
D3634									BR Darlington		11/58	30A	11/71	36A
D3635									BR Darlington		11/58	30A	12/68	30A
D3636									BR Darlington		11/58	30A	04/68	30A
D3637									BR Darlington		11/58	40B	11/68	36A
D3638									BR Darlington		11/58	36C	11/70	52A
D3639									BR Darlington		11/58	36C	07/69	36A
D3640									BR Darlington		11/58	36C	12/68	36C
D3641									BR Darlington		12/58	36C	12/71	36C
D3642									BR Darlington		12/58	36C	06/69	36C
D3643									BR Darlington		12/58	36C	10/68	36C
D3644									BR Darlington		12/58	36C	06/70	36C
D3645									BR Darlington		12/58	36C	03/69	36C
D3646									BR Darlington		12/58	36C	05/71	36A
D3647									BR Darlington		12/58	36C	03/70	36C
D3648									BR Darlington		01/59	36C	01/71	52A
D3649									BR Darlington		01/59	36C	07/69	36A
D3650									BR Darlington		01/59	36A	12/71	36A
D3651									BR Darlington		02/59	36A	12/71	36A
D3652			08497	02/74					BR Doncaster		03/58	55B	10/85	NL
D3653			08498	04/74					BR Doncaster		03/58	55B		
D3654			08499	01/74					BR Doncaster		03/58	55B		
D3655			08500	02/74					BR Doncaster		04/58	55B		
D3656			08501	02/74					BR Doncaster		04/58	55B	07/85	CF
D3657			08502	02/74					BR Doncaster		05/58	55B	10/88	TE
D3658			08503	02/74					BR Doncaster		05/58	55B	09/88	TE
D3659			08504	02/74					BR Doncaster		05/58	41A	10/88	TE
D3660			08505	07/74					BR Doncaster		05/58	41A	02/61	GM
D3661			08506	02/74					BR Doncaster		06/58	41A		
D3662			08507	02/74					BR Doncaster		06/58	41A		
D3663			08508	03/74					BR Doncaster		06/58	41A	10/91	IM
D3664			08509	02/74					BR Doncaster		07/58	41A		
D3665			09001	12/73					BR Darlington		02/59	71A		
D3666			09002	12/73					BR Darlington		02/59	71A	09/92	SU
D3667			09003	01/74					BR Darlington		02/59	71A		
D3668			09004	12/73					BR Darlington		02/59	71A		
D3669			09005	02/74					BR Darlington		02/59	75C		
D3670			09006	01/74					BR Darlington		03/59	73F		
D3671			09007	01/74					BR Darlington		03/59	73C		
D3672			08510	02/74					BR Darlington		05/58	51A		
D3673			08511	02/74					BR Darlington		05/58	52E		
D3674			08512	02/74					BR Darlington		05/58	52E		
D3675			08513	04/74					BR Darlington		06/58	53C	03/78	BG
D3676			08514	04/74					BR Darlington		06/58	53C		
D3677			08515	11/74					BR Darlington		06/58	51A	01/92	HT
D3678			08516	02/74					BR Darlington		08/58	52E		
D3679			08517	02/74					BR Darlington		08/58	52E		
D3680			08518	05/74					BR Doncaster		07/58	30A	05/86	SF
D3681			08519	05/74					BR Doncaster		07/58	30A		
D3682			08520	03/74					BR Doncaster		08/58	30A	01/86	SF
D3683			08521	03/74					BR Doncaster		08/58	30A		
D3684			08522	03/74					BR Doncaster		09/58	30A	09/85	CA
D3685			08523	02/74					BR Doncaster		09/58	41A		
D3686			08524	04/74					BR Doncaster		09/58	41A	01/86	LO
D3687			08525	01/74			Percy The Pilot	11/84-12/84	BR Darlington		02/59	34B		
D3688			08526	03/74					BR Darlington		02/59	34B		
D3689			08527	03/74					BR Darlington		03/59	34A		

BR 1957 No.	Disposal Code	Disposal Detail	Date Cut Up	Notes
D3612	C	G Cohen, Kettering	07/69	
D3613	C	NCB Moor Green	03/85	PI
D3614	C	Steelbreaking & Dismantling Co Chesterfield	08/69	
D3615	C	G Cohen, Kettering	07/69	
D3616	C	Steelbreaking & Dismantling Co Chesterfield	11/69	
D3617	C	Steelbreaking & Dismantling Co Chesterfield	11/69	
D3618	C	NCB Moor Green	04/85	PI
D3619	C	NCB Moor Green	03/85	PI
D3620	C	Slag Reduction, Ickles	08/67	
D3621	C	C F Booth, Rotherham	11/69	
D3622	C	C F Booth, Rotherham	07/67	
D3623	C	C F Booth, Rotherham	07/69	
D3624	C	C F Booth, Rotherham	08/69	
D3625	C	C F Booth, Rotherham	01/69	
D3626	C	C F Booth, Rotherham	01/69	
D3627	C	C F Booth, Rotherham	01/69	
D3628	C	C F Booth, Rotherham	09/68	
D3629	C	G Cohen, Kettering	10/69	
D3630	C	Cox & Danks, Park Royal	07/69	
D3631	C	BR Stratford, by G Cohen	10/67	
D3632	C	C F Booth, Rotherham	01/69	
D3633	C	G Cohen, Kettering	06/69	
D3634	C	C F Booth, Rotherham	04/72	
D3635	C	G Cohen, Kettering	05/69	
D3636	C	BR Canning Town, by G Cohen	06/69	
D3637	C	G Cohen, Kettering	04/69	
D3638	C	NCB Ashington	09/75	Withdrawn: 07/69, R/I: 03/70, PI
D3639	E	Guinea, West Africa [03/70]	-	
D3640	C	G Cohen, Kettering	05/69	
D3641	C	C F Booth, Rotherham	12/72	
D3642	C	BSC Appleby-Frodingham	10/78	PI
D3643	C	G Cohen, Kettering	07/69	
D3644	C	C F Booth, Rotherham	05/71	
D3645	C	C F Booth, Rotherham	07/69	
D3646	C	C F Booth, Rotherham	12/71	
D3647	C	C F Booth, Rotherham	09/70	Withdrawn: 07/69, R/I: 08/69
D3648	C	NCB Bates Colliery	03/77	PI
D3649	E	Guinea, West Africa [03/70]	-	
D3650	C	C F Booth, Rotherham	04/72	
D3651	C	C F Booth, Rotherham	04/72	
D3652	C	BREL Doncaster	02/86	
D3653				Stored: [U] 11/81, R/I: 08/82
D3654				
D3655				Stored: [U] 03/83, R/I: 11/84
D3656	C	M C Processors, Glasgow	08/91	Stored: [U] 09/83, R/I: 10/83
D3657	S	ICI Wilton	-	
D3658	S	ICI Wilton	-	
D3659	C	Booth-Roe, Rotherham	10/92	Stored: [U] 08/77, R/I: 09/77
D3660	C	BREL Swindon	03/83	
D3661				
D3662				
D3663	A	BR Frodingham		Worked after withdrawal
D3664				
D3665				
D3666	A	BR Selhurst		
D3667				
D3668				
D3669				
D3670				
D3671				
D3672				
D3673				Stored: [U] 02/88
D3674				Stored: [U] 09/77, R/I: 12/77
D3675	C	BREL Doncaster	10/78	
D3676				Stored: [U] 01/77, R/I: 09/77
D3677	A	BR Gateshead		
D3678				
D3679				
D3680	A	BR March		Stored: [U] 05/86
D3681				
D3682	C	BREL Doncaster	03/86	
D3683				
D3684	C	BREL Doncaster	12/85	
D3685				
D3686	C	C F Booth, Rotherham	12/87	
D3687				Stored: [U] 01/84, R/I: 02/84
D3688				
D3689				Withdrawn: 05/89, R/I: 06/89

BR 1957 No.	Date Re No.	BR 1948 No.	First TOPS No.	Date Re No.	Second TOPS No.	Date Re No.	Name	Name Date	Built By	Works No.	Date Introduced	Depot of First Allocation	Date Withdrawn	Depot of Final Allocation
D3690			08528	01/74					BR Darlington		03/59	34B		
D3691			08529	01/74					BR Darlington		03/59	34B		
D3692			08530	07/74					BR Darlington		03/59	34B		
D3693			08531	06/74					BR Darlington		04/59	34B		
D3694			08532	02/74					BR Darlington		04/59	41A		
D3695			08533	08/74					BR Darlington		04/59	41A	10/92	SF
D3696			08534	02/74					BR Darlington		04/59	41A		
D3697									BR Darlington		05/59	41A	02/65	41C
D3698									BR Darlington		05/59	41A	02/65	41A
D3699			08535	06/74					BR Darlington		05/59	41A		
D3700			08536	02/74					BR Darlington		05/59	41A		
D3701			08537	02/74					BR Darlington		05/59	41A	05/92	BS
D3702			08538	03/74					BR Darlington		05/59	41A		
D3703			08539	05/74					BR Darlington		06/59	41A	04/92	CA
D3704			08540	02/74					BR Darlington		06/59	34B		
D3705			08541	05/74					BR Darlington		06/59	34B		
D3706			08542	02/74					BR Darlington		06/59	34B		
D3707			08543	02/74					BR Darlington		06/59	40B		
D3708			08544	01/74					BR Darlington		06/59	40B	07/92	HT
D3709			08545	02/74					BR Darlington		07/59	40B	09/80	FP
D3710			08546	02/74					BR Darlington		08/59	34E	07/85	SF
D3711			08547	01/74					BR Darlington		08/59	34E	12/81	SF
D3712			08548	11/73					BR Darlington		08/59	34B	12/82	HS
D3713			08549	03/74					BR Darlington		09/59	34B	11/85	MR
D3714			08550	02/74					BR Darlington		09/59	41F	03/83	SF
D3715			08551	01/74					BR Darlington		09/59	41F	10/82	FP
D3716			08552	01/74					BR Darlington		09/59	36A	08/82	SF
D3717			08553	02/74					BR Darlington		10/59	36A	05/81	CA
D3718			08554	03/74					BR Darlington		10/59	36A	08/82	SF
D3719			09008	01/74			Sheffield Childrens Hospital	09/90	BR Darlington		10/59	73C		
D3720			09009	12/73			Three Bridges CED	08/90	BR Darlington		10/59	73C		
D3721			09010	01/74					BR Darlington		10/59	73C		
D3722			08555	01/74					BR Darlington		10/59	34A	08/82	FP
D3723			08556	02/74					BR Darlington		10/59	34A	06/90	WN
D3724			08557	03/74					BR Darlington		11/59	34A	12/82	FP
D3725			08558	02/74					BR Darlington		11/59	34A	01/86	MR
D3726			08559	02/74					BR Darlington		12/59	41A	12/81	YK
D3727			08560	02/74					BR Darlington		12/59	41A	05/81	SB
D3728			08561	04/74					BR Crewe		01/59	64A		
D3729			08562	11/73			The Doncaster Postman	11/85	BR Crewe		01/59	64A		
D3730			08563	09/74					BR Crewe		01/59	64A	04/82	ML
D3731			08564	09/74					BR Crewe		02/59	64A	11/85	TJ
D3732			08565	11/73					BR Crewe		02/59	64A		
D3733			08566	09/74					BR Crewe		04/59	64A	10/75	ED
D3734			08567	03/74					BR Crewe		04/59	64A		
D3735			08568	07/74					BR Crewe		04/59	64A		
D3736			08569	02/74					BR Crewe		05/59	64B		
D3737			08570	04/75					BR Crewe		05/59	64C	01/92	ED
D3738			08571	04/74					BR Crewe		06/59	64C		
D3739			08572	09/74					BR Crewe		05/59	64C	02/81	HA
D3740			08573	03/74					BR Crewe		06/59	64C		
D3741			08574	02/74					BR Crewe		06/59	64C	03/81	HM
D3742			08575	09/74					BR Crewe		06/59	64C		
D3743			08576	04/74					BR Crewe		06/59	86B		
D3744			08577	03/74					BR Crewe		06/59	86B		
D3745			08578	05/74					BR Crewe		06/59	86B		
D3746			08579	04/74					BR Crewe		07/59	86B	11/88	HM
D3747			08580	04/74					BR Crewe		07/59	86B		
D3748			08581	05/74					BR Crewe		07/59	86B		
D3749			08582	04/74					BR Crewe		08/59	86B		
D3750			08583	03/74					BR Crewe		08/59	87C		
D3751			08584	02/74					BR Crewe		08/59	87C	08/91	LA
D3752			08585	03/74					BR Crewe		08/59	87C		
D3753			08586	05/74					BR Crewe		11/59	87C		
D3754			08587	03/74					BR Crewe		11/59	87C		
D3755			08588	04/74					BR Crewe		11/59	87C		
D3756			08589	03/74					BR Crewe		10/59	87C	11/92	CF
D3757			08590	04/74					BR Crewe		10/59	87C		
D3758			08591	05/74					BR Crewe		10/59	87C		
D3759			08592	04/74	08993	07/85	Ashburnham	01/86	BR Crewe		10/59	87C		
D3760			08593	12/73					BR Crewe		12/59	87C		
D3761			08594	05/74					BR Crewe		12/59	87C		
D3762			08595	03/74					BR Crewe		12/59	87C		
D3763			08596	05/74					BR Derby		04/59	24K	03/77	OC
D3764			08597	04/74					BR Derby		04/59	6A		
D3765			08598	04/74					BR Derby		05/59	9A	11/86	CD
D3766			08599	05/74					BR Derby		05/59	9A		
D3767			08600	12/73					BR Derby		05/59	9A		

BR 1957 No.	Disposal Code	Disposal Detail	Date Cut Up	Notes
D3690				
D3691				
D3692				
D3693				
D3694				
D3695	A	BR Colchester		Stored: [U] 07/85, R/I: 04/86
D3696				
D3697	R	Rebuilt as Class 13 No D4502 (S)	-	
D3698	R	Rebuilt as Class 13 No D4500 (S)	-	
D3699				
D3700				
D3701	A	BR Bescot		Stored: [U] 12/91
D3702				
D3703	A	ABB Transportation, Crewe		
D3704				
D3705				
D3706				
D3707				
D3708	A	BR Heaton		
D3709	C	BREL Swindon	11/83	
D3710	C	BREL Doncaster	01/86	
D3711	C	BREL Swindon	05/82	Withdrawn: 08/81, R/I: 08/81
D3712	C	BREL Doncaster	11/83	
D3713	C	Booth-Roe, Rotherham	10/92	
D3714	C	BREL Swindon	05/86	
D3715	C	BREL Doncaster	11/83	
D3716	C	BREL Doncaster	01/83	
D3717	C	BREL Swindon	03/83	
D3718	C	BREL Doncaster	02/83	
D3719				
D3720				
D3721				
D3722	C	BREL Doncaster	11/83	
D3723	A	BR Willesden		Withdrawn: 05/85, R/I: 07/85
D3724	C	BREL Doncaster	12/83	
D3725	C	V Berry, Leicester	05/89	
D3726	C	BREL Swindon	11/83	Withdrawn: 04/81, R/I: 08/81
D3727	C	BREL Swindon	02/82	
D3728				Stored: [U] 07/84, R/I: 12/84
D3729				Withdrawn: 12/81, R/I: [S] 01/83, R/I: 02/85
D3730	C	D Christie, Camlackie	04/87	Stored: [U] 04/82
D3731	C	J McWilliams, Shettleston	05/86	
D3732				Stored: [U] 07/81, Withdrawn: 10/81, R/I: [S] 01/83, R/I: 09/84
D3733	C	BREL Glasgow	11/77	
D3734				
D3735				Stored: [U] 01/85, R/I: 10/85
D3736				Stored: [U] 11/84, R/I: 11/85
D3737	A	BR Motherwell		Stored: [U] 03/84, R/I: 05/84
D3738				Stored: [U] 07/84, R/I: 12/84
D3739	C	BREL Swindon	10/83	
D3740				Stored: [U] 05/82, R/I: 11/84
D3741	C	BREL Swindon	08/86	
D3742				Stored: [U] 12/84, R/I: 07/85
D3743				
D3744				Stored: [U] 02/83, R/I: 01/84
D3745				Withdrawn: 12/81, R/I: [S] 01/83, R/I: 01/84
D3746	A	BR Knottingley		
D3747				
D3748				
D3749				Withdrawn: 01/82, R/I: 04/82
D3750				Stored: [U] 12/80, Withdrawn: 10/81, R/I: [S] 01/83, R/I: 01/84
D3751	A	ABB Transportation, Crewe		
D3752				Withdrawn: 10/81, R/I: [S] 01/83, R/I: 01/84
D3753				
D3754				Withdrawn: 12/81, R/I: [S] 01/83, R/I: 02/84
D3755				Stored: [U] 12/84, R/I: 05/85
D3756	A	BR Cardiff		
D3757				Stored: [U] 07/82, R/I: 08/82, Stored: [U] 07/84, R/I: 12/84
D3758				Stored: [U] 02/83, R/I: 01/84
D3759				
D3760				Withdrawn: 12/81, R/I: [S] 01/83, R/I: 01/84
D3761				
D3762				
D3763	S	Balfour-Beatty, Cheriton	-	Stored: [U] 12/76
D3764				Stored: [U] 03/83, R/I: 04/84
D3765	S	BCOE, Gwaun-Cae-Gurwan	-	
D3766				Stored: [U] 02/85, R/I: 10/85
D3767				To Departmental Stock - 97800: 05/79 - 03/90

BR 1957 No.	Date Re No.	BR 1948 No.	First TOPS No.	Date Re No.	Second TOPS No.	Date Re No.	Name	Name Date	Built By	Works No.	Date Introduced	Depot of First Allocation	Date Withdrawn	Depot of Final Allocation
D3768			08601	05/74			[Sceptre]	10/86-11/90	BR Derby		05/59	9A		
D3769			08602	03/74					BR Derby		05/59	9A	03/86	KD
D3770			08603	04/74					BR Derby		05/59	9A		
D3771			08604	02/74					BR Derby		06/59	9A		
D3772			08605	03/74					BR Derby		06/59	9B		
D3773			08606	04/74					BR Derby		06/59	9B	02/88	SF
D3774			08607	03/74					BR Derby		07/59	17B		
D3775			08608	03/74					BR Derby		07/59	17C	11/88	GD
D3776			08609	02/74					BR Derby		07/59	17C		
D3777			08610	02/74					BR Derby		07/59	15B		
D3778			08611	02/74					BR Derby		08/59	15B		
D3779			08612	02/74					BR Derby		08/59	15B	09/90	OC
D3780			08613	09/74					BR Derby		08/59	24D		
D3781			08614	04/74					BR Derby		08/59	24D		
D3782			08615	02/74					BR Derby		09/59	24B		
D3783			08616	02/74					BR Derby		09/59	24B		
D3784			08617	09/73					BR Derby		09/59	26A		
D3785			08618	04/74					BR Derby		09/59	26A	11/90	GD
D3786			08619	07/74					BR Derby		09/59	15C		
D3787			08620	07/74	09205	12/92			BR Derby		09/59	15C		
D3788			08621	03/74					BR Derby		10/59	15C	06/88	LE
D3789			08622	12/73					BR Derby		10/59	15C		
D3790			08623	02/74					BR Derby		10/59	15C		
D3791			08624	03/74					BR Derby		10/59	15C		
D3792			08625	02/74					BR Derby		10/59	18C		
D3793			08626	03/74					BR Derby		10/59	8D	04/88	AN
D3794			08627	12/73					BR Derby		11/59	8D		
D3795			08628	02/74					BR Derby		11/59	8B		
D3796			08629	06/74					BR Derby		11/59	8B		
D3797			08630	01/74					BR Derby		11/59	8B		
D3798			08631	04/74			[Eagle CURC]	05/88-10/92	BR Derby		11/59	5D	12/92	MR
D3799			08632	02/74					BR Derby		11/59	5D		
D3800			08633	02/74			The Sorter	10/91	BR Derby		12/59	5D		
D3801			08634	03/74					BR Derby		12/59	5D		
D3802			08635	01/74					BR Derby		12/59	5D		
D3803			08636	09/74					BR Horwich		12/58	82B	07/80	LE
D3804			08637	06/74					BR Horwich		12/58	82B	08/91	AF
D3805			08638	05/74					BR Horwich		02/58	82B	12/92	RG
D3806			08639	03/74					BR Horwich		12/58	82B	10/88	DR
D3807			08640	02/74					BR Horwich		12/58	82B	11/88	RG
D3808			08641	04/74					BR Horwich		01/59	86A		
D3809			08642	01/74					BR Horwich		01/59	82B	09/92	EH
D3810			08643	02/74					BR Horwich		01/59	86A		
D3811			08644	09/74					BR Horwich		02/59	86A		
D3812			08645	03/74					BR Horwich		02/59	86A		
D3813			08646	04/74					BR Horwich		02/59	86A		
D3814			08647	03/74			Crimpsall	06/89	BR Horwich		02/59	86B		
D3815			08648	01/74					BR Horwich		03/59	86B		
D3816			08649	09/74					BR Horwich		03/59	86B		
D3817			08650	12/73					BR Horwich		03/59	86B	08/89	EH
D3818			08651	04/74					BR Horwich		03/59	86B		
D3819			08652	03/74					BR Horwich		03/59	86B	06/92	CF
D3820			08653	02/74					BR Horwich		04/59	86B		
D3821			08654	03/74					BR Horwich		04/59	86B	10/91	CF
D3822			08655	12/73					BR Horwich		04/59	86B		
D3823			08656	04/74					BR Horwich		04/59	86B		
D3824			08657	03/74					BR Horwich		04/59	86B	11/91	KY
D3825			08658	04/74					BR Horwich		05/59	87C		
D3826			08659	02/74					BR Horwich		05/59	87C	06/91	KY
D3827			08660	03/74					BR Horwich		05/59	87C	11/91	CF
D3628			08661	05/74					BR Horwich		05/59	87C		
D3829			08662	03/74					BR Horwich		05/59	87C		
D3830			08663	03/74					BR Horwich		06/59	87C		
D3831			08664	02/74					BR Horwich		06/59	87C		
D3832			08665	05/74					BR Crewe		01/60	6B		
D3833			08666	05/74					BR Crewe		01/60	6B		
D3834			08667	06/74					BR Crewe		01/60	8B	11/92	NL
D3835			08668	02/74					BR Crewe		01/60	8B		
D3836			08669	02/74					BR Crewe		01/60	8F	05/89	LO
D3837			08670	02/74					BR Crewe		01/60	8F		
D3838			08671	04/74					BR Crewe		01/60	24L	11/88	GD
D3839			08672	04/74					BR Crewe		01/60	24L	09/92	BS
D3840			08673	03/74					BR Crewe		01/60	24L		
D3841			08674	01/74					BR Crewe		02/60	24J	07/81	BS
D3842			08675	03/74					BR Horwich		07/59	26F		
D3843			08676	03/74					BR Horwich		07/59	26F		
D3844			08677	03/74					BR Horwich		07/59	26F		
D3845			08678	02/74					BR Horwich		08/59	26F	10/88	TI
D3846			08679	05/74					BR Horwich		08/59	26F	06/76	AN
D3847			08680	01/74					BR Horwich		09/59	9E	07/92	ED
D3848			08681	04/74					BR Horwich		09/59	9E	10/85	DY
D3849			08682	05/74					BR Horwich		09/59	9E		

BR 1957 No.	Disposal Code	Disposal Detail	Date Cut Up	Notes
D3768				
D3769	S	RFS Industries, Doncaster	-	Hire Loco
D3770				Withdrawn: 12/81, R/I: [S] 01/83, R/I: 02/84
D3771				Withdrawn: 12/81, R/I: [S] 01/81, R/I: 02/84
D3772				Stored: [S] 01/84, R/I: 11/84
D3773	C	V Berry, Leicester	09/89	
D3774				
D3775	A	BR Gateshead		
D3776				Stored: [U] 04/82, R/I: 03/84
D3777				Stored: [U] 02/85, R/I: 11/85
D3778				Stored: [U] 05/82, R/I: 08/82
D3779	C	M C Processors, Glasgow	03/91	Stored: [U] 04/82, R/I: 03/84
D3780				
D3781				Stored: [U] 04/84, R/I: 08/85
D3782				Stored: [U] 06/82, R/I: 03/84
D3783				Stored: [U] 05/85, R/I: 11/85
D3784				Withdrawn: 10/81, R/I: [S] 01/83, R/I: 04/84
D3785	A	BR Gateshead		Stored: [U] 05/84, R/I: 07/84
D3786				Withdrawn: 09/81, R/I: [S] 01/83, R/I: 03/84
D3787				Stored: [U] 11/82, R/I: 11/83
D3788	C	V Berry, Leicester	10/89	
D3789				Withdrawn: 12/81, R/I: [S] 01/83, R/I: 04/84
D3790				Stored: [U] 11/84, R/I: 01/85
D3791				Stored: [U] 07/84, R/I: 01/85
D3792				
D3793	A	BR Allerton		Stored: [U] 03/88
D3794				Stored: [U] 01/87, R/I: 03/87
D3795				Withdrawn: 10/81, R/I: [S] 01/83, R/I: 04/84
D3796				Stored: [U] 08/81, Withdrawn: 11/81, R/I: 01/83
D3797				
D3798	A	BR Cambridge		Stored: [U] 03/85, R/I: 07/85
D3799				Stored: [U] 08/78, R/I: 11/83
D3800				Stored: [U] 07/84, R/I: 12/84
D3801				
D3802				Withdrawn: 12/81, R/I: [S] 01/84, R/I: 05/84
D3803	C	BREL Swindon	05/83	
D3804	C	BR Eastleigh	08/92	
D3805	A	BR Reading		Withdrawn: 12/81, R/I: [S] 01/83, R/I: 05/84
D3806	C	V Berry, Leicester	10/89	
D3807	C	M C Processors, Glasgow	03/91	
D3808				
D3809	A	BR Eastleigh		
D3810				
D3811				
D3812				
D3813				
D3814				
D3815				
D3816				
D3817	S	Foster Yeoman, Grain		
D3818				
D3819	A	BR Cardiff		
D3820				
D3821	A	BR Cardiff		Stored: [U] 07/91
D3822				
D3823				
D3824	A	BR York		Stored: [U] 07/90
D3825				
D3826	A	BR Healey Mills		
D3827	A	BR Cardiff		Stored: [S] 08/91
D3828				
D3829				
D3830				Withdrawn: 05/90, R/I: 06/90
D3831				
D3832				Stored: [U] 03/85, R/I: 06/85
D3833				
D3834	A	BR Neville Hill		Stored: [U] 01/84, R/I: 05/85
D3835				
D3836	S	Trafford Park Estates	-	
D3837				Stored: [U] 10/81, R/I: 03/82, Stored: [S] 12/92
D3838	A	BR Gateshead		
D3839	A	BR Bescot		
D3840				Stored: [U] 09/82, R/I: 06/84
D3841	C	BREL Swindon	07/83	Stored: [U] 07/81
D3842				Stored: [U] 05/85, R/I: 12/85
D3843				Stored: [U] 03/83, R/I: 06/84
D3844				Stored: [U] 03/85, R/I: 11/85, Stored: [U] 06/90
D3845	S	Glaxochem, Ulverston	-	
D3846	C	C F Booth, Rotherham	04/86	PI
D3847	A	BR Motherwell		Stored: [U] 10/81, R/I: 07/82
D3848	C	BREL Doncaster	09/86	Stored: [U] 04/84
D3849				Stored: [U] 07/84, R/I: 12/84

BR 1957 No.	Date Re No.	BR 1948 No.	First TOPS No.	Date Re No.	Second TOPS No.	Date Re No.	Name	Name Date	Built By	Works No.	Date Introduced	Depot of First Allocation	Date Withdrawn	Depot of Final Allocation
D3850			08683	05/74					BR Horwich		09/59	9E		
D3851			08684	04/74					BR Horwich		10/59	9E	12/86	BY
D3852			08685	02/74					BR Horwich		10/59	9E		
D3853			08686	02/74					BR Horwich		10/59	9E	11/91	AN
D3854			08687	02/74	08995	09/87	Kidwelly	10/87	BR Horwich		10/59	9E		
D3855			08688	11/73					BR Horwich		10/59	27F	03/91	AN
D3856			08689	02/74					BR Horwich		10/59	27F		
D3857			08690	01/74					BR Horwich		10/59	27F		
D3858			08691	12/74					BR Horwich		10/59	27E		
D3859			08692	01/75					BR Horwich		11/59	16D		
D3860			08693	01/74					BR Horwich		11/59	16D		
D3861			08694	12/73					BR Horwich		11/59	16D		
D3862			08695	05/74					BR Horwich		11/59	17A		
D3863			08696	06/74					BR Horwich		11/59	17A		
D3864			08697	03/74					BR Horwich		12/59	17A		
D3865			08698	04/74					BR Horwich		12/59	17A		
D3866			08699	07/74					BR Horwich		01/60	5D		
D3867			08700	06/74					BR Horwich		01/60	24J	12/92	BS
D3868			08701	03/74			Gateshead MPD 1852-1991	07/91	BR Horwich		02/60	26A		
D3869			08702	04/74					BR Horwich		02/60	14A		
D3870			08703	06/74					BR Horwich		02/60	5D		
D3871			08704	05/74					BR Horwich		02/60	24J	11/90	BY
D3872			08705	03/74					BR Crewe		03/60	50A		
D3873			08706	02/74					BR Crewe		03/60	51L		
D3874			08707	12/73					BR Crewe		03/60	50A		
D3875			08708	02/74					BR Crewe		03/60	51L	10/92	SF
D3876			08709	03/74					BR Crewe		04/60	51L		
D3877			08710	09/74					BR Crewe		04/60	64A		
D3878			08711	09/74					BR Crewe		05/60	64A		
D3879			08712	02/74					BR Crewe		05/60	61A	12/92	ML
D3880			08713	02/74					BR Crewe		05/60	64A		
D3881			08714	04/74					BR Crewe		05/60	64A		
D3882			08715	07/74					BR Crewe		05/60	64A		
D3883			08716	03/74					BR Crewe		05/60	64A	12/88	CA
D3884			08717	09/74	09204	12/92			BR Crewe		05/60	64A		
D3885									BR Crewe		05/60	64A	02/72	64H
D3886			08718	07/75					BR Crewe		06/60	64A		
D3887			08719	12/73					BR Crewe		06/60	64A	11/91	BY
D3888			08720	04/74					BR Crewe		06/60	64A		
D3889			08721	04/74					BR Crewe		06/60	64A		
D3890			08722	04/74					BR Crewe		07/60	64A	03/87	GD
D3891			08723	09/74					BR Crewe		07/60	64A		
D3892			08724	06/74					BR Crewe		07/60	64A		
D3893			08725	05/74					BR Crewe		07/60	64G	08/92	ED
D3894			08726	11/73					BR Crewe		07/60	65C	01/87	HA
D3895			08727	09/74					BR Crewe		08/60	65C		
D3896			08728	09/74					BR Crewe		08/60	65C	09/87	ML
D3897			08729	03/74					BR Crewe		08/60	67C	02/90	DR
D3898			08730	09/74					BR Crewe		08/60	65C		
D3899			08731	04/74					BR Crewe		09/60	65C		
D3900			08732	04/74	09202	10/92			BR Crewe		09/60	65C		
D3901			08733	04/74					BR Crewe		09/60	65C		
D3902			08734	04/74					BR Crewe		09/60	65E		
D3903			08735	06/74					BR Crewe		10/60	65E		
D3904			08736	09/74					BR Crewe		10/60	65E	09/87	ML
D3905			08737	05/74					BR Crewe		10/60	66A		
D3906			08738	05/74					BR Crewe		11/60	66A		
D3907			08739	03/74					BR Crewe		11/60	66A		
D3908			08740	04/74					BR Crewe		11/60	66A		
D3909			08741	08/74					BR Crewe		11/60	66A	04/91	DY
D3910			08742	04/74					BR Crewe		11/60	66A		
D3911			08743	03/74					BR Crewe		11/60	66A		
D3912			08744	03/74					BR Crewe		11/60	66A	03/92	LO
D3913			08745	03/74					BR Crewe		11/60	66A		
D3914			08746	06/74					BR Crewe		12/60	66A		
D3915			08747	03/74					BR Crewe		12/60	66A	06/90	GD
D3916			08748	02/74					BR Crewe		12/60	66A		
D3917			08749	02/74	09104	01/93			BR Crewe		12/60	66A		
D3918			08750	05/74					BR Crewe		12/60	66D		
D3919			08751	02/74					BR Crewe		12/60	66D		
D3920			08752	03/74					BR Crewe		12/60	66D		
D3921			08753	10/73					BR Crewe		12/60	66D		
D3922			08754	04/74					BR Horwich		01/61	67A		
D3923			08755	03/74					BR Horwich		01/61	67A		
D3924			08756	12/73					BR Horwich		01/61	67A		
D3925			08757	04/74					BR Horwich		02/61	67A		
D3926			08758	02/74					BR Horwich		02/61	67A		
D3927			08759	11/73	09106	01/93			BR Horwich		02/61	67D		
D3928			08760	01/74					BR Horwich		02/61	67D	02/92	EH
D3929			08761	12/74					BR Horwich		03/61	67D	03/92	ED

BR 1957 No.	Disposal Code	Disposal Detail	Date Cut Up	Notes
D3850				
D3851	S	W Smith Ltd, Wakefield	-	Stored: [U] 05/85, R/I: 12/85
D3852				Stored: [U] 03/84, R/I: 04/84
D3853	A	BR Allerton		
D3854				Stored: [U] 11/86, R/I: 09/87,
D3855	A	BR Allerton		
D3856				
D3857				Withdrawn: 10/81, R/I: 07/84
D3858				Stored: [U] 07/84, R/I: 11/84
D3859				Stored: [U] 04/84, R/I: 03/85
D3860				Stored: [U] 12/83, R/I: 02/84
D3861				
D3862				
D3863				
D3864				Stored: [U] 01/84, R/I: 05/84
D3865				Stored: [U] 10/81, R/I: 03/82
D3866				Stored: [U] 04/84, R/I: 03/85
D3867	A	BR Bescot		
D3868				Stored: [U] 11/84, R/I: 05/85
D3869				Stored: [U] 07/84, R/I: 12/84
D3870				
D3871	P	Nene Valley Railway	-	Withdrawn: 11/81, R/I: 01/83, Stored: [U] 08/83, R/I: 11/83, Stored: [S] 12/89, Withdrawn: 05/90, R/I: 06/90
D3872				
D3873				
D3874				
D3875	A	BR Colchester		
D3876				
D3877				
D3878				Stored: [U] 12/84, R/I: 05/85
D3879	A	BR Motherwell		Stored: [U] 12/83, R/I: 02/84
D3880				
D3881				Stored: [U] 12/84
D3882				Withdrawn: 07/90, R/I: 07/90
D3883	C	V Berry, Leicester	04/90	
D3884				Withdrawn: 10/81, R/I: [S] 01/83, R/I: 04/83
D3885	C	BREL Glasgow	03/72	
D3886				Stored: [U] 01/83, R/I: 05/83
D3887	A	BR Bletchley		Stored: [U] 01/87, R/I: 03/87
D3888				
D3889				Stored: [U] 03/85, R/I: 05/85
D3890	C	V Berry, Leicester	07/88	Stored: [U] 03/85, R/I: 07/85, Stored: [U] 01/87
D3891				Stored: [U] 05/84, R/I: 10/84
D3892				
D3893	A	BR Motherwell		Withdrawn: 11/81, R/I: 12/81, Stored: [U] 01/82, R/I: 01/82
D3894	C	BR Thornton Junction, by V Berry	05/87	Stored: [U] 12/80, R/I: 05/81
D3895				Stored: [U] 05/84
D3896	S	Deanside Transit, Glasgow	-	
D3897	A	BRML Doncaster		
D3898				Stored: [U] 05/84, R/I: 07/84
D3899				
D3900				
D3901				Stored: [U] 07/83, R/I: 12/83
D3902				
D3903				Stored: [U] 05/83, R/I: 02/84
D3904	S	Deanside Transit, Glasgow	-	Stored: [U] 01/81, R/I: 06/81
D3905				
D3906				
D3907				
D3908				
D3909	A	BRML Doncaster		
D3910				
D3911				
D3912	A	ABB Transportation, Crewe		
D3913				
D3914				
D3915	A	BR Doncaster		
D3916				
D3917				
D3918				
D3919				
D3920				
D3921				
D3922				
D3923				
D3924				
D3925				
D3926				
D3927				Stored: [U] 01/87,
D3928	C	BR Eastleigh	09/92	Stored: [U] 06/88, R/I: 07/88
D3929	A	BR Motherwell		

BR 1957 No.	Date Re No.	BR 1948 No.	First TOPS No.	Date Re No.	Second TOPS No.	Date Re No.	Name	Name Date	Built By	Works No.	Date Introduced	Depot of First Allocation	Date Withdrawn	Depot of Final Allocation
D3930			08762	01/74					BR Horwich		03/61	61A		
D3931			08763	12/73					BR Horwich		03/61	61A	10/89	HA
D3932			08764	11/74					BR Horwich		03/61	61A	05/88	HA
D3933			08765	05/74					BR Horwich		03/61	61A		
D3934			08766	10/73	09103	12/92			BR Horwich		04/61	61A		
D3935			08767	02/74					BR Horwich		04/61	61A		
D3936			08768	03/74					BR Horwich		04/61	61A		
D3937			08769	03/74					BR Derby		03/60	55D	05/89	LE
D3938			08770	02/74					BR Derby		03/60	52E		
D3939			08771	02/74					BR Derby		03/60	52E	03/92	HT
D3940			08772	02/74			Camulddunum	05/88	BR Derby		03/60	50A		
D3941			08773	02/74					BR Derby		03/60	55D		
D3942			08774	04/74					BR Derby		04/60	52E	09/88	TE
D3943			08775	02/74					BR Derby		04/60	52E		
D3944			08776	03/74					BR Derby		04/60	50B		
D3945			08777	03/74					BR Derby		04/60	50B	11/91	KY
D3946			08778	01/74					BR Derby		04/60	50A	09/92	CF
D3947			08779	01/74					BR Derby		04/60	81A	02/88	AF
D3948			08780	02/74					BR Derby		04/60	81A		
D3949			08781	05/74	09203	10/92			BR Derby		04/60	81A		
D3950			08782	01/74					BR Derby		04/60	81A		
D3951			08783	02/74					BR Derby		04/60	81A		
D3952			08784	04/74					BR Derby		04/60	81A		
D3953			08785	02/74					BR Derby		04/60	81A	03/89	CF
D3954			08786	02/74					BR Derby		04/60	81A		
D3955			08787	02/74					BR Derby		04/60	81A	02/91	CF
D3956			08788	02/74					BR Derby		04/60	81A		
D3957			08789	04/74					BR Derby		05/60	81A		
D3958			08790	04/74					BR Derby		05/60	81A		
D3959			08791	02/74					BR Derby		05/60	81A	08/92	ED
D3960			08792	03/74					BR Derby		05/60	81A		
D3961			08793	02/74					BR Derby		05/60	81A		
D3962			08794	02/74					BR Derby		05/60	81A	11/92	NL
D3963			08795	01/74					BR Derby		05/60	81A		
D3964			08796	03/74					BR Derby		05/60	81A	05/90	AF
D3965			08797	02/74					BR Derby		05/60	81A	11/90	GD
D3966			08798	02/74					BR Derby		05/60	84E		
D3967			08799	02/74					BR Derby		06/60	83E		
D3968			08800	04/74					BR Derby		06/60	83E		
D3969			08801	02/74					BR Derby		06/60	83E		
D3970			08802	03/74					BR Derby		06/60	83E		
D3971			08803	02/74			[Isis]	01/92-12/92	BR Derby		06/60	83E	12/92	RG
D3972			08804	04/74					BR Derby		06/60	83E		
D3973			08805	03/74					BR Derby		06/60	84E		
D3974			08806	04/74					BR Derby		06/60	86A		
D3975			08807	04/74					BR Derby		07/60	86A		
D3976			08808	04/74					BR Derby		07/60	86A	05/90	CL
D3977			08809	03/74					BR Derby		07/60	86A		
D3978			08810	04/74					BR Derby		07/60	86A		
D3979			08811	03/74					BR Derby		07/60	86A		
D3980			08812	04/74					BR Derby		07/60	86A	01/78	EJ
D3981			08813	02/74					BR Derby		08/60	84E		
D3982			08814	04/74					BR Derby		08/60	84E	11/92	DY
D3983			08815	04/74					BR Derby		08/60	84E		
D3984			08816	03/74					BR Derby		08/60	84E	02/86	TE
D3985			08817	11/74					BR Derby		08/60	84E		
D3986			08818	02/74					BR Derby		08/60	84E		
D3987			08819	03/74					BR Derby		08/60	84E		
D3988			08820	02/74					BR Derby		08/60	84E	02/90	LO
D3989			08821	08/74					BR Derby		09/60	84E	12/92	LA
D3990			08822	03/74					BR Derby		09/60	84E		
D3991			08823	04/74					BR Derby		09/60	84E		
D3992			08824	06/74					BR Derby		09/60	84E		
D3993			08825	03/74					BR Derby		09/60	84E		
D3994			08826	03/74					BR Derby		09/60	84E		
D3995			08827	03/74					BR Derby		09/60	84E		
D3996			08828	07/74					BR Derby		10/60	84E		
D3997			08829	07/74					BR Derby		10/60	84E		
D3998			08830	12/73					BR Derby		11/60	84E		
D3999			08831	02/74					BR Derby		11/60	84E		
D4000			08832	03/74	09102	08/92			BR Derby		11/60	84E		
D4001			08833	11/73	09101	08/92			BR Derby		11/60	84E		
D4002			08834	06/74					BR Derby		11/60	86A		
D4003			08835	02/74					BR Derby		11/60	84E		
D4004			08836	06/74					BR Derby		12/60	84E		
D4005			08837	02/74					BR Derby		12/60	81A		
D4006			08838	10/73					BR Derby		12/60	81A	05/90	DY
D4007			08839	02/74					BR Derby		12/60	81A	01/92	LA
D4008			08840	02/74					BR Derby		12/60	81A	06/91	AN
D4009			08841	03/74					BR Derby		12/60	81C	06/90	BS
D4010			08842	03/74					BR Derby		12/60	81B		
D4011			08843	03/74					BR Horwich		04/61	DG	02/90	LO

BR 1957 No.	Disposal Code	Disposal Detail	Date Cut Up	Notes
D3930				
D3931	C	BR Eastfield	10/90	Stored: [U] 08/89
D3932	S	RFS Industries Doncaster [003]	-	Hire Loco
D3933				
D3934				
D3935				
D3936				
D3937	S	Fire Training College, Moreton in Marsh	-	
D3938				
D3939	A	BR Heaton		
D3940				Stored: [U] 05/77, R/I: 09/77
D3941				Withdrawn: 08/82, R/I: 10/82
D3942	C	A V Dawson, Middlesbrough	09/88	
D3943				Withdrawn: 08/82, R/I: 10/82
D3944				
D3945	A	BR Botanic Gardens, Hull		
D3946	A	BR Cardiff		
D3947	C	M C Processors, Glasgow	10/91	
D3948				
D3949				
D3950				
D3951				Stored: [U] 04/84, R/I: 04/84
D3952				Stored: [U] 07/84, R/I: 01/85
D3953	S	RFS Locomotives, Doncaster [004]	-	Hire Loco
D3954				Withdrawn: 12/81, R/I: [U] 01/82, R/I: 05/82
D3955	A	ABB Transportation, Crewe		
D3956				
D3957				Stored: [U] 11/80, R/I: [S] 01/81, R/I: 05/81
D3958				Stored: [U] 11/84, R/I: 03/85
D3959	A	BR Millerhill		
D3960				
D3961				
D3962	A	BR Neville Hill		
D3963				
D3964	C	BR Eastleigh	09/92	
D3965	A	BR Thornaby		
D3966				
D3967				
D3968				
D3969				
D3970				
D3971	A	BR Reading		
D3972				
D3973				
D3974				
D3975				Stored: [U] 02/85, R/I: 03/85
D3976	C	Booth-Roe, Rotherham	03/92	Stored: [U] 01/84, R/I: 03/84
D3977				
D3978				
D3979				Stored: [S] 12/92
D3980	C	BREL Swindon	03/78	
D3981				Withdrawn: 09/81, R/I: 03/82, Stored: [U] 10/86, R/I: 12/86
D3982	A	BR Derby		
D3983				
D3984	S	Cobra, Middlesbrough	-	
D3985				Stored: [U] 02/84, R/I: 07/84
D3986				
D3987				
D3988	C	M C Processors, Glasgow	10/91	
D3989	A	BR Laira		
D3990				
D3991				
D3992				
D3993				
D3994				
D3995				
D3996				Withdrawn: 03/86, R/I: 05/86
D3997				
D3998				
D3999				Withdrawn: 04/92, R/I: 02/93
D4000				
D4001				Withdrawn: 11/91, R/I: [S] 12/91
D4002				
D4003				
D4004				
D4005				
D4006	A	BR Derby		
D4007	A	BR Laira		
D4008	A	BR Allerton		
D4009	A	ABB Transportation, Crewe		
D4010				
D4011	C	M C Processors, Glasgow	09/91	

BR 1957 No.	Date Re No.	BR 1948 No.	First TOPS No.	Date Re No.	Second TOPS No.	Date Re No.	Name	Name Date	Built By	Works No.	Date Introduced	Depot of First Allocation	Date Withdrawn	Depot of Final Allocation
D4012			08844	03/74					BR Horwich		04/61	DG		
D4013			08845	12/73					BR Horwich		05/61	83D		
D4014			08846	03/74					BR Horwich		05/61	83A	09/89	AN
D4015			08847	12/73					BR Horwich		05/61	83B		
D4016			08848	03/74					BR Horwich		05/61	82B	05/92	CF
D4017			08849	02/74					BR Horwich		05/61	82B		
D4018			08850	02/74					BR Horwich		06/61	82B	12/92	RG
D4019			08851	04/74					BR Horwich		06/61	82B	02/90	ML
D4020			08852	05/74					BR Horwich		06/61	82B	08/88	ML
D4021			08853	03/74					BR Horwich		06/61	82B		
D4022			08854	02/74					BR Horwich		06/61	82B		
D4023			08855	04/74					BR Horwich		06/61	82B		
D4024			08856	03/74					BR Horwich		07/61	82B		
D4025			08857	03/74					BR Horwich		07/61	82B	10/91	BS
D4026			08858	03/74					BR Horwich		08/61	82B	11/92	AN
D4027			08859	03/74					BR Horwich		08/61	82B	12/92	MR
D4028			08860	02/74					BR Darlington		05/60	41A	03/81	TE
D4029			08861	10/73					BR Darlington		05/60	41B	10/81	TI
D4030			08862	11/73					BR Darlington		06/60	41B	12/80	LA
D4031			08863	12/73					BR Darlington		07/60	41A	12/80	LA
D4032			08864	03/74					BR Darlington		08/60	41A	08/81	TE
D4033			08865	02/74					BR Darlington		09/60	41A		
D4034			08866	02/74					BR Darlington		10/60	41A		
D4035			08867	02/74					BR Darlington		11/60	41B		
D4036			08868	02/74					BR Darlington		11/60	41B	12/92	MR
D4037			08869	02/74				The Canary	05/87	BR Darlington		11/60	41B		
D4038			08870	02/74					BR Darlington		11/60	41B		
D4039			08871	02/74					BR Darlington		12/60	41B	10/90	TI
D4040			08872	02/74					BR Darlington		12/60	41B		
D4041			08873	02/74					BR Darlington		12/60	41B		
D4042			08874	02/74					BR Darlington		12/60	41B	02/92	NL
D4043			08875	02/74					BR Darlington		12/60	41B	05/91	TE
D4044			08876	03/74					BR Darlington		12/60	41B	10/91	TI
D4045			08877	05/74					BR Darlington		01/61	41A		
D4046			08878	02/74					BR Darlington		01/61	41B		
D4047			08879	03/74					BR Darlington		01/61	41B		
D4048			08880	02/74					BR Darlington		01/61	41B		
D4049									BR Darlington		01/61	41A	01/72	36A
D4050									BR Darlington		01/61	41A	12/71	40B
D4051									BR Darlington		01/61	41A	07/71	40B
D4052									BR Darlington		01/61	41A	10/70	36A
D4053									BR Darlington		01/61	41A	12/71	40B
D4054									BR Darlington		02/61	41A	06/72	51L
D4055									BR Darlington		02/61	41A	12/71	36A
D4056									BR Darlington		02/61	41A	06/72	40B
D4057									BR Darlington		02/61	41A	01/72	41J
D4058									BR Darlington		02/61	41A	01/72	36A
D4059									BR Darlington		03/61	41A	12/71	40B
D4060									BR Darlington		03/61	41A	12/71	36A
D4061									BR Darlington		03/61	41A	01/72	40A
D4062									BR Darlington		03/61	41A	02/72	40B
D4063									BR Darlington		04/61	41A	06/72	40A
D4064									BR Darlington		04/61	41A	08/68	16B
D4065									BR Darlington		04/61	41A	11/71	40A
D4066									BR Darlington		04/61	41A	06/72	40B
D4067									BR Darlington		05/61	41A	12/70	41J
D4068									BR Darlington		05/61	41A	06/72	40B
D4069									BR Darlington		05/61	41A	04/72	41J
D4070									BR Darlington		06/61	41A	04/72	41J
D4071									BR Darlington		06/61	41A	07/68	30A
D4072									BR Darlington		07/61	41A	04/72	31B
D4073									BR Darlington		08/61	41A	06/72	40B
D4074									BR Darlington		08/61	41A	04/72	31B
D4075									BR Darlington		09/61	34E	06/72	40A
D4076									BR Darlington		10/61	34E	07/68	34E
D4077									BR Darlington		10/61	34E	11/70	31B
D4078									BR Darlington		11/61	34G	06/72	36A
D4079									BR Darlington		11/61	36A	06/72	40B
D4080									BR Darlington		12/61	36A	02/68	36A
D4081									BR Darlington		01/62	36A	06/68	36A
D4082									BR Darlington		01/62	36A	06/68	36A
D4083									BR Darlington		01/62	34G	07/68	34E
D4084									BR Darlington		02/62	34E	07/68	34E
D4085									BR Darlington		02/62	34G	06/68	41J
D4086									BR Darlington		02/62	40F	08/68	16B
D4087									BR Darlington		02/62	40B	09/68	34E
D4088									BR Darlington		03/63	40B	08/68	40B
D4089									BR Darlington		03/62	36A	08/68	40B
D4090									BR Darlington		03/62	9G	08/68	40B
D4091									BR Darlington		04/62	41A	08/68	36A
D4092									BR Darlington		05/62	41A	09/68	34E
D4093									BR Darlington		05/62	41A	08/68	40B

102

BR 1957 No.	Disposal Code	Disposal Detail	Date Cut Up	Notes
D4012				Stored: [U] 09/92
D4013				
D4014	S	ABB Transportation, Derby Litchurch	-	Withdrawn: 09/86, R/I: 09/86
D4015				Withdrawn: 06/92, R/I: 07/92
D4016	A	BR Cardiff		
D4017				
D4018	A	BR Reading		
D4019	C	M C Processors, Glasgow	08/91	
D4020	C	M C Processors, Glasgow	12/89	
D4021				
D4022				
D4023				
D4024				
D4025	A	ABB Transportation, Crewe		
D4026	A	BR Allerton		
D4027	A	BR March		
D4028	C	BREL Swindon	06/81	
D4029	C	BREL Swindon	03/82	
D4030	C	BREL Swindon	10/81	Stored: [U] 09/80
D4031	C	BREL Swindon	10/81	Stored: [U] 09/80
D4032	C	BREL Swindon	03/82	
D4033				Stored: [U] 01/87
D4034				Stored: [U] 05/83, R/I: 03/84
D4035				
D4036	A	BR March		
D4037				Withdrawn: 11/81, R/I: [S] 05/83, R/I: 01/84
D4038				
D4039	S	Humberside Sea & Land Services, Grimsby	-	
D4040				
D4041				
D4042	A	RFS Locomotives, Kilnhurst		
D4043	S	RFS Locomotives, Kilnhurst		
D4044	A	RFS Locomotives, Kilnhurst		
D4045				
D4046				
D4047				
D4048				
D4049	C	G Cohen, Kettering	08/72	
D4050	C	C F Booth, Rotherham	08/72	
D4051	C	C F Booth, Rotherham	06/72	
D4052	C	C F Booth, Rotherham	02/71	
D4053	C	C F Booth, Rotherham	08/72	
D4054	C	G Cohen, Kettering	03/73	
D4055	C	C F Booth, Rotherham	04/72	
D4056	C	NCB Shilbottle	03/83	PI
D4057	C	C F Booth, Rotherham	10/72	
D4058	C	G Cohen, Kettering	08/72	
D4059	C	G Cohen, Kettering	08/72	
D4060	C	C F Booth, Rotherham	04/72	
D4061	C	C F Booth, Rotherham	09/72	
D4062	C	C F Booth, Rotherham	09/72	
D4063	C	G Cohen, Kettering	04/73	
D4064	C	C F Booth, Rotherham	01/69	
D4065	C	C F Booth, Rotherham	05/72	
D4066	C	G Cohen, Kettering	02/73	
D4067	P	Great Central Railway, Loughborough	-	PI
D4068	C	NCB Whittle Colliery	10/85	PI
D4069	C	NCB Whittle Colliery	12/85	PI
D4070	C	NCB Whittle Colliery	10/85	PI
D4071	C	G Cohen, Kettering	01/69	
D4072	C	NCB Philadelphia	11/85	PI
D4073	C	G Cohen, Kettering	02/73	
D4074	C	NCB Philadelphia	08/78	PI
D4075	C	G Cohen, Kettering	04/73	
D4076	C	C F Booth, Rotherham	05/69	
D4077	C	C F Booth, Rotherham	04/71	
D4078	C	G Cohen, Kettering	03/73	
D4079	C	G Cohen, Kettering	02/73	
D4080	C	C F Booth, Rotherham	11/68	
D4081	C	C F Booth, Rotherham	12/68	
D4082	C	C F Booth, Rotherham	11/68	
D4083	C	C F Booth, Rotherham	05/69	
D4084	C	C F Booth, Rotherham	05/69	
D4085	C	C F Booth, Rotherham	01/69	
D4086	C	G Cohen, Kettering	01/69	
D4087	C	G Cohen, Kettering	01/69	
D4088	C	C F Booth, Rotherham	01/69	
D4089	C	C F Booth, Rotherham	01/69	
D4090	C	G Cohen, Kettering	01/69	
D4091	C	C F Booth, Rotherham	01/69	
D4092	P	South Yorkshire Railway	-	PI
D4093	C	G Cohen, Kettering	02/69	

BR 1957 No.	Date Re No.	BR 1948 No.	First TOPS No.	Date Re No.	Second TOPS No.	Date Re No.	Name	Name Date	Built By	Works No.	Date Introduced	Depot of First Allocation	Date Withdrawn	Depot of Final Allocation
D4094									BR Darlington		06/62	41A	08/68	40B
D4095			08881	09/74					BR Horwich		08/61	63B		
D4096			08882	10/73					BR Horwich		09/61	65A		
D4097			08883	05/74					BR Horwich		09/61	63A		
D4098			08884	02/74					BR Horwich		09/61	63A		
D4099			09011	12/73					BR Horwich		10/61	73C		
D4100			09012	01/74			Dick Hardy	04/88	BR Horwich		10/61	73F		
D4101			09013	02/74					BR Horwich		10/61	73F		
D4102			09014	01/74					BR Horwich		11/61	75C		
D4103			09015	01/74					BR Horwich		11/61	75C		
D4104			09016	12/73					BR Horwich		11/61	75C		
D4105			09017	01/74					BR Horwich		12/61	6A	10/87	SU
D4106			09018	01/74					BR Horwich		12/61	12A		
D4107			09019	12/73					BR Horwich		12/61	12B		
D4108			09020	02/74					BR Horwich		12/61	5D		
D4109			09021	01/74					BR Horwich		12/61	5D		
D4110			09022	12/73					BR Horwich		12/61	5D		
D4111			09023	02/74					BR Horwich		12/61	2B		
D4112			09024	02/74					BR Horwich		12/61	2B		
D4113			09025	02/74					BR Horwich		01/62	9A		
D4114			09026	01/74			William Pearson	03/89	BR Horwich		01/62	24C		
D4115			08885	02/74					BR Horwich		01/62	24C		
D4116			08886	05/74					BR Horwich		01/62	14A		
D4117			08887	02/74					BR Horwich		02/62	14A		
D4118			08888	02/74			Postmans Pride	04/91	BR Horwich		02/62	82C		
D4119			08889	05/74					BR Horwich		02/62	82C	02/92	MR
D4120			08890	10/74					BR Horwich		02/62	82C		
D4121			08891	06/74					BR Horwich		03/62	82C		
D4122			08892	12/73					BR Horwich		03/62	82C		
D4123			08893	09/74					BR Horwich		03/62	82C		
D4124			08894	03/74					BR Horwich		04/62	82C		
D4125			08895	02/74					BR Horwich		04/62	82C		
D4126			08896	06/74					BR Horwich		04/62	86A		
D4127			08897	04/74					BR Horwich		04/62	86A		
D4128			08898	05/74					BR Horwich		05/62	86A	11/88	LE
D4129			08899	02/74					BR Horwich		05/62	83C		
D4130			08900	02/74					BR Horwich		05/62	83C		
D4131			08901	04/74					BR Horwich		05/62	14A		
D4132			08902	05/74					BR Horwich		96/62	14A		
D4133			08903	04/74					BR Horwich		06/62	14A		
D4134			08904	08/74					BR Horwich		06/62	14E		
D4135			08905	05/74					BR Horwich		07/62	14E		
D4136			08906	02/74					BR Horwich		07/62	14A		
D4137			08907	04/74					BR Horwich		07/62	17B		
D4138			08908	04/74					BR Horwich		07/62	14A		
D4139			08909	09/74					BR Horwich		07/62	24L		
D4140			08910	05/74					BR Horwich		09/62	24L		
D4141			08911	06/74					BR Horwich		09/62	24L		
D4142			08912	06/74					BR Horwich		09/62	24L		
D4143			08913	02/74					BR Horwich		09/62	5B		
D4144			08914	02/74					BR Horwich		09/62	2B		
D4145			08915	02/74					BR Horwich		10/62	9B		
D4146			08916	02/74					BR Horwich		10/62	8C	10/92	AN
D4147			08917	02/74					BR Horwich		10/62·	8C	11/91	AN
D4148			08918	02/74					BR Horwich		10/62	8C		
D4149			08919	02/74					BR Horwich		10/62	17B		
D4150			08920	03/74					BR Horwich		10/62	17B		
D4151			08921	05/74					BR Horwich		11/62	17B		
D4152			08922	05/74					BR Horwich		11/62	8C		
D4153			08923	02/74					BR Horwich		12/62	8C		
D4154			08924	02/74					BR Horwich		12/62	8A		
D4155			08925	02/74					BR Horwich		12/62	8A		
D4156			08926	05/74					BR Horwich		12/62	24J		
D4157			08927	05/74					BR Horwich		12/62	24J		
D4158			08928	02/74					BR Darlington		04/62	83E		
D4159			08929	12/73					BR Darlington		04/62	83D	12/91	OC
D4160			08930	03/74					BR Darlington		04/62	83A	02/90	SF
D4161			08931	01/74					BR Darlington		05/62	83G		
D4162			08932	04/74					BR Darlington		05/62	83C		
D4163			08933	02/74					BR Darlington		05/62	83B		
D4164			08934	05/74					BR Darlington		05/62	83B		
D4165			08935	04/74					BR Darlington		05/62	83B		
D4166			08936	02/74					BR Darlington		05/62	83E	12/92	MR
D4167			08937	03/74					BR Darlington		05/62	83D		
D4168			08938	06/74					BR Darlington		05/62	82C		
D4169			08939	07/74					BR Darlington		05/62	82A		
D4170			08940	04/74					BR Darlington		06/62	86A		
D4171			08941	02/74					BR Darlington		06/62	86A		
D4172			08942	07/74					BR Darlington		07/62	86A		
D4173			08943	06/74					BR Darlington		07/62	86A	07/88	WN
D4174			08944	03/74					BR Darlington		08/62	86A		

BR 1957 No.	Disposal Code	Disposal Detail	Date Cut Up	Notes
D4094	C	G Cohen, Kettering	01/69	
D4095				
D4096				
D4097				
D4098				
D4099				
D4100				Names off: 10/90 - 09/91
D4101				
D4102				
D4103				
D4104				
D4105	D	To Departmental Stock - 97806		
D4106				
D4107				
D4108				
D4109				
D4110				
D4111				
D4112				
D4113				
D4114				
D4115				
D4116				
D4117				
D4118				
D4119	A	BR Peterborough		
D4120				
D4121				
D4122				
D4123				
D4124				Stored: [U] 08/82, R/I: 10/82
D4125				
D4126				
D4127				
D4128	A	BR Bescot		
D4129				
D4130				
D4131				
D4132				
D4133				
D4134				
D4135				
D4136				
D4137				Stored: [U] 11/86
D4138				
D4139				
D4140				
D4141				
D4142				
D4143				
D4144				
D4145				
D4146	A	BR Allerton		
D4147	A	BR Allerton		
D4148				
D4149				
D4150				
D4151				
D4152				
D4153				
D4154				
D4155				
D4156				
D4157				
D4158				
D4159	A	BR Old Oak Common		
D4160	C	M C Processors, Glasgow	06/91	
D4161				
D4162				
D4163				
D4164				
D4165				
D4166	A	BR March		
D4167				
D4168				
D4169				
D4170				Stored: [S] 07/88, R/I: 09/88
D4171				
D4172				
D4173	S	ABB Transportation, Crewe		
D4174				

BR 1957 No.	Date Re No.	BR 1948 No.	First TOPS No.	Date Re No.	Second TOPS No.	Date Re No.	Name	Name Date	Built By	Works No.	Date Introduced	Depot of First Allocation	Date Withdrawn	Depot of Final Allocation
D4175			08945	05/74					BR Darlington		08/62	86A		
D4176			08946	02/74					BR Darlington		09/62	86A		
D4177			08947	01/74					BR Darlington		09/62	86A		
D4178			08948	03/74					BR Darlington		09/62	86A		
D4179			08949	02/74					BR Darlington		09/62	86A		
D4180			08950	02/74					BR Darlington		10/62	86A		
D4181			08951	02/74					BR Darlington		10/62	86A		
D4182			08952	04/74					BR Darlington		10/62	86A		
D4183			08953	09/74					BR Darlington		01/63	86A		
D4184			08954	02/74					BR Darlington		01/63	86A		
D4185			08955	03/74					BR Darlington		01/63	86A		
D4186			08956	05/74					BR Darlington		07/62	30A		
D4187									BR Darlington		07/62	30A	02/65	41A
D4188									BR Darlington		07/62	30A	02/65	41A
D4189									BR Darlington		07/62	30A	02/65	41A
D4190									BR Darlington		08/62	30A	02/65	41A
D4191			08957	03/74					BR Darlington		09/62	30A		
D4192			08958	03/74					BR Darlington		09/62	30A		

BR 0-6-0+0-6-0 'MASTER & SLAVE' SHUNTERS CLASS 13

BR 1957 No.	TOPS No.	Date Re No.	Former Nos Master	Former Nos Slave	Rebuilt By	Works No.	Date Introduced	Depot of First Allocation	Date Withdrawn	Depot of Final Allocation	Disposal Code
D4500	13003	02/74	D4188	D3698	BR Darlington	-	05/65	41A	01/85	TI	C
D4501	13001	02/74	D4190	D4189	BR Darlington	-	06/65	41A	01/85	TI	C
D4502	13002	02/74	D4187	D3697	BR Darlington	-	07/65	41A	06/81	TI	C

BR TYPE 2 Bo-Bo CLASS 24

BR 1957 No.	TOPS No.	Date Re No.	Built By	Works No.	Date Introduced	Depot of First Allocation	Date Withdrawn	Depot of Final Allocation
D5000	24005	11/73	BR Derby	-	07/58	5B	01/76	LO
D5001	24001	04/74	BR Derby	-	10/58	5B	10/75	LO
D5002	24002	04/74	BR Derby	-	12/58	5B	10/75	ED
D5003	24003	04/74	BR Derby	-	12/58	5B	08/75	ED
D5004	24004	02/74	BR Derby	-	12/58	5B	10/75	ED
D5005	-	-	BR Derby	-	01/59	5B	01/69	DO5
D5006	24006	04/74	BR Derby	-	01/59	5B	08/75	ED
D5007	24007	04/74	BR Derby	-	02/59	5B	10/75	ED
D5008	24008	07/74	BR Derby	-	02/59	5B	08/75	ED
D5009	24009	04/74	BR Derby	-	03/59	5B	07/76	HA
D5010	24010	04/74	BR Derby	-	03/59	5B	10/75	ED
D5011	24011	11/73	BR Derby	-	03/59	5B	10/75	ED
D5012	24012	05/74	BR Derby	-	04/59	5B	08/75	ED
D5013	24013	11/73	BR Derby	-	04/59	5B	10/75	ED
D5014	24014	03/74	BR Derby	-	05/59	5B	10/75	ED
D5015	24015	04/74	BR Derby	-	05/59	5B	08/75	ED
D5016	24016	04/74	BR Derby	-	05/59	5B	08/75	ED
D5017	24017	04/74	BR Derby	-	06/59	5B	10/75	ED
D5018	24018	04/74	BR Derby	-	06/59	5B	08/75	ED
D5019	24019	05/74	BR Derby	-	07/59	5B	10/75	ED
D5020	24020	04/74	BR Derby	-	08/59	32B	08/75	LO
D5021	24021	03/74	BR Derby	-	08/59	32B	08/75	LO
D5022	24022	04/74	BR Derby	-	09/59	32A	01/76	LO
D5023	24023	03/74	BR Derby	-	09/59	32A	09/78	CD
D5024	24024	02/74	BR Derby	-	10/59	32B	08/75	LO
D5025	24025	05/74	BR Derby	-	10/59	32B	01/76	CD
D5026	24026	03/74	BR Derby	-	10/59	32B	08/75	CD
D5027	24027	05/74	BR Derby	-	10/59	32B	07/76	CD
D5028	-	-	BR Derby	-	11/59	32B	06/72	DO5
D5029	24029	03/74	BR Derby	-	11/59	32B	08/75	CD
D5030	24030	06/74	BR Crewe	-	06/59	31B	07/76	CD
D5031	24031	02/74	BR Crewe	-	06/59	31B	10/75	CD
D5032	24032	03/74	BR Crewe	-	07/59	31B	07/76	CD
D5033	24033	03/74	BR Crewe	-	08/59	31B	10/75	CD
D5034	24034	03/74	BR Crewe	-	09/59	31B	01/76	CD
D5035	24035	04/74	BR Crewe	-	08/59	31B	10/78	CD
D5036	24036	05/74	BR Crewe	-	09/59	31B	11/77	CD
D5037	24037	01/74	BR Crewe	-	09/59	31B	07/76	CD
D5038	24038	02/74	BR Crewe	-	09/59	32B	07/76	CD
D5039	24039	06/74	BR Crewe	-	09/59	32B	07/76	CD
D5040	24040	03/74	BR Crewe	-	10/59	32B	01/76	CD

BR 1957 No.	Disposal Code	Disposal Detail	Date Cut Up	Notes
D4175				
D4176				
D4177				
D4178				
D4179				
D4180				
D4181				
D4182				
D4183				
D4184				
D4185				
D4186				
D4187	R	Rebuilt as Class 13, No. D4502 (M)	-	
D4188	R	Rebuilt as Class 13, No. D4500 (M)	-	
D4189	R	Rebuilt as Class 13, No. D4501 (M)	-	
D4190	R	Rebuilt as Class 13, No. D4501 (S)	-	
D4191				
D4192				

BR 1957 No.	Disposal Detail	Date Cut Up	Notes
D4500	BREL Doncaster	09/86	
D4501	BREL Swindon	05/85	
D4502	BREL Swindon	10/82	Stored: [U] 04/81

BR 1957 No.	Disposal Code	Disposal Detail	Date Cut Up	Notes
D5000	C	BREL Swindon	04/77	Withdrawn: 01/69, R/I: 10/69, Stored: [S] 07/75
D5001	C	BREL Doncaster	11/77	Stored: [U] 01/69, R/I: 10/69, Stored: [S] 07/75
D5002	C	BREL Glasgow	07/77	Stored: [U] 01/69, R/I: 03/69, Stored: [S] 07/75
D5003	C	BREL Doncaster	07/76	Stored: [U] 07/75
D5004	C	BREL Glasgow	04/77	Stored: [U] 07/75
D5005	C	BREL Derby	09/69	
D5006	C	BREL Glasgow	12/80	Stored: [U] 07/75
D5007	C	BREL Doncaster	01/78	Stored: [U] 01/69, R/I: 08/69, Stored: [S] 07/75
D5008	C	BREL Doncaster	05/76	Stored: [U] 07/75
D5009	C	BREL Doncaster	11/77	Stored: [U] 08/75, R/I: 09/75, Stored: [U] 04/76
D5010	C	BREL Doncaster	02/77	Withdrawn: 01/69, R/I: 09/75
D5011	C	BREL Glasgow	04/77	Stored: [U] 01/69, R/I: 05/69, Stored: [S] 07/75
D5012	C	BREL Doncaster	02/76	Stored: [U] 01/69, R/I: 10/69, Stored: [S] 07/75
D5013	C	BREL Doncaster	02/78	Stored: [U] 04/69, R/I: 10/69, Stored: [S] 07/75
D5014	C	BREL Doncaster	03/78	Stored: [S] 07/75
D5015	C	BREL Doncaster	01/77	Stored: [S] 07/75
D5016	C	BREL Doncaster	03/77	Stored: [U] 03/69, R/I: 06/69, Stored: [U] 07/75
D5017	C	BREL Doncaster	03/77	Stored: [U] 01/69, R/I: 07/69, Stored: [S] 07/75
D5018	C	BREL Doncaster	06/76	Stored: [S] 07/75
D5019	C	BREL Doncaster	05/78	Withdrawn: 01/69, R/I: 10/69, Stored: [S] 07/75
D5020	C	BREL Swindon	04/77	Stored: [S] 07/75
D5021	C	BREL Swindon	04/77	Stored: [S] 07/75
D5022	C	BREL Doncaster	08/78	Stored: [S] 07/75
D5023	C	BREL Doncaster	11/78	Stored: [U] 11/77, R/I: 11/77
D5024	C	BREL Swindon	04/77	Stored: [U] 02/69, R/I: 03/69, Stored: [S] 07/75
D5025	C	BREL Swindon	07/77	Stored: [S] 07/75
D5026	C	BREL Swindon	01/77	Stored: [S] 07/75
D5027	C	BREL Swindon	05/77	Stored: [S] 07/75, R/I: 12/75, Stored: [S] 02/76
D5028	C	BREL Crewe	09/72	
D5029	C	BREL Swindon	03/77	Stored: [S] 07/75
D5030	C	BREL Swindon	05/77	Stored: [S] 02/76
D5031	C	BREL Swindon	12/76	Stored: [S] 07/75
D5032	P	North Yorkshire Moors Railway, Grosmont	-	Stored: [S] 05/76
D5033	C	BREL Swindon	03/77	Stored: [S] 08/75
D5034	C	BREL Swindon	05/77	Stored: [S] 08/75
D5035	C	BREL Doncaster	01/79	Stored: [S] 11/77, R/I: 11/77
D5036	C	BREL Doncaster	06/78	Stored: [U] 10/77
D5037	C	BREL Swindon	06/77	Stored: [S] 08/75, R/I: 08/75, Stored: [S] 05/76
D5038	C	BREL Swindon	06/77	Stored: [S] 08/75, R/I: 12/75, Stored: [S] 02/76
D5039	C	BREL Swindon	07/78	Stored: [S] 05/76
D5040	C	BREL Swindon	02/77	Stored: [U] 12/75

BR 1957 No.	TOPS No.	Date Re No.	Built By	Works No.	Date Introduced	Depot of First Allocation	Date Withdrawn	Depot of Final Allocation
D5041	24041	09/73	BR Crewe	-	09/59	32B	07/76	CD
D5042	24042	01/74	BR Crewe	-	10/59	32B	08/75	CD
D5043	-	-	BR Crewe	-	10/59	32B	08/69	DO5
D5044	24044	03/74	BR Crewe	-	10/59	32B	01/76	CD
D5045	24045	05/74	BR Crewe	-	11/59	31B	08/75	CD
D5046	24046	11/73	BR Crewe	-	10/59	31B	07/76	CD
D5047	24047	02/74	BR Crewe	-	11/59	31B	11/78	CD
D5048	24048	02/74	BR Crewe	-	11/59	31B	08/75	CD
D5049	24049	02/74	BR Crewe	-	04/60	31B	01/76	CD
D5050	24050	03/74	BR Crewe	-	11/59	31B	10/75	CD
D5051	-	-	BR Crewe	-	12/59	31B	11/67	ED
D5052	24052	01/74	BR Crewe	-	12/59	31B	12/76	CD
D5053	24053	11/73	BR Crewe	-	12/59	31B	01/76	CD
D5054	24054	02/74	BR Crewe	-	12/59	31B	07/76	CD
D5055	24055	02/74	BR Crewe	-	12/59	31B	10/75	CD
D5056	24056	02/74	BR Crewe	-	12/59	31B	10/75	CD
D5057	24057	03/74	BR Crewe	-	12/59	31B	01/78	CD
D5058	24058	02/74	BR Crewe	-	01/60	31B	10/75	CD
D5059	24059	02/74	BR Crewe	-	12/59	31B	10/75	CD
D5060	24060	03/74	BR Crewe	-	01/60	31B	10/75	CD
D5061	24061	03/74	BR Crewe	-	01/60	31B	01/75	CD
D5062	24062	02/74	BR Crewe	-	01/60	31B	10/75	CD
D5063	24063	03/74	BR Crewe	-	01/60	31B	04/79	CD
D5064	24064	02/74	BR Crewe	-	01/60	31B	01/76	CD
D5065	24065	10/73	BR Crewe	-	02/60	31B	12/76	ED
D5066	24066	03/74	BR Derby	-	12/59	31B	02/76	HA
D5067	-	-	BR Derby	-	12/59	31B	10/72	HA
D5068	-	-	BR Derby	-	12/59	31B	10/72	HA
D5069	24069	12/73	BR Derby	-	01/60	31B	12/76	HA
D5070	24070	12/73	BR Derby	-	01/60	31B	02/76	HA
D5071	24071	04/74	BR Derby	-	02/60	31B	08/75	HA
D5072	24072	04/74	BR Derby	-	02/60	31B	10/75	HA
D5073	24073	12/73	BR Derby	-	03/60	31B	09/78	CD
D5074	24074	03/74	BR Derby	-	03/60	31B	10/75	CD
D5075	24075	02/74	BR Derby	-	03/60	31B	01/76	CD
D5076	24076	04/74	BR Crewe	-	02/60	31B	10/75	CD
D5077	24077	04/74	BR Crewe	-	02/60	31B	07/76	CD
D5078	24078	03/74	BR Crewe	-	02/60	31B	04/76	CD
D5079	24079	06/74	BR Crewe	-	02/60	31B	07/76	CD
D5080	24080	04/74	BR Crewe	-	03/60	31B	09/76	CD
D5081	24081	12/74	BR Crewe	-	03/60	31B	10/80	CD
D5082	24082	10/73	BR Crewe	-	03/60	31B	03/79	CD
D5083	24083	03/74	BR Crewe	-	04/60	31B	03/76	CD
D5084	24084	03/74	BR Crewe	-	04/60	31B	07/76	CD
D5085	24085	05/74	BR Crewe	-	05/60	31B	07/76	CD
D5086	24086	02/74	BR Crewe	-	05/60	31B	01/76	CD
D5087	24087	02/74	BR Crewe	-	06/60	31B	02/76	CD
D5088	-	-	BR Crewe	-	06/60	31B	07/70	DO5
D5089	24089	03/74	BR Crewe	-	06/60	31B	01/76	CD
D5090	24090	04/74	BR Crewe	-	06/60	31B	02/76	ED
D5091	24091	04/74	BR Crewe	-	06/60	31B	11/77	CD
D5092	24092	04/74	BR Crewe	-	07/60	31B	10/75	CD
D5093	-	-	BR Crewe	-	07/60	31B	08/69	DO5
D5094	24094	04/74	BR Darlington	-	02/60	31B	12/76	HA
D5095	24095	10/73	BR Darlington	-	03/60	31B	08/75	HA
D5096	24096	04/74	BR Darlington	-	04/60	52A	08/75	ED
D5097	24097	01/74	BR Darlington	-	05/60	52A	02/76	ED
D5098	24098	02/74	BR Darlington	-	05/60	52A	08/75	ED
D5099	24099	05/74	BR Darlington	-	06/60	52A	02/76	ED
D5100	24100	10/73	BR Darlington	-	06/60	52A	02/76	ED
D5101	24101	05/74	BR Darlington	-	07/60	52A	02/76	ED
D5102	24102	11/73	BR Darlington	-	07/60	52A	02/76	IS
D5103	24103	02/74	BR Darlington	-	08/60	52A	12/76	HA
D5104	24104	03/74	BR Darlington	-	08/60	52A	12/76	HA
D5105	24105	02/74	BR Darlington	-	09/60	52A	10/75	ED
D5106	24106	02/74	BR Darlington	-	09/60	52A	12/76	HA
D5107	24107	01/74	BR Darlington	-	10/60	52A	12/76	HA
D5108	24108	11/73	BR Darlington	-	10/60	52A	97/76	HA
D5109	24109	02/74	BR Darlington	-	11/60	52A	02/76	HA
D5110	24110	10/73	BR Darlington	-	11/60	52A	12/76	HA
D5111	24111	12/74	BR Darlington	-	11/60	52A	02/76	HA
D5112	24112	07/74	BR Darlington	-	12/60	52A	12/76	HA
D5113	24113	07/74	BR Darlington	-	01/61	52A	12/76	HA
D5114	-	-	BR Derby	-	04/60	61A	10/72	60A
D5115	24115	07/74	BR Derby	-	04/60	61A	12/76	HA
D5116	24116	04/74	BR Derby	-	05/60	61A	09/76	HA
D5117	24117	07/74	BR Derby	-	05/60	60A	02/76	HA
D5118	24118	07/74	BR Derby	-	05/60	60A	12/76	HA
D5119	24119	03/74	BR Derby	-	05/60	60A	07/76	HA
D5120	24120	07/74	BR Derby	-	06/60	60A	12/76	HA

BR 1957 No.	Disposal Code	Disposal Detail	Date Cut Up	Notes
D5041	C	BREL Swindon	06/78	Stored: [S] 08/75, R/I: 08/75, Stored: [S] 10/75 R/I: 12/75, Stored: [U] 12/75, R/I: 01/76, Stored: [S] 02/76
D5042	C	BREL Swindon	02/76	Stored: [S] 08/75
D5043	C	J Cashmore, Great Bridge	06/70	Stored: [U] 07/69
D5044	C	BREL Swindon	01/77	Stored: [U] 01/76
D5045	C	BREL Swindon	12/76	Stored: [U] 08/75
D5046	C	BREL Swindon	07/77	Stored: [U] 08/75, R/I: 08/75, Stored: [S] 10/75, R/I: 12/75, Stored: [S] 02/76
D5047	C	BREL Doncaster	02/79	Stored: [S] 11/77, R/I: 11/77
D5048	C	BREL Swindon	03/76	
D5049	C	BREL Swindon	11/76	Stored: [S] 08/75, R/I: 08/75, Stored: [U] 01/76
D5050	C	BREL Swindon	12/76	Stored: [U] 03/79, R/I: 04/69, Stored: [U] 08/75
D5051	C	BR Workshops, Inverurie	08/68	
D5052	C	BREL Swindon	05/76	Stored: [S] 07/76
D5053	C	BREL Swindon	05/76	Stored: [S] 01/76
D5054	D	To Departmental Stock - 968008	-	Stored: [S] 05/76
D5055	C	BREL Swindon	12/76	
D5056	C	BREL Swindon	11/76	
D5057	C	BREL Doncaster	05/78	
D5058	C	BREL Swindon	10/76	
D5059	C	BREL Swindon	03/77	
D5060	C	BREL Swindon	01/77	Stored: [U] 02/69, R/I: 04/69
D5061	D	To Departmental Stock - 97201	-	Stored: [U] 08/75
D5062	C	BREL Swindon	10/76	
D5063	C	BREL Doncaster	07/79	Stored: [U] 11/77, R/I: 11/77
D5064	C	BREL Swindon	12/76	Stored: [S] 01/76
D5065	C	BREL Swindon	06/77	Stored: [S] 07/75, R/I: 12/75, Stored: [U] 10/76
D5066	C	BREL Doncaster	08/78	Stored: [S] 08/75
D5067	C	BREL Glasgow	10/73	
D5068	C	BREL Glasgow	04/73	
D5069	C	BREL Doncaster	09/77	Stored: [S] 08/75, R/I: 12/75, Stored: [U] 09/76
D5070	C	BREL Doncaster	06/76	Stored: [S] 08/75
D5071	C	BREL Doncaster	01/77	Stored: [U] 07/75
D5072	C	BREL Doncaster	12/77	Stored: [S] 08/75
D5073	C	BREL Doncaster	12/78	Stored: [U] 08/75
D5074	C	J Cashmore, Great Bridge	05/76	
D5075	C	BREL Swindon	07/76	Stored: [U] 01/76
D5076	C	BREL Swindon	01/77	
D5077	C	BREL Swindon	06/78	Stored: [S] 02/76
D5078	C	BREL Swindon	06/78	Stored: [U] 03/76
D5079	C	BREL Swindon	09/78	Stored: [S] 05/76
D5080	C	BREL Doncaster	06/78	
D5081	P	Steamport, Southport	-	Stored: [U] 11/77, R/I: 11/77
D5082	C	BREL Doncaster	04/79	Stored: [U] 07/76, R/I: 10/76
D5083	C	BREL Swindon	05/77	Stored: [U] 02/76
D5084	C	BREL Swindon	05/78	Stored: [S] 05/76
D5085	C	BREL Swindon	10/78	Stored: [S] 05/76
D5086	C	BREL Doncaster	01/77	Stored: [U] 11/75
D5087	C	BREL Doncaster	08/78	Stored: [S] 07/76
D5088	C	G Cohen, Kettering	10/71	
D5089	C	BREL Swindon	07/77	Stored: [U] 12/75
D5090	C	BREL Doncaster	01/78	Stored: [S] 08/69, R/I: 11/69, Stored: [S] 08/75
D5091	C	BREL Doncaster	06/78	Stored: [U] 10/77
D5092	C	BREL Swindon	12/76	
D5093	C	J Cashmore, Great Bridge	04/70	Stored: [U] 07/69
D5094	C	BREL Doncaster	09/77	Stored: [S] 08/75, R/I: 12/75, Stored: [U] 09/76
D5095	C	BREL Doncaster	06/76	Stored: [U] 07/75
D5096	C	BREL Doncaster	08/76	Stored: [S] 08/75
D5097	C	BREL Doncaster	02/77	Stored: [S] 08/75
D5098	C	BREL Doncaster	05/76	Stored: [S] 08/75
D5099	C	BREL Doncaster	08/77	Stored: [S] 08/75
D5100	C	BREL Doncaster	04/76	Stored: [S] 08/75
D5101	C	BREL Doncaster	03/76	Stored: [S] 08/75
D5102	C	BREL Doncaster	04/78	Stored: [S] 10/75
D5103	C	BREL Doncaster	10/77	Stored: [S] 10/75, R/I: 04/76, Stored: [U] 07/76
D5104	C	BREL Doncaster	09/77	Stored: [S] 08/75, R/I: 04/76, Stored: [S] 10/76
D5105	C	BREL Doncaster	03/78	Stored: [S] 08/75
D5106	C	BREL Doncaster	10/77	Stored: [S] 10/75, R/I: 11/75, Stored: [S] 10/76
D5107	C	BREL Swindon	06/77	Stored: [U] 10/76
D5108	C	BREL Doncaster	02/78	Stored: [S] 10/75
D5109	C	BREL Doncaster	03/78	Stored: [S] 10/75
D5110	C	BREL Doncaster	04/77	Stored: [U] 10/76
D5111	C	BREL Doncaster	02/78	Stored: [S] 10/75
D5112	C	BREL Doncaster	10/77	Stored: [S] 10/76
D5113	C	BREL Doncaster	10/77	Stored: [S] 10/76
D5114	C	BREL Glasgow	10/73	
D5115	C	BREL Swindon	06/77	Stored: [S] 10/76, [Worked during storage]
D5116	C	BREL Doncaster	07/76	Stored: [U] 09/76
D5117	C	BREL Doncaster	01/77	Stored: [U] 01/76
D5118	C	BREL Doncaster	06/78	Stored: [U] 10/76
D5119	C	BREL Doncaster	05/77	Stored: [U] 04/76
D5120	C	BREL Doncaster	08/77	Stored: [U] 10/76

BR 1957 No.	TOPS No.	Date Re No.	Built By	Works No.	Date Introduced	Depot of First Allocation	Date Withdrawn	Depot of Final Allocation
D5121	24121	05/74	BR Derby	-	06/60	60A	12/76	HA
D5122	-	-	BR Derby	-	06/60	60A	09/68	60A
D5123	24123	04/74	BR Derby	-	06/60	60A	07/76	HA
D5124	24124	04/74	BR Derby	-	07/60	60A	12/76	HA
D5125	24125	04/74	BR Derby	-	07/60	60A	03/76	HA
D5126	24126	07/74	BR Derby	-	07/60	60A	02/76	HA
D5127	24127	10/74	BR Derby	-	07/60	60A	02/76	HA
D5128	24128	05/74	BR Derby	-	08/60	60A	07/76	HA
D5129	24129	04/74	BR Derby	-	08/60	60A	12/76	HA
D5130	24130	04/74	BR Derby	-	09/60	60A	12/76	HA
D5131	-	-	BR Derby	-	09/60	60A	09/71	60A
D5132	24132	10/74	BR Derby	-	09/60	60A	02/76	IS
D5133	24133	02/74	BR Derby	-	09/60	9A	03/78	CD
D5134	24134	02/74	BR Derby	-	10/60	9A	12/76	CD
D5135	24135	04/74	BR Derby	-	10/60	9A	01/76	CD
D5136	24136	04/74	BR Derby	-	10/60	9A	10/75	CD
D5137	24137	04/74	BR Derby	-	10/60	9A	07/76	CD
D5138	-	-	BR Derby	-	10/60	9A	08/69	DO5
D5139	-	-	BR Derby	-	10/60	9A	08/69	DO5
D5140	24140	02/74	BR Derby	-	11/60	9A	01/76	CD
D5141	24141	02/74	BR Derby	-	11/60	9A	07/76	CD
D5142	24142	04/74	BR Derby	-	11/60	9A	07/76	CD
D5143	24143	03/74	BR Derby	-	11/60	1A	01/76	CD
D5144	24144	05/74	BR Derby	-	12/60	1A	01/76	CD
D5145	24145	05/74	BR Derby	-	12/60	1A	01/76	CD
D5146	24146	03/74	BR Derby	-	12/60	1A	01/76	CD
D5147	24147	04/74	BR Derby	-	12/60	52A	07/76	HA
D5148	24148	04/74	BR Derby	-	12/60	52A	10/75	ED
D5149	-	-	BR Derby	-	01/61	52A	10/72	65A
D5150	24150	04/74	BR Derby	-	01/61	52A	12/76	HA

BR TYPE 2 Bo-Bo CLASS 25

BR 1957 No.	TOPS No.	Date Re No.	TOPS Re No.	Date Re No.	Built By	Works No.	Date Introduced	Depot of First Allocation	Date Withdrawn	Depot of Final Allocation
D5151	25001	04/74	-	-	BR Darlington	-	04/61	51L	09/80	ED
D5152	25002	03/74	-	-	BR Darlington	-	05/61	51L	12/80	ED
D5153	25003	03/74	-	-	BR Darlington	-	05/61	51L	08/76	ED
D5154	25004	04/74	-	-	BR Darlington	-	05/61	51L	09/76	ED
D5155	25005	01/74	-	-	BR Darlington	-	06/61	51L	12/80	ED
D5156	25006	03/74	-	-	BR Darlington	-	06/61	51L	12/80	HA
D5157	25007	05/74	-	-	BR Darlington	-	07/61	51L	12/80	HA
D5158	25008	09/74	-	-	BR Darlington	-	07/61	51L	06/80	HA
D5159	25009	04/74	-	-	BR Darlington	-	07/61	51L	07/80	HA
D5160	25010	04/74	-	-	BR Darlington	-	08/61	51L	12/80	HA
D5161	25011	10/73	-	-	BR Darlington	-	09/61	51L	12/80	HA
D5162	25012	02/74	-	-	BR Darlington	-	09/61	51L	02/77	HA
D5163	25013	03/74	-	-	BR Darlington	-	09/61	51L	09/80	HA
D5164	25014	05/74	-	-	BR Darlington	-	10/61	51L	07/77	HA
D5165	25015	05/74	-	-	BR Darlington	-	10/61	51L	12/75	TI
D5166	25016	02/74	-	-	BR Darlington	-	11/61	51L	01/76	TI
D5167	25017	02/74	-	-	BR Darlington	-	11/61	51L	01/76	TI
D5168	25018	02/74	-	-	BR Darlington	-	12/61	51L	11/76	ED
D5169	25019	11/73	-	-	BR Darlington	-	12/61	51L	09/80	HA
D5170	25020	03/74	-	-	BR Darlington	-	12/61	51L	01/76	TI
D5171	25021	02/74	-	-	BR Darlington	-	01/62	51L	09/80	HA
D5172	25022	03/74	-	-	BR Darlington	-	02/62	51L	01/76	TI
D5173	25023	02/74	-	-	BR Darlington	-	03/62	51L	09/80	HA
D5174	25024	03/74	-	-	BR Darlington	-	03/62	51L	01/76	TI
D5175	25025	02/74	-	-	BR Darlington	-	04/62	51L	04/77	ED
D5176	25026	02/74	-	-	BR Darlington	-	01/63	55A	11/80	TO
D5177	25027	10/73	-	-	BR Darlington	-	02/63	55A	05/83	CD
D5178	25028	04/74	-	-	BR Darlington	-	02/63	55A	12/80	HA
D5179	25029	04/74	-	-	BR Darlington	-	02/63	52A	08/77	TO
D5180	25030	02/74	-	-	BR Darlington	-	02/63	52A	08/76	CW
D5181	25031	11/73	-	-	BR Darlington	-	03/63	52A	12/77	ED
D5182	25032	03/74	-	-	BR Darlington	-	03/63	52A	03/86	CD
D5183	25033	02/74	-	-	BR Darlington	-	04/63	18A	04/83	CD
D5184	25034	01/74	-	-	BR Darlington	-	05/63	18A	12/86	CD
D5185	25035	02/74	-	-	BR Darlington	-	05/63	18A	03/87	CD
D5186	25036	03/74	-	-	BR Derby	-	03/63	18A	12/82	BS
D5187	25037	03/74	-	-	BR Derby	-	03/63	18A	01/87	CD
D5188	25038	02/74	-	-	BR Derby	-	03/63	18A	05/81	KD
D5189	25039	02/74	-	-	BR Derby	-	03/63	18A	05/81	LO
D5190	25040	04/74	-	-	BR Derby	-	04/63	18A	11/80	LO
D5191	25041	02/74	-	-	BR Derby	-	05/63	18A	05/81	KD

BR 1957 No.	Disposal Code	Disposal Detail	Date Cut Up	Notes
D5121	C	BREL Doncaster	05/78	Stored: [S] 10/76, R/I: 10/76
D5122	C	BREL Glasgow	03/71	
D5123	C	BREL Doncaster	07/77	Stored: [U] 06/76
D5124	C	BREL Swindon	05/77	Stored: [U] 07/76
D5125	C	BREL Doncaster	07/77	Stored: [U] 11/75
D5126	C	BREL Doncaster	07/77	Stored: [U] 12/75
D5127	C	BREL Doncaster	03/77	Stored: [U] 12/75
D5128	C	BREL Doncaster	10/77	Stored: [U] 04/76
D5129	C	BREL Doncaster	04/77	Stored: [S] 10/76
D5130	C	BREL Doncaster	08/77	Stored: [S] 10/76
D5131	C	BREL Glasgow	11/71	
D5132	C	BREL Doncaster	03/76	Stored: [U] 10/75
D5133	C	BREL Doncaster	09/78	
D5134	C	BREL Swindon	09/78	Stored: [U] 10/76
D5135	C	BREL Swindon	07/77	Stored: [U] 12/75
D5136	C	BREL Swindon	01/77	
D5137	C	BREL Doncaster	10/78	Stored: [U] 03/69, R/I: 04/69, Stored: [U] 02/76, R/I: 03/76, Stored: [U] 05/76
D5138	C	J Cashmore, Great Bridge	04/70	Stored: [U] 07/69
D5139	C	J Cashmore, Great Bridge	04/70	Stored: [U] 07/69
D5140	C	BREL Swindon	09/76	Stored: [U] 01/76
D5141	C	BREL Swindon	06/78	Stored: [U] 02/76
D5142	D	To Departmental Stock - 968009	-	Stored: [U] 02/76
D5143	C	BREL Swindon	05/76	Stored: [U] 01/76
D5144	C	BREL Swindon	09/76	Stored: [U] 01/76
D5145	C	BREL Swindon	06/76	Stored: [U] 01/76
D5146	C	BREL Swindon	10/76	Stored: [S] 01/76
D5147	C	BREL Doncaster	11/77	Stored: [U] 04/76
D5148	C	BREL Doncaster	12/77	Stored: [S] 08/75
D5149	C	BREL Glasgow	04/73	
D5150	C	BREL Doncaster	08/77	Stored: [U] 10/76

BR 1957 No.	Disposal Code	Disposal Detail	Date Cut Up	Notes
D5151	C	BREL Swindon	12/80	Stored: [U] 10/77
D5152	C	BREL Swindon	04/81	Stored: [U] 06/77, R/I: 06/77
D5153	C	BREL Glasgow	01/79	
D5154	C	BREL Glasgow	09/77	
D5155	C	BREL Swindon	08/81	
D5156	C	BREL Swindon	05/83	Stored: [U] 09/80
D5157	C	BREL Swindon	07/82	Stored: [U] 09/80
D5158	C	BREL Glasgow	09/80	
D5159	C	BREL Glasgow	10/80	Stored: [S] 02/76, R/I: 06/76
D5160	C	BREL Swindon	05/81	
D5161	C	BREL Swindon	03/81	
D5162	C	BREL Glasgow	09/77	Stored: [S] 02/76, R/I: 05/76
D5163	C	BREL Swindon	01/81	Stored: [S] 02/76, R/I: 06/76
D5164	C	BREL Glasgow	09/77	Stored: [S] 02/76, R/I: 06/76, Stored: [U] 06/77
D5165	C	BREL Doncaster	01/77	
D5166	C	BREL Swindon	12/76	Stored: [S] 01/76
D5167	C	BREL Swindon	07/76	Stored: [S] 01/76
D5168	C	BREL Glasgow	09/76	Stored: [U] 11/76
D5169	C	BREL Swindon	01/81	
D5170	C	BREL Swindon	08/76	Stored: [S] 01/76
D5171	C	BREL Swindon	12/80	
D5172	C	BREL Glasgow	01/79	
D5173	C	BREL Swindon	03/83	
D5174	C	BREL Glasgow	01/77	Stored: [S] 01/76
D5175	C	BREL Glasgow	06/78	
D5176	C	BREL Swindon	03/81	
D5177	C	V Berry, Leicester	06/87	
D5178	C	V Berry, Leicester	06/87	Stored: [U] 10/80
D5179	C	BREL Glasgow	01/78	
D5180	C	BREL Derby	05/80	
D5181	C	BREL Glasgow	05/78	Stored: [U] 10/77
D5182	C	V Berry, Leicester	12/88	Stored: [U] 10/77
D5183	C	V Berry, Leicester	03/86	
D5184	C	V Berry, Leicester	06/87	
D5185	P	Northampton Steam Railway	-	Stored: [U] 06/79, R/I: 06/80, Withdrawn: 08/86, R/I: 08/86
D5186	C	BREL Swindon	06/85	
D5187	C	V Berry, Leicester	06/87	Stored: [U] 05/86, R/I: 05/86
D5188	C	BREL Derby	11/82	
D5189	C	BREL Swindon	12/81	
D5190	C	BREL Swindon	02/82	
D5191	C	BREL Swindon	05/83	

BR 1957 No.	TOPS No.	Date Re No.	TOPS Re No.	Date Re No.	Built By	Works No.	Date Introduced	Depot of First Allocation	Date Withdrawn	Depot of Final Allocation
D5192	25042	02/74	-	-	BR Derby	-	05/63	18A	05/86	CD
D5193	25043	04/74	-	-	BR Derby	-	05/63	18A	02/81	CD
D5194	25044	03/74	-	-	BR Derby	-	05/63	18A	07/85	KD
D5195	25045	02/74	-	-	BR Derby	-	05/63	18A	10/75	SP
D5196	25046	02/74	-	-	BR Derby	-	05/63	18A	02/81	HA
D5197	25047	03/74	-	-	BR Derby	-	05/63	18A	09/80	SP
D5198	25048	02/74	-	-	BR Derby	-	05/63	18A	02/86	CD
D5199	25049	04/74	-	-	BR Derby	-	05/63	18A	01/84	CD
D5200	25050	03/74	-	-	BR Derby	-	06/63	18A	04/83	CW
D5201	25051	12/74	-	-	BR Derby	-	05/63	18A	09/85	CD
D5202	25052	02/74	-	-	BR Derby	-	06/63	18A	10/80	LA
D5203	25053	02/74	-	-	BR Derby	-	06/63	18A	12/80	BS
D5204	25054	03/74	-	-	BR Derby	-	06/63	18A	05/85	CD
D5205	25055	03/74	-	-	BR Derby	-	06/63	18A	11/80	KD
D5206	25056	03/74	-	-	BR Derby	-	06/63	18A	08/82	KD
D5207	25057	04/74	-	-	BR Derby	-	06/63	18A	03/87	CD
D5208	25058	02/74	-	-	BR Derby	-	06/63	18A	02/87	CD
D5209	25059	02/74	-	-	BR Derby	-	06/63	18A	03/87	CD
D5210	25060	03/74	-	-	BR Derby	-	06/63	18A	11/85	CD
D5211	25061	05/74	-	-	BR Derby	-	06/63	18A	11/80	CD
D5212	25062	02/74	-	-	BR Derby	-	06/63	18A	12/82	KD
D5213	25063	02/74	-	-	BR Derby	-	07/63	18A	11/80	CD
D5214	25064	03/74	-	-	BR Derby	-	07/63	18A	11/85	CD
D5215	25065	03/74	-	-	BR Derby	-	07/63	18A	02/81	HA
D5216	25066	03/74	-	-	BR Derby	-	07/63	18A	06/81	CW
D5217	25067	03/74	-	-	BR Derby	-	08/63	18A	12/82	CW
D5218	25068	03/74	-	-	BR Derby	-	08/63	14A	07/80	ED
D5219	25069	02/74	-	-	BR Derby	-	08/63	18A	12/83	CW
D5220	25070	03/74	-	-	BR Derby	-	09/63	14B	11/80	SP
D5221	25071	04/74	-	-	BR Derby	-	08/63	14A	06/81	CW
D5222	25072	03/74	-	-	BR Derby	-	10/63	14B	11/85	CD
D5223	25073	03/74	-	-	BR Darlington	-	07/63	18A	09/81	TO
D5224	25074	03/74	-	-	BR Darlington	-	07/63	18A	09/80	TO
D5225	25075	08/74	-	-	BR Darlington	-	07/63	18A	03/83	CW
D5226	25076	03/74	-	-	BR Darlington	-	08/63	18A	11/84	KD
D5227	25077	04/74	-	-	BR Darlington	-	09/63	18A	05/78	ED
D5228	25078	02/74	-	-	BR Darlington	-	09/63	16A	09/85	CD
D5229	25079	02/74	-	-	BR Darlington	-	10/63	14B	08/83	LO
D5230	25080	01/74	-	-	BR Darlington	-	10/63	14B	09/85	CD
D5231	25081	02/74	-	-	BR Darlington	-	12/63	16A	02/82	CW
D5232	25082	04/74	-	-	BR Darlington	-	11/63	16A	05/81	HA
D5233	25083	02/74	-	-	BR Derby	-	12/63	16A	07/84	CD
D5234	25084	02/74	-	-	BR Derby	-	12/63	16A	12/83	KD
D5235	25085	04/74	-	-	BR Derby	-	12/63	16A	03/82	KD
D5236	25086	02/74	-	-	BR Derby	-	12/63	16A	10/83	CD
D5237	25087	02/74	-	-	BR Derby	-	12/63	16A	09/80	HA
D5238	25088	04/74	-	-	BR Derby	-	12/63	16A	08/81	LO
D5239	25089	04/74	-	-	BR Derby	-	12/63	16A	02/86	CD
D5240	25090	02/74	-	-	BR Derby	-	01/64	16A	05/83	BS
D5241	25091	04/74	-	-	BR Derby	-	01/64	16A	10/78	HA
D5242	25092	02/74	-	-	BR Derby	-	01/64	16A	05/80	CW
D5243	25093	04/74	-	-	BR Derby	-	01/64	16A	11/82	BS
D5244	25094	04/74	-	-	BR Derby	-	01/64	16A	02/81	TO
D5245	25095	04/74	-	-	BR Derby	-	01/64	16A	07/86	CD
D5246	25096	04/74	-	-	BR Derby	-	01/64	16A	12/77	ED
D5247	25097	05/74	-	-	BR Derby	-	02/64	16A	11/83	TO
D5248	25098	02/74	-	-	BR Derby	-	02/64	16A	10/78	ED
D5249	25099	03/74	-	-	BR Derby	-	02/64	16A	12/80	CW
D5250	25100	04/74	-	-	BR Derby	-	02/64	16A	02/81	LO
D5251	25101	12/73	-	-	BR Derby	-	02/64	16A	01/83	BS
D5252	25102	04/74	-	-	BR Derby	-	02/64	16A	05/80	LO
D5253	25103	03/74	-	-	BR Derby	-	02/64	16A	09/80	SP
D5254	25104	04/74	-	-	BR Derby	-	02/64	16A	09/82	LO
D5255	25105	02/74	-	-	BR Derby	-	03/64	16A	04/82	LO
D5256	25105	05/74	-	-	BR Derby	-	03/64	16A	11/83	KD
D5257	25107	03/74	-	-	BR Derby	-	03/64	16A	05/81	TO
D5288	25108	03/74	-	-	BR Derby	-	03/64	16A	07/80	ED
D5259	25109	02/74	-	-	BR Derby	-	03/64	16A	03/87	CD
D5260	25110	02/74	-	-	BR Derby	-	04/64	16A	11/80	SP
D5261	25111	03/74	-	-	BR Derby	-	04/64	16A	03/80	CW
D5262	25112	04/74	-	-	BR Derby	-	04/64	16A	11/80	CW
D5263	25113	04/74	-	-	BR Derby	-	04/64	16A	06/83	CD
D5264	25114	02/74	-	-	BR Derby	-	04/64	16A	02/81	CW
D5265	25115	04/74	-	-	BR Derby	-	04/64	16A	10/83	CW
D5266	25116	02/74	-	-	BR Derby	-	05/64	16A	11/80	CW
D5267	25117	02/74	-	-	BR Derby	-	05/64	16A	01/84	CW
D5268	25118	03/74	-	-	BR Derby	-	05/64	16A	01/81	TO
D5269	25119	02/74	-	-	BR Derby	-	05/64	16A	06/85	CD
D5270	25120	03/74	-	-	BR Derby	-	05/64	16A	11/83	CD
D5271	25121	02/74	-	-	BR Derby	-	05/64	16A	11/80	TO
D5272	25122	01/74	-	-	BR Derby	-	05/64	16A	11/80	TO
D5273	25123	05/74	-	-	BR Derby	-	05/64	16A	05/83	CD

BR 1957 No.	Disposal Code	Disposal Detail	Date Cut Up	Notes
D5192	C	V Berry, Leicester	07/87	
D5193	C	BREL Derby	08/81	
D5194	C	BREL Doncaster	07/86	
D5195	C	BREL Derby	08/79	Stored: [U] 08/75
D5196	C	V Berry, Leicester	07/87	
D5197	C	BREL Swindon	01/81	
D5198	C	V Berry, Leicester	01/87	
D5199	C	BREL Swindon	06/85	
D5200	C	BREL Swindon	03/85	
D5201	C	V Berry, Leicester	03/87	
D5202	C	BREL Swindon	11/80	
D5203	C	BREL Swindon	11/81	Stored: [U] 11/80
D5204	C	BREL Doncaster	09/86	
D5205	C	BREL Swindon	12/81	
D5206	C	BREL Swindon	03/85	
D5207	P	North Norfolk Railway	-	
D5208	C	V Berry, Leicester	07/87	
D5209	P	Keighley & Worth Valley Railway	-	
D5210	C	V Berry, Leicester	10/86	
D5211	C	BREL Swindon	03/83	
D5212	C	BREL Swindon	05/85	
D5213	C	BREL Swindon	02/83	
D5214	C	V Berry. Leicester	02/87	
D5215	C	BREL Swindon	02/82	
D5216	C	BREL Derby	01/83	
D5217	P	Mid Hants Railway	-	
D5218	C	BREL Glasgow	08/81	
D5219	C	BREL Swindon	06/86	
D5220	C	BREL Swindon	05/83	
D5221	C	BREL Swindon	01/83	
D5222	P	Swindon & Cricklade Railway	-	
D5223	C	BREL Swindon	08/82	
D5224	C	BREL Swindon	05/82	
D5225	C	V Berry, Leicester	06/87	
D5226	C	BREL Swindon	08/86	
D5227	C	BREL Glasgow	09/78	
D5228	C	V Berry, Leicester	10/86	
D5229	C	BREL Swindon	12/83	
D5230	A	BR Crewe		
D5231	C	BREL Swindon	06/82	
D5232	C	BREL Swindon	01/83	
D5233	P	The Railway Age Crewe	-	
D5234	C	BREL Swindon	06/86	
D5235	C	BREL Derby	06/83	
D5236	C	BREL Swindon	03/86	
D5237	C	BREL Swindon	04/81	
D5238	C	BR Derby, by V Berry	04/86	
D5239	C	V Berry, Leicester	07/87	
D5240	C	BREL Swindon	03/85	
D5241	C	BREL Glasgow	11/79	
D5242	C	BREL Derby	06/82	
D5243	C	V Berry, Leicester	06/87	
D5244	C	BREL Derby	04/82	
D5245	C	V Berry, Leicester	01/89	
D5246	C	BREL Glasgow	05/78	Stored: [U] 10/77
D5247	C	BREL Swindon	01/85	
D5248	C	BREL Glasgow	07/79	
D5249	C	BREL Swindon	02/81	Stored: [U] 11/80
D5250	C	BREL Swindon	02/83	
D5251	C	BREL Swindon	12/85	
D5252	C	BREL Swindon	10/80	
D5253	C	BREL Swindon	03/83	
D5254	C	BREL Swindon	12/86	
D5255	C	BREL Swindon	11/85	
D5256	C	BREL Swindon	02/84	
D5257	C	BREL Swindon	09/81	
D5258	C	BREL Glasgow	02/81	
D5259	C	V Berry, Leicester	07/87	
D5260	C	BREL Swindon	03/82	
D5261	C	BREL Swindon	10/80	
D5262	C	BREL Swindon	03/82	
D5263	C	BREL Swindon	10/83	
D5264	C	BREL Swindon	07/81	
D5265	C	BREL Swindon	12/85	
D5266	C	BREL Swindon	07/82	
D5267	C	BREL Swindon	05/84	
D5268	C	BREL Swindon	08/83	
D5269	C	BREL Doncaster	07/86	
D5270	C	BREL Swindon	05/84	
D5271	C	BREL Swindon	08/82	
D5272	C	BREL Swindon	05/83	
D5273	C	V Berry, Leicester	07/87	

BR 1957 No.	TOPS No.	Date Re No.	TOPS Re No.	Date Re No.	Built By	Works No.	Date Introduced	Depot of First Allocation	Date Withdrawn	Depot of Final Allocation
D5274	25124	10/73	-	-	BR Derby	-	05/64	16A	11/83	CW
D5275	25125	04/74	-	-	BR Derby	-	06/64	16A	12/81	TO
D5276	25126	04/74	-	-	BR Derby	-	06/64	16A	11/82	CD
D5277	25127	03/74	-	-	BR Derby	-	06/64	16A	11/80	TO
D5278	-	-	-	-	BR Derby	-	06/64	16A	05/71	DO9
D5279	25129	04/74	-	-	BR Derby	-	06/64	16A	02/82	TO
D5280	25130	03/74	-	-	BR Derby	-	06/64	16A	12/82	CW
D5281	25131	04/74	-	-	BR Derby	-	06/64	16A	12/82	TO
D5282	25132	02/74	-	-	BR Derby	-	07/64	16A	12/82	LO
D5283	25133	10/73	-	-	BR Derby	-	07/64	16A	08/83	KD
D5284	25134	05/74	-	-	BR Derby	-	07/64	16A	12/82	BS
D5285	25135	03/74	-	-	BR Derby	-	07/64	16A	01/83	BS
D5286	25136	03/74	-	-	BR Derby	-	08/64	16A	03/83	CW
D5287	25137	04/74	-	-	BR Derby	-	08/64	16A	10/80	TO
D5288	25138	02/74	-	-	BR Derby	-	08/64	16A	12/83	CD
D5289	25139	02/74	-	-	BR Derby	-	08/64	16A	11/82	TO
D5290	25140	03/74	-	-	BR Derby	-	08/64	16A	12/83	KD
D5291	25141	03/74	-	-	BR Derby	-	08/64	16A	09/82	LO
D5292	25142	03/74	-	-	BR Derby	-	08/64	16A	10/81	KD
D5293	25143	02/74	-	-	BR Derby	-	09/64	16A	11/82	TO
D5294	25144	05/74	-	-	BR Derby	-	09/64	16A	04/83	BS
D5295	25145	02/74	-	-	BR Derby	-	09/64	16A	01/86	KD
D5296	25146	11/73	-	-	BR Derby	-	09/64	16A	08/83	KD
D5297	25147	03/74	-	-	BR Derby	-	09/64	16A	03/80	KD
D5298	25148	02/74	-	-	BR Derby	-	10/64	16A	09/81	CW
D5299	25149	09/73	-	-	BR Derby	-	10/65	D16	01/82	KD

Class continued from D7500

BIRMINGHAM RCW TYPE 2 Bo-Bo CLASS 26

BR 1957 No.	TOPS No.	Date Re No.	Name	Name Date	Built By	Works No.	Date Introduced	Depot of First Allocation	Date Withdrawn	Depot of Final Allocation
D5300	26007	04/74			BRCW	DEL45	07/58	34B		
D5301	26001	04/74	Eastfield	07/91	BRCW	DEL46	09/58	34B		
D5302	26002	04/74			BRCW	DEL47	10/58	34B	10/92	IS
D5303	26003	05/74			BRCW	DEL48	10/58	34B		
D5304	26004	04/74			BRCW	DEL49	10/58	34B	11/92	IS
D5305	26005	04/74			BRCW	DEL50	11/58	34B		
D5306	26006	12/73			BRCW	DEL51	11/58	34B		
D5307	26020	04/74			BRCW	DEL52	12/58	34B	01/77	IS
D5308	26008	04/74			BRCW	DEL53	12/58	34B		
D5309	26009	04/74			BRCW	DEL54	12/58	34B	01/77	IS
D5310	26010	04/74			BRCW	DEL55	01/59	34B	12/92	IS
D5311	26011	04/74			BRCW	DEL56	01/59	34B	11/92	IS
D5312	26012	12/73			BRCW	DEL57	01/59	34B	01/82	HA
D5313	26013	04/74			BRCW	DEL58	01/59	34B	03/85	HA
D5314	26014	04/74			BRCW	DEL59	01/59	34B	10/92	IS
D5315	26015	04/74			BRCW	DEL60	02/59	34B	06/91	ED
D5316	26016	09/74			BRCW	DEL61	02/59	34B	10/75	HA
D5317	26017	11/73			BRCW	DEL62	02/59	34B	08/77	IS
D5318	26018	01/74			BRCW	DEL63	03/59	34B	01/82	HA
D5319	26019	04/74			BRCW	DEL64	03/59	34B	03/85	HA
D5320	26028	04/74			BRCW	DEL65	04/59	64H	10/91	ED
D5321	26021	09/74			BRCW	DEL66	04/59	64H	10/91	ED
D5322	26022	03/74			BRCW	DEL67	04/59	64B	02/81	HA
D5323	26023	02/74			BRCW	DEL68	05/59	64B	10/90	ED
D5324	26024	04/74			BRCW	DEL69	05/59	64B	10/92	IS
D5325	26025	04/74			BRCW	DEL70	05/59	64B		
D5326	26026	02/74			BRCW	DEL71	05/59	64B	11/92	IS
D5327	26027	07/74			BRCW	DEL72	06/59	64B	08/91	ED
D5328	-	-			BRCW	DEL73	05/59	64B	07/72	HA
D5329	26029	07/74			BRCW	DEL74	06/59	64B	10/88	ED
D5330	26030	07/74			BRCW	DEL75	06/59	64B	03/85	IS
D5331	26031	04/74			BRCW	DEL76	06/59	64B	09/89	ED
D5332	26032	04/74			BRCW	DEL77	07/59	64B		
D5333	26033	10/73			BRCW	DEL78	07/59	64B	03/85	HA
D5334	26034	07/74			BRCW	DEL79	07/59	64B	09/89	ED
D5335	26035	11/73			BRCW	DEL80	07/59	64B	12/92	IS
D5336	26036	10/73			BRCW	DEL81	08/59	64B		
D5337	26037	09/74			BRCW	DEL82	08/59	64B		
D5338	26038	07/74			BRCW	DEL83	08/59	64B	10/92	IS
D5339	26039	09/74			BRCW	DEL84	09/59	64B	10/90	ED
D5340	26040	10/73			BRCW	DEL85	09/59	64B	12/92	IS
D5341	26041	06/74			BRCW	DEL86	09/59	64B	11/92	IS
D5342	26042	04/74			BRCW	DEL87	09/59	64B	10/92	IS
D5343	26043	01/74			BRCW	DEL88	10/59	64B	01/93	IS
D5344	26044	05/74			BRCW	DEL89	10/59	64B	01/84	IS
D5345	26045	04/74			BRCW	DEL90	10/59	64B	07/83	IS
D5346	26046	01/74			BRCW	DEL91	10/59	64B	04/91	ED

BR 1957 No.	Disposal Code	Disposal Detail	Date Cut Up	Notes
D5274	C	BREL Swindon	12/86	
D5275	C	BREL Swindon	03/83	
D5276	C	V Berry, Leicester	06/87	
D5277	C	BREL Swindon	04/83	
D5278	C	BR Peak Forest, by G Cohen	10/71	
D5279	C	BREL Swindon	03/85	
D5280	C	BREL Swindon	01/87	
D5281	D	To Departmental Stock - 97202	-	
D5282	C	BREL Swindon	11/84	
D5283	C	V Berry, Leicester	07/87	
D5284	C	V Berry, Leicester	06/87	
D5285	C	BREL Swindon	07/86	
D5286	C	BREL Swindon	08/86	
D5287	C	BREL Swindon	04/83	
D5288	C	BREL Swindon	11/86	
D5289	C	BREL Swindon	10/86	
D5290	C	BREL Swindon	08/86	
D5291	C	BREL Swindon	01/87	
D5292	C	BREL Swindon	11/82	
D5293	C	BREL Swindon	10/84	
D5294	C	V Berry, Leicester	06/87	
D5295	C	V Berry, Leicester	01/87	
D5296	C	BREL Swindon	04/85	
D5297	C	BREL Swindon	09/80	
D5298	C	BREL Swindon	05/82	
D5299	C	BREL Swindon	05/83	

BR 1957 No.	Disposal Code	Disposal Detail	Date Cut Up	Notes
D5300				
D5301				
D5302	A	BR Inverness		Stored: [U] 10/92
D5303				Stored: [U] 10/87, R/I: 11/87
D5304	A	BR Inverness		
D5305				
D5306				
D5307	C	BREL Glasgow	03/78	Stored: [U] 12/76
D5308				
D5309	C	BREL Glasgow	04/78	Stored: [U] 01/77
D5310	A	BR Inverness		Stored: [U] 10/87, R/I: 11/87
D5311	A	BR Motherwell		
D5312	C	BREL Glasgow	08/82	Stored: [U] 02/71, R/I: 09/72
D5313	C	V Berry, Leicester	06/87	Stored: [U] 04/80, R/I: 10/80, Stored: [U] 09/82
D5314	A	BR Perth		Stored: [U] 09/82, R/I: 06/84, Stored: [U] 10/92
D5315	A	BR Inverness		Stored: [U] 06/81, R/I: 05/82
D5316	C	BREL Glasgow	09/78	
D5317	C	BREL Glasgow	02/78	Stored: [U] 08/77
D5318	C	BREL Glasgow	05/82	
D5319	C	BR Thornton, by V Berry	05/87	Stored: [U] 02/83
D5320	C	M C Processors, Glasgow	11/92	Stored: [U] 06/82, R/I: 06/83
D5321	A	BR Inverness		
D5322	C	BREL Glasgow	03/81	
D5323	C	M C Processors, Glasgow	01/91	
D5324	A	BR Motherwell		Stored: [U] 10/92
D5325				Stored: [U] 09/87, R/I: 11/87
D5326	A	BR Perth		
D5327	A	BR Perth		Stored: [U] 11/82, R/I: 06/83, Stored: [U] 04/84, R/I: 10/84
D5328	C	BREL Glasgow	12/72	
D5329	C	M C Processors, Glasgow	11/89	
D5330	C	BR Thornton, by V Berry	03/87	Stored: [U] 10/83
D5331	C	M C Processors, Glasgow	11/90	
D5332				
D5333	C	BR Thornton, by V Berry	03/87	Stored: [U] 10/82
D5334	C	M C Processors, Glasgow	09/90	
D5335	A	BR Inverness		Stored: [U] 08/83
D5336				
D5337				
D5338	A	BR Inverness		
D5339	A	M C Processors, Glasgow		Stored: [U] 04/84, R/I: 10/84, Stored: [U] 10/87, R/I: 11/87
D5340	A	BR Perth		Stored: [S] 08/92, R/I: 09/92, Stored: [U] 10/92
D5341	A	BR Inverness		
D5342	A	BR Inverness		Stored: [U] 10/92
D5343	A	BR Perth		
D5344	C	BR Thornton, by V Berry	01/87	
D5345	C	V Berry, Leicester	07/87	Stored: [U] 09/82
D5346	A	M C Processors, Glasgow		

BIRMINGHAM RCW TYPE 2 Bo-Bo CLASS 27

BR 1957 No.	First TOPS No.	Date Re No.	Second TOPS No.	Date Re No.	Third TOPS No.	Date Re No.	Built By	Works No.	Date Introduced	Depot of First Allocation	Date Withdrawn	Depot of Final Allocation
D5347	27001	04/74	-	-	-	-	BRCW	DEL190	06/61	62A	07/87	ED
D5348	27002	04/74	-	-	-	-	BRCW	DEL191	07/61	62A	01/86	HA
D5349	27003	11/73	-	-	-	-	BRCW	DEL192	07/61	65A	01/87	ED
D5350	27004	04/74	-	-	-	-	BRCW	DEL193	08/61	65A	05/86	HA
D5351	27005	10/74	-	-	-	-	BRCW	DEL194	08/61	65A	07/87	ED
D5352	27006	09/74	-	-	-	-	BRCW	DEL195	09/61	65A	01/76	ED
D5353	27007	04/74	-	-	-	-	BRCW	DEL196	09/61	65A	01/85	IS
D5354	27008	12/73	-	-	-	-	BRCW	DEL197	09/61	65A	08/87	ED
D5355	27009	04/74	-	-	-	-	BRCW	DEL198	09/61	65A	07/80	ED
D5356	27010	11/73	-	-	-	-	BRCW	DEL199	10/61	65A	04/86	HA
D5357	27011	04/74	-	-	-	-	BRCW	DEL200	10/61	65A	03/81	ED
D5358	27012	05/74	-	-	-	-	BRCW	DEL201	10/61	65A	05/86	HA
D5359	27013	04/74	-	-	-	-	BRCW	DEL202	11/61	65A	07/76	ED
D5360	27014	02/74	-	-	-	-	BRCW	DEL203	11/61	65A	09/86	HA
D5361	27015	05/74	-	-	-	-	BRCW	DEL204	11/61	65A	01/77	ED
D5362	27016	04/74	-	-	-	-	BRCW	DEL205	11/61	65A	04/84	ED
D5363	27017	04/74	-	-	-	-	BRCW	DEL206	12/61	65A	05/86	HA
D5364	27018	04/74	-	-	-	-	BRCW	DEL207	12/61	65A	05/86	HA
D5365	27019	04/74	-	-	-	-	BRCW	DEL208	12/61	65A	04/84	ED
D5366	27020	05/74	-	-	-	-	BRCW	DEL209	01/62	65A	04/86	ED
D5367	27021	04/74	-	-	-	-	BRCW	DEL210	01/62	65A	05/85	IS
D5368	27022	11/73	-	-	-	-	BRCW	DEL211	01/62	65A	01/85	ED
D5369	27023	02/74	-	-	-	-	BRCW	DEL212	01/62	51L	05/86	HA
D5370	27024	04/74	-	-	-	-	BRCW	DEL213	01/62	51L	06/87	ED
D5371	27025	09/74	-	-	-	-	BRCW	DEL214	01/62	51L	07/87	ED
D5372	27026	09/74	-	-	-	-	BRCW	DEL215	01/62	51L	07/87	ED
D5373	27027	04/74	-	-	-	-	BRCW	DEL216	01/62	51L	06/83	ED
D5374	27101	04/74	27045	09/84	-	-	BRCW	DEL217	02/62	51L	05/86	ED
D5375	27028	12/73	-	-	-	-	BRCW	DEL218	02/62	51L	08/84	ED
D5376	27029	04/74	-	-	-	-	BRCW	DEL219	02/62	51L	01/86	ED
D5377	27030	09/74	-	-	-	-	BRCW	DEL220	02/62	51L	04/86	ED
D5378	27031	04/74	-	-	-	-	BRCW	DEL221	03/62	51L	05/78	ED
D5379	27032	04/74	-	-	-	-	BRCW	DEL222	03/62	14A	05/85	IS
D5380	27102	10/73	27046	04/83	-	-	BRCW	DEL223	04/62	14A	07/87	ED
D5381	27033	09/74	-	-	-	-	BRCW	DEL224	04/62	14A	01/86	HA
D5382	27034	04/74	-	-	-	-	BRCW	DEL225	04/62	14A	07/84	ED
D5383	-	-	-	-	-	-	BRCW	DEL226	04/62	14A	01/66	66B
D5384	27035	04/74	-	-	-	-	BRCW	DEL227	05/62	14A	09/76	ED
D5385	27036	11/73	-	-	-	-	BRCW	DEL228	05/62	14A	04/86	HA
D5386	27103	04/74	27212	03/75	27066	01/82	BRCW	DEL229	05/62	14A	07/87	ED
D5387	27104	04/74	27048	09/83	-	-	BRCW	DEL230	05/62	14A	05/86	IS
D5388	27105	04/74	27049	08/83	-	-	BRCW	DEL231	06/62	14A	04/87	ED
D5389	27037	10/73	-	-	-	-	BRCW	DEL232	06/62	14A	03/86	ED
D5390	27038	04/74	-	-	-	-	BRCW	DEL233	06/62	14A	02/87	ED
D5391	27119	06/73	27201	05/74	-	-	BRCW	DEL234	06/62	14A	01/79	HA
D5392	27120	08/74	27202	08/74	-	-	BRCW	DEL235	06/62	14A	08/80	ED
D5393	27121	07/74	27203	11/74	27057*	-	BRCW	DEL236	06/62	14A	05/83	ED
D5394	27106	04/74	27050	11/82	-	-	BRCW	DEL237	06/62	14A	07/87	ED
D5395	27107	04/74	27051	09/82	-	-	BRCW	DEL238	06/62	14A	07/87	ED
D5396	27108	04/74	27052	12/84	-	-	BRCW	DEL239	06/62	14A	07/87	ED
D5397	27109	05/74	27053	06/83	-	-	BRCW	DEL240	08/62	14A	05/87	ED
D5398	27039	12/73	-	-	-	-	BRCW	DEL241	07/62	14A	10/75	ED
D5399	27110	04/74	27054	03/83	-	-	BRCW	DEL242	07/62	14A	07/87	ED
D5400	27111	04/74	27055	05/83	-	-	BRCW	DEL243	07/62	14A	07/87	ED
D5401	27112	02/74	27056	08/82	-	-	BRCW	DEL244	07/62	14A	05/87	ED
D5402	27040	04/74	-	-	-	-	BRCW	DEL245	07/62	14A	01/86	HA
D5403	27122	04/74	27204	04/75	27058	03/86	BRCW	DEL246	07/62	14A	05/86	ED
D5404	27113	02/74	27207	10/75	27061*	-	BRCW	DEL247	07/62	14A	05/86	ED
D5405	27041	04/74	-	-	-	-	BRCW	DEL248	07/62	14A	05/86	ED
D5406	27042	01/74	-	-	-	-	BRCW	DEL249	08/62	14A	07/87	ED
D5407	27114	01/74	27208	12/75	27062*	-	BRCW	DEL250	08/62	14A	02/86	ED
D5408	27115	01/74	27209	02/75	27063	09/82	BRCW	DEL251	08/62	14A	07/87	ED
D5409	27116	03/73	27210	06/74	27064	07/84	BRCW	DEL252	08/62	14A	05/86	IS
D5410	27123	11/73	27205	12/74	27059	11/82	BRCW	DEL253	09/62	14A	07/87	ED
D5411	27117	03/74	27211	03/76	27065	02/85	BRCW	DEL254	09/62	14A	05/86	IS
D5412	27124	06/74	27206	07/74	27060*	-	BRCW	DEL255	09/62	14A	03/86	ED
D5413	27118	11/73	27103	04/75	27047	07/85	BRCW	DEL256	09/62	14A	04/86	IS
D5414	27043	04/74	-	-	-	-	BRCW	DEL257	09/62	14A	04/80	ED
D5415	27044	04/74	-	-	-	-	BRCW	DEL258	10/62	14A	07/80	ED

BR 1957 No.	Disposal Code	Disposal Detail	Date Cut Up	Notes
D5347	P	Class 27 Loco Group at Bo'ness & Kinneil Railway	-	Stored: [U] 09/82, R/I: 06/83
D5348	C	V Berry, Leicester	12/87	Stored: [U] 02/84, R/I: 05/84
D5349	C	M C Processors, Glasgow	12/87	
D5350	C	V Berry, Leicester	07/87	
D5351	P	Bo'ness and Kinneil Railway	-	
D5352	C	BREL Glasgow	01/77	
D5353	P	Mid Hants Railway	-	
D5354	C	M C Processors, Glasgow	11/87	
D5355	C	BREL Glasgow	01/82	
D5356	C	V Berry, Leicester	07/87	
D5357	C	BREL Derby	08/82	
D5358	C	V Berry, Leicester	12/87	Stored: [U] 07/75, R/I: 09/75
D5359	C	BREL Glasgow	07/77	
D5360	C	V Berry, Leicester	10/87	
D5361	C	BREL Glasgow	04/77	Stored: [U] 01/77
D5362	C	V Berry, Leicester	07/87	Stored: [U] 04/83
D5363	C	V Berry, Leicester	07/87	
D5364	C	M C Processors, Glasgow	08/89	
D5365	C	BREL Swindon	04/85	Stored: [U] 07/83, R/I: 12/83, Stored: [U] 03/84
D5366	C	BR Thornton, By V Berry	02/87	
D5367	C	BR Thornton, By V Berry	04/87	
D5368	C	V Berry, Leicester	09/87	
D5369	C	V Berry, Leicester	08/87	
D5370	D	To Departmental Stock 060020	-	
D5371	C	V Berry, Leicester	08/87	
D5372	C	V Berry, Leicester	07/87	
D5373	C	BR Thornton, By V Berry	04/87	Stored: [U] 09/82
D5374	C	V Berry, Leicester	01/89	
D5375	C	BREL Swindon	08/85	Stored: [U] 03/83
D5376	C	BR Thornton, By V Berry	04/87	
D5377	C	V Berry, Leicester	10/87	Stored: [U] 10/84, R/I: 12/84
D5378	C	BREL Glasgow	09/78	
D5379	C	BR Thornton, By V Berry	03/87	
D5380	C	M C Processors, Glasgow	05/88	
D5381	C	BR Thornton, By V Berry	02/87	
D5382	C	BREL Swindon	12/85	
D5383	C	J Cashmore, Great Bridge	06/67	
D5384	C	BREL Glasgow	03/77	
D5385	C	BR Thornton, By V Berry	02/87	Stored: [U] 12/83, R/I: 06/84
D5386	P	North Norfolk Railway	-	
D5387	C	V Berry, Leicester	08/87	
D5388	C	V Berry, Leicester	03/88	
D5389	C	V Berry, Leicester	08/87	
D5390	C	V Berry, Leicester	08/87	
D5391	C	BREL Glasgow	09/79	
D5392	C	BREL Glasgow	01/82	
D5393	C	V Berry, Leicester	02/86	Stored: [U] 12/82
D5394	P	Strathspey Railway	-	
D5395	C	V Berry, Leicester	10/87	
D5396	C	V Berry, Leicester	07/87	
D5397	C	V Berry, Leicester	08/87	
D5398	C	BREL Glasgow	04/77	Stored: [U] 09/75
D5399	C	M C Processors, Glasgow	06/88	
D5400	C	V Berry, Leicester	07/87	
D5401	P	Northampton Steam Railway	-	
D5402	C	BR Thornton, By V Berry	04/87	
D5403	C	V Berry, Leicester	08/87	
D5404	D	To Departmental Stock - 968025	-	Stored: [U] 08/81, R/I: 09/83
D5405	C	V Berry, Leicester	07/87	Stored: [U] 01/83, R/I: 04/83
D5406	C	V Berry, Leicester	07/87	
D5407	C	V Berry, Leicester	01/89	
D5408	C	V Berry, Leicester	07/87	
D5409	C	V Berry, Leicester	08/87	
D5410	P	Severn Valley Railway	-	
D5411	C	V Berry, Leicester	07/87	
D5412	C	V Berry, Leicester	09/87	
D5413	C	V Berry, Leicester	07/87	
D5414	X	Used as Landfill, Patersons Tip, Mount Vernon	11/85	
D5415	C	BREL Glasgow	02/82	

BRUSH TYPE 2 A1A-A1A CLASS 31

BR 1957 No.	TOPS No.	Date Re No.	First TOPS Re No.	Date Re No.	Second TOPS Re No.	Date Re No.	Third TOPS Re No.	Date Re No.	Name	Name Date	Built By	Works No.	Date Introduced
D5500	31018	01/74									Brush	71	10/57
D5501	31001	02/74									Brush	72	11/57
D5502	31002	02/74									Brush	73	12/57
D5503	31003	03/74									Brush	74	01/58
D5504	31004	02/74									Brush	75	01/58
D5505	31005	02/74									Brush	76	02/58
D5506	31006	04/74									Brush	77	03/58
D5507	31007	02/74									Brush	78	04/58
D5508	31008	02/74									Brush	79	04/58
D5509	31009	03/74									Brush	80	05/58
D5510	31010	04/74									Brush	81	05/58
D5511	31011	04/74									Brush	82	06/58
D5512	31012	04/74									Brush	83	06/58
D5513	31013	02/74									Brush	84	07/58
D5514	31014	02/74									Brush	85	07/58
D5515	31015	03/74									Brush	86	07/58
D5516	31016	05/74									Brush	87	09/58
D5517	31017	04/74									Brush	88	09/58
D5518	31101	02/74									Brush	89	10/58
D5519	31019	03/74									Brush	90	12/58
D5520	31102	02/74								Cricklewood	01/90	Brush	119	02/59
D5521	31103	03/74										Brush	120	03/59
D5522	31104*	-	31418	11/73							Brush	121	03/59
D5523	31105	02/74								Bescot TMD	08/92	Brush	122	03/59
D5524	31106	02/74								The Blackcountryman	08/92	Brush	123	03/59
D5525	31107	02/74								John H. Carless VC	08/92	Brush	124	03/59
D5526	31108	02/74										Brush	125	04/59
D5527	31109	02/74										Brush	126	04/59
D5528	31110	02/74										Brush	127	04/59
D5529	31111	03/74										Brush	128	05/59
D5530	31112	05/74										Brush	129	05/59
D5531	31113	02/74										Brush	130	05/59
D5532	31114	02/74	31453	10/84	31553	05/90					Brush	131	06/59
D5533	31115	03/74	31466	03/85							Brush	132	06/59
D5534	31116	03/74							Rail 1981-1991	10/91	Brush	133	06/59
D5535	31117	03/74									Brush	134	06/59
D5536	31118	02/74									Brush	135	06/59
D5537	31119	02/74									Brush	136	06/59
D5538	31120	02/74									Brush	137	07/59
D5539	31121	02/74									Brush	138	07/59
D5540	31122	02/74									Brush	139	07/59
D5541	31123	01/74									Brush	140	07/59
D5542	31124	02/74									Brush	141	07/59
D5543	31125	03/74									Brush	142	07/59
D5544	31126	03/74									Brush	143	08/59
D5545	31127	02/74									Brush	144	10/59
D5546	31128	02/74									Brush	145	08/59
D5547	31129	02/74	31461	01/85							Brush	146	08/59
D5548	31130	05/74							Calder Hall Power Station	07/92	Brush	147	09/59
D5549	31131	01/74									Brush	148	09/59
D5550	31132	03/74									Brush	149	09/59
D5551	31133	02/74	31450	09/84							Brush	150	09/59
D5552	31134	03/74									Brush	151	09/59
D5553	31135	02/74									Brush	152	09/59
D5554	31136	05/74									Brush	153	10/59
D5555	31137	02/74	31444	05/84	31544	05/90			Keighley & Worth Valley Railway	08/88	Brush	154	10/59
D5556	31138	02/74									Brush	155	10/59
D5557	31139	10/73	31438	03/84							Brush	156	10/59
D5558	31140	03/74	31421	11/74							Brush	157	10/59
D5559	31141	12/74									Brush	158	10/59
D5560	31142	03/74									Brush	159	10/59
D5561	31143	03/74									Brush	160	10/59
D5562	31144	02/74									Brush	161	11/59
D5563	31145	03/74									Brush	162	11/59
D5564	31146	03/74							Brush Veteran	08/92	Brush	163	11/59
D5565	31147	02/74									Brush	164	11/59
D5566	31148	02/74	31448	08/84	31548	05/90					Brush	165	11/59
D5567	31149	02/74									Brush	166	11/59
D5568	31150	02/74									Brush	167	12/59
D5569	31151	02/74	31436	03/84							Brush	168	12/59
D5570	31152	12/73									Brush	169	12/59
D5571	31153	02/74	31432	11/83							Brush	170	12/59
D5572	31154	02/74									Brush	171	12/59
D5573	31155	02/74									Brush	172	12/59
D5574	31156	02/74									Brush	173	12/59
D5575	31157	01/74	31424	01/75	31524	05/90					Brush	174	12/59
D5576	31158	02/74									Brush	175	12/59

BR 1957 No.	Depot of First Allocation	Date Withdrawn	Depot of Final Allocation	Disposal Code	Disposal Detail	Date Cut Up	Notes
D5500	30A	07/76	SF	P	National Railway Museum, York	-	Stored: [U] 05/76
D5501	30A	07/76	SF	C	BREL Doncaster	01/77	Stored: [S] 05/76
D5502	30A	01/80	SF	D	To Departmental Stock - 968014	-	
D5503	30A	02/80	SF	C	BREL Doncaster	04/80	
D5504	30A	10/80	SF	C	BREL Swindon	06/81	
D5505	30A	02/80	SF	C	BREL Doncaster	03/80	
D5506	30A	01/80	SF	C	BREL Doncaster	12/80	
D5507	30A	11/76	SF	C	BREL Doncaster	06/78	Stored: [S] 05/76
D5508	30A	10/80	SF	D	To Departmental Stock - 968016	-	
D5509	30A	07/76	SF	C	BREL Doncaster	11/77	Stored: [S] 05/76
D5510	30A	07/76	SF	C	BREL Doncaster	12/76	Stored: [S] 05/76
D5511	30A	07/76	SF	C	BREL Doncaster	01/77	Stored: [S] 05/76
D5512	30A	11/76	SF	C	BREL Doncaster	10/76	Stored: [S] 05/76
D5513	30A	03/79	SF	D	To Departmental Stock - 968013	-	
D5514	30A	11/76	SF	D	To Departmental Stock - 968015	-	Stored: [S] 05/76
D5515	30A	05/80	SF	C	BREL Doncaster	10/80	
D5516	30A	07/76	SF	C	BREL Doncaster	12/76	Stored: [S] 05/76
D5517	30A	05/80	SF	C	BREL Swindon	01/83	
D5518	30A	01/93	BS	A	BR Tinsley		Withdrawn: 05/81, R/I: [S] 09/82, R/I: 11/82
D5519	30A	10/80	SF	C	BREL Swindon	08/81	
D5520	30A						
D5521	32B	10/80	SF	C	BREL Swindon	02/83	
D5522	30A						
D5523	30A						
D5524	30A						
D5525	31B						
D5526	30A	09/91	IM	A	BR Scunthorpe		Stored: [U] 06/91
D5527	30A	03/88	SF	C	Booth-Roe, Rotherham	08/89	
D5528	32A						
D5529	31B	06/83	IM	C	BREL Swindon	06/86	
D5530	31B						
D5531	30A						
D5532	32A						
D5533	32A						
D5534	32A						
D5535	32A	03/87	CW	C	BRML Doncaster, by Booth-Roe	09/88	
D5536	30A	02/89	TE	C	M C Processors, Glasgow	01/90	
D5537	30A						
D5538	32B	09/91	CD	A	Booth-Roe, Rotherham		Stored: [U] 08/90
D5539	32B	09/88	BS	C	M C Processors, Glasgow	07/90	
D5540	32B	01/87	KD	C	BR Stratford	08/87	Withdrawn: 05/81, R/I: 01/82
D5541	32B	03/92	BS	A	BR Bescot		
D5542	32B	05/90	IM	C	M C Processors, Glasgow	11/91	
D5543	32B						
D5544	32B						Withdrawn: 04/91, R/I: 04/91
D5545	32B	06/89	TE	C	V Berry, Leicester	11/89	
D5546	32B						
D5547	32B						
D5548	32B						
D5549	32B	03/89	TE	C	V Berry, Leicester	04/90	
D5550	32B						
D5551	32B						
D5552	32B						
D5553	32B						
D5554	32B	08/80	BR	C	BREL Swindon	11/80	Stored: [U] 08/80
D5555	32B						
D5556	32B	02/89	BS	C	M C Processors, Glasgow	01/90	
D5557	32B						Stored: [U] 03/91, R/I: 04/91
D5558	32B						
D5559	32B	02/89	CD	C	M C Processors, Glasgow	10/89	
D5560	32B						
D5561	32B	12/88	TI	C	Booth-Roe, Rotherham	08/89	Stored: [U] 10/88
D5562	32B						
D5563	31B						
D5564	31B						
D5565	31B						
D5566	30A						
D5567	34B						
D5568	34B	10/75	MR	C	BREL Doncaster	01/76	
D5569	34B	03/86	IM	C	BREL Doncaster	12/86	
D5570	31B	06/89	SF	C	V Berry, Leicester	04/90	
D5571	31B						
D5572	31B						
D5573	31B						
D5574	32A	02/92	CD	A	BR Scunthorpe		Stored: [U] 08/91
D5575	32A						
D5576	32A						

BR 1957 No.	TOPS No.	Date Re No.	First TOPS Re No.	Date Re No.	Second TOPS Re No.	Date Re No.	Third TOPS Re No.	Date Re No.	Name	Name Date	Built By	Works No.	Date Introduced
D5577	31159	02/74									Brush	176	12/59
D5578	31160	02/74									Brush	177	01/60
D5579	31161	12/74	31400	05/88					[North Yorkshire Moors Railway]	04/91-10/91	Brush	178	01/60
D5580	31162	11/73									Brush	180	01/60
D5581	31163	02/74									Brush	181	01/60
D5582	31164	02/74									Brush	182	01/60
D5583	31165	03/74							Stratford Major Depot	03/91	Brush	183	01/60
D5584	31166	02/74									Brush	184	01/60
D5585	31167	03/74									Brush	185	01/60
D5586	31168	02/74									Brush	186	02/60
D5587	31169	02/74	31457	11/84							Brush	187	02/60
D5588	31170	11/73									Brush	188	02/60
D5589	31401	02/74									Brush	189	02/60
D5590	31171	04/74									Brush	190	02/60
D5591	31172	03/74	31420	10/74							Brush	191	02/60
D5592	31402	02/74									Brush	192	02/60
D5593	31173	02/74									Brush	193	03/60
D5594	31174	05/74									Brush	194	03/60
D5595	31175	04/74									Brush	195	03/60
D5596	31403	03/74									Brush	196	03/60
D5597	31176	03/74									Brush	197	03/60
D5598	31177	02/74	31443	05/84							Brush	198	03/60
D5599	31178	02/74									Brush	199	03/60
D5600	31179	02/74	31435	02/84							Brush	200	03/60
D5601	31180	04/74									Brush	201	03/60
D5602	31181	03/74									Brush	202	03/60
D5603	31182	02/74	31437	03/84	31537	05/90					Brush	203	03/60
D5604	31183	02/74									Brush	204	04/60
D5605	31404	03/74									Brush	205	04/60
D5606	31405	03/74							Mappa Mundi	05/91	Brush	206	04/60
D5607	31184	02/74									Brush	207	04/60
D5608	31185	03/74									Brush	208	04/60
D5609	31186	03/74									Brush	209	04/60
D5610	31187	03/74									Brush	210	04/60
D5611	31188	03/74									Brush	211	04/60
D5612	31189	03/74									Brush	212	05/60
D5613	31190	03/74									Brush	213	05/60
D5614	31191	03/74									Brush	214	05/60
D5615	31192	03/74									Brush	215	05/60
D5616	31406	02/74									Brush	216	05/60
D5617	31193	06/74	31426	05/83	31526	05/90					Brush	217	05/60
D5618	31194	03/74	31427	05/83							Brush	218	05/60
D5619	31195	02/74									Brush	219	05/60
D5620	31196	03/74									Brush	220	06/60
D5621	31197	03/74	31423	01/75					Jerome K Jerome	05/90	Brush	221	06/60
D5622	31198	03/74									Brush	222	06/60
D5623	31199	11/73									Brush	223	06/60
D5624	31200	03/74									Brush	224	06/60
D5625	31201	03/74							[Fina Energy]	03/89-09/91	Brush	225	06/60
D5626	31202	02/74									Brush	226	06/60
D5627	31203	04/74									Brush	227	06/60
D5628	31204	03/74	31440	04/84							Brush	228	06/60
D5629	31205	02/74									Brush	229	06/60
D5630	31206	03/74									Brush	230	07/60
D5631	31207	02/74									Brush	231	07/60
D5632	31208	10/73									Brush	232	07/60
D5633	31209	03/74									Brush	233	07/60
D5634	31210	03/74									Brush	234	07/60
D5635	31211	02/74	31428	06/83					[North Yorkshire Moors Railway]	04/88-03/91	Brush	235	07/60
D5636	31212	02/74									Brush	236	07/60
D5637	31213	03/74	31465	02/85	31565	05/90	31465	12/92			Brush	237	07/60
D5638	31214	03/74									Brush	238	08/60
D5639	31215	01/74									Brush	239	08/60
D5640	31407	02/74	31507	05/90	31407	08/91					Brush	240	08/60
D5641	31216	03/74	31467	03/85							Brush	241	08/60
D5642	31217	02/74									Brush	242	08/60
D5643	31218	02/74									Brush	243	08/60
D5644	31219	03/74									Brush	244	08/60
D5645	31220	01/74	31441	04/84	31541	05/90					Brush	245	09/60
D5646	31408	01/74									Brush	246	09/60
D5647	31221	02/74									Brush	247	09/60
D5648	31222	02/74									Brush	248	09/60
D5649	31223	03/74									Brush	249	09/60
D5650	31224	03/74									Brush	250	09/60

BR 1957 No.	Depot of First Allocation	Date Withdrawn	Depot of Final Allocation	Disposal Code	Disposal Detail	Date Cut Up	Notes
D5577	32A						
D5578	30A						
D5579	30A	10/91	CD	A	Booth-Roe, Rotherham		Withdrawn: 03/88, R/I: 04/88
D5580	32A	05/91	BS	A	BR Immingham		
D5581	32A						Stored: [U] 06/91, R/I: 07/91
D5582	32A						
D5583	30A						Reverted to No. D5583 03/91
D5584	31B						
D5585	31B	09/88	BS	C	V Berry, Leicester	01/89	Stored: [U] 09/83, R/I: 11/83
D5586	34B	12/91	BS	A	BR Bescot		
D5587	34B						
D5588	34B	01/90	IM	C	M C Processors, Glasgow	12/91	
D5589	34B	03/88	OC	C	BRML Doncaster by Booth-Roe	09/88	
D5590	34B						Stored: [U] 03/87, R/I: 04/87
D5591	34B						
D5592	34B	03/92	BS	A	BR Bescot		
D5593	34B	04/89	SF	C	M C Processors, Glasgow	12/91	
D5594	34B						
D5595	34B	03/87	TI	C	BR Carlisle	09/88	
D5596	34B						
D5597	34B	04/87	TI	C	V Berry, Leicester	03/88	Stored: [U] 03/87
D5598	34B	07/89	CD	C	V Berry, Leicester	04/90	
D5599	34B						
D5600	34B						
D5601	34B						
D5602	34B						
D5603	34B						
D5604	34B	07/88	TI	C	Booth-Roe, Doncaster	08/89	Stored: [U] 05/88
D5605	34B	03/91	CD	A	BRML Doncaster		Stored: [U] 02/91
D5606	34B						
D5607	34B						
D5608	34B						Stored: [U] 06/91, R/I: 07/91
D5609	34B						
D5610	34G						
D5611	34G						
D5612	34G	05/89	SF	C	V Berry, Leicester	11/89	
D5613	34G						
D5614	34G						
D5615	34G	10/82	IM	C	BREL Doncaster	01/83	
D5616	30A	01/90	BS	C	M C Processors, Glasgow	12/91	Stored: [U] 07/90, Withdrawn: 10/90, R/I: 11/90
D5617	30A						Stored: [U] 05/81, R/I: 09/82, Stored: [U] 11/82, R/I: 05/83
D5618	30A						Stored: [U] 05/81, R/I: 09/82, Stored: [U] 10/82, R/I: 05/83
D5619	30A	02/88	IM	C	BRML Doncaster, by Booth-Roe	09/88	
D5620	31B						
D5621	31B						
D5622	32B	03/90	SF	C	M C Processors, Glasgow	09/91	
D5623	32B						
D5624	32B						
D5625	32B						Stored: [U] 07/91, R/I: 08/91
D5626	32B	10/88	SF	C	BR & V Berry, Cricklewood & Leicester	11/88	
D5627	32B						
D5628	32B	03/87	KD	C	V Berry, Leicester	03/88	
D5629	32A						Stored: [U] 03/91, R/I: 04/91
D5630	32A						
D5631	32A						Stored: [S] 07/90
D5632	30A	03/90	IM	C	M C Processors, Glasgow	02/92	
D5633	30A						
D5634	30A	02/92	IM	A	BR Scunthorpe		Stored: [U] 08/91
D5635	30A	03/91	CD	A	BR Crewe Basford Hall		Stored: [U] 05/81, R/I: 09/82, Stored: [U] 1/91
D5636	30A	12/91	IM	A	BR Scunthorpe		Stored: [S] 05/90, R/I: 06/90
D5637	30A						
D5638	30A	05/83	MR	C	BREL Doncaster	10/83	
D5639	34G						
D5640	34G						
D5641	34G						
D5642	34G						
D5643	34G	04/88	SF	C	Booth-Roe, Rotherham	08/89	
D5644	34G						
D5645	34G						
D5646	34G						
D5647	34G	08/91	IM	A	BR Scunthorpe		Stored: [S] 05/90, R/I: 06/90
D5648	34G	10/88	BS	C	V Berry, Leicester	02/90	
D5649	34G	12/91	IM	A	BR Scunthorpe		Stored: [S] 05/90, R/I: 06/90
D5650	34G						

BR 1957 No.	TOPS No.	Date Re No.	First TOPS Re No.	Date Re No.	Second TOPS Re No.	Date Re No.	Third TOPS Re No.	Date Re No.	Name	Name Date	Built By	Works No.	Date Introduced
D5651	31225	02/74									Brush	251	09/60
D5652	31226	03/74									Brush	252	09/60
D5653	31227	02/74									Brush	253	09/60
D5654	31228	03/74	31454	10/84	31554	05/90					Brush	254	09/60
D5655	31229	02/74									Brush	255	09/60
D5656	31409	02/74									Brush	256	10/60
D5657	31230	03/74									Brush	257	10/60
D5658	31231	12/73									Brush	258	10/60
D5659	31232	02/74									Brush	259	10/60
D5660	31233	03/74							[Phillips-Imperial]	03/90-11/92	Brush	260	10/60
D5661	31234	03/74									Brush	261	10/60
D5662	31235	03/74									Brush	262	10/60
D5663	31236	03/74	31433	01/84	31533	06/90					Brush	263	11/60
D5664	31237	02/74									Brush	264	11/60
D5665	31238	02/74									Brush	265	11/60
D5666	31239	03/74	31439	04/84					North Yorkshire Moors Railway	09/92	Brush	266	11/60
D5667	31240	02/74									Brush	267	11/60
D5668	31241	03/74									Brush	268	11/60
D5669	31410	02/74							Granada Telethon	06/92	Brush	269	11/60
D5670	31242	01/74									Brush	270	01/61
D5671	31243	01/74									Brush	271	10/60
D5672	31244	02/74									Brush	272	11/60
D5673	31245	02/74									Brush	273	12/60
D5674	31246	02/74	31455	11/84	31555	05/90	31455	12/92	'Our Eli'	01/93	Brush	274	12/60
D5675	31247	02/74									Brush	275	12/60
D5676	31248	02/74									Brush	276	12/60
D5677	31249	02/74									Brush	277	12/60
D5678	31250	02/74									Brush	278	12/60
D5679	31251	02/74	31442	05/84							Brush	279	12/60
D5680	31252	01/74									Brush	281	12/60
D5681	31253	01/74	31431	11/83	31531	10/90					Brush	282	01/61
D5682	31254	02/74									Brush	283	01/61
D5683	31255	04/74									Brush	284	01/61
D5684	31256	02/74	31459	12/84							Brush	285	01/61
D5685	31257	03/74									Brush	286	02/61
D5686	31258	02/74	31434	02/84							Brush	287	02/61
D5687	31259	02/74									Brush	288	02/61
D5688	31260	02/74									Brush	289	02/61
D5689	31261	03/74									Brush	290	03/61
D5690	31262	02/74									Brush	291	03/61
D5691	31411	02/74	31511	05/90	31411	05/92					Brush	292	03/61
D5692	31412	02/74	31512	10/90							Brush	293	03/61
D5693	31263	02/74									Brush	294	03/61
D5694	31264	03/74									Brush	295	04/61
D5695	31265	04/74	31430	08/83	31530	05/90			Sister Dora	10/88	Brush	296	04/61
D5696	31266	01/74	31460	12/84							Brush	297	05/61
D5697	31267*	-	31419	04/74	31519	05/90					Brush	299	05/61
D5698	31268	02/74									Brush	300	05/61
D5699	31269	02/74	31429	08/83							Brush	298	04/61

Class continued from D5800

METROPOLITAN-VICKERS TYPE 2 Co-Bo CLASS 28

BR 1957 No.	Built By	Works No.	Date Introduced	Depot of First Allocation	Date Withdrawn	Depot of Final Allocation	Disposal Code
D5700	MV	-	07/58	17A	12/67	12B	C
D5701	MV	-	08/58	17A	09/68	D10	C
D5702	MV	-	09/58	17A	09/68	D10	C
D5703	MV	-	10/58	17A	12/67	12B	C
D5704	MV	-	11/58	17A	12/67	12B	C
D5705	MV	-	12/58	17A	09/68	D10	D
D5706	MV	-	12/58	17A	09/68	D10	C
D5707	MV	-	12/58	17A	09/68	D10	C
D5708	MV	-	01/59	17A	09/68	D10	C
D5709	MV	-	01/59	17A	12/69	12B	C
D5710	MV	-	02/59	17A	12/67	12B	C
D5711	MV	-	02/59	17A	09/68	D10	C
D5712	MV	-	02/59	17A	09/68	D10	C
D5713	MV	-	03/59	17A	12/67	12B	C
D5714	MV	-	03/59	17A	09/68	D10	C
D5715	MV	-	04/59	17A	05/68	12B	C
D5716	MV	-	05/59	17A	09/68	D10	C
D5717	MV	-	06/59	17A	09/68	D10	C
D5718	MV	-	07/59	17A	04/68	12B	C
D5719	MV	-	10/59	17A	09/68	D10	C

BR 1957 No.	Depot of First Allocation	Date Withdrawn	Depot of Final Allocation	Disposal Code	Disposal Detail	Date Cut Up	Notes
D5651	34G	03/89	IM	C	M C Processors, Glasgow	10/91	
D5652	34G	01/89	SF	C	M C Processors, Glasgow	09/91	
D5653	34G	12/88	IM	C	V Berry, Leicester	01/89	
D5654	34G						Stored: [U] 05/81, R/I: 02/82
D5655	31B						
D5656	31B	01/90	BS	C	M C Processors, Glasgow	03/92	
D5657	31B						
D5658	31B	01/90	SF	C	M C Processors, Glasgow	03/91	
D5659	31B						
D5660	31B						
D5661	31B						
D5662	31B						
D5663	31B						
D5664	31B						
D5665	31B						Stored: [U] 06/91, R/I: 07/91
D5666	31B						Withdrawn: 02/92, R/I: 04/92
D5667	31B	01/90	SF	A	BR Stratford		
D5668	31B	03/82	OC	C	BREL Swindon	05/82	Stored: [U] 01/82
D5669	31B						
D5670	31B						
D5671	41A	04/91	IM	A	BR Stratford		Stored: [U] 09/82
D5672	34G	05/83	MR	C	BREL Doncaster	09/83	
D5673	34G	01/87	IM	C	BR Stratford Major	03/88	Stored: [U] 12/86
D5674	34G						
D5675	34G						
D5676	34G						
D5677	34G	12/91	IM	A	BR Scunthorpe		Stored: [S] 05/90, R/I: 06/90
D5678	34G						
D5679	34G						Withdrawn: 03/92, R/I: 04/92
D5680	41A						
D5681	41A						Stored: [U] 08/83, R/I: 10/83
D5682	41A	12/79	OC	C	BR Old Oak Common	07/80	
D5683	41A						
D5684	41A						
D5685	41A	04/91	IM	C	Booth-Roe, Rotherham	04/92	
D5686	41A						
D5687	41A	03/89	TI	C	V Berry, Leicester	04/90	Stored: [U] 09/83, R/I: 10/88
D5688	41A	01/90	TE	C	M C Processors, Glasgow	12/91	
D5689	41A	01/87	IM	C	BR Stratford Major	03/88	Stored: [U] 12/86
D5690	41A	05/83	MR	C	BREL Doncaster	08/83	
D5691	41A						
D5692	41A						
D5693	41A						
D5694	30A	04/91	TE	A	BR Thornaby		Stored: [U] 10/80, Withdrawn: 12/80, R/I: 01/83
D5695	30A						Stored: [U] 10/80, Withdrawn: 12/80, R/I: 08/83
D5696	30A						Stored: [U] 12/92
D5697	30A						
D5698	30A						Stored: [U] 09/82, R/I: 10/82
D5699	30A	12/91	CD	A	BR Crewe		Stored: [U] 12/91

BR 1957 No.	Disposal Detail	Date Cut Up	Notes
D5700	J McWilliam, Shettleston	09/68	Stored: [S] 01/61, R/I: 02/62
D5701	J Cashmore, Great Bridge	10/69	Stored: [S] 02/61, R/I: 02/62
D5702	J Cashmore, Great Bridge	10/69	Stored: [S] 02/61, R/I: 02/62
D5703	J McWilliam, Shettleston	06/68	Stored: [S] 02/61, R/I: 02/62
D5704	J McWilliam, Shettleston	08/68	Stored: [S] 02/61, R/I: 02/62, Stored: [U] 04/66, R/I: 12/66
D5705	To Departmental Stock - 15705	-	Stored: [S] 02/61, R/I: 02/62
D5706	J Cashmore, Great Bridge	10/69	Stored: [S] 02/61, R/I: 02/62
D5707	J Cashmore, Geeat Bridge	10/69	Stored: [S] 02/61, R/I: 02/62
D5708	J Cashmore, Great Bridge	10/69	Stored: [S] 02/61, R/I: 02/62
D5709	J McWilliam, Shettleston	10/68	Stored: [S] 01/61, R/I: 02/62, Stored: [U] 11/65, R/I: 12/66
D5710	J McWilliam, Shettleston	09/68	Stored: [S] 02/61, R/I: 02/62, Stored: [U] 04/66, R/I: 12/66
D5711	J Cashmore, Great Bridge	10/69	Stored: [S] 02/61, R/I: 02/62
D5712	J Cashmore, Great Bridge	12/69	Stored: [S] 02/61, R/I: 02/62
D5713	J McWilliam, Shettleston	10/68	Stored: [S] 01/61, R/I: 02/62, Stored: [U] 01/66, R/I: 12/66
D5714	J Cashmore, Great Bridge	09/69	Stored: [S] 02/61, R/I: 02/62
D5715	J McWilliam, Shettleston	09/68	Stored: [S] 02/61, R/I: 02/62, Stored: [U] 07/66, R/I: 02/68
D5716	J Cashmore, Great Bridge	12/69	Stored: [S] 02/61, R/I: 02/62
D5717	J Cashmore, Great Bridge	10/69	Stored: [S] 02/61, R/I: 02/62, Stored: [U] 02/64, R/I: 12/66
D5718	J McWilliam, Shettleston	09/68	Stored: [S] 02/61, R/I: 02/62
D5719	J Cashmore, Great Bridge	09/69	Stored: [S] 02/61, R/I: 12/61, Stored: [U] 04/64, R/I: 12/66

BRUSH TYPE 2 A1A-A1A CLASS 31

Class continued from D5699

BR 1957 No.	TOPS No.	Date Re No.	First TOPS Re No.	Date Re No.	Second TOPS Re No.	Date Re No.	Third TOPS Re No.	Date Re No.	Name	Name Date	Built By	Works No.	Date Introduced
D5800	31270	02/74									Brush	301	06/61
D5801	31271	02/74									Brush	302	06/61
D5802	31272	04/74									Brush	303	06/61
D5803	31273	11/73									Brush	304	06/61
D5804	31274	03/74	31425	08/83							Brush	305	06/61
D5805	31275	03/74									Brush	306	07/61
D5806	31276	03/74							[Calder Hall Power Station]	12/88-04/92	Brush	307	07/61
D5807	31277	02/74	31469	02/88	31569	10/90					Brush	308	07/61
D5808	31278	02/74									Brush	309	07/61
D5809	31279	02/74	31452	10/84	31552	05/90					Brush	310	07/61
D5810	31280	02/74									Brush	311	08/61
D5811	31281	02/74									Brush	312	08/61
D5812	31413	03/74							Severn Valley Railway	04/88	Brush	313	08/61
D5813	31282	02/74									Brush	314	09/61
D5814	31414	02/74	31514	10/90							Brush	315	09/61
D5815	31283	02/74									Brush	316	09/61
D5816	31284	02/74									Brush	317	09/61
D5817	31285	03/74									Brush	318	10/61
D5818	31286	05/74									Brush	319	10/61
D5819	31287	02/74									Brush	320	10/61
D5820	31288	02/74									Brush	321	10/61
D5821	31289	01/74									Brush	322	11/61
D5822	31290	02/74									Brush	323	11/61
D5823	31291	11/73	31456	11/84	31556	05/90					Brush	324	11/61
D5824	31415	03/74									Brush	325	11/61
D5825	31292	03/74									Brush	326	11/61
D5826	31293	04/74									Brush	362	12/61
D5827	31294	02/74									Brush	363	12/61
D5828	31295	12/73	31447	07/84	31547	05/90					Brush	364	12/61
D5829	31296	02/74							[Amlwch Freighter/ Tren Nwyddau Amlwych]	09/86-03/90	Brush	365	01/62
D5830	31297	05/74	31463	02/85	31563	10/90					Brush	366	01/62
D5831	31298	02/74									Brush	367	01/62
D5832	31299	02/74									Brush	368	02/62
D5833	31300	12/73	31445	06/84	31545	06/90					Brush	369	02/62
D5834	31301	05/74									Brush	371	03/62
D5835	31302	02/74									Brush	370	04/62
D5836	31303	03/74	31458	12/84							Brush	372	04/62
D5837	31304	05/74									Brush	373	04/62
D5838	31305	02/74									Brush	374	04/62
D5839	31306	02/74									Brush	375	05/62
D5840	31307	02/74	31449	08/84	31549	06/90					Brush	376	05/62
D5841	31308	12/73									Brush	377	05/62
D5842	31416	09/73	31516	11/90							Brush	378	05/62
D5843	31309	10/73							[Cricklewood]	05/87-12/89	Brush	379	05/62
D5844	31310	02/74	31422	12/74	31522	05/90	31422	05/91			Brush	380	06/62
D5845	31311	03/74									Brush	381	06/62
D5846	31312	02/74									Brush	382	06/62
D5847	31313	02/74									Brush	383	06/62
D5848	31314	05/74									Brush	384	07/62
D5849	31315	02/74	31462	01/85							Brush	385	07/62
D5850	31316	03/74	31446	06/84	31546	05/90					Brush	386	07/62
D5851	31317	03/74									Brush	387	07/62
D5852	31318	10/73	31451	09/84	31551	05/90					Brush	388	08/62
D5853	31319	03/74									Brush	389	08/62
D5854	31320	01/74									Brush	390	08/62
D5855	31321	12/74	31468	04/85	31568	05/90			The Enginemen's Fund	05/90	Brush	391	08/62
D5856	31417	02/74									Brush	392	08/62
D5857	31322	02/74									Brush	393	09/62
D5858	31323	11/73									Brush	394	09/62
D5859	31324	03/74									Brush	395	10/62
D5860	31325	03/74	31464	02/85							Brush	396	10/62
D5861	31326	10/73	31970	07/89							Brush	397	10/62
D5862	31327	02/74							[Phillips Imperial]	05/87-09/90	Brush	398	10/62

BR 1957 No.	Depot of First Allocation	Date Withdrawn	Depot of Final Allocation	Disposal Code	Disposal Detail	Date Cut Up	Notes
D5800	31B						
D5801	30A						
D5802	31B						
D5803	31B						
D5804	41A	12/91	CD	A	Booth-Roe, Rotherham		Stored [U]: 10/80, Withdrawn: 10/80, R/I: 09/82, Stored: [U] 12/91
D5805	41A						
D5806	41A						
D5807	41A						Stored: 03/81, R/I: 04/87
D5808	41A	03/89	SF	C	M C Processors, Glasgow	12/89	
D5809	41A						
D5810	41A	10/88	SF	C	Booth-Roe, Rotherham	01/91	Withdrawn: 05/81, R/I: 01/82, Stored: [U] 03/87, R/I: 04/87
D5811	41A	05/89	TE	C	V Berry, Leicester	02/90	
D5812	41A						
D5813	41A						
D5814	41A						
D5815	41A	11/89	TE	A	BR Stratford		Withdrawn: 11/89, R/I: 04/91
D5816	41A	08/89	TI	C	M C Processors, Glasgow	12/91	
D5817	41A						
D5818	41A	12/91	BS	A	BR Bescot		
D5819	41A	09/87	BS	C	M C Processors, Glasgow	12/89	
D5820	41A	04/91	IM	C	Booth-Roe, Rotherham	04/92	
D5821	41A	03/92	TI	A	BR Bescot		
D5822	41A						
D5823	41A						
D5824	41A						
D5825	41A	02/90	BS	C	M C Processors, Glasgow	09/90	
D5826	41A	10/90	BS	A	BR Stratford		
D5827	41A						
D5828	41A						
D5829	41A						
D5830	41A						
D5831	41A	07/86	LO	D	To Departmental Stock - 97203	-	Stored: [U] 05/81, R/I: 02/82
D5832	41A	04/91	IM	A	BR Stratford		Withdrawn: 03/90, R/I: 05/90
D5833	41A						
D5834	41A						
D5835	41A						
D5836	41A						
D5837	41A						
D5838	41A	11/91	BS	A	BR Bescot		Stored: [U] 05/81, R/I: 02/82
D5839	41A						
D5840	41A						
D5841	41A						
D5842	41A						
D5843	41A	04/91	IM	C	Booth-Roe, Rotherham	04/92	Stored: [U] 06/81, R/I: 02/82, Stored: [S] 05/90
D5844	41A						
D5845	41A	02/89	TO	C	M C Processors, Glasgow	07/90	
D5846	41A						
D5847	41A	05/83	MR	C	BREL Doncaster	05/83	
D5848	41A	09/82	IM	C	BREL Doncaster	01/83	
D5849	41A						
D5850	41A						
D5851	41A						
D5852	41A						
D5853	41A						
D5854	41A	04/90	TE	A	BR Stratford		Withdrawn: 03/89, R/I: 06/89,
D5855	41A						
D5856	41A						
D5857	41A	07/89	IM	C	V Berry, Leicester		
D5858	41A	06/89	SF	C	M C Processors, Glasgow	09/91	
D5859	41A						
D5860	41A	01/90	BS	C	M C Processors, Glasgow	11/91	
D5861	41A	09/91	CD	A	ABB Transportation, Crewe		Stored: [U] 05/81, R/I: 01/82, As departmental No. 97204 05/87-07/89 Stored: [U] 12/90
D5862	41A						

ENGLISH ELECTRIC TYPE 2 Bo-Bo CLASS 23

BR 1957 No.	Built By	Works No.	Date Introduced	Depot of First Allocation	Date Withdrawn	Depot of Final Allocation	Disposal Code
D5900	EE.VF	2377/D417	05/59	34B	11/68	34G	C
D5901	EE.VF	2378/D418	05/59	34B	12/69	34G	D
D5902	EE.VF	2379/D419	05/59	34B	11/69	34G	C
D5903	EE.VF	2380/D420	04/59	34B	11/68	34G	C
D5904	EE.VF	2381/D421	04/59	34B	01/69	34G	C
D5905	EE.VF	2382/D422	05/59	34B	02/71	34G	C
D5906	EE.VF	2382/D423	05/59	34B	09/68	34G	C
D5907	EE.VF	2384/D424	05/59	34B	10/68	34G	C
D5908	EE.VF	2385/D425	05/59	34B	03/69	34G	C
D5909	EE.VF	2386/D426	06/59	34B	03/71	34G	C

NORTH BRITISH TYPE 2 Bo-Bo CLASS 21/29

BR 1957 No.	Built By	Works No.	Date Introduced	Depot of First Allocation	Date Refurbished To Class 29	Date Withdrawn	Depot of Final Allocation	Disposal Code
D6100	NBL	27681	12/58	34B	09/67	10/71	65A	C
D6101	NBL	27682	12/58	34B	12/65	08/71	65A	C
D6102	NBL	27683	12/58	34B	11/65	10/71	65A	C
D6103	NBL	27684	05/59	34B	10/65	10/71	65A	C
D6104	NBL	27685	02/59	34B	-	12/67	65A	C
D6105	NBL	27686	03/59	34B	-	06/68	65A	C
D6106	NBL	27687	03/59	34B	11/65	07/71	65A	C
D6107	NBL	27688	03/59	34B	02/67	10/71	65A	C
D6108	NBL	27689	04/59	34B	01/67	05/69	65A	C
D6109	NBL	27690	04/59	34B	-	04/68	65A	C
D6110	NBL	27840	05/59	30A	-	04/68	65A	C
D6111	NBL	27841	05/59	30A	-	08/68	65A	C
D6112	NBL	27842	05/59	30A	07/66	12/71	65A	C
D6113	NBL	27843	05/59	30A	08/66	10/71	65A	C
D6114	NBL	27844	05/59	30A	06/66	10/71	65A	C
D6115	NBL	27845	06/59	30A	-	06/68	65A	C
D6116	NBL	27846	06/59	30A	05/66	12/71	65A	C
D6117	NBL	27847	06/59	30A	-	08/68	65A	C
D6118	NBL	27848	06/59	30A	-	12/67	65A	C
D6119	NBL	27849	06/59	30A	02/67	12/71	65A	C
D6120	NBL	27850	07/59	32B	-	12/67	65A	C
D6121	NBL	27851	08/59	32B	09/66	10/71	65A	C
D6122	NBL	27852	08/59	32B	-	12/67	65A	C
D6123	NBL	27853	09/59	32B	06/63	09/71	65A	C
D6124	NBL	27854	09/59	32B	03/67	10/71	65A	C
D6125	NBL	27855	09/59	32B	-	12/67	65A	C
D6126	NBL	27856	10/59	32B	-	04/68	65A	C
D6127	NBL	27857	10/59	32B	-	12/67	65A	C
D6128	NBL	27858	10/59	32B	-	12/67	65A	C
D6129	NBL	27859	10/59	32B	04/67	10/71	65A	C
D6130	NBL	27860	10/59	32B	04/66	10/71	65A	C
D6131	NBL	27861	11/59	32B	-	12/67	65A	C
D6132	NBL	27682	11/59	32B	03/66	10/71	65A	C
D6133	NBL	27863	11/59	32B	07/66	12/71	65A	C
D6134	NBL	27864	12/59	32B	-	12/67	65A	C
D6135	NBL	27865	12/59	32B	-	12/67	65A	C
D6136	NBL	27866	12/59	32B	-	12/67	65A	C
D6137	NBL	27867	12/59	32B	02/67	04/71	65A	C
D6138	NBL	27942	02/60	64A	-	12/67	61B	C
D6139	NBL	27943	02/60	61A	-	12/67	61B	C
D6140	NBL	27944	02/60	61A	-	12/67	65A	C
D6141	NBL	27945	03/60	61A	-	12/67	61B	C
D6142	NBL	27946	03/60	61A	-	12/67	61B	C
D6143	NBL	27947	03/60	61A	-	12/67	61B	C
D6144	NBL	27948	03/60	61A	-	12/67	61B	C
D6145	NBL	27949	04/60	61A	-	12/67	61B	C
D6146	NBL	27950	04/60	61A	-	12/67	61B	C
D6147	NBL	27951	05/60	61A	-	12/67	61B	C
D6148	NBL	27952	05/60	61A	-	12/67	61B	C
D6149	NBL	27953	05/60	61A	-	12/67	61B	C
D6150	NBL	27954	06/60	61A	-	12/67	61B	C
D6151	NBL	27955	06/60	61A	-	12/67	61B	C
D6152	NBL	27956	06/60	61A	-	08/68	65A	C
D6153	NBL	27957	07/60	61A	-	12/67	61B	C
D6154	NBL	27958	07/60	61A	-	12/67	61B	C
D6155	NBL	27959	08/60	61A	-	12/67	61B	C
D6156	NBL	27960	09/60	61A	-	12/67	61B	C
D6157	NBL	27961	12/60	61A	-	12/67	61B	C

BR 1957 No.	Disposal Detail	Date Cut Up	Notes
D5900	G Cohen, Kettering	06/69	
D5901	To Departmental Stock - D5901	-	
D5902	G Cohen, Kettering	08/70	
D5903	G Cohen, Kettering	06/69	
D5904	G Cohen, Kettering	07/69	
D5905	G Cohen, Kettering	08/73	
D5906	G Cohen, Kettering	07/69	Stored: [U] 06/68
D5907	G Cohen, Kettering	07/69	Stored: [U] 04/62, R/I: 07/62
D5908	J Cashmore, Great Bridge	12/69	
D5909	G Cohen, Kettering	08/73	

BR 1957 No.	Disposal Detail	Date Cut Up	Notes
D6100	BREL Glasgow	11/72	Stored: [U] 02/60, R/I: 04/60, Stored: [U] 10/64, R/I: 09/67, Stored: [U] 08/70
D6101	BREL Glasgow	10/72	Stored: [U] 02/60, R/I: 04/60, Stored: [U] 04/62, R/I: 12/65, Stored: [U] 09/68
D6102	BREL Glasgow	08/72	Stored: [U] 02/60, R/I: 04/60, Stored: [U] 05/70, R/I: 09/70
D6103	BREL Glasgow	11/72	Stored: [U] 02/60, R/I: 04/60
D6104	Barnes & Bell, Coatbridge	05/68	Stored: [U] 02/60, R/I: 04/60, Stored: [U] 04/62, R/I: 05/63, Stored: [U] 09/67
D6105	J McWilliam, Shettleston	08/68	Stored: [U] 02/60, R/I: 04/60
D6106	BREL Glasgow	07/72	Stored: [U] 02/60, R/I: 04/60, Stored: [U] 04/62, R/I: 11/65, Stored: [U] 02/71
D6107	BREL Glasgow	10/72	Stored: [U] 02/60, R/I: 04/60, Stored: [U] 10/64, R/I: 02/67, Stored: [U] 02/71
D6108	J McWilliam, Shettleston	07/71	Stored: [U] 02/60, R/I: 04/60, Stored: [U] 04/69
D6109	J McWilliam, Shettleston	01/69	Stored: [U] 02/60, R/I: 04/60
D6110	J McWilliam, Shettleston	02/69	
D6111	J McWilliam, Shettleston	09/69	
D6112	BREL Glasgow	06/72	Stored: [U] 04/62, R/I: 07/66, Stored: [U] 06/71
D6113	BREL Glasgow	11/72	Stored: [U] 10/64, R/I: 08/66
D6114	BREL Glasgow	09/72	
D6115	J McWilliam, Shettleston	08/68	
D6116	BREL Glasgow	05/72	
D6117	J McWilliam, Shettleston	11/68	Stored: [U] 08/64
D6118	J McWilliam, Shettleston	05/68	Stored: [U] 10/65
D6119	BREL Glasgow	08/72	Stored: [U] 08/64, R/I: 02/67
D6120	J McWilliam, Shettleston	09/68	Stored: [U] 09/67
D6121	BREL Glasgow	09/72	
D6122	D Woodham, Barry	06/80	Stored: [U] 09/67
D6123	BREL Glasgow	10/72	Stored: [U] 04/62, R/I: 06/63
D6124	BREL Glasgow	09/72	Stored: [U] 09/67, R/I: 12/67
D6125	Barnes & Bell, Coatbridge	04/68	Stored: [U] 03/64, R/I: 01/66, Stored: [U] 09/67
D6126	J McWilliam, Shettleston	02/69	Stored: [U] 09/67
D6127	Barnes & Bell, Coatbridge	04/68	Stored: [U] 04/62
D6128	J McWilliam, Shettleston	05/68	Stored: [U] 08/67
D6129	BREL Glasgow	06/72	
D6130	BREL Glasgow	10/72	Stored: [U] 03/64, R/I: 04/66
D6131	J McWilliam, Shettleston	06/68	Stored: [U] 09/67
D6132	BREL Glasgow	09/72	
D6133	BREL Glasgow	07/72	Stored: [U] 04/62, R/I: 07/66, Stored: [U] 10/71
D6134	Barnes & Bell, Coatbridge	06/68	Stored: [U] 08/67
D6135	J McWilliam, Shettleston	05/68	
D6136	Barnes & Bell, Coatbridge	04/68	Stored: [U] 03/67
D6137	BREL Glasgow	08/72	
D6138	J McWilliam, Shettleston	06/68	Stored: [U] 09/67
D6139	J McWilliam, Shettleston	06/68	Stored: [U] 09/67
D6140	J McWilliam, Shettleston	05/68	
D6141	J McWilliam, Shettleston	08/68	Stored: [U] 08/67
D6142	J McWilliam, Shettleston	07/68	Stored: [U] 09/67
D6143	Barnes & Bell, Coatbridge	06/68	Stored: [U] 08/67
D6144	J McWilliam, Shettleston	05/68	Stored: [U] 08/67
D6145	J McWilliam, Shettleston	06/68	Stored: [U] 07/67
D6146	J McWilliam, Shettleston	08/68	Stored: [U] 07/67
D6147	J McWilliam, Shettleston	06/68	Stored: [U] 07/67
D6148	J McWilliam, Shettleston	06/68	Stored: [U] 07/67
D6149	J McWilliam, Shettleston	05/68	Stored: [U] 07/67
D6150	J McWilliam, Shettleston	06/68	Stored: [U] 09/67
D6151	J McWilliam, Shettleston	05/68	Stored: [U] 07/67
D6152	J McWilliam, Shettleston	02/69	Stored: [U] 08/64, R/I: 08/65
D6153	J McWilliam, Shettleston	05/68	Stored: [U] 07/67
D6154	J McWilliam, Shettleston	05/68	Stored: [U] 07/67
D6155	J McWilliam, Shettleston	06/68	Stored: [U] 08/64, R/I: 09/67, Stored: [U] 10/67
D6156	J McWilliam, Shettleston	05/68	Stored: [U] 07/67
D6157	J McWilliam, Shettleston	06/68	Stored: [U] 08/64

NORTH BRITISH TYPE 2 B-B CLASS 22

BR 1957 No.	Built By	Works No.	Date Introduced	Depot of First Allocation	Date Withdrawn	Depot of Final Allocation	Disposal Code
D6300	NBL	27665	01/59	82C	05/68	84A	C
D6301	NBL	27666	02/59	82C	12/67	84A	C
D6302	NBL	27667	02/59	82C	05/68	84A	C
D6303	NBL	27668	05/59	83D	05/68	84A	C
D6304	NBL	27669	06/69	83D	05/68	84A	C
D6305	NBL	27670	01/60	83D	05/68	84A	C
D6306	NBL	27679	10/59	83D	12/68	84A	C
D6307	NBL	27680	10/59	83D	03/71	84A	C
D6308	NBL	27881	01/60	83D	09/71	84A	C
D6309	NBL	27882	01/60	83D	05/71	84A	C
D6310	NBL	27883	01/60	82D	03/71	82A	C
D6311	NBL	27884	01/60	81D	09/68	82A	C
D6312	NBL	27885	01/60	83D	05/71	84A	C
D6313	NBL	27886	01/60	83D	08/68	84A	C
D6314	NBL	27887	01/60	83D	04/69	84A	C
D6315	NBL	27888	01/60	83D	05/71	84A	C
D6316	NBL	27889	03/60	83D	03/68	82A	C
D6317	NBL	27890	03/60	83D	09/68	84A	C
D6318	NBL	27891	03/60	83D	05/71	84A	C
D6319	NBL	27892	04/60	83D	09/71	84A	C
D6320	NBL	27893	03/60	83D	05/71	82A	C
D6321	NBL	27894	04/60	83D	08/68	82A	C
D6322	NBL	27895	04/60	83D	10/71	84A	C
D6323	NBL	27896	04/60	83D	05/71	84A	C
D6324	NBL	27897	06/60	83D	09/68	82A	C
D6325	NBL	27898	06/60	83D	10/68	82A	C
D6326	NBL	27899	05/60	83D	10/71	84A	C
D6327	NBL	27900	08/60	83D	05/71	82A	C
D6328	NBL	27901	06/60	83D	07/71	84A	C
D6329	NBL	27902	06/60	83D	11/68	82A	C
D6330	NBL	27903	06/60	83D	10/71	84A	C
D6331	NBL	27904	07/60	83D	03/71	82A	C
D6332	NBL	27905	07/60	83D	05/71	81A	C
D6333	NBL	27906	08/60	83D	01/72	83A	C
D6334	NBL	27907	12/60	83D	10/71	84A	C
D6335	NBL	27908	02/61	83D	09/68	81A	C
D6336	NBL	27909	07/61	83A	01/72	84A	C
D6337	NBL	27910	03/62	83A	10/71	84A	C
D6338	NBL	27911	03/62	83A	01/72	84A	C
D6339	NBL	27912	04/62	83D	01/72	84A	C
D6340	NBL	27913	04/62	83D	05/71	84A	C
D6341	NBL	27914	05/62	83D	11/68	81A	C
D6342	NBL	27915	05/62	83D	12/68	82A	C
D6343	NBL	27916	05/62	83D	10/71	84A	C
D6344	NBL	27917	05/62	83D	09/68	81A	C
D6345	NBL	27918	05/62	83D	09/68	81A	C
D6346	NBL	27919	06/62	83D	04/69	81A	C
D6347	NBL	27920	06/62	82A	03/68	81A	C
D6348	NBL	27921	06/62	83D	07/71	84A	C
D6349	NBL	27922	06/62	83D	09/68	82A	C
D6350	NBL	27923	06/62	83D	09/68	81A	C
D6351	NBL	27924	06/62	83D	11/68	81A	C
D6352	NBL	27925	07/62	83D	05/71	82A	C
D6353	NBL	27926	07/62	83D	09/68	81A	C
D6354	NBL	27927	08/62	83D	05/71	82A	C
D6355	NBL	27928	08/62	82A	09/68	81A	C
D6356	NBL	27929	09/62	82A	10/71	84A	C
D6357	NBL	27930	11/62	82A	12/68	81A	C

BIRMINGHAM RCW TYPE 3 Bo-Bo CLASS 33

BR 1957 No.	TOPS No.	Date Re No.	First TOPS Re No.	Date Re No.	Second TOPS Re No.	Date Re No.	Name	Name Date	Built By	Works No.	Date Introduced	Depot of First Allocation	Date Withdrawn
D6500	33001	02/74							BRCW	DEL92	01/60	73C	03/88
D6501	33002	02/74					Sea King	11/91	BRCW	DEL93	02/60	73C	
D6502	-	-							BRCW	DEL94	03/60	73C	05/64
D6503	33003	02/74							BRCW	DEL95	03/60	73C	06/87
D6504	33004	01/74							BRCW	DEL96	03/60	73C	06/91
D6505	33005	02/74							BRCW	DEL97	04/60	73C	06/87
D6506	33006	12/73							BRCW	DEL98	04/60	73C	08/91
D6507	33007	02/74							BRCW	DEL99	05/60	73C	12/86
D6508	33008	04/74					Eastleigh	04/80	BRCW	DEL100	05/60	73C	
D6509	33009	03/74					[Walrus]	09/91-02/92	BRCW	DEL101	05/60	73C	03/92
D6510	33010	02/74							BRCW	DEL102	06/60	73C	04/88
D6511	33101	02/74							BRCW	DEL103	06/60	73C	
D6512	33011	01/74							BRCW	DEL104	06/60	73C	03/89

BR 1957 No.	Disposal Detail	Date Cut Up	Notes
D6300	J Cashmore, Newport	01/69	
D6301	G Cohen, Morriston	08/68	Stored: [U] 05/67, R/I: 06/67, Stored: [U] 11/67
D6302	J Cashmore, Newport	11/68	
D6303	J Cashmore, Newport	12/68	
D6304	J Cashmore, Newport	12/68	
D6305	J Cashmore, Newport	11/68	
D6306	J Cashmore, Newport	05/69	
D6307	BREL Swindon	12/71	
D6308	BREL Swindon	05/72	Stored: [U] 06/69, R/I: 06/70, Stored: [U] 08/71
D6309	BREL Swindon	12/71	Stored: [U] 05/71
D6310	BREL Swindon	05/72	
D6311	J Cashmore, Newport	05/69	Stored: [U] 09/68
D6312	BREL Swindon	01/72	Stored: [U] 05/71
D6313	J Cashmore, Newport	11/68	Stored: [U] 06/68
D6314	J Cashmore, Newport	07/69	Stored: [U] 04/69
D6315	BREL Swindon	01/72	Stored: [U] 04/69, R/I: 05/69, Stored: [U] 04/71
D6316	J Cashmore, Newport	12/68	Stored: [U] 03/68
D6317	J Cashmore, Newport	06/69	Stored: [U] 09/68
D6318	BREL Swindon	03/72	Stored: [U] 04/71
D6319	BREL Swindon	11/72	Withdrawn: 03/71, R/I: 05/71
D6320	BREL Swindon	06/72	Stored: [U] 05/71
D6321	J Cashmore, Newport	06/69	Stored: [U] 08/68
D6322	BREL Swindon	05/72	
D6323	BREL Swindon	08/72	Stored: [U] 04/71
D6324	J Cashmore, Newport	05/69	Stored: [U] 08/68
D6325	J Cashmore, Newport	05/69	Stored: [U] 09/68
D6326	BREL Swindon	03/72	Stored: [U] 09/71
D6327	BREL Swindon	07/72	Stored: [U] 05/71
D6328	BREL Swindon	05/72	Stored: [U] 07/71
D6329	J Cashmore, Newport	05/69	Stored: [U] 10/68
D6330	BREL Swindon	06/72	Withdrawn: 09/68, R/I: 09/69
D6331	BREL Swindon	03/72	Stored: [U] 12/68
D6332	BREL Swindon	12/71	Stored: [U] 04/71
D6333	BREL Swindon	08/72	Stored: [U] 08/68, R/I: 05/71, Stored: [U] 12/71
D6334	BREL Swindon	04/72	Stored: [U] 12/68, R/I: 05/71
D6335	J Cashmore, Newport	06/69	Stored: [U] 08/68
D6336	BREL Swindon	06/72	Stored: [U] 05/67, R/I: 07/67, Withdrawn: 10/71, R/I: 10/71, Stored: [U] 12/71
D6337	BREL Swindon	05/72	
D6338	BREL Swindon	02/72	Stored: [U] 05/67, R/I: 06/67, Stored: [U] 12/71
D6339	BREL Swindon	06/72	Withdrawn: 10/71, R/I: 10/71, Stored: [U] 12/71
D6340	BREL Swindon	04/72	Stored: [U] 11/68, R/I: 05/71
D6341	J Cashmore, Newport	05/69	Stored: [U] 05/67, R/I: 06/67, Stored: [U] 10/68
D6342	J Cashmore, Newport	05/69	
D6343	BREL Swindon	01/72	
D6344	J Cashmore, Newport	05/69	Stored: [U] 08/68
D6345	J Cashmore, Newport	08/69	Stored: [U] 08/68
D6346	J Cashmore, Newport	07/69	Stored: [U] 12/68
D6347	J Cashmore, Newport	11/68	
D6348	BREL Swindon	05/72	Stored: [U] 05/69, R/I: 06/69, Stored: [U] 02/70, R/I: 03/70, Stored: [U] 07/71
D6349	BREL Swindon	10/71	Stored: [U] 08/68
D6350	J Cashmore, Newport	05/69	Stored: [U] 08/68
D6351	J Cashmore, Newport	05/69	Stored: [U] 11/68
D6352	BREL Swindon	11/71	
D6353	J Cashmore, Newport	08/69	Stored: [U] 08/68
D6354	BREL Swindon	02/72	Stored: [U] 05/71
D6355	J Cashmore, Newport	05/69	Stored: [U] 08/68
D6356	BREL Swindon	01/72	Stored: [U] 09/69, R/I: 02/70
D6357	J Cashmore, Newport	05/69	Stored: [U] 12/68

BR 1957 No.	Depot of Final Allocation	Disposal Code	Disposal Detail	Date Cut Up	Notes
D6500	EH	C	BR Eastleigh, by V Berry	03/89	
D6501					
D6502	73C	C	BR Workshops, Eastleigh	07/64	
D6503	EH	C	BR Eastleigh, by V Berry	05/89	
D6504	SL	C	BR Eastleigh	08/92	
D6505	EH	C	V Berry, Leicester	10/90	
D6506	SL	A	BR Eastleigh		
D6507	EH	C	BR Eastleigh	05/87	
D6508					
D6509	EH	A	BR Eastleigh		
D6510	EH	C	BR Eastleigh, by V Berry	03/89	
D6511					
D6512	EH	C	V Berry, Leicester	11/90	

BR 1957 No.	TOPS No.	Date Re No.	First TOPS Re No.	Date Re No.	Second TOPS Re No.	Date Re No.	Name	Name Date	Built By	Works No.	Date Introduced	Depot of First Allocation	Date Withdrawn
D6513	33102	12/73							BRCW	DEL105	06/60	73C	11/92
D6514	33103	02/74							BRCW	DEL106	07/60	73C	
D6515	33012	01/74							BRCW	DEL107	07/60	73C	
D6516	33104	02/74							BRCW	DEL108	07/60	73C	12/85
D6517	33105	12/73							BRCW	DEL109	07/60	73C	10/87
D6518	33013	01/74							BRCW	DEL110	08/60	73C	03/85
D6519	33106	01/74							BRCW	DEL111	08/60	73C	10/90
D6520	33107	12/73							BRCW	DEL112	09/60	73C	
D6521	33108	01/74							BRCW	DEL113	09/60	73C	07/89
D6522	33014	02/74							BRCW	DEL114	09/60	73C	
D6523	33015	01/74							BRCW	DEL115	09/60	73C	02/86
D6524	33016	03/74							BRCW	DEL116	10/60	73C	07/89
D6525	33109	02/74							BRCW	DEL117	10/60	73C	09/89
D6526	33017	02/74							BRCW	DEL118	10/60	73C	
D6527	33110	02/74							BRCW	DEL119	10/60	73C	01/88
D6528	33111	03/74							BRCW	DEL120	10/60	73C	09/92
D6529	33112	12/73					[Templecombe]	09/87-11/88	BRCW	DEL121	11/60	73C	03/91
D6530	33018	12/73							BRCW	DEL122	11/60	73C	11/88
D6531	33113	12/73							BRCW	DEL123	11/60	73C	02/88
D6532	33114	12/73					Ashford 150	05/92	BRCW	DEL124	11/60	73C	10/92
D6533	33115	02/74							BRCW	DEL125	12/60	73C	06/89
D6534	33019	03/74					Griffon	12/91	BRCW	DEL126	12/60	73C	
D6535	33116	03/74							BRCW	DEL127	12/60	73C	
D6536	33117	02/74							BRCW	DEL128	12/60	73C	
D6537	33020	01/74							BRCW	DEL129	12/60	73C	
D6538	33118	12/73							BRCW	DEL130	01/61	73C	
D6539	33021	01/74							BRCW	DEL131	01/61	73C	
D6540	33022	02/74							BRCW	DEL132	01/61	73C	12/89
D6541	33023	02/74							BRCW	DEL133	01/61	73C	
D6542	33024	02/74							BRCW	DEL134	02/61	73C	02/86
D6543	33025	12/75					Sultan	08/81	BRCW	DEL135	02/61	73C	
D6544	33026	01/74					Seafire	08/91	BRCW	DEL136	02/61	73C	
D6545	33027	01/74					[Earl Mountbatten of Burma]	09/80-08/89	BRCW	DEL137	03/61	73C	07/91
D6546	33028	02/74							BRCW	DEL138	03/61	73C	10/88
D6547	33029	01/74							BRCW	DEL139	03/61	73C	
D6548	33030	01/74							BRCW	DEL140	04/61	73C	
D6549	33031	02/74							BRCW	DEL141	04/61	73C	02/89
D6550	33032	02/74							BRCW	DEL142	04/61	73C	03/87
D6551	33033	01/74							BRCW	DEL143	04/61	73C	
D6552	33034	01/74							BRCW	DEL144	04/61	73C	01/88
D6553	33035	02/74							BRCW	DEL145	05/61	73C	
D6554	33036	01/74							BRCW	DEL146	05/61	73C	07/79
D6555	33037	12/73							BRCW	DEL147	05/61	73C	09/87
D6556	33038	02/74							BRCW	DEL148	06/61	73C	10/88
D6557	33039	01/74							BRCW	DEL149	06/61	73C	05/89
D6558	33040	12/73							BRCW	DEL150	06/61	73C	
D6559	33041	01/74							BRCW	DEL151	06/61	73C	11/75
D6560	33042	12/73							BRCW	DEL152	07/61	73C	
D6561	33043	01/74							BRCW	DEL153	07/61	73C	09/87
D6562	33044	12/73							BRCW	DEL154	07/61	73C	09/87
D6563	33045	12/73							BRCW	DEL155	07/61	73C	10/87
D6564	33046	12/73					Merlin	09/91	BRCW	DEL156	07/61	73C	
D6565	33047	01/74					[Spitfire]	04/91-01/92	BRCW	DEL169	08/61	73C	
D6566	33048	12/73							BRCW	DEL170	08/61	73C	
D6567	33049	01/74							BRCW	DEL171	09/61	73C	03/88
D6568	33050	12/73					Isle of Grain	05/88	BRCW	DEL172	09/61	73C	
D6569	33051	12/73					Shakespeare Cliff	05/88	BRCW	DEL173	09/61	73C	
D6570	33052	02/74					Ashford	05/80	BRCW	DEL174	09/61	73C	
D6571	33053	01/74							BRCW	DEL175	10/61	73C	
D6572	33054	02/74							BRCW	DEL176	10/61	73C	02/86
D6573	33055	12/73							BRCW	DEL177	11/61	73C	12/89
D6574	33056	12/73					[The Burma Star]	09/80-02/91	BRCW	DEL178	10/61	73C	02/91
D6575	33057	12/73					Seagull	09/91	BRCW	DEL179	11/61	73C	
D6576	-	-							BRCW	DEL180	11/61	73C	11/68
D6577	33058	02/74							BRCW	DEL181	11/61	73C	06/91
D6578	33059	02/74							BRCW	DEL182	11/61	73C	09/88
D6579	33060	01/74							BRCW	DEL183	12/61	73C	10/90
D6580	33119	02/74							BRCW	DEL184	12/61	73C	10/89
D6581	33061	12/74							BRCW	DEL185	12/61	73C	06/87
D6582	33062	01/74							BRCW	DEL186	12/61	73C	09/87
D6583	33063	12/73							BRCW	DEL187	01/62	73C	
D6584	33064	12/73							BRCW	DEL188	01/62	73C	
D6585	33065	01/74					Sealion	08/91	BRCW	DEL189	01/62	73C	
D6586	33201	01/74							BRCW	DEL157	02/62	73C	
D6587	33202	12/73					The Burma Star	03/91	BRCW	DEL158	02/62	73C	
D6588	33203	01/74	33301*						BRCW	DEL159	02/62	73C	04/91
D6589	33204	02/74							BRCW	DEL160	02/62	73C	
D6590	33205	01/74	33302	07/88	33205	10/88			BRCW	DEL161	02/62	73C	07/92
D6591	33206	01/74	33303*						BRCW	DEL162	03/62	73C	
D6592	33207	02/74					Earl Mountbatten of Burma	09/89	BRCW	DEL163	03/62	73C	

BR 1957 No.	Depot of Final Allocation	Disposal Code	Disposal Detail	Date Cut Up	Notes
D6513	EH	A	BR Eastleigh		
D6514					
D6515					
D6516	EH	C	BR Slade Green	01/86	Stored: [U] 09/85
D6517	EH	C	V Berry, Leicester	11/90	
D6518	EH	C	M C Processors, Glasgow	06/91	
D6519	EH	C	BR Eastleigh	09/92	
D6520	EH	C	M C Processors, Glasgow	05/91	
D6521					
D6522	EH	C	BR Eastleigh	11/86	
D6523	EH	C	V Berry, Leicester	11/90	
D6524	SL	C	M C Processors, Glasgow	02/92	Stored: [U] 09/89
D6525					
D6526	EH	C	V Berry, Leicester	11/90	
D6527	SL	A	BR Eastleigh		Withdrawn: 09/89, R/I: 09/89
D6528	EH	P	Class 33/1 Preservation Group		
D6529	EH	C	BR Eastleigh	09/92	
D6530	EH	D	To Departmental Stock - 968030	-	
D6531	SL	A	BR Stewarts Lane		
D6532					Named: 'Sultan' 04/88-01/89
D6533	EH	R	Rebuilt to vehicle No. 83301		
D6534					
D6535					
D6536					Stored: [U] 04/91
D6537					
D6538					
D6539					
D6540	SL	C	M C Processors, Glasgow	09/91	
D6541					
D6542	EH	C	BR Eastleigh	05/86	
D6543					Withdrawn: 02/88, R/I: 05/88
D6544					
D6545	SL	C	BR Eastleigh East	09/92	
D6546	EH	C	BRML Eastleigh	09/89	
D6547					
D6548					
D6549	EH	C	M C Processors, Glasgow	06/91	
D6550	EH	C	BREL Eastleigh	07/87	
D6551					
D6552	EH	P	Private at Ludgershall	-	
D6553					Withdrawn: 10/89, R/I: 01/90
D6554	HG	C	BR Slade Green	10/79	
D6555	SL	C	M C Processors, Glasgow	00/91	
D6556	SL	A	BR Stratford		
D6557	SL	C	M C Processors, Glasgow	05/91	
D6558					
D6559	HG	C	BR Selhurst	06/76	
D6560					
D6561	SL	C	M C Processors, Glasgow	05/91	
D6562	SL	C	V Berry, Leicester	10/90	
D6563	SL	C	V Berry, Leicester	11/90	
D6564					
D6565					
D6566					
D6567	SL	C	BR Eastleigh, by V Berry	03/89	
D6568					
D6569					Stored: [U] 02/87, R/I: 03/87
D6570					Withdrawn: 07/91, R/I: 08/91
D6571					
D6572	SL	C	BRML Eastleigh	08/87	
D6573	SL	C	M C Processors, Glasgow	09/91	
D6574	SL	P	BRCW Type 3 Preservation Group		
D6575					
D6576	HG	C	BR Workshops, Eastleigh	03/69	
D6577	SL	A	BR Eastleigh		Stored: [U] 02/88, R/I: 03/88
D6578	SL	C	M C Processors, Glasgow	07/91	
D6579	SL	C	BR Eastleigh	07/92	Stored: [U] 05/90
D6580	EH	C	M C Processors, Glasgow	04/92	Stored: [U] 09/89
D6581	SL	C	BREL Eastleigh	11/87	
D6582	SL	C	V Berry, Leicester	11/90	
D6583					
D6584					
D6585					
D6586					
D6587					Withdrawn: 02/88, R/I: 12/88
D6588	SL	P	BRCW Type 3 Preservation Group		
D6589					
D6590	SL	A	BR Stewarts Lane		Withdrawn: 09/87, R/I: 12/87
D6591					
D6592					

BR 1957 No.	TOPS No.	Date Re No.	First TOPS Re No.	Date Re No.	Second TOPS Re No.	Date Re No.	Name	Name Date	Built By	Works No.	Date Introduced	Depot of First Allocation	Date Withdrawn
D6593	33208	12/73							BRCW	DEL164	03/62	73C	
D6594	33209	12/73							BRCW	DEL165	03/62	73C	12/88
D6595	33210	01/74							BRCW	DEL166	04/62	73C	08/87
D6596	33211	02/74							BRCW	DEL167	04/62	73C	
D6597	33212	02/74							BRCW	DEL168	05/62	73C	09/87

ENGLISH ELECTRIC TYPE 3 Co-Co CLASS 37

BR 1957 No.	TOPS No.	Date Re No.	First TOPS Re No.	Date Re No.	Second TOPS Re No.	Date Re No.	Third TOPS Re No.	Date Re No.	Name	Name Date	Built By	Works No.	Date Introduced
D6600	37300	11/73	37429	03/86					Eisteddfod Genedlaethol	08/87	EE.VF	3560/D989	08/65
D6601	37301	05/74	37412	11/85					[Loch Lomond]	04/87-06/86	EE.VF	3561/D990	09/65
D6602	37302	03/74	37416	11/85							EE.VF	3562/D991	09/65
D6603	37303	03/74	37271	01/89							EE.VF	3563/D992	09/65
D6604	37304	02/74	37272	01/89							EE.VF	3564/D993	09/65
D6605	37305	03/74	37407	08/85					Loch Long	06/86	EE.VF	3565/D994	10/65
D6606	37306	04/74	37273	02/89							EE.VF	3566/D995	10/65
D6607	37307	03/74	37403	06/85					Glendarroch	11/88	EE.VF	3567/D996	10/65
D6608	37308	03/74	37274	02/89							EE.VF	3568/D997	11/65
D6700	37119	02/74	37350	03/88							EE.VF	2863/D579	12/60
D6701	37001	03/74	37707	12/87							EE.VF	2864/D580	12/60
D6702	37002	02/74	37351	05/89							EE.VF	2865/D581	12/60
D6703	37003	02/74							[First East Anglian Regiment]	03/64-03/64	EE.VF	2866/D582	12/60
D6704	37004	03/74									EE.VF	2867/D583	01/61
D6705	37005	02/74	37501	05/86					[Teeside Steelmaster]	03/87-01/91	EE.VF	2868/D584	01/61
D6706	37006	03/74	37798	09/86							EE.VF	2869/D585	01/61
D6707	37007	02/74	37506	04/86					British Steel Skinningrove	03/87	EE.VF	2870/D586	02/61
D6708	37008	03/74	37352	06/88	37008	07/89					EE.VF	2871/D587	02/61
D6709	37009	03/74									EE.VF	2872/D588	02/61
D6710	37010	01/74									EE.VF	2874/D590	02/61
D6711	37011	02/74									EE.VF	2875/D591	03/61
D6712	37012	02/74							[Loch Rannoch]	03/82-06/86	EE.VF	2873/D589	03/61
D6713	37013	02/74									EE.VF	2876/D592	03/61
D6714	37014	02/74	37709	02/88							EE.VF	2877/D593	03/61
D6715	37015	12/73									EE.VF	2878/D594	05/61
D6716	37016	03/74	37706	11/87					Conidae	05/91	EE.VF	2879/D595	06/61
D6717	37017	02/74	37503	03/86					British Steel Shelton	04/87	EE.VF	2880/D596	05/61
D6718	37018	02/74	37517	04/87							EE.VF	2881/D597	06/61
D6719	37019	05/74									EE.VF	2882/D598	06/61
D6720	37020	02/74	37702	12/86					Taff Merthyr	11/89	EE.VF	2883/D599	07/61
D6721	37021	02/74	37715	08/88					British Steel Teesside	08/92	EE.VF	2884/D600	07/61
D6722	37022	02/74	37512	01/87					Thornaby Demon	05/87	EE.VF	2885/D601	07/61
D6723	37023	03/74							Stratford	11/91	EE.VF	2886/D602	07/61
D6724	37024	02/74	37714	10/88					[Thornaby TMD]	09/92-02/93	EE.VF	2887/D603	08/61
D6725	37025	02/74									EE.VF	2888/D604	08/61
D6726	37026	02/74	37320	07/86	37026	09/89			Shap Fell	07/86	EE.VF	2889/D605	09/61
D6727	37027	02/74	37519	06/87					[Loch Eil]	10/81-02/87	EE.VF	2890/D606	09/61
D6728	37028	02/74	37505	04/86					[British Steel Workington]	03/87-04/91	EE.VF	2891/D607	09/61
D6729	37029	11/73									EE.VF	2892/D608	10/61
D6730	37030	02/74	37701	12/86							EE.VF	2893/D609	10/61
D6731	37031	02/74									EE.VF	2894/D610	10/61
D6732	37032	02/74	37353	06/88	37032	06/89					EE.VF	2895/D611	03/62
D6733	37033	10/73	37719	03/89							EE.VF	2896/D612	03/62
D6734	37034	03/74	37704	01/87							EE.VF	2897/D613	03/62
D6735	37035	02/74									EE.VF	2898/D614	04/62
D6736	37036	03/74	37507	04/86					[Hartlepool Pipe Mill]	12/87-06/92	EE.VF	2899/D615	04/62
D6737	37037	02/74	37321	07/86	37037	11/88			[Gartcosh]	07/86-05/92	EE.VF	2900/D616	05/62
D6738	37038	02/74									EE.VF	2901/D617	05/62
D6739	37039	02/74	37504	03/86					[British Steel Corby]	03/87-08/91	EE.VF	2902/D618	05/62
D6740	37040	02/74									EE.VF	2903/D619	06/62
D6741	37041	03/74	37520	09/87							EE.VF	2904/D620	06/62
D6742	37042	02/74									EE.VF	3034/D696	06/62
D6743	37043	02/74	37354	06/88	37043	06/92			[Loch Lomond]	10/81-06/86	EE.VF	3035/D697	06/62
D6744	37044	02/74	37710	02/88							EE.VF	3036/D698	06/62
D6745	37045	02/74	37355	06/88							EE.VF	3037/D699	07/62
D6746	37046	02/74									EE.VF	3038/D700	07/62
D6747	37047	03/74									EE.VF	3039/D701	07/62
D6748	37048	03/74									EE.VF	3040/D702	08/62
D6749	37049	01/74	37322	07/86	37049	06/88			Imperial	07/86	EE.VF	3041/D703	08/62
D6750	37050	11/73	37717	01/89					Stainless Pioneer	08/92	EE.VF	3042/D704	08/62
D6751	37051	02/74									EE.VF	3043/D705	08/62
D6752	37052	02/74	37713	08/88					British Steel Workington	08/92	EE.VF	3044/D706	09/62
D6753	37053	02/74									EE.VF	3045/D707	09/62
D6754	37054	02/74									EE.VF	3046/D708	09/62

BR 1957 No.	Depot of Final Allocation	Disposal Code	Disposal Detail	Date Cut Up	Notes
D6593					
D6594	SL	C	BRML Eastleigh	10/89	
D6595	SL	C	V Berry, Leicester	10/90	
D6596					
D6597	SL	C	M C Processors, Glasgow	05/91	

BR 1957 No.	Depot of First Allocation	Date Withdrawn	Depot of Final Allocation	Disposal Code	Disposal Detail	Date Cut Up	Notes
D6600	86A						Named: 'Sir Dyfed/County of Dyfed' 04/87-07/87
D6601	87E						
D6602	86A						
D6603	87E						
D6604	87E						
D6605	86A						
D6606	87E	09/91	CF	A	BR Cardiff		Stored: [U] 08/91
D6607	87E						
D6608	87E						Named: 'Isle of Mull' 01/86-11/88
D6700	30A						
D6701	30A						Stored: [U] 09/92
D6702	30A						Stored: [U] 09/92
D6703	30A						Nameplate covered
D6704	30A						
D6705	31B						
D6706	32A						
D6707	31B						
D6708	32B	02/92	TI	A	BR Stratford		
D6709	30A						
D6710	30A						
D6711	30A	08/87	ED	C	Rollesons, Wellington	08/89	
D6712	30A						
D6713	30A						
D6714	30A						
D6715	30A						
D6716	30A						
D6717	30A						
D6718	30A						
D6719	30A						
D6720	30A						
D6721	32A						
D6722	32B						
D6723	31B						
D6724	31B						
D6725	30A						
D6726	30A						Named: 'Loch Awe' 10/81-07/86
D6727	30A						
D6728	30A						
D6729	30A						
D6730	50B						
D6731	50B						
D6732	50B						
D6733	50B						
D6734	50B						
D6735	50B						
D6736	50B						
D6737	50B						Stored: [U] 08/91
D6738	50B						
D6739	50B						
D6740	50B						
D6741	50B						
D6742	41A						
D6743	41A						
D6744	41A						Stored: [U] 02/88, R/I: 02/88
D6745	41A						
D6746	41A						
D6747	41A						
D6748	41A						
D6749	41A						
D6750	41A						
D6751	41A						
D6752	41A						
D6753	41A						
D6754	41A						

BR 1957 No.	TOPS No.	Date Re No.	First TOPS Re No.	Date Re No.	Second TOPS Re No.	Date Re No.	Third TOPS Re No.	Date Re No.	Name	Name Date	Built By	Works No.	Date Introduced
D6755	37055	02/74									EE.VF	3047/D709	09/62
D6756	37056	02/74	37513	01/87							EE.VF	3048/D710	09/62
D6757	37057	02/74									EE.VF	3049/D711	10/62
D6758	37058	03/74									EE.VF	3050/D712	10/62
D6759	37059	11/73							Port of Tilbury	09/88	EE.VF	3051/D713	10/62
D6760	37060	07/74	37705	10/87							EE.VF	3052/D714	10/62
D6761	37061	02/74	37799	08/66					Sir Dyfed/County of Dyfed	11/87	EE.VF	3053/D715	10/62
D6762	37062	02/74							[British Steel Corby]	09/85-03/87	EE.VF	3054/D716	10/62
D6763	37063	02/74									EE.VF	3055/D717	11/62
D6764	37064	02/74	37515	03/87							EE.VF	3056/D718	11/62
D6765	37065	02/74									EE.VF	3057/D719	11/62
D6766	37066	02/74							[British Steel Worksington]	09/85-03/87	EE.VF	3058/D720	11/62
D6767	37067	02/74	37703	01/87							EE.VF	3059/D721	11/62
D6768	37068	01/74	37356	05/88	37068	06/89			Grainflow	09/87	EE.VF	3060/D722	11/62
D6769	37069	02/74							[Thornaby TMD]	09/86-05/92	EE.RSH	3061/8315	07/62
D6770	37070	01/74									EE.RSH	3062/8316	08/62
D6771	37071	02/74							[British Steel Skinningrove]	09/85-04/87	EE.RSH	3063/8317	08/62
D6772	37072	02/74									EE.RSH	3064/8318	09/62
D6773	37073	02/74							Fort William/An Gearasdan	08/91	EE.RSH	3065/8319	09/62
D6774	37074	02/74									EE.RSH	3066/8320	09/62
D6775	37075	02/74									EE.RSH	3067/8321	09/62
D6776	37076	10/73	37518	06/87							EE.RSH	3068/8322	10/62
D6777	37077	02/74							[British Steel Shelton]	09/85-07/87	EE.RSH	3069/8323	10/62
D6778	37078	02/74							[Teesside Steelmaster]	06/84-03/87	EE.RSH	3070/8324	10/62
D6779	37079	03/74	37357	06/88	37079	03/90			[Medite]	06/91-11/92	EE.RSH	3206/8325	11/62
D6780	37080	02/74									EE.RSH	3207/8326	11/62
D6781	37081	02/74	37797	10/86					[Loch Long]	10/81-05/86	EE.RSH	3208/8327	11/62
D6782	37082	02/74	37502	03/86					[British Steel Teesside]	03/87-01/91	EE.RSH	3209/8328	11/62
D6783	37083	03/74									EE.RSH	3210/8329	12/62
D6784	37084	10/73	37718	01/89					Hartlepool Pipe Mill	07/92	EE.RSH	3211/8330	12/62
D6785	37085	03/74	37711	06/88					Tremorfa Steel Works	11/88	EE.RSH	3212/8331	12/62
D6786	37086	02/74	37516	04/87							EE.RSH	3213/8332	12/62
D6787	37087	02/74									EE.RSH	3214/8333	12/62
D6788	37088	02/74	37323	07/86	37088	09/89			Clydesdale	07/86	EE.RSH	3215/8334	01/63
D6789	37089	02/74	37708	02/88							EE.RSH	3216/8335	01/63
D6790	37090	05/74	37508	05/86							EE.RSH	3217/8336	01/63
D6791	37091	01/74	37358	03/88					P & O Containers	04/88	EE.RSH	3218/8337	01/63
D6792	37092	02/74									EE.RSH	3219/8338	02/63
D6793	37093	02/74	37509	05/86							EE.RSH	3220/8339	02/63
D6794	37094	11/73	37716	12/88					British Steel Corby	07/92	EE.RSH	3221/8341	02/63
D6795	37095	02/74							[British Steel Teesside]	09/85-03/87	EE.RSH	3222/8342	03/63
D6796	37096	03/74									EE.VF	3225/D750	11/62
D6797	37097	02/74									EE.VF	3226/D751	12/62
D6798	37098	03/74									EE.VF	3227/D752	12/62
D6799	37099	03/74	37324	07/86	37099	08/89			Clydebridge	07/86	EE.VF	3228/D753	12/62
D6800	37100	02/74									EE.VF	3229/D754	12/62
D6801	37101	02/74									EE.VF	3230/D755	12/62
D6802	37102	02/74	37712	08/88					Teesside Steelmaster	09/92	EE.VF	3231/D756	01/63
D6803	37103	02/74	37511	07/86					Stockton Haulage	02/88	EE.VF	3232/D757	01/63
D6804	37104	02/74									EE.VF	3233/D758	01/63
D6805	37105	02/74	37796	11/86							EE.VF	3234/D759	01/63
D6806	37106	02/74									EE.VF	3235/D760	01/63
D6807	37107	01/74									EE.VF	3236/D761	01/63
D6808	37108	05/74	37325	07/86	37108	09/89			[Lanarkshire Steel]	08/86-02/92	EE.VF	3237/D762	01/63
D6809	37109	02/74									EE.VF	3238/D763	02/63
D6810	37110	02/74									EE.VF	3239/D764	02/63
D6811	37111	03/74	37326	07/86	37111	07/89			Glengarnock	07/86	EE.VF	3240/D765	02/63
D6812	37112	02/74	37510	06/86							EE.VF	3241/D766	02/63
D6813	37113	02/74							Radio Highland	09/89	EE.VF	3242/D767	02/63
D6814	37114	02/74							Dunrobin Castle	06/85	EE.VF	3243/D768	02/63
D6815	37115	05/74	37514	03/87							EE.VF	3244/D769	02/63
D6816	37116	02/74									EE.VF	3245/D770	03/63
D6817	37117	02/74	37521	04/88							EE.VF	3246/D771	03/63
D6818	37118	02/74	37359	11/88							EE.VF	3247/D772	03/63
D6819	37283	02/74	37895	02/87							EE.RSH	3264/8379	03/63
D6820	37120	03/74	37887	01/88					Castell Caerffilli/ Caerphilly Castle	09/92	EE.RSH	3265/8380	03/63
D6821	37121	02/74	37677	05/87							EE.RSH	3266/8381	04/63
D6822	37122	02/74	37692	01/87							EE.RSH	3267/8382	05/63
D6823	37123	03/74	37679	05/87							EE.RSH	3268/8383	04/63
D6824	37124	04/74	37894	01/87							EE.RSH	3269/8384	04/63
D6825	37125	02/74	37904	04/87							EE.RSH	3270/8385	05/63
D6826	37126	05/74	37676	05/87							EE.RSH	3271/8386	05/63
D6827	37127	05/74	37370	07/88							EE.RSH	3272/8387	05/64
D6828	37128	04/74									EE.RSH	3273/8388	06/63
D6829	37129	03/74	37669	08/87							EE.VF	3274/D803	03/63
D6830	37130	03/74	37681	04/87							EE.VF	3275/D804	03/63
D6831	37131	02/74									EE.VF	3276/D805	03/63
D6832	37132	02/74	37673	07/87							EE.VF	3277/D806	04/63
D6833	37133	05/74									EE.VF	3278/D807	04/63
D6834	37134	02/74	37684	03/87					Peak National Park	09/91	EE.VF	3279/D808	04/63
D6835	37135	05/74	37888	12/87					Petrolea	05/88	EE.VF	3280/D809	04/63

BR 1957 No.	Depot of First Allocation	Date Withdrawn	Depot of Final Allocation	Disposal Code	Disposal Detail	Date Cut Up	Notes
D6755	51L						
D6756	51L						
D6757	51L						
D6758	51L						
D6759	51L						
D6760	51L						
D6761	51L						
D6762	51L	03/89	TI	C	V Berry, Leicester	04/90	
D6763	51L						
D6764	51L						
D6765	51L						
D6766	51L						
D6767	51L						
D6768	51L						
D6769	51L						
D6770	51L						
D6771	51L						
D6772	51L						
D6773	51L						
D6774	51L						
D6775	51L						
D6776	51L						
D6777	51L						
D6778	51L						
D6779	50B						
D6780	50B						
D6781	50B						
D6782	50B						
D6783	50B						
D6784	52A						
D6785	52A						
D6786	52A						
D6787	52A						
D6788	52A						
D6789	52A						
D6790	52A						
D6791	52A						
D6792	52A						
D6793	52A						
D6794	52A						
D6795	52A						
D6796	41A	03/91	TI	C	M C Processors, at Doncaster	08/91	Stored: [U] 09/89, R/I: 11/89, Stored: [U] 03/91
D6797	41A						
D6798	41A						
D6799	41A						
D6800	41A						
D6801	41A						
D6802	41A						Named: 'The Cardiff Rod Mill' 11/88-06/92
D6803	41A						
D6804	41A						
D6805	41A						
D6806	41A						
D6807	41A						
D6808	41A						
D6809	41A						
D6810	41A						
D6811	41A						Named: 'Loch Eil Outward Bound' 04/75-07/86
D6812	41A						
D6813	41A						Withdrawn: 03/89, R/I: 08/89
D6814	41A						
D6815	41A						
D6816	41A						
D6817	41A						
D6818	41A						
D6819	88A						
D6820	88A						
D6821	88A						
D6822	88A						
D6823	88A						
D6824	88A						
D6825	88A						
D6826	88A						
D6827	88A						
D6828	88A						
D6829	88A						
D6830	88A	08/92	IM	A	ABB Transportation, Crewe		
D6831	88A						
D6832	88A						
D6833	88A						
D6834	88A						
D6835	88A						

BR 1957 No.	TOPS No.	Date Re No.	First TOPS Re No.	Date Re No.	Second TOPS Re No.	Date Re No.	Third TOPS Re No.	Date Re No.	Name	Name Date	Built By	Works No.	Date Introduced
D6836	37136	02/74	37905	12/86					Vulcan Enterprise	02/87	EE.VF	3281/D810	04/63
D6837	37137	04/74	37312	07/86	37137	03/89			Clyde Iron	07/86	EE.VF	3282/D811	04/63
D6838	37138	04/74									EE.VF	3283/D812	04/63
D6839	37139	05/74									EE.VF	3314/D813	05/63
D6840	37140	12/73									EE.VF	3315/D814	05/63
D6841	37141	06/74									EE.VF	3316/D815	05/63
D6842	37142	03/74									EE.VF	3317/D816	05/63
D6843	37143	05/74	37800	09/86					Glo Cymru	09/86	EE.VF	3318/D817	05/63
D6844	37144	04/74									EE.VF	3319/D818	06/63
D6845	37145	04/74	37313	08/86	37145	09/89	37382	06/92			EE.VF	3320/D819	06/63
D6846	37146	01/74									EE.VF	3321/D820	06/63
D6847	37147	09/74	37371	10/88							EE.VF	3322/D821	06/63
D6848	37148	04/74	37902	10/86					[British Steel Llanwern]	06/91-09/92	EE.VF	3323/D822	06/63
D6849	37149	09/74	37892	09/87					Ripple Lane	10/87	EE.VE	3324/D823	06/63
D6850	37150	04/74	37901	10/86					Mirrlees Pioneer	12/86	EE.VF	3325/D824	07/63
D6851	37151	05/74	37667	06/88					[Wensleydale]	10/88-12/90	EE.VF	3326/D825	07/63
D6852	37152	04/74	37310	03/86	37152	09/89			[British Steel Ravenscraig]	03/86-06/90	EE.VF	3327/D826	07/63
D6853	37153	04/74									EE.VF	3328/D827	07/63
D6854	37154	09/74							Johnson Stevens Agencies	05/92	EE.VF	3329/D828	07/63
D6855	37155	04/74	37897	12/86							EE.VF	3330/D829	07/63
D6856	37156	05/74	37311	03/86	37156	09/89			British Steel Hunterston	03/86	EE.VF	3331/D830	07/63
D6857	37157	04/74	37695	06/86							EE.VF	3332/D831	07/63
D6858	37158	03/74									EE.VF	3333/D832	08/63
D6859	37159	03/74	37372	06/88							EE.RSH	3337/8390	06/63
D6860	37160	02/74	37373	07/88							EE.RSH	3338/8391	07/63
D6861	37161	02/74	37899	12/86					County of West Glamorgan/Sir Gorllewin Morgannwg	04/91	EE.RSH	3339/8392	07/63
D6862	37162	04/74									EE.RSH	3340/8393	07/63
D6863	37163	02/74	37802	10/86							EE.RSH	3341/8394	08/63
D6864	37164	02/74	37675	05/87					William Cookworthy	05/87	EE.RSH	3342/8395	08/63
D6865	37165	02/74	37374	02/89	37165	07/89					EE.RSH	3343/8396	08/63
D6866	37166	02/74	37891	10/87							EE.RSH	3344/8397	09/63
D6867	37167	02/74									EE.RSH	3345/8398	09/63
D6868	37168	02/74	37890	10/87							EE.RSH	3346/8399	10/63
D6869	37169	05/74	37674	06/87							EE.VF	3347/D833	08/63
D6870	37170	05/74									EE.VF	3348/D834	08/63
D6871	37171	03/74	37690	02/87							EE.VF	3349/D835	09/63
D6872	37172	02/74	37686	02/87							EE.VF	3350/D836	09/63
D6873	37173	02/74	37801	09/86					Aberthaw/Aberddawan	09/86	EE.VF	3351/D837	09/63
D6874	37174	02/74									EE.VF	3352/D838	09/63
D6875	37175	03/74									EE.VF	3353/D839	09/63
D6876	37176	04/74	37883	08/88							EE.VF	3354/D840	10/63
D6877	37177	03/74	37885	10/88							EE.VF	3355/D841	10/63
D6878	37178	05/74									EE.VF	3356/D842	10/63
D6879	37179	12/73	37691	02/87							EE.RSH	3357/8400	10/63
D6880	37180	11/73	37886	09/88					[Sir Dyfed/County of Dyfed]	05/81-01/87	EE.RSH	3358/8401	10/63
D6881	37181	02/74	37687	03/87							EE.RSH	3359/8402	10/63
D6882	37182	03/74	37670	08/87							EE.RSH	3360/8403	10/63
D6883	37183	03/74	37884	10/88					Gartcosh	07/92	EE.RSH	3361/8404	11/63
D6884	37184	02/74									EE.RSH	3362/8405	11/63
D6885	37185	03/74									EE.RSH	3363/8406	12/63
D6886	37186	05/74	37898	12/86							EE.RSH	3364/8407	11/63
D6887	37187	02/74	37683	03/87							EE.RSH	3365/8408	01/64
D6888	37188	04/74							[Jimmy Shand]	05/85-07/91	EE.RSH	3366/8409	01/64
D6889	37189	04/74	37672	07/87					Freight Transport Association	09/87	EE.RSH	3367/8410	01/64
D6890	37190	02/74	37314	07/86	37190	09/88			[Dalzell]	07/86-06/92	EE.RSH	3368/8411	01/64
D6891	37191	02/74							[International Youth Year 1985]	01/85-11/85	EE.RSH	3369/8412	02/64
D6892	37192	03/74	37694	06/86					The Lass O'Ballochmyle	10/90	EE.RSH	3370/8413	02/64
D6893	37193	03/74	37375	09/88							EE.RSH	3371/8414	02/64
D6894	37194	03/74							British International Freight Association	09/90	EE.RSH	3372/8415	03/64
D6895	37195	02/74	37689	02/87							EE.RSH	3373/8416	03/64
D6896	37196	04/74							[Tre Pol and Pen]	07/85-05/87	EE.RSH	3374/8417	04/64
D6897	37197	02/74									EE.RSH	3375/8418	04/64
D6898	37198	02/74									EE.RSH	3376/8419	05/64
D6899	37199	02/74	37376	06/88							EE.VF	3377/D843	10/63
D6900	37200	02/74	37377	05/88							EE.VF	3378/D844	10/63
D6901	37201	10/73									EE.VF	3379/D845	10/63
D6902	37202	01/74									EE.VF	3380/D846	10/63
D6903	37203	05/74									EE.VF	3381/D847	10/63
D6904	37204	04/74	37378	05/88							EE.VF	3382/D848	11/63
D6905	37205	05/74	37688	02/87					Great Rocks	06/88	EE.VF	3383/D849	11/63
D6906	37206	06/74	37906	12/86							EE.VF	3384/D850	11/63
D6907	37207	05/74							[William Cookworthy]	05/82-05/87	EE.VF	3385/D851	11/63
D6908	37208	02/74	37803	10/86							EE.VF	3386/D852	11/63
D6909	37209	12/73									EE.VF	3387/D853	12/63
D6910	37210	02/74	37693	07/86					Sir William Arrol	03/90	EE.VF	3388/D854	11/63
D6911	37211	02/74									EE.VF	3389/D855	12/63
D6912	37212	02/74									EE.VF	3390/D856	01/64

BR 1957 No.	Depot of First Allocation	Date Withdrawn	Depot of Final Allocation	Disposal Code	Disposal Detail	Date Cut Up	Notes
D6836	88A						
D6837	87E						Stored: [U] 03/87
D6838	88A						
D6839	88A						
D6840	88A						
D6841	88A						
D6842	88A						
D6843	88A						
D6844	88A						
D6845	88A						
D6846	88A						
D6847	88A						
D6848	88A						Stored: [U] 09/92
D6849	88A						
D6850	88A						
D6851	88A						
D6852	87E						
D6853	87E						
D6854	87E						
D6855	87E						
D6856	87E						
D6857	88A						
D6858	87E						
D6859	88A						
D6860	88A						
D6861	87E						
D6862	87E						
D6863	87E						
D6864	87E						
D6865	87E						
D6866	86A						
D6867	86A						
D6868	86A						
D6869	87E						
D6870	87E						
D6871	88A						
D6872	86A						
D6873	87E						
D6874	86A						
D6875	86A						
D6876	86A						
D6877	86A						
D6878	86A						
D6879	86A						
D6880	87E						
D6881	86A						
D6882	86A						
D6883	87E						
D6884	87E						
D6885	87E						
D6886	87E						
D6887	87E						
D6888	87E						
D6889	86A						
D6890	87E						
D6891	87E						
D6892	87E						
D6893	87E						
D6894	87E						
D6895	87E						
D6896	87E						
D6897	87E						
D6898	87E						
D6899	86A						
D6900	86A						
D6901	86A						
D6902	86A						
D6903	86A						
D6904	86A						
D6905	87E						
D6906	87E						
D6907	87E						
D6908	87E						
D6909	87E						
D6910	87E						
D6911	87E						
D6912	87E						

BR 1957 No.	TOPS No.	Date Re No.	First TOPS Re No.	Date Re No.	Second TOPS Re No.	Date Re No.	Third TOPS Re No.	Date Re No.	Name	Name Date	Built By	Works No.	Date Introduced
D6913	37213	03/74									EE.VF	3391/D857	01/64
D6914	37214	04/74									EE.VF	3392/D858	01/64
D6915	37215	03/74									EE.VF	3393/D859	01/64
D6916	37216	03/74							Great Eastern	03/92	EE.VF	3394/D860	01/64
D6917	37217	04/74									EE.VF	3395/D861	01/64
D6918	37218	04/74									EE.VF	3396/D862	01/64
D6919	37219	02/74									EE.VF	3405/D863	01/64
D6920	37220	01/74							[Westerleigh]	06/90-11/92	EE.VF	3406/D864	01/64
D6921	37221	02/74									EE.VF	3407/D865	01/64
D6922	37222	05/74									EE.VF	3408/D866	01/64
D6923	37223	03/74									EE.VF	3409/D867	02/64
D6924	37224	05/74	37680	04/87							EE.VF	3410/D868	01/64
D6925	37225	05/74									EE.VF	3411/D869	02/64
D6926	37226	05/74	37379	07/88							EE.VF	3412/D870	02/64
D6927	37227	06/74									EE.VF	3413/D871	02/64
D6928	37228	02/74	37696	06/86							EE.VF	3414/D872	02/64
D6929	37229	03/74							[The Cardiff Rod Mill]	05/84-10/88	EE.VF	3415/D873	02/64
D6930	37230	04/74									EE.VF	3416/D874	03/64
D6931	37231	05/74	37896	12/86							EE.VF	3417/D875	03/64
D6932	37232	03/74							The Institution of Railway Signal Engineers	12/90	EE.VF	3418/D876	03/64
D6933	37233	04/74	37889	10/87							EE.VF	3419/D877	04/64
D6934	37234	03/74	37685	03/87							EE.VF	3420/D878	04/64
D6935	37235	02/74							[The Coal Merchants Association of Scotland]	11/87-03/91	EE.VF	3421/D879	04/64
D6936	37236	03/74	37682	03/87							EE.VF	3422/D880	04/64
D6937	37237	09/74	37893	10/87							EE.VF	3423/D881	05/64
D6938	37238	02/74									EE.VF	3424/D882	06/64
D6939	37239	05/74							The Coal Merchants Association of Scotland	03/91	EE.VF	3496/D927	08/64
D6940	37240	02/74									EE.VF	3497/D928	08/64
D6941	37241	04/74									EE.VF	3498/D929	09/64
D6942	37242	02/74									EE.VF	3499/D930	09/64
D6943	37243	02/74	37697	05/86							EE.VF	3500/D931	09/64
D6944	37244	02/74									EE.VF	3501/D932	09/64
D6945	37245	03/74									EE.VF	3502/D933	10/64
D6946	37246	03/74	37698	02/86					Coedbach	09/88	EE.VF	3503/D934	10/64
D6947	37247	04/74	37671	07/87					Tre Pol and Pen	07/87	EE.VF	3504/D935	10/64
D6948	37248	04/74									EE.VF	3505/D936	10/64
D6949	37249	02/74	37903	02/87							EE.VF	3506/D937	12/64
D6950	37250	02/74									EE.VF	3507/D938	12/64
D6951	37251	02/74							The Northern Lights	11/92	EE.VF	3508/D939	12/64
D6952	37252	02/74									EE.VF	3509/D940	01/65
D6953	37253	02/74	37699	12/85							EE.VF	3510/D941	01/65
D6954	37254	04/74									EE.VF	3511/D942	01/65
D6955	37255	04/74									EE.VF	3512/D943	01/65
D6956	37256	04/74	37678	05/87							EE.VF	3513/D944	01/65
D6957	37257	04/74	37668	05/88					[Leyburn]	10/88-12/90	EE.VF	3514/D945	01/65
D6958	37258	02/74									EE.VF	3515/D946	01/65
D6959	37259	03/74	37380	06/88							EE.VF	3519/D948	01/65
D6960	37260	12/73							[Radio Highland]	07/84-09/89	EE.VF	3520/D949	01/65
D6961	37261	11/73							Caithness	06/85	EE.VF	3521/D950	01/65
D6962	37262	02/74							Dounreay	06/85	EE.VF	3522/D951	01/65
D6963	37263	03/74									EE.VF	3523/D952	01/65
D6964	37264	01/74									EE.VF	3524/D953	01/65
D6965	37265	03/74	37430	03/86					Cwmbran	05/86	EE.VF	3525/D954	02/65
D6966	37266	03/74	37422	01/86							EE.VF	3526/D955	02/65
D6967	37267	05/74	37421	12/85					Strombidae	03/91	EE.VF	3527/D956	02/65
D6968	37268	02/74	37401	06/85					Mary Queen of Scots	11/85	EE.VF	3528/D957	02/65
D6969	37269	02/74	37417	11/85					Highland Region	12/85	EE.VF	3529/D958	02/65
D6970	37270	02/74	37409	10/85					Loch Awe	08/86	EE.VF	3530/D959	03/65
D6971	37271	03/74	37418	11/85					Pectinidae	03/91	EE.VF	3531/D960	03/65
D6972	37272	03/74	37431	04/86					Bullidae	05/91	EE.VF	3532/D961	03/65
D6973	37273	03/74	37410	10/85					Aluminium 100	09/86	EE.VF	3533/D962	04/65
D6974	37274	03/74	37402	06/85					Oor Wullie	12/85	EE.VF	3534/D963	04/65
D6975	37275	02/74							[Stainless Pioneer]	12/88-05/92	EE.VF	3535/D964	04/65
D6976	37276	03/74	37413	10/85					Loch Eil Outward Bound	11/86	EE.VF	3536/D965	04/65
D6977	37277	05/74	37415	11/85							EE.VF	3537/D966	04/65
D6978	37278	03/74									EE.VF	3538/D967	04/65
D6979	37279	04/74	37424	01/86					Isle of Mull	03/90	EE.VF	3539/D968	05/65
D6980	37280	02/74									EE.VF	3540/D969	05/65
D6981	37281	05/74	37428	02/86					David Lloyd George	05/87	EE.VF	3541/D970	05/65
D6982	37282	03/74	37405	08/85					Strathclyde Region	04/86	EE.VF	3542/D971	05/65
D6983											EE.VF	3543/D972	05/65
D6984	37284	05/74	37381	06/88							EE.VF	3544/D973	05/65
D6985	37285	03/74									EE.VF	3545/D974	05/65
D6986	37286	03/74	37404	12/85					Ben Cruachan	01/86	EE.VF	3546/D975	06/65
D6987	37287	05/74	37414	11/85							EE.VF	3547/D976	06/65
D6988	37288	03/74	37427	02/86					Bont Y Bermo	04/86	EE.VF	3548/D977	06/65
D6989	37289	02/74	37408	08/85					Loch Rannoch	09/86	EE.VF	3549/D978	06/65

BR 1957 No.	Depot of First Allocation	Date Withdrawn	Depot of Final Allocation	Disposal Code	Disposal Detail	Date Cut Up	Notes
D6913	87E						
D6914	86A						
D6915	87E						
D6916	87E						Reverted to No. D6916 in 03/92
D6917	87E						
D6918	87E						
D6919	86A						
D6920	86A						
D6921	87E						
D6922	87E						
D6923	87E						
D6924	87E						
D6925	87E						
D6926	87E						
D6927	87E						
D6928	87E						
D6929	87E						
D6930	87E						
D6931	87E						
D6932	87E						
D6933	87E						
D6934	87E						
D6935	87E						
D6936	87E						
D6937	86A						
D6938	86A						
D6939	85A						
D6940	85A						
D6941	85A						
D6942	85A						
D6943	86A						
D6944	86A						
D6945	86A						Stored: [U] 04/87, R/I: 05/87
D6946	86A						
D6947	86A						
D6948	86A						
D6949	86A						Stored: [U] 08/91
D6950	86A						
D6951	86A						
D6952	86A						
D6953	86A						
D6954	86A						
D6955	86A						
D6956	86A						
D6957	86A						
D6958	86A						
D6959	41A						
D6960	41A	09/89	IS	C	M C Processors, at Doncaster	09/91	
D6961	41A						
D6962	41A						
D6963	41A						
D6964	41A						
D6965	41A						
D6966	41A						
D6967	41A						
D6968	41A						
D6969	86A						
D6970	86A						
D6971	86A						Named: 'An Comunn Gaidhealach' 10/86-02/91
D6972	86A						Named: 'Sir Powys/County of Powis' 06/87 - 05/91
D6973	86A						
D6974	86A						
D6975	86A						
D6976	86A						
D6977	86A						
D6978	86A						
D6979	86A						Named: 'Glendarroch' 12/87 - 11/88
D6980	86A						
D6981	86A						
D6982	86A						
D6983	86A	04/66	86A	C	R S Hayes, Bridgend	06/66	
D6984	86A						
D6985	86A						
D6986	86A						
D6987	86A						
D6988	86A						
D6989	86A						

BR 1957 No.	TOPS No.	Date Re No.	First TOPS Re No.	Date Re No.	Second TOPS Re No.	Date Re No.	Third TOPS Re No.	Date Re No.	Name	Name Date	Built By	Works No.	Date Introduced
D6990	37290	03/74	37411	10/85					[Institution of Railway Signal Engineers]	05/87-10/90	EE.VF	3550/D979	06/65
D6991	37291	03/74	37419	12/85							EE.VF	3551/D980	06/65
D6992	37292	02/74	37425	04/86					Sir Robert McAlpine/ Concrete Bob	10/86	EE.VF	3552/D981	07/65
D6993	37293	02/74									EE.VF	3553/D982	07/65
D6994	37294	11/73									EE.VF	3554/D983	07/65
D6995	37295	05/74	37406	08/85					The Saltire Society	06/86	EE.VF	3555/D984	07/65
D6996	37296	01/74	37423	01/86					Sir Murray Morrison 1874-1948	05/88	EE.VF	3556/D985	07/65
D6997	37297	03/74	37420	12/85					The Scottish Hosteller	06/86	EE.VF	3557/D986	07/65
D6998	37298	05/74									EE.VF	3558/D987	08/65
D6999	37299	02/74	37426	02/86					[Y Lein Fach/ Vale of Rheidol]	05/86-07/91	EE.VF	3559/D988	08/65

BEYER PEACOCK TYPE 3 B-B CLASS 35

BR 1957 No.	Built By	Works No.	Date Introduced	Depot of First Allocation	Date Withdrawn	Depot of Final Allocation	Disposal Code
D7000	B.Peacock	7894	05/61	82A	07/73	OC	C
D7001	B.Peacock	7895	07/61	82B	03/74	OC	C
D7002	B.Peacock	7896	07/61	82A	10/71	82A	C
D7003	B.Peacock	7897	08/61	82A	01/72	82A	C
D7004	B.Peacock	7898	08/61	82A	06/72	82A	C
D7005	B.Peacock	7899	09/61	82A	07/72	82A	C
D7006	B.Peacock	7900	10/61	82A	09/71	82A	C
D7007	B.Peacock	7901	10/61	82D	04/72	82A	C
D7008	B.Peacock	7902	10/61	82A	01/72	82A	C
D7009	B.Peacock	7903	11/61	82A	05/73	82A	C
D7010	B.Peacock	7904	11/61	82A	01/72	82A	C
D7011	B.Peacock	7905	12/61	82A	03/75	OC	C
D7012	B.Peacock	7906	12/61	82A	01/72	82A	C
D7013	B.Peacock	7907	12/61	82A	01/72	82A	C
D7014	B.Peacock	7908	12/61	82A	01/72	82A	C
D7015	B.Peacock	7909	12/61	82A	06/72	81A	C
D7016	B.Peacock	7910	01/62	82A	07/74	OC	C
D7017	B.Peacock	7911	01/62	82A	03/75	OC	P
D7018	B.Peacock	7912	01/62	82A	03/75	OC	P
D7019	B.Peacock	7913	02/62	82A	09/72	81A	C
D7020	B.Peacock	7914	02/62	82A	01/72	82A	C
D7021	B.Peacock	7915	02/62	82A	01/72	82A	C
D7022	B.Peacock	7916	02/62	88A	03/75	OC	C
D7023	B.Peacock	7917	02/62	82A	05/73	OC	C
D7024	B.Peacock	7918	03/62	88A	01/72	82A	C
D7025	B.Peacock	7919	03/62	88A	01/72	82A	C
D7026	B.Peacock	7920	03/62	82A	10/74	OC	C
D7027	B.Peacock	7921	04/62	82A	11/71	81A	C
D7028	B.Peacock	7922	04/62	88A	01/75	OC	C
D7029	B.Peacock	7923	04/62	88A	02/75	OC	P
D7030	B.Peacock	7924	04/62	88A	05/73	81A	C
D7031	B.Peacock	7925	04/62	88A	05/73	81A	C
D7032	B.Peacock	7926	05/62	88A	05/73	81A	C
D7033	B.Peacock	7927	05/62	88A	01/72	81A	C
D7034	B.Peacock	7928	05/62	88A	01/72	81A	C
D7035	B.Peacock	7929	06/62	88A	01/72	81A	C
D7036	B.Peacock	7930	06/62	88A	06/72	81A	C
D7037	B.Peacock	7931	06/62	88A	09/72	81A	C
D7038	B.Peacock	7932	06/62	88A	07/72	82A	C
D7039	B.Peacock	7933	06/62	88A	06/72	82A	C
D7040	B.Peacock	7934	07/62	82A	01/72	82A	C
D7041	B.Peacock	7935	07/62	82A	01/72	82A	C
D7042	B.Peacock	7936	07/62	82A	01/72	82A	C
D7043	B.Peacock	7937	07/62	82A	01/72	82A	C
D7044	B.Peacock	7938	08/62	82A	05/73	81A	C
D7045	B.Peacock	7949	08/62	82A	11/72	82A	C
D7046	B.Peacock	7950	08/62	82A	01/72	81A	C
D7047	B.Peacock	7951	08/62	82A	01/72	82A	C
D7048	B.Peacock	7952	09/62	82A	01/72	81A	C
D7049	B.Peacock	7953	10/62	82A	01/72	81A	C
D7050	B.Peacock	7954	10/62	82A	11/72	82A	C
D7051	B.Peacock	7955	10/62	82A	01/72	81A	C
D7052	B.Peacock	7956	10/62	82A	11/72	86A	C
D7053	B.Peacock	7957	10/62	82A	01/72	81A	C
D7054	B.Peacock	7958	11/62	82A	12/72	81A	C
D7055	B.Peacock	7959	11/62	82A	04/73	82A	C
D7056	B.Peacock	7960	11/62	88A	01/72	82A	C
D7057	B.Peacock	7961	11/62	88A	01/72	82A	C
D7058	B.Peacock	7962	11/62	88A	10/71	86A	C
D7059	B.Peacock	7963	11/62	88A	10/71	86A	C
D7060	B.Peacock	7964	12/62	88A	10/71	86A	C

BR 1957 No.	Depot of First Allocation	Date Withdrawn	Depot of Final Allocation	Disposal Code	Disposal Detail	Date Cut Up	Notes
D6990	86A						
D6991	86A						
D6992	86A						
D6993	86A						
D6994	86A						
D6995	86A						
D6996	86A						
D6997	86A						
D6998	86A						
D6999	86A						

BR 1957 No.	Disposal Detail	Date Cut Up	Notes
D7000	BREL Swindon	10/75	
D7001	G Cohen, Kettering	05/75	
D7002	BREL Swindon	07/72	
D7003	BREL Swindon	08/72	
D7004	BREL Swindon	08/72	Withdrawn: 01/72, R/I: 01/72
D7005	BREL Swindon	10/72	Withdrawn: 01/72, R/I: 01/72
D7006	BREL Swindon	09/72	Stored: [U] 09/71
D7007	BREL Swindon	06/72	Withdrawn: 01/72, R/I: 01/72
D7008	BREL Swindon	09/72	
D7009	BREL Swindon	10/74	
D7010	BREL Swindon	11/72	
D7011	Marple and Gillot, Sheffield	03/77	Stored: [U] 01/75
D7012	BREL Swindon	07/72	
D7013	BREL Swindon	08/72	
D7014	BREL Swindon	08/72	
D7015	BREL Swindon	09/72	
D7016	BREL Swindon	06/75	
D7017	D&EPG on West Somerset Railway	-	Withdrawn: 05/73, R/I: 05/73
D7018	D&EPG on West Somerset Railway	-	
D7019	BREL Swindon	10/72	Withdrawn: 07/72, R/I: 09/72
D7020	BREL Swindon	09/72	
D7021	BREL Swindon	06/72	
D7022	G Cohen, Kettering	01/77	Stored: [U] 01/75, R/I: 02/75
D7023	BREL Swindon	05/75	
D7024	BREL Swindon	11/72	
D7025	BREL Swindon	07/72	
D7026	G Cohen, Kettering	01/77	
D7027	BREL Swindon	08/72	Stored: [U] 11/71
D7028	G Cohen, Kettering	02/77	Stored: [U] 10/74, R/I: 10/74, Stored: [U] 11/74
D7029	Diesel Traction Group on NYMR	-	Stored: [U] 01/75, R/I: 02/75
D7030	Birds, Long Marston	03/74	
D7031	BREL Swindon	09/75	
D7032	BREL Swindon	07/75	
D7033	BREL Swindon	11/72	
D7034	BREL Swindon	09/72	
D7035	BREL Swindon	08/72	
D7036	BREL Swindon	10/72	Withdrawn: 01/72, R/I: 04/72
D7037	BREL Swindon	11/72	Stored: [U] 08/75
D7038	BREL Swindon	06/73	Withdrawn: 01/72, R/I: 01/72
D7039	BREL Swindon	08/72	
D7040	BREL Swindon	08/72	
D7041	BREL Swindon	09/72	
D7042	BREL Swindon	07/72	
D7043	BREL Swindon	08/72	
D7044	Birds, Long Marston	03/74	
D7045	BREL Swindon	08/73	
D7046	BREL Swindon	07/72	
D7047	BREL Swindon	08/72	
D7048	BREL Swindon	08/72	
D7049	BREL Swindon	07/72	
D7050	BREL Swindon	00/73	
D7051	BREL Swindon	07/72	
D7052	BREL Swindon	05/73	Withdrawn: 01/72, R/I: 05/72, Withdrawn: 07/72, R/I: 10/72
D7053	BREL Swindon	08/72	
D7054	BREL Swindon	04/75	Withdrawn: 01/72, R/I: 05/72, Withdrawn: 07/72, R/I: 09/72
D7055	BREL Swindon	11/75	Allocated Dptl No. 968004*
D7056	BREL Swindon	07/72	
D7057	BREL Swindon	10/72	
D7058	BREL Swindon	06/72	
D7059	BREL Swindon	07/72	
D7060	BREL Swindon	10/72	

BR 1957 No.	Built By	Works No.	Date Introduced	Depot of First Allocation	Date Withdrawn	Depot of Final Allocation	Disposal Code
D7061	B.Peacock	7965	12/62	88A	01/72	81A	C
D7062	B.Peacock	7966	01/63	88A	10/71	86A	C
D7063	B.Peacock	7967	12/62	88A	10/71	86A	C
D7064	B.Peacock	7968	01/63	88A	10/71	86A	C
D7065	B.Peacock	7969	01/63	88A	01/72	81A	C
D7066	B.Peacock	7970	01/63	88A	11/71	81A	C
D7067	B.Peacock	7971	02/63	88A	10/71	86A	C
D7068	B.Peacock	7972	02/63	88A	12/72	82A	C
D7069	B.Peacock	7973	02/63	82A	10/71	86A	C
D7070	B.Peacock	7974	03/63	82A	09/72	86A	C
D7071	B.Peacock	7975	03/63	82A	01/72	81A	C
D7072	B.Peacock	7976	03/63	82A	10/71	86A	C
D7073	B.Peacock	7977	03/63	82A	12/71	86A	C
D7074	B.Peacock	7978	03/63	82A	12/72	82A	C
D7075	B.Peacock	7979	03/63	83A	05/73	82A	C
D7076	B.Peacock	7980	05/63	81A	05/73	82A	D
D7077	B.Peacock	7981	12/63	82A	07/72	82A	C
D7078	B.Peacock	7982	05/63	81A	10/71	86A	C
D7079	B.Peacock	7983	12/63	82A	10/71	86A	C
D7080	B.Peacock	7984	12/63	86A	11/72	82A	C
D7081	B.Peacock	7985	12/63	82A	09/71	86A	C
D7082	B.Peacock	7986	06/63	88A	04/72	86A	C
D7083	B.Peacock	7987	06/63	88A	10/71	86A	C
D7084	B.Peacock	7988	06/63	88A	10/72	86A	C
D7085	B.Peacock	7989	06/63	88A	10/72	81A	C
D7086	B.Peacock	7990	07/63	88A	01/72	86A	C
D7087	B.Peacock	7991	07/63	88A	10/72	82A	C
D7088	B.Peacock	7992	10/63	88A	01/72	86A	C
D7089	B.Peacock	7993	07/63	88A	05/73	82A	D
D7090	B.Peacock	7994	09/63	86A	06/72	86A	C
D7091	B.Peacock	7995	09/63	86A	08/72	86A	C
D7092	B.Peacock	7996	12/63	86A	06/72	86A	C
D7093	B.Peacock	7997	12/63	82A	11/74	OC	C
D7094	B.Peacock	7998	12/63	86A	11/72	86A	C
D7095	B.Peacock	7999	12/63	86A	10/72	86A	C
D7096	B.Peacock	8000	12/63	86A	12/72	82A	D
D7097	B.Peacock	8001	12/63	86A	12/72	82A	C
D7098	B.Peacock	8002	01/64	83A	12/72	86A	C
D7099	B.Peacock	8003	01/64	83A	10/72	86A	C
D7100	B.Peacock	8004	02/64	83A	11/72	82A	C

BR TYPE 2 Bo-Bo CLASS 25

Class continued from D5299

BR 1957 No.	TOPS No.	Date Re No.	TOPS Re No.	Date Re No.	Built By	Works No.	Date Introduced	Depot of First Allocation	Date Withdrawn	Depot of Final Allocation
D7500	25150	02/74	-	-	BR Derby	-	10/64	16A	06/82	LO
D7501	25151	02/74	-	-	BR Derby	-	10/64	16A	09/82	LO
D7502	25152	04/74	-	-	BR Derby	-	10/64	16A	01/84	CW
D7503	25153	04/74	-	-	BR Derby	-	10/64	16A	04/83	CD
D7504	25154	03/74	-	-	BR Derby	-	10/64	16A	03/85	BS
D7505	25155	04/74	-	-	BR Derby	-	10/64	16A	12/80	SP
D7506	25156	07/74	-	-	BR Derby	-	10/64	16A	12/81	CD
D7507	25157	05/74	-	-	BR Derby	-	10/64	16A	10/82	CD
D7508	25158	05/74	-	-	BR Derby	-	11/64	16A	05/83	CD
D7509	25159	03/74	-	-	BR Derby	-	11/64	16A	11/80	CD
D7510	25160	04/74	-	-	BR Derby	-	11/64	16A	10/82	CD
D7511	25161	04/74	-	-	BR Derby	-	11/64	16A	11/84	CD
D7512	25162	04/74	-	-	BR Derby	-	11/64	16A	05/81	CD
D7513	25163	03/74	-	-	BR Derby	-	11/64	16A	11/80	CD
D7514	25164	04/74	-	-	BR Derby	-	11/64	16A	08/83	CD
D7515	25165	04/74	-	-	BR Derby	-	12/64	16A	11/78	CD
D7516	25166	04/74	-	-	BR Derby	-	12/64	16A	11/80	CD
D7517	25167	02/74	-	-	BR Derby	-	12/64	16A	05/83	CD
D7518	25168	03/74	-	-	BR Derby	-	12/64	16A	05/83	CD
D7519	25169	04/73	-	-	BR Derby	-	01/65	16A	09/81	CD
D7520	25170	07/74	-	-	BR Derby	-	12/64	16A	08/82	CD
D7521	25171	06/74	-	-	BR Derby	-	01/65	16A	10/78	HA
D7522	25172	05/74	-	-	BR Derby	-	01/65	16A	02/81	SP
D7523	25173	04/74	-	-	BR Derby	-	01/65	16A	03/87	CD
D7524	25174	03/74	-	-	BR Derby	-	01/65	16A	09/76	CW
D7525	25175	03/74	-	-	BR Derby	-	01/65	16A	11/85	CD
D7526	25176	04/74	-	-	BR Derby	-	01/65	16A	03/87	CD
D7527	25177	03/74	-	-	BR Derby	-	01/65	16A	11/82	CW
D7528	25178	03/74	-	-	BR Derby	-	02/65	16A	04/85	CD
D7529	25179	04/74	-	-	BR Derby	-	02/65	16A	11/82	CW
D7530	25180	04/74	-	-	BR Derby	-	02/65	16A	11/82	CW
D7531	25181	04/74	-	-	BR Derby	-	02/65	16A	08/86	CD
D7532	25182	02/74	-	-	BR Derby	-	02/65	16A	01/85	CD

BR 1957 No.	Disposal Detail	Date Cut Up	Notes
D7061	BREL Swindon	08/72	
D7062	BREL Swindon	08/72	
D7063	BREL Swindon	11/72	
D7064	BREL Swindon	09/72	
D7065	BREL Swindon	09/72	
D7066	BREL Swindon	09/72	Stored: [U] 11/71
D7067	BREL Swindon	08/72	
D7068	BREL Swindon	04/75	
D7069	BREL Swindon	08/72	
D7070	BREL Swindon	10/72	Withdrawn: 01/72, R/I: 05/72, Withdrawn: 07/72, R/I: 08/72
D7071	BREL Swindon	09/72	
D7072	BREL Swindon	09/72	Stored: [U] 09/71
D7073	BREL Swindon	10/72	Stored: [U] 11/71
D7074	BREL Swindon	08/75	
D7075	Birds, Long Marston	02/74	
D7076	To Departmental Stock - D7076	-	
D7077	BREL Swindon	10/72	Completed: 05/63
D7078	BREL Swindon	05/72	
D7079	BREL Swindon	08/72	Completed: 05/63
D7080	BREL Swindon	05/73	Completed: 05/63
D7081	BREL Swindon	08/72	Completed: 05/63, Stored: [U] 08/71
D7082	BREL Swindon	10/72	
D7083	BREL Swindon	09/72	
D7084	BREL Swindon	11/72	
D7085	BREL Swindon	03/73	Withdrawn: 07/72, R/I: 09/72
D7086	BREL Swindon	09/72	
D7087	BREL Swindon	09/73	
D7088	BREL Swindon	10/72	
D7089	To Departmental Stock - 968005	-	
D7090	BREL Swindon	09/72	
D7091	BREL Swindon	10/72	
D7092	BREL Swindon	09/72	
D7093	G Cohen, Kettering	02/77	Withdrawn: 05/73, R/I: 08/73, Stored: [U] 10/74
D7094	BREL Swindon	07/73	
D7095	BREL Swindon	12/72	
D7096	To Departmental Stock - D7096	-	
D7097	BREL Swindon	03/75	
D7098	BREL Swindon	03/75	
D7099	BREL Swindon	10/72	
D7100	BREL Swindon	12/74	Stored: [U] 10/72

BR 1957 No.	Disposal Code	Disposal Detail	Date Cut Up	Notes
D7500	C	BREL Swindon	06/85	
D7501	C	BR Toton, by V Berry	11/87	
D7502	C	BREL Swindon	02/85	
D7503	C	BREL Swindon	02/87	
D7504	C	V Berry, Leicester	07/87	
D7505	C	BREL Swindon	09/81	Stored: [U] 11/80
D7506	C	BREL Swindon	05/82	
D7507	C	BREL Swindon	03/87	
D7508	C	V Berry, Leicester	06/87	
D7509	C	BREL Swindon	03/81	
D7510	C	V Berry, Leicester	07/87	
D7511	C	V Berry, Leicester	03/88	
D7512	C	BREL Swindon	06/82	
D7513	C	BREL Swindon	03/83	
D7514	C	V Berry, Leicester	07/87	
D7515	C	BREL Derby	10/79	
D7516	C	BREL Swindon	07/81	
D7517	C	BREL Swindon	04/84	
D7518	C	BREL Swindon	04/84	Withdrawn: 05/81, R/I: 05/81
D7519	C	BREL Swindon	02/82	
D7520	C	BREL Derby	04/83	Withdrawn: 04/82, R/I: 06/82
D7521	C	BR Arbroath	01/79	
D7522	C	BREL Swindon	06/83	
D7523	P	The Railway Age, Crewe	-	
D7524	C	BREL Derby	07/78	
D7525	C	V Berry, Leicester	01/87	
D7526	C	V Berry, Leicester	07/88	
D7527	C	BREL Swindon	06/86	
D7528	C	V Berry, Leicester	12/88	
D7529	C	BREL Swindon	02/87	
D7530	C	V Berry, Leicester	06/87	
D7531	C	BR Eastleigh, by V Berry	07/88	
D7532	C	BREL Swindon	08/85	

BR 1957 No.	TOPS No.	Date Re No.	TOPS Re No.	Date Re No.	Built By	Works No.	Date Introduced	Depot of First Allocation	Date Withdrawn	Depot of Final Allocation
D7533	25183	02/74	-	-	BR Derby	-	02/65	16A	12/80	CW
D7534	25184	07/74	-	-	BR Derby	-	03/65	16A	08/83	LO
D7535	25185	02/74	-	-	BR Derby	-	03/65	16A	11/84	CD
D7536	25186	04/74	-	-	BR Derby	-	03/65	16A	11/82	TO
D7537	25187	04/74	-	-	BR Derby	-	03/65	16A	11/82	LO
D7538	25188	04/74	-	-	BR Derby	-	03/65	16A	08/82	LO
D7539	25189	02/74	-	-	BR Derby	-	03/65	16A	07/85	CD
D7540	25190	06/74	-	-	BR Derby	-	04/65	16A	01/87	CD
D7541	25191	04/74	-	-	BR Derby	-	04/65	16A	03/87	CD
D7542	25192	02/74	-	-	BR Derby	-	04/65	16A	05/85	CD
D7543	25193	02/74	-	-	BR Derby	-	04/65	16A	11/84	CD
D7544	25194	02/74	-	-	BR Derby	-	05/65	16A	06/85	CD
D7545	25195	02/74	-	-	BR Derby	-	05/65	16A	06/85	CD
D7546	25196	02/74	-	-	BR Derby	-	05/65	16A	03/86	CD
D7547	25197	02/74	-	-	BR Derby	-	05/65	D16	12/80	BS
D7548	25198	03/74	-	-	BR Derby	-	05/65	D16	02/86	KD
D7549	25199	03/74	-	-	BR Derby	-	05/65	D16	02/87	CD
D7550	25200	04/74	-	-	BR Derby	-	06/65	D16	02/86	CD
D7551	25201	03/74	-	-	BR Derby	-	06/65	D16	01/87	CD
D7552	25202	02/74	-	-	BR Derby	-	06/65	D16	06/86	CD
D7553	25203	02/74	-	-	BR Derby	-	06/65	D16	12/80	CW
D7554	25204	05/74	-	-	BR Derby	-	07/65	D16	07/80	CW
D7555	25205	04/74	-	-	BR Derby	-	07/65	D16	05/86	KD
D7556	25206	05/74	-	-	BR Derby	-	07/65	D16	05/86	CD
D7557	25207	02/74	-	-	BR Derby	-	08/65	D16	11/84	BS
D7558	25208	09/73	-	-	BR Derby	-	08/65	D16	05/84	LO
D7559	25209	04/74	-	-	BR Derby	-	08/65	D16	12/85	CD
D7560	25210	12/73	-	-	BR Derby	-	09/65	D16	04/85	CD
D7561	25211	11/73	-	-	BR Derby	-	10/65	D16	07/86	CD
D7562	25212	02/74	-	-	BR Derby	-	12/65	D16	05/85	CD
D7563	25213	11/73	-	-	BR Derby	-	11/65	D16	03/87	CD
D7564	25214	05/74	-	-	BR Derby	-	12/65	D16	11/82	CD
D7565	25215	12/73	-	-	BR Derby	-	12/65	D16	06/83	CD
D7566	25216	04/74	-	-	BR Derby	-	01/66	D16	12/80	CD
D7567	25217	04/74	-	-	BR Derby	-	01/66	D16	02/81	CD
D7568	25218	03/74	-	-	BR Derby	-	09/63	16A	01/85	KD
D7569	25219	05/74	-	-	BR Derby	-	10/63	14B	04/83	CD
D7570	25220	02/74	-	-	BR Derby	-	10/63	16A	06/82	CD
D7571	25221	02/74	-	-	BR Derby	-	10/63	14B	02/84	LO
D7572	25222	03/74	-	-	BR Derby	-	10/63	14B	12/80	LO
D7573	25223	04/74	-	-	BR Derby	-	10/63	14B	10/80	LA
D7574	25224	02/74	-	-	BR Derby	-	10/63	14B	05/86	CD
D7575	25225	02/74	-	-	BR Derby	-	11/63	16A	10/80	LA
D7576	25226	09/74	-	-	BR Derby	-	11/63	16A	06/86	CD
D7577	25227	06/74	-	-	BR Derby	-	11/63	16A	06/83	CD
D7578	25228	04/74	-	-	BR Darlington	-	11/63	16A	03/84	LO
D7579	25229	04/74	-	-	BR Darlington	-	12/63	16A	05/85	CD
D7580	25230	03/74	-	-	BR Darlington	-	12/63	16A	08/86	CD
D7581	25231	09/74	-	-	BR Darlington	-	12/63	16A	08/85	CD
D7582	25232	02/74	-	-	BR Darlington	-	01/64	16A	12/80	ED
D7583	25233	03/74	-	-	BR Darlington	-	01/64	16A	03/83	LO
D7584	25234	02/74	-	-	BR Darlington	-	02/64	16A	03/85	CD
D7585	25235	11/73	-	-	BR Darlington	-	02/64	16A	04/85	CD
D7586	25236	03/74	-	-	BR Darlington	-	02/64	16A	12/84	KD
D7587	25237	11/73	-	-	BR Darlington	-	03/64	16A	06/85	CD
D7588	25238	05/74	-	-	BR Darlington	-	03/64	16A	10/80	ED
D7589	25239	04/74	-	-	BR Darlington	-	04/64	16A	11/84	CD
D7590	25240	02/74	-	-	BR Darlington	-	04/64	16A	08/83	CD
D7591	25241	04/74	-	-	BR Darlington	-	06/64	16A	05/81	ED
D7592	25242	02/74	-	-	BR Darlington	-	05/64	16A	05/84	CD
D7593	25243	03/74	-	-	BR Darlington	-	05/64	16A	09/83	CW
D7594	25244	10/73	-	-	BR Darlington	-	06/64	16A	07/86	CD
D7595	25245	03/74	-	-	BR Darlington	-	06/64	16A	06/85	CD
D7596	25246	02/74	-	-	BR Darlington	-	07/64	16A	10/81	ED
D7597	25247	03/74	-	-	BR Darlington	-	08/64	16A	06/83	KD
D7598	25248	02/74	-	-	BR Derby	-	02/66	41A	11/82	TO
D7599	25249	02/74	-	-	BR Derby	-	02/66	41A	01/87	CD
D7600	25250	02/74	-	-	BR Derby	-	02/66	41A	05/84	CD
D7601	25251	04/74	-	-	BR Derby	-	02/66	41A	01/85	CD
D7602	25252	03/74	-	-	BR Derby	-	02/66	41A	03/80	BS
D7603	25253	03/74	-	-	BR Derby	-	02/66	41A	09/83	CD
D7604	25254	04/74	-	-	BR Derby	-	03/66	41A	09/86	CD
D7605	-	-	-	-	BR Derby	-	03/66	41A	03/72	D16
D7606	25256	03/74	-	-	BR Derby	-	03/66	41A	04/85	CD
D7607	25257	04/74	-	-	BR Derby	-	03/66	41A	12/85	CD
D7608	25258	04/74	-	-	BR Derby	-	04/66	41A	12/84	CD
D7609	25259	03/74	-	-	BR Derby	-	04/66	41A	08/86	CD
D7610	25260	03/74	-	-	BR Derby	-	04/66	41A	12/82	BS
D7611	25261	04/74	-	-	BR Derby	-	04/66	65A	01/81	BS
D7612	25262	03/74	25901	12/85	BR Derby	-	04/66	65A	03/87	KD
D7613	25263	05/74	-	-	BR Derby	-	04/66	65A	11/80	BS

BR 1957 No.	Disposal Code	Disposal Detail	Date Cut Up	Notes
D7533	C	BREL Swindon	09/81	Stored: [U] 11/80
D7534	C	BREL Swindon	02/84	
D7535	P	Paignton & Dartmouth Railway	-	
D7536	C	BREL Swindon	01/87	Stored: [U] 08/82
D7537	C	BREL Swindon	01/87	
D7538	C	BREL Swindon	02/87	
D7539	C	BREL Doncaster	05/86	
D7540	C	V Berry, Leicester	06/87	
D7541	P	North Yorkshire Moors Railway	-	
D7542	C	V Berry, Leicester	07/87	
D7543	C	V Berry, Leicester	07/87	
D7544	A	BR Bescot		
D7545	C	V Berry, Leicester	06/87	
D7546	C	V Berry, Leicester	07/87	
D7547	C	BREL Swindon	02/83	Stored: [U] 11/80
D7548	C	V Berry, Leicester	12/88	
D7549	C	V Berry, Leicester	06/87	
D7550	C	V Berry, Leicester	01/87	
D7551	C	V Berry, Leicester	07/87	
D7552	C	V Berry, Leicester	07/89	
D7553	C	BREL Swindon	02/82	Stored: [U] 11/80
D7554	C	BREL Swindon	10/80	
D7555	A	BR Bescot		
D7556	A	BRML Doncaster		
D7557	C	V Berry, Leicester	06/87	
D7558	C	V Berry, Leicester	07/87	
D7559	C	M C Processors, Glasgow	03/92	
D7560	C	BREL Doncaster	04/86	
D7561	A	BR Bescot		
D7562	C	V Berry, Leicester	07/87	
D7563	C	V Berry, Leicester	12/90	
D7564	C	BREL Swindon	09/86	
D7565	C	BREL Swindon	12/83	
D7566	C	BREL Derby	01/83	
D7567	C	BREL Derby	01/82	
D7568	C	BREL Swindon	05/85	
D7569	C	BREL Swindon	02/87	
D7570	C	BREL Swindon	04/85	
D7571	C	BREL Swindon	12/86	
D7572	C	BREL Swindon	12/81	
D7573	C	BREL Swindon	11/80	
D7574	C	V Berry, Leicester	03/87	Withdrawn: 01/83, R/I: 03/83
D7575	C	BREL Swindon	11/80	
D7576	C	BREL Doncaster	08/86	
D7577	C	BREL Swindon	12/83	
D7578	C	V Berry, Leicester	07/87	
D7579	C	BREL Doncaster	09/86	
D7580	C	V Berry, Leicester	07/87	
D7581	C	V Berry, Leicester	01/87	
D7582	C	BREL Swindon	04/83	
D7583	C	BREL Swindon	04/85	
D7584	C	V Berry, Leicester	05/89	
D7585	P	Bo'ness & Kinneil Railway		
D7586	C	BREL Swindon	11/86	
D7587	C	BREL Doncaster	04/86	Stored: [U] 01/84, R/I: 01/84
D7588	C	BREL Swindon	07/83	
D7589	C	BREL Swindon	09/86	
D7590	C	BREL Swindon	12/83	
D7591	C	BREL Swindon	10/81	
D7592	C	BREL Swindon	11/84	
D7593	C	BREL Swindon	01/84	
D7594	P	Nene Valley Railway	-	
D7595	C	BREL Doncaster	08/86	
D7596	C	BREL Swindon	01/83	
D7597	C	BREL Swindon	12/83	
D7598	C	BREL Swindon	04/86	
D7599	C	V Berry, Leicester	06/87	
D7600	C	BREL Swindon	01/85	
D7601	C	BREL Swindon	12/86	
D7602	C	BREL Swindon	09/80	
D7603	C	BREL Swindon	10/85	
D7604	C	V Berry, Leicester	06/89	
D7605	C	BREL Derby	06/72	
D7606	C	BREL Swindon	10/86	
D7607	C	V Berry, Leicester	10/86	Withdrawn: 12/83, R/I: 01/84, Withdrawn: 01/84, R/I: 07/84
D7608	C	BREL Swindon	02/86	Stored: [U] 10/85
D7609	A	BR Bescot		
D7610	C	V Berry, Leicester	06/87	
D7611	C	BREL Derby, by V Berry	12/85	
D7612	P	East Lancashire Railway	-	
D7613	C	BREL Swindon	04/83	

BR 1957 No.	TOPS No.	Date Re No.	TOPS Re No.	Date Re No.	Built By	Works No.	Date Introduced	Depot of First Allocation	Date Withdrawn	Depot of Final Allocation
D7614	25264	03/74	-	-	BR Derby	-	05/66	65A	12/80	TO
D7615	25265	03/74	-	-	BR Derby	-	05/66	65A	03/87	CD
D7616	25266	12/73	-	-	BR Derby	-	05/66	65A	09/86	CD
D7617	25267	02/74	-	-	BR Derby	-	08/66	65A	02/81	TO
D7618	25268	05/74	25902	12/85	BR Derby	-	08/66	65A	03/87	KD
D7619	25269	04/74	-	-	BR Derby	-	08/66	65A	04/86	KD
D7620	25270	06/74	-	-	BR Derby	-	08/66	65A	11/82	TO
D7621	25271	04/74	-	-	BR Derby	-	08/66	65A	10/81	BS
D7622	25272	05/74	-	-	BR Derby	-	09/66	65A	07/81	BS
D7623	25273	03/74	-	-	BR Derby	-	09/66	65A	02/81	BS
D7624	25274	05/74	-	-	B.Peacock	8034	07/65	41A	05/82	BS
D7625	25275	04/74	-	-	B.Peacock	8035	07/65	41A	04/82	BS
D7626	25276	04/74	25903	01/86	B.Peacock	8036	09/65	41A	03/87	KD
D7627	25277	02/74	-	-	B.Peacock	8037	09/65	41A	04/84	BS
D7628	25278	01/74	-	-	B.Peacock	8038	09/65	41A	03/86	CD
D7629	25279	01/74	-	-	B.Peacock	8039	09/65	41A	03/87	CD
D7630	25280	04/74	-	-	B.Peacock	8040	09/65	41A	11/81	TO
D7631	25281	04/74	-	-	B.Peacock	8041	09/65	41A	02/81	SP
D7632	25282	09/73	-	-	B.Peacock	8042	10/65	41A	03/86	CD
D7633	25283	02/74	25904	11/85	B.Peacock	8043	10/65	41A	03/87	KD
D7634	25284	10/73	-	-	B.Peacock	8044	10/65	41A	01/85	CD
D7635	25285	04/74	-	-	B.Peacock	8045	11/65	41A	03/86	CD
D7636	25286	04/74	25905	11/85	B.Peacock	8046	11/65	41A	09/86	CD
D7637	25287	03/74	-	-	B.Peacock	8047	11/65	41A	12/85	CD
D7638	25288	10/73	-	-	B.Peacock	8048	11/65	41A	03/87	CD
D7639	25289	02/74	-	-	B.Peacock	8049	11/65	41A	01/84	BS
D7640	25290	03/74	-	-	B.Peacock	8050	11/65	41A	10/81	CD
D7641	25291	02/74	-	-	B.Peacock	8051	01/66	41A	05/81	CD
D7642	25292	02/74	-	-	B.Peacock	8052	01/66	41A	10/81	CD
D7643	25293	02/74	-	-	B.Peacock	8053	02/66	41A	02/81	CD
D7644	25294	05/74	-	-	B.Peacock	8054	02/66	41A	10/82	CD
D7645	25295	02/74	-	-	B.Peacock	8055	03/66	41A	06/78	CW
D7646	25296	02/74	25906	12/85	B.Peacock	8056	03/66	41A	10/86	KD
D7647	25297	12/73	25907	11/85	B.Peacock	8057	04/66	41A	09/86	CD
D7648	25298	04/74	-	-	B.Peacock	8058	04/66	41A	03/85	CD
D7649	25299	02/74	-	-	B.Peacock	8059	04/66	41A	10/81	CW
D7650	25300	11/73	-	-	B.Peacock	8060	05/66	D16	12/85	CD
D7651	25301	03/74	-	-	B.Peacock	8061	05/66	D16	12/83	CD
D7652	25302	02/74	-	-	B.Peacock	8062	05/66	D16	06/85	CD
D7653	25303	02/74	-	-	B.Peacock	8063	06/66	LMML	02/86	CD
D7654	25304	04/74	-	-	B.Peacock	8064	06/66	LMML	11/82	CD
D7655	25305	03/74	-	-	B.Peacock	8065	07/66	LMML	07/83	BS
D7656	25306	05/74	-	-	B.Peacock	8066	07/66	LMML	02/85	CD
D7657	25307	05/74	25908	12/85	B.Peacock	8067	07/66	LMML	10/86	KD
D7658	25308	04/74	-	-	B.Peacock	8068	07/66	LMML	10/83	TO
D7659	25309	04/74	25909	12/85	B.Peacock	8069	07/66	LMML	09/86	CD
D7660	25310	02/74	-	-	BR Derby	-	12/66	DO1	10/82	CW
D7661	25311	02/74	-	-	BR Derby	-	12/66	DO1	03/86	CD
D7662	25312	04/74	-	-	BR Derby	-	12/66	DO1	06/82	CD
D7663	25313	03/74	-	-	BR Derby	-	11/66	DO1	03/87	CD
D7664	25314	03/74	-	-	BR Derby	-	11/66	DO1	03/83	CW
D7665	25315	03/74	25910	12/85	BR Derby	-	12/66	DO1	03/87	KD
D7666	25316	03/74	25911	12/85	BR Derby	-	12/66	DO1	09/86	CD
D7667	25317	02/74	-	-	BR Derby	-	01/67	DO1	04/83	CD
D7668	25318	03/74	-	-	BR Derby	-	01/67	DO1	07/82	SP
D7669	25319	03/74	-	-	BR Derby	-	03/67	DO1	03/83	CD
D7670	25320	03/74	-	-	BR Derby	-	01/67	DO1	12/83	CO
D7671	25321	05/74	-	-	BR Derby	-	02/67	DO1	09/86	CD
D7672	25322	04/74	25912	11/85	BR Derby	-	02/67	D16	04/91	HO
D7673	25323	11/73	-	-	BR Derby	-	03/67	D16	03/86	CD
D7674	25324	04/74	-	-	BR Derby	-	03/67	D16	12/85	CD
D7675	25325	02/74	-	-	BR Derby	-	04/67	D16	12/85	CD
D7676	25326	02/74	-	-	BR Derby	-	04/67	D16	01/85	CD
D7677	25327	05/74	-	-	BR Derby	-	04/67	D16	02/84	BS

ENGLISH ELECTRIC TYPE 1 Bo-Bo CLASS 20

BR 1957 No.	TOPS No.	Date Re No.	First TOPS Re No.	Date Re No.	Second TOPS Re No.	Date Re No.	Name	Name Date	Built By	Works No.	Date Introduced
D8000	20050	02/74							EE.VF	2347/D375	06/57
D8001	20001	03/74							EE.VF	2348/D376	07/57
D8002	20002	09/74							EE.VF	2349/D377	07/57
D8003	20003	05/74							EE.VF	2350/D378	08/57
D8004	20004	03/74							EE.VF	2351/D379	08/57
D8005	20005	02/74							EE.VF	2352/D380	09/57

BR 1957 No.	Disposal Code	Disposal Detail	Date Cut Up	Notes
D7614	C	BREL Swindon	06/83	
D7615	P	PeakRail, Matlock	-	
D7616	C	V Berry, Leicester	01/89	
D7617	C	BREL Derby	02/82	
D7618	C	V Berry, Leicester	07/87	
D7619	C	V Berry, Leicester	02/87	
D7620	C	BREL Swindon	05/86	
D7621	C	BREL Derby	10/82	
D7622	C	BREL Derby	08/82	
D7623	C	BREL Swindon	04/82	
D7624	C	BREL Swindon	04/85	
D7625	C	BREL Swindon	02/85	
D7626	C	V Berry, Leicester	06/89	
D7627	C	BREL Swindon	10/85	
D7628	P	North Yorkshire Moors Railway	-	
D7629	P	Llangollen Railway	-	
D7630	C	BREL Swindon	07/83	
D7631	C	BREL Swindon	05/81	
D7632	C	V Berry, Leicester	12/88	
D7633	P	Severn Valley Railway	-	
D7634	C	BREL Swindon	10/86	
D7635	C	V Berry, Leicester	07/87	Stored: [U] 08/86
D7636	C	M C Processors, Glasgow	02/90	
D7637	C	V Berry, Leicester	10/86	Withdrawn: 07/76, R/I: 09/76, Stored: [U] 05/85
D7638	C	V Berry, Leicester	03/88	
D7639	C	BREL Swindon	10/84	
D7640	C	BREL Derby	04/83	
D7641	C	BREL Swindon	11/82	
D7642	C	BREL Swindon	02/83	
D7643	C	BREL Swindon	08/81	
D7644	C	BREL Swindon	04/85	
D7645	C	BREL Derby	11/81	
D7646	C	V Berry, Leicester	06/87	
D7647	C	V Berry, Leicester	07/89	Stored: [U] 06/85, R/I: 08/85
D7648	C	BREL Doncaster	02/86	
D7649	C	BREL Derby	05/82	
D7650	C	BREL Doncaster	05/85	
D7651	C	BREL Swindon	08/84	
D7652	C	BREL Doncaster	07/86	
D7653	C	V Berry, Leicester	01/87	
D7654	C	BREL Swindon	06/85	
D7655	D	To Departmental Stock - 97251	-	Withdrawn: 06/81, R/I: 08/82
D7656	C	V Berry, Leicester	06/87	
D7657	D	To Departmental Stock 968026	-	
D7658	C	BREL Swindon	11/84	
D7659	P	East Lancashire Railway	-	Stored: [U] 08/86
D7660	D	To Departmental Stock - 97250	-	
D7661	C	V Berry, Leicester	12/88	Withdrawn: 03/85, R/I: 10/85
D7662	C	BREL Swindon	08/85	
D7663	P	Llangollen Railway	-	
D7664	D	To Departmental Stock - 97252	-	
D7665	C	V Berry, Leicester	07/87	
D7666	C	V Berry, Leicester	12/88	Withdrawn: 02/84, R/I: 04/84, Stored: [U] 08/86
D7667	C	BREL Swindon	05/86	
D7668	C	V Berry, Leicester	05/87	
D7669	C	BREL Swindon	12/85	
D7670	C	BREL Swindon	05/85	
D7671	P	Midland Railway Centre, Butterley	-	
D7672	P	North Staffordshire Railway	-	Withdrawn: 02/84, R/I: 05/84, Named: Tamworth Castle In Departmental Stock - ADB968027 03/87 - 09/90
D7673	C	V Berry, Leicester	07/87	
D7674	C	V Berry, Leicester	07/87	
D7675	C	M C Processors, Glasgow	02/90	
D7676	C	BREL Swindon	03/86	
D7677	C	V Berry, Leicester	06/87	

BR 1957 No	Depot of First Allocation	Date Withdrawn	Depot of Final Allocation	Disposal Code	Disposal Detail	Date Cut Up	Notes
D8000	1D	12/80	TO	P	National Railway Museum, York	-	Stored: [U] 08/75, R/I: 11/75, Stored: [U] 10/80
D8001	1D	04/88	TO		Class 20 Society, Butterley	-	Stored: [U] 08/75, R/I: 11/75
D8002	1D	02/88	IM	C	M C Processors, Glasgow	08/90	
D8003	1D	12/82	IM	C	BREL Crewe	01/84	Stored: [U] 08/81, R/I: 09/82
D8004	1D	10/90	TO	C	M C Processors, Glasgow	08/91	Stored: [U] 08/75, R/I: 11/75, Stored: [U] 01/83, R/I: 12/83
D8005	1D	07/89	TO	C	M C Processors, Glasgow	12/90	Stored: [U] 09/83, R/I: 11/83, Stored: [U] 01/84, R/I: 04/84

BR 1957 No.	TOPS No.	Date Re No.	First TOPS Re No.	Date Re No.	Second TOPS Re No.	Date Re No.	Name	Name Date	Built By	Works No.	Date Introduced
D8006	20006	03/74							EE.VF	2353/D381	09/57
D8007	20007	04/74							EE.VF	2354/D382	09/57
D8008	20008	03/74							EE.VF	2355/D383	10/57
D8009	20009	11/73							EE.VF	2356/D384	10/57
D8010	20010	04/74							EE.VF	2357/D385	10/57
D8011	20011	04/74							EE.VF	2358/D386	11/57
D8012	20012	03/74							EE.VF	2359/D387	11/57
D8013	20013	04/74							EE.VF	2360/D388	11/57
D8014	20014	02/74							EE.VF	2361/D389	12/57
D8015	20015	09/74							EE.VF	2362/D390	12/57
D8016	20016	04/74							EE.VF	2363/D391	01/58
D8017	20017	04/74							EE.VF	2364/D392	01/58
D8018	20018	09/74							EE.VF	2365/D393	02/58
D8019	20019	03/74							EE.VF	2366/D394	03/58
D8020	20020	04/74							EE.RSH	2742/8052	10/59
D8021	20021	12/74							EE.RSH	2743/8053	10/59
D8022	20022	03/74							EE.RSH	2744/8054	10/59
D8023	20023	03/74	20301	04/86	20023	11/86			EE.RSH	2745/8055	11/59
D8024	20024	03/74							EE.RSH	2746/8056	11/59
D8025	20025	02/74							EE.RSH	2747/8057	11/59
D8026	20026	03/74							EE.RSH	2748/8058	12/59
D8027	20027	03/74							EE.RSH	2749/8059	12/59
D8028	20028	02/74							EE.RSH	2750/8060	12/59
D8029	20029	02/74							EE.RSH	2751/8061	12/59
D8030	20030	04/74							EE.RSH	2752/8062	12/59
D8031	20031	02/74							EE.RSH	2753/8063	01/60
D8032	20032	09/73							EE.RSH	2754/8064	01/60
D8033	20033	04/74							EE.RSH	2755/8065	02/60
D8034	20034	09/74							EE.RSH	2756/8066	03/60
D8035	20035	03/74							EE.VF	2757/D482	09/59
D8036	20036	04/74							EE.VF	2758/D483	10/59
D8037	20037	02/74							EE.VF	2759/D484	10/59
D8038	20038	05/74							EE.VF	2760/D485	10/59
D8039	20039	04/74							EE.VF	2761/D486	10/59
D8040	20040	03/74							EE.VF	2762/D487	10/59
D8041	20041	04/74	20901	03/89			Nancy	05/89	EE.VF	2763/D488	11/59
D8042	20042	03/74							EE.VF	2764/D489	11/59
D8043	20043	03/74							EE.VF	2765/D490	11/59
D8044	20044	04/74							EE.VF	2766/D491	11/59
D8045	20045	03/74							EE.VF	2767/D492	12/59
D8046	20046	05/74							EE.VF	2768/D493	12/59
D8047	20047	04/74							EE.VF	2769/D494	12/59
D8048	20048	05/74							EE.VF	2770/D495	12/59
D8049	20049	02/74							EE.VF	2771/D496	12/59
D8050	20128	11/73							EE.RSH	2956/8208	03/61
D8051	20051	02/74							EE.RSH	2957/8209	03/61
D8052	20052	02/74							EE.RSH	2958/8210	03/61
D8053	20053	01/74							EE.RSH	2959/8211	04/61
D8054	20054	02/74							EE.RSH	2960/8212	06/61
D8055	20055	04/74							EE.RSH	2961/8213	04/61
D8056	20056	02/74							EE.RSH	2962/8214	04/61
D8057	20057	02/74							EE.RSH	2963/8215	05/61
D8058	20058	03/74							EE.RSH	2964/8216	05/61
D8059	20059	04/74	20302	04/86	20059	11/86			EE.RSH	2965/8217	05/61
D8060	20060	10/73	20902	05/89			Lorna	05/89	EE.RSH	2966/8218	05/61
D8061	20061	05/74							EE.RSH	2967/8219	05/61
D8062	20062	03/74							EE.RSH	2968/8220	05/61
D8063	20063	02/74							EE.RSH	2969/8221	06/61
D8064	20064	02/74							EE.RSH	2970/8222	06/61
D8065	20065	04/74							EE.RSH	2971/8223	06/61
D8066	20066	02/74							EE.RSH	2972/8224	06/61
D8067	20067	03/74							EE.RSH	2973/8225	06/61

BR 1957 No.	Depot of First Allocation	Date Withdrawn	Depot of Final Allocation	Disposal Code	Disposal Detail	Date Cut Up	Notes
D8006	1D	10/90	TO	C	M C Processors, Glasgow	06/91	Stored: [U] 08/75, R/I: 11/75, Stored: [U] 10/81, R/I:03/82, Stored: [U] 10/82, R/I: 07/84
D8007	1D						Stored: [U] 04/81, R/I: 09/81, Stored: [U] 02/83, R/I:05/84, Stored: [S] 03/92
D8008	1D	02/89	TE	A	BR Thornaby		Stored: [U] 11/75, R/I: 04/76
D8009	1D	07/89	TE	A	BR Thornaby		Withdrawn: 06/89, R/I: [S] 07/89
D8010	1D	12/91	TO	A	BR Toton		
D8011	1D	02/87	TI	A	BR Derby		
D8012	1D	03/76	TO	C	BREL Glasgow	01/77	Stored: [U] 11/75
D8013	1D	12/91	TO	A	BR Toton		Stored: [U] 11/75, R/I: 08/76, Stored: [U] 10/81, R/I: 03/82, Stored: [U] 05/83, R/I: 05/84
D8014	1D	03/76	TO	C	BREL Glasgow	01/77	Stored: [U] 11/75
D8015	1D	07/87	IM	C	BR Thornaby, by V Berry	09/88	
D8016	1D						Stored: [U] 11/75, R/I: 11/75, Stored: [U] 01/83, R/I:01/84
D8017	1D	12/82	TI	C	BREL Crewe	03/85	
D8018	1D	12/76	TO	C	BREL Glasgow	05/78	Stored: [U] 04/76
D8019	1D	11/91	TO	A	BR Toton		Stored: [U] 11/83, R/I: 01/84, Stored: [U] 10/91
D8020	14B	10/90	TO	P	Scottish RPS, Bo'ness	-	Stored: [U] 04/81, R/I: 09/81, Stored: [U] 02/82, Withdrawn: 01/83, R/I: 06/84
D8021	34B	03/91	TO	C	M C Processors, Glasgow	03/92	
D8022	34B	08/88	IM	C	V Berry, Leicester	04/90	Stored: [U] 12/82, R/I: 02/83
D8023	34B	05/91	TO	C	M C Processors, Glasgow	04/92	
D8024	34B	05/77	ED	C	BREL Glasgow	03/78	
D8025	34B	09/91	TO	A	BR Frodingham		Stored: [U] 12/82, R/I: 04/83, Stored: [U] 09/91
D8026	34B	10/90	TO	C	M C Processors, Glasgow	09/91	Stored: [U] 08/83, R/I: 12/83
D8027	34B	12/82	HA	C	BREL Glasgow	08/86	Stored: [U] 09/82
D8028	64H	11/91	TO	A	BR Toton		Stored: [U] 04/83, R/I: 04/84
D8029	64H	07/91	TE	A	BR Thornaby		Stored: [U] 08/81, R/I: 09/82
D8030	64H	10/90	TO	C	M C Processors, Glasgow	08/91	Stored: [U] 02/81, R/I: 09/82, Stored: [U] 01/83, R/I: 07/83
D8031	61C	09/90	TO	P	Private on KWVR	-	Stored: [U] 12/82, R/I: 09/83, Withdrawn: 10/89, R/I: 11/89
D8032	60A						Withdrawn: 02/90. R/I: 07/90, Withdrawn: 10/90, R/I: 11/90 Stored: [S] 03/92
D8033	60A	11/87	TI	C	BREL Crewe	01/80	
D8034	60A	10/90	TO	C	M C Processors, Glasgow	07/91	
D8035	32A	07/91	BS	E	CFD Industries, Paris (2001)	-	
D8036	32A	06/84	ED	C	BREL Glasgow	08/86	
D8037	32A	01/87	BS	C	V Berry, Leicester	05/88	Stored: [U] 08/76, R/I: 10/76, Stored: [U] 10/81, R/I: 03/82
D8038	32A	03/76	TO	C	BREL Glasgow	06/77	Stored: [U] 11/75
D8039	32A	09/86	BS	C	V Berry, Leicester	06/88	
D8040	32A	09/91	TO	C	M C Processors, Glasgow	02/92	Stored: [U] 10/81, R/I: 03/82, Stored: [U] 02/83, R/I: 05/84, Stored: [S] 08/91. Sold to RFS for spares
D8041	32A	11/88	TO	S	Hunslet-Barclay, Kilmarnock	-	
D8042	32A	06/91	TE	A	BR Frodingham		Withdrawn: 02/81, R/I: 05/81, Stored: [U] 10/81, R/I: 03/83
D8043	32A	06/91	TE	A	BR Frodingham		Stored: [U] 10/81, R/I: 03/82, Stored: [U] 08/82, R/I: 01/83
D8044	32A	09/89	TO	C	M C Processors, Glasgow	09/91	Stored: [U] 10/81, R/I: 03/82, Stored: [U] 10/82, R/I: 01/83
D8045	34B	10/90	TO	C	M C Processors, Glasgow	10/91	Stored: [U] 10/81, R/I: 03/82, Stored: [U] 04/83, R/I: 08/83
D8046	34B						Stored: [U] 11/82, R/I: 10/83 Stored: [S] 01/92
D8047	34B	09/91	TO	S	RFS Locomotives, No. 2004	-	Stored: [U] 08/91
D8048	34B	11/90	BS	P#	Peak Rail, Darley Dale	-	Stored: [U] 02/81, R/I: 09/82
D8049	34B	05/87	TO	C	V Berry, Leicester	07/88	Stored: [U] 03/83, R/I: 01/84, Stored: [U] 03/87, R/I: 04/87
D8050	41A						Stored: [S] 03/92
D8051	41A	02/91	TO	S	RFS Locomotives	-	Stored: [U] 09/89
D8052	41A	11/90	TO	C	M C Processors, Glasgow	07/91	
D8053	41A	11/90	TO	C	M C Processors, Glasgow	09/91	
D8054	41A	09/89	TO	C	M C Processors, Glasgow	10/91	Stored: [U] 12/82, R/I: 05/83
D8055	41A						Stored: [U] 09/89, R/I: 11/89, Stored: [U] 01/92
D8056	41A	10/90	TO	P#	Caledonian Railway, Brechin	-	
D8057	41A						
D8058	41A	02/92	TO	A	BR Toton		
D8059	41A						
D8060	41A	11/88	TO	S	Hunslet-Barclay, Kilmarnock	-	
D8061	41A	01/91	TE	A	BR Frodingham		Stored: [U] 06/81, R/I: 09/82
D8062	41A	09/76	TO	C	BREL Derby	08/79	
D8063	41A	07/91	BS	E	CFD Industries, Paris (2002)	-	Stored: [U] 10/81, R/I: 03/82, Stored: [U] 10/82, R/I: 12/82
D8064	41A	09/90	TO	C	M C Processors, Glasgow	10/91	
D8065	41A	12/90	TO	C	M C Processors, Glasgow	01/91	Stored: [U] 09/89
D8066	41A						Stored: [U] 12/90, R/I: 01/91
D8067	41A	01/87	TO	C	V Berry, Leicester	06/88	Stored: [U] 10/81, R/I: 03/82

BR 1957 No.	TOPS No.	Date Re No.	First TOPS Re No.	Date Re No.	Second TOPS Re No.	Date Re No.	Name	Name Date	Built By	Works No.	Date Introduced
D8068	20068	03/74							EE.RSH	2974/8226	06/61
D8069	20069	02/74							EE.RSH	2975/8227	06/61
D8070	20070	05/74							EE.RSH	2976/8228	06/61
D8071	20071	05/74							EE.RSH	2977/8229	07/61
D8072	20072	02/74							EE.RSH	2978/8230	07/61
D8073	20073	03/74							EE.RSH	2979/8231	07/61
D8074	20074	04/74							EE.RSH	2980/8232	07/61
D8075	20075	02/74							EE.RSH	2981/8233	07/61
D8076	20076	04/74							EE.RSH	2982/8234	07/61
D8077	20077	04/74							EE.RSH	2983/8235	07/61
D8078	20078	04/74							EE.RSH	2984/8236	07/61
D8079	20079	04/74							EE.RSH	2985/8237	08/61
D8080	20080	10/73							EE.RSH	2986/8238	08/61
D8081	20081	04/74							EE.RSH	2987/8239	08/61
D8082	20082	03/74							EE.RSH	2988/8240	08/61
D8083	20083	03/74	20903	05/89			Alison	05/89	EE.RSH	2989/8241	08/61
D8084	20084	03/74							EE.RSH	2990/8242	09/61
D8085	20085	04/74							EE.RSH	2991/8243	09/61
D8086	20086	09/74							EE.RSH	2992/8244	09/61
D8087	20087	06/74							EE.RSH	2993/8245	09/61
D8088	20088	02/74							EE.RSH	2994/8246	09/61
D8089	20089	04/74							EE.RSH	2995/8247	10/61
D8090	20090	03/74							EE.RSH	2996/8248	09/61
D8091	20091	05/74							EE.RSH	2997/8249	10/61
D8092	20092	04/74							EE.RSH	2998/8250	10/61
D8093	20093	04/74							EE.RSH	2999/8251	10/61
D8094	20094	04/74							EE.RSH	3000/8252	10/61
D8095	20095	11/73							EE.RSH	3001/8253	10/61
D8096	20096	04/74							EE.RSH	3002/8254	10/61
D8097	20097	04/74							EE.RSH	3004/8256	11/61
D8098	20098	05/74							EE.RSH	3003/8255	11/61
D8099	20099	04/74							EE.RSH	3005/8257	11/61
D8100	20100	02/74							EE.RSH	3006/8258	11/61
D8101	20101	04/74	20904	03/89			Janis	05/89	EE.RSH	3007/8259	11/61
D8102	20102	01/74							EE.RSH	3008/8260	12/61
D8103	20103	04/74							EE.RSH	3009/8261	12/61
D8104	20104	12/73							EE.RSH	3010/8262	12/61
D8105	20105	04/74							EE.RSH	3011/8263	12/61
D8106	20106	12/73							EE.RSH	3012/8264	12/61
D8107	20107	04/74							EE.RSH	3013/8265	12/61
D8108	20108	04/74							EE.RSH	3014/8266	12/61
D8109	20109	04/74							EE.RSH	3015/8267	12/61
D8110	20110	12/73							EE.RSH	3016/8268	01/62
D8111	20111	09/74							EE.RSH	3017/8269	01/62
D8112	20112	05/74							EE.RSH	3018/8270	01/62
D8113	20113	04/74							EE.RSH	3019/8271	02/62
D8114	20114	01/74							EE.RSH	3020/8272	02/62
D8115	20115	04/74							EE.RSH	3021/8273	02/62
D8116	20116	04/74							EE.RSH	3022/8274	02/62
D8117	20117	05/74							EE.RSH	3023/8275	02/62
D8118	20118	08/74					[Saltburn by the Sea]	08/87-08/90	EE.RSH	3024/8276	03/62
D8119	20119	02/74							EE.RSH	3025/8277	03/62
D8120	20120	01/74							EE.RSH	3026/8278	03/62
D8121	20121	04/74							EE.RSH	3027/8279	03/62
D8122	20122	04/74					[Cleveland Potash]	06/87-09/90	EE.RSH	3028/8280	03/62
D8123	20123	11/73							EE.RSH	3029/8281	03/62
D8124	20124	04/74							EE.RSH	3030/8282	03/62
D8125	20125	04/74							EE.RSH	3031/8283	04/62
D8126	20126	04/74							EE.RSH	3032/8284	05/62
D8127	20127	04/74							EE.RSH	3033/8285	07/62
D8128	20228	11/73							EE.VF	3599/D998	01/66
D8129	20129	03/74							EE.VF	3600/D999	02/66
D8130	20130	02/74							EE.VF	3601/D1000	02/66
D8131	20131	05/74							EE.VF	3602/D1001	02/66
D8132	20132	02/74							EE.VF	3603/D1002	03/66
D8133	20133	02/74							EE.VF	3604/D1003	03/66
D8134	20134	02/74	20303	04/86	20134	12/86			EE.VF	3605/D1004	03/66
D8135	20135	10/73							EE.VF	3606/D1005	03/66
D8136	20136	05/74							EE.VF	3607/D1006	04/66
D8137	20137	05/74					[Murry B Hofmeyr]	06/87-07/91	EE.VF	3608/D1007	04/66

BR 1957 No.	Depot of First Allocation	Date Withdrawn	Depot of Final Allocation	Disposal Code	Disposal Detail	Date Cut Up	Notes
D8068	41A	06/87	IM	A	BR Immingham	-	Stored: [U] 11/81, R/I: 09/83
D8069	41A	05/91	TE	P	Type 1 Fund, County School Stn	-	Stored: [U] 02/81, R/I: 06/82
D8070	65A	04/91	TO	C	M C Processors, Glasgow	12/92	
D8071	65A						Stored: [S] 03/92
D8072	65A						Stored: [U] 01/92
D8073	65A						
D8074	65A	01/76	TO	C	BREL Glasgow	11/76	
D8075	65A						Stored: [S] 03/92
D8076	65A	05/88	IM	C	BR Thornaby, by V Berry	10/88	
D8077	65A	05/88	TO	C	V Berry, Leicester	04/90	
D8078	65A	01/92	TO	A	BR Toton	-	Stored: [S] 10/80, Withdrawn 10/80, R/I: 03/81, Stored [S] 01/82, R/I: 01/82, Stored: [U] 09/82, R/I: 11/82, Stored: [U] 11/82, R/I:12/84
D8079	65A	04/77	TI	C	BREL Derby	11/78	
D8080	65A	07/90	TO	A	BRML Doncaster	-	Stored: [U] 04/81, R/I: 04/82, Stored: [U] 08/82, R/I: 06/84
D8081	65A						Stored: [U] 10/81, R/I: 03/82
D8082	65A						Stored: [S] 02/92
D8083	65A	01/89	TO	S	Hunslet-Barclay, Kilmarnock	-	Stored: [U] 03/89
D8084	65A	09/91	TO	S	RFS Locomotives, No. 2002	-	Stored: [S] 06/91
D8085	65A	03/91	TO	S	RFS Locomotives	-	Stored: [U] 04/81, R/I: 09/81, Stored: [U] 10/83, R/I: 04/84
D8086	65A	09/88	TO	C	MC Processors, Glasgow	02/91	
D8087	65A						Stored: [U] 02/83, R/I: 04/84
D8088	65A	09/91	TO	S	RFS Locomotives, No. 2017	-	Stored: [U] 11/81, R/I: 04/83, Stored: [U] 05/83, R/I: 06/83, Stored: [U] 08/91
D8089	65A	04/87	IM	A	BR Immingham		
D8090	65A						Stored: [U] 04/83, R/I: 03/83, Stored: [S] 01/92
D8091	65A	09/78	ED	C	BREL Glasgow	11/78	Stored: [U] 08/78
D8092	65A						Stored: [U] 02/82, R/I: 04/83
D8093	65A	05/91	TE	C	M C Processors, Glasgow	05/92	Stored: [U] 05/82, R/I: 04/83
D8094	65A						Stored: [U] 12/82, R/I: 11/83
D8095	65A	01/91	TE	S	RFS Locomotives, No. 2020	-	Stored: [U] 12/82, R/I: 03/83, Stored: [U] 01/91
D8096	65A						Stored: [U] 12/82, R/I: 02/83, Stored; [U] 06/92
D8097	65A	09/89	TO	C	M C Processors, Glasgow	10/91	Stored: [U] 02/81, R/I: 06/82
D8098	65A	06/91	TE	P	Type 1 Assoc, on G C Railway	-	Stored: [U] 12/82, R/I: 08/83
D8099	65A	02/91	TO	A	BR Toton	-	Stored: [U] 10/83, R/I: 05/84, Stored: [U] 09/89
D8100	65A	08/89	TO	C	M C Processors, Glasgow	02/91	Stored: [U] 04/81, R/I: 05/81, Stored: [U] 08/87, R/I: 04/83
D8101	65A	11/88	TO	S	Hunslet-Barclay, Kilmarnock	-	Stored: [U] 10/83
D8102	65A	09/91	TO	S	RFS Locomotives, No. 2008	-	Stored: [S] 06/91
D8103	65A	03/91	TO	C	M C Processors, Glasgow	06/92	Stored: [U] 10/82, R/I: 03/85, Stored: [U] 09/92
D8104	65A						Stored: [U] 12/83, R/I: 02/84, Stored: [U] 06/91
D8105	65A	09/91	TO	S	RFS Locomotives, No. 2016	-	Stored: [U] 12/82, R/I: 05/84, Stored: [U] 01/92
D8106	65A	12/92	HQ	A	BRML Doncaster	-	Stored: [U] 12/82, R/I: 06/83
D8107	65A	01/91	TE	P#	Private, on East Lancs Railway	-	Stored: [U] 04/81, R/I: 03/81, Stored: [U] 02/83, R/I: 01/84, Stored: [U] 06/91
D8108	65A	09/91	TO	S	RFS Locomotives, No. 2001	-	Stored: [U] 04/81, R/I: 05/81
D8109	65A	02/82	HA	C	BREL Glasgow	12/82	Stored: [U] 04/81, R/I: 07/81, Stored: [U] 08/81, R/I: 04/83
D8110	65A	09/90	TO	P#	Private on South Devon Railway	-	
D8111	65A	04/87	IM	C	V Berry, Leicester	03/88	Stored: [U] 11/82, R/I: 06/83
D8112	65A	01/91	TE	A	BR Thornaby	-	Stored: [U] 08/91
D8113	65A	09/91	TO	S	RFS Locomotives, No. 2003	-	
D8114	65A	01/90	TO	C	M C Processors, Glasgow	07/91	
D8115	65A	01/87	BS	C	V Berry, Leicester	05/88	
D8116	66A	01/87	TI	C	V Berry, Leicester	07/88	Stored: [U] 10/82, R/I: 05/84, Stored: [S] 03/92
D8117	66A						Stored: [U] 12/92
D8118	66A						Stored: [U] 04/81, R/I: 01/83, Stored: [U] 06/92
D8119	66A	09/92	TO	A		-	Stored: [U] 07/83, R/I: 09/84, Stored: [U] 06/91
D8120	66A	09/91	TO	S	RFS Locomotives, No. 2009	-	Stored: [U] 10/82, R/I: 01/84, Stored: [S] 03/92
D8121	66A						Stored: [U] 01/91
D8122	66A	07/91	HQ	A	M C Processors, Glasgow	-	
D8123	66A	04/87	BS	C	V Berry, Leicester	07/88	
D8124	66A	06/91	ED	A	M C Processors, Glasgow	-	
D8125	66A	09/86	BS	C	V Berry, Leicester	05/88	Stored: [U] 04/81, R/I: 08/81, Stored: [U] 12/82, R/I: 04/83
D8126	66A	09/89	IM	C	M C Processors, Glasgow	05/92	Stored: [U] 06/81, R/I: 06/82
D8127	66A	01/91	TE	S	RFS Locomotives, No. 2018	-	Stored: [U] 11/82, R/I: 01/83
D8128	41A	07/91	BS	E	CFD Industries, Paris (2004)	-	Stored: [U] 10/82, R/I: 08/83
D8129	41A	10/90	TO	C	M C Processors, Glasgow	09/91	Stored: [U] 12/82, R/I: 04/83
D8130	41A	10/90	TO	C	M C Processors, Glasgow	10/91	Stored: [U] 08/83, R/I: 04/84, Stored: [U] 01/92
D8131	41A						Stored: [U] 08/83, R/I: 04/84, Stored: [U] 01/92
D8132	41A						Stored: [U] 04/83, R/I: 02/84
D8133	41A	09/91	TO	S	RFS Locomotives, No. 2005	-	Stored: [S] 08/91
D8134	2F	07/89	TO	C	M C Processors, Glasgow	08/91	
D8135	2F						Stored: [S] 02/92
D8136	2F	10/90	TO	C	M C Processors, Glasgow	10/91	
D8137	DO2						

BR 1957 No.	TOPS No.	Date Re No.	First TOPS Re No.	Date Re No.	Second TOPS Re No.	Date Re No.	Name	Name Date	Built By	Works No.	Date Introduced
D8138	20138	03/74							EE.VF	3609/D1008	04/66
D8139	20139	03/74							EE.VF	3610/D1009	04/66
D8140	20140	05/74							EE.VF	3611/D1010	05/66
D8141	20141	03/74							EE.VF	3612/D1011	05/66
D8142	20142	03/74							EE.VF	3614/D1013	05/66
D8143	20143	03/74							EE.VF	3613/D1012	07/66
D8144	20144	03/74							EE.VF	3615/D1014	07/66
D8145	20145	03/74							EE.VF	3616/D1015	06/66
D8146	20146	03/74							EE.VF	3617/D1016	06/66
D8147	20147	02/74							EE.VF	3618/D1017	06/66
D8148	20148	02/74							EE.VF	3619/D1018	07/66
D8149	20149	02/74							EE.VF	3620/D1019	07/66
D8150	20150	03/74							EE.VF	3621/D1020	07/66
D8151	20151	03/74							EE.VF	3622/D1021	07/66
D8152	20152	02/74							EE.VF	3623/D1022	08/66
D8153	20153	05/74							EE.VF	3624/D1023	08/66
D8154	20154	04/74							EE.VF	3625/D1024	08/66
D8155	20155	03/74							EE.VF	3626/D1025	08/66
D8156	20156	02/74							EE.VF	3627/D1026	08/66
D8157	20157	04/74							EE.VF	3628/D1027	09/66
D8158	20158	03/74							EE.VF	3629/D1028	09/66
D8159	20159	02/74							EE.VF	3630/D1029	09/66
D8160	20160	03/74							EE.VF	3631/D1030	12/66
D8161	20161	10/73							EE.VF	3632/D1031	09/66
D8162	20162	02/74							EE.VF	3633/D1032	09/66
D8163	20163	05/74							EE.VF	3634/D1033	09/66
D8164	20164	03/74							EE.VF	3635/D1034	10/66
D8165	20165	02/74					[Henry Pease]	08/87-09/90	EE.VF	3636/D1035	10/66
D8166	20166	04/74							EE.VF	3637/D1036	10/66
D8167	20167	04/74							EE.VF	3638/D1037	10/66
D8168	20168	04/74	20304	04/86	20168	12/86			EE.VF	3639/D1038	10/66
D8169	20169	02/74							EE.VF	3640/D1039	10/66
D8170	20170	02/74							EE.VF	3641/D1040	10/66
D8171	20171	11/73							EE.VF	3642/D1041	10/66
D8172	20172	02/74	20305	04/86	20172	11/86			EE.VF	3643/D1042	10/66
D8173	20173	02/74	20306	04/86	20173	11/86			EE.VF	3644/D1043	11/66
D8174	20174	03/74							EE.VF	3645/D1044	11/66
D8175	20175	02/74							EE.VF	3646/D1045	11/66
D8176	20176	06/74							EE.VF	3647/D1046	11/66
D8177	20177	04/74							EE.VF	3648/D1047	11/66
D8178	20178	05/74							EE.VF	3649/D1048	12/66
D8179	20179	05/74							EE.VF	3660/D1055	12/66
D8180	20180	03/74							EE.VF	3661/D1056	12/66
D8181	20181	05/74							EE.VF	3662/D1057	12/66
D8182	20182	03/74							EE.VF	3663/D1058	12/66
D8183	20183	03/74							EE.VF	3664/D1059	12/66
D8184	20184	04/74							EE.VF	3665/D1060	12/66
D8185	20185	09/73							EE.VF	3666/D1061	02/67
D8186	20186	03/74							EE.VF	3667/D1062	01/67
D8187	20187	03/74							EE.VF	3668/D1063	01/67
D8188	20188	02/74							EE.VF	3669/D1064	01/67
D8189	20189	05/74							EE.VF	3670/D1065	01/67
D8190	20190	05/74							EE.VF	3671/D1066	01/67
D8191	20191	05/74							EE.VF	3672/D1067	02/67
D8192	20192	02/74							EE.VF	3673/D1068	02/67
D8193	20193	03/74							EE.VF	3674/D1069	02/67
D8194	20194	03/74	20307	04/86	20194	11/86			EE.VF	3675/D1070	02/67
D8195	20195	12/73							EE.VF	3676/D1071	02/67
D8196	20196	06/74	20308	04/86	20196	11/86			EE.VF	3677/D1072	02/67
D8197	20197	02/74							EE.VF	3678/D1073	02/67
D8198	20198	10/73							EE.VF	3679/D1074	04/67
D8199	20199	11/73							EE.VF	3680/D1075	04/67

Class continued from D8300

BRITISH THOMSON-HOUSTON TYPE 1 Bo-Bo CLASS 15

BR 1957 No.	Built By	Works No.	Date Introduced	Depot of First Allocation	Date Withdrawn
D8200	YEC	2642	11/57	1D	03/71
D8201	YEC	2643	02/58	1D	03/71
D8202	YEC	2644	02/58	1D	06/68
D8203	YEC	2645	03/58	1D	03/69
D8204	YEC	2646	04/58	1D	03/71
D8205	YEC	2647	05/58	1D	09/68
D8206	YEC	2648	06/58	1D	09/68
D8207	YEC	2649	07/58	1D	03/71
D8208	YEC	2650	07/58	1D	09/68
D8209	YEC	2651	11/58	1D	03/71
D8210	BTH Clayton	1108	10/59	30A	03/71

BR 1957 No.	Depot of First Allocation	Date Withdrawn	Depot of Final Allocation	Disposal Code	Disposal Detail	Date Cut Up	Notes
D8138	DO2						Stored: [S] 12/89, R/I: 01/90
D8139	DO2	07/91	BS	E	CFD Industries, Paris (2003)	-	Stored: [U] 06/86, R/I: 07/82
D8140	DO2						Stored: [U] 12/82, R/I: 01/84, Stored: [U] 01/92
D8141	DO2	11/91	TO	A	BR Toton		Stored: [U] 09/91
D8142	DO2						Stored: [U] 02/83, R/I: 01/84, Stored; [S] 02/92
D8143	DO2	01/92	TO	A	BR Toton		
D8144	D16	01/90	TE	A	BR Thornaby		
D8145	D16	05/91	TE	S	RFS Locomotives, No. 2019	-	
D8146	D16	12/88	ED	C	M C Processors, Glasgow	03/89	
D8147	D16	05/89	TO	C	M C Processors, Glasgow	10/91	
D8148	D16	11/91	TO	A	BR Toton		Stored: [U] 02/91
D8149	D16	11/87	TO	C	V Berry, Leicester	09/88	Stored: [U] 05/87
D8150	D16	11/87	TI	C	V Berry, Leicester	04/90	Stored: [U] 05/87
D8151	D16						Stored: [U] 02/83, R/I: 01/84, Stored: [U] 01/92
D8152	D16	04/88	HA	C	M C Processors, Glasgow	08/88	
D8153	D16	11/87	TI	C	V Berry, Leicester	03/88	Stored: [U] 05/87
D0154	D16						
D8155	D16	11/87	TI	C	V Berry, Leicester	11/87	Stored: [U] 05/87
D8156	D16	07/91	HQ	A	M C Processors, Glasgow		Stored: [U] 07/91
D8157	D16	10/90	TO	C	M C Processors, Glasgow	06/92	
D8158	D16	05/89	TO	C	M C Processors, Glasgow	03/91	
D8159	D16	09/91	TO	S	RFS Locomotives, No. 2010	-	Stored: [S] 08/91
D8160	D16	12/90	BS	A	BR Bescot		
D8161	D16	02/88	TO	C	V Berry, Leicester	04/90	
D8162	D16	11/87	TI	C	V Berry, Leicester	11/87	Stored: [U] 05/87
D8163	D16						
D8164	D16	11/87	TI	C	V Berry, Leicester	07/88	Stored: [U] 05/87
D8165	D16						
D8166	D16	05/91	TO	P#	Bodmin & Wenford Railway	-	Stored: [U] 02/83, R/I: 01/84
D8167	D16	11/87	TI	C	V Berry, Leicester	08/88	Stored: [U] 05/87
D8168	D16						
D8169	D16						Stored: [S] 03/92
D8170	D16	11/91	TO	A	BR Toton		Stored: [U] 09/91
D8171	D16	12/89	TO	C	M C Processors, Glasgow	08/91	
D8172	D16	10/90	TO	A	BR Toton		
D8173	D16	02/91	TO	C	M C Processors, Glasgow	02/92	Stored: [U] 09/89, Sold to RFS for spares
D8174	D16	04/88	TE	A	BR Thornaby		
D8175	D16	09/91	TO	S	RFS Locomotives, No. 2007	-	Stored: [U] 08/91
D8176	D16	10/91	TE	A	BR Frodingham		Stored: [U] 01/83, R/I: 03/83
D8177	D16						Stored: [U] 01/92
D8178	D16	12/89	TE	C	M C Processors, Glasgow	06/92	
D8179	D16	02/89	TO	C	M C Processors, Glasgow	02/91	
D8180	D16	03/88	TO	C	V Berry, Leicester	04/90	
D8181	D16	11/87	TO	A	BR Bescot		Stored: [U] 03/87
D8182	D16	03/91	TO	A	BR Toton		Stored: [U] 02/83, R/I: 01/84
D8183	D16	01/90	TO	C	V Berry, Leicester	03/90	Stored: [U] 04/89
D8184	D16	10/86	TO	C	V Berry, Leicester	07/88	
D8185	D16	10/92	TO	A	BR Toton		Stored: [U] 02/91
D8186	D16						Stored: [U] 01/92
D8187	D16						Stored: [U] 01/92
D8188	D16	01/90	TO	A	M C Processors, Glasgow		Stored: [U] 02/83, R/I: 04/83
D8189	D16	09/90	TO	S	M C Processors, Glasgow (Pilot)		Stored: [S] 03/92
D8190	D16						
D8191	D16	11/87	TO	C	V Berry, Leicester	11/87	Stored: [U] 05/87
D8192	D16	09/89	TO	C	M C Processors, Glasgow	10/91	Stored: [U] 03/89, R/I: 08/89,
D8193	D16	07/89	TO	C	M C Processors, Glasgow	01/91	
D8194	D16	09/91	TO	S	RFS Locomotives, No. 2006	-	Stored: [U] 02/83, R/I: 04/84 Stored: [U] 08/91
D8195	D16						Stored: [U] 01/92
D8196	D16						Stored: [U] 04/83, R/I: 04/84, Stored: [U] 01/92
D8197	D16	11/91	TO	A	BR Toton		
D0198	D16	07/91	ED	A	M C Processors, Glasgow		Stored: [U] 05/89, R/I: 12/89
D8199	D16	01/90	TO	A	M C Processors, Glasgow		

BR 1957 No.	Depot of Final Allocation	Disposal Code	Disposal Detail	Date Cut Up	Notes
D8200	30A	C	BREL Crewe	03/72	
D8201	30A	C	BREL Crewe	02/72	
D8202	30A	C	G Cohen, Kettering	04/69	
D8203	30A	D	To Departmental Stock - 968003	-	
D8204	30A	C	BREL Crewe	04/72	
D8205	30A	C	G Cohen, Kettering	05/69	
D8206	30A	C	D Woodham, Barry	02/70	
D8207	30A	C	BREL Crewe	04/72	
D8208	30A	C	G Cohen, Kettering	04/69	
D8209	30A	C	BREL Crewe	04/72	
D8210	30A	C	BREL Crewe	03/72	

153

BR 1957 No.	Built By	Works No.	Date Introduced	Depot of First Allocation	Date Withdrawn
D8211	BTH Clayton	1109	11/59	30A	03/71
D8212	BTH Clayton	1110	11/59	31B	09/68
D8213	BTH Clayton	1111	12/59	31B	12/68
D8214	BTH Clayton	1112	12/59	31B	12/70
D8215	BTH Clayton	1113	01/60	31B	12/70
D8216	BTH Clayton	1114	01/60	31B	03/71
D8217	BTH Clayton	1115	01/60	31B	03/68
D8218	BTH Clayton	1116	02/60	31B	03/71
D8219	BTH Clayton	1117	02/60	31B	09/68
D8220	BTH Clayton	1118	03/60	30A	03/71
D8221	BTH Clayton	1119	03/60	30A	03/71
D8222	BTH Clayton	1120	03/60	30A	03/71
D8223	BTH Clayton	1121	04/60	32B	12/68
D8224	BTH Clayton	1122	04/60	32B	03/71
D8225	BTH Clayton	1123	05/60	32B	03/71
D8226	BTH Clayton	1124	05/60	32B	03/71
D8227	BTH Clayton	1125	05/60	32B	09/68
D8228	BTH Clayton	1126	06/60	32B	03/71
D8229	BTH Clayton	1127	06/60	32B	03/71
D8230	BTH Clayton	1128	07/60	32A	03/71
D8231	BTH Clayton	1129	07/60	32A	03/71
D8232	BTH Clayton	1130	08/60	32A	03/71
D8233	BTH Clayton	1131	08/60	30A	02/69
D8234	BTH Clayton	1132	09/60	30A	03/71
D8235	BTH Clayton	1133	09/60	32B	10/68
D8236	BTH Clayton	1134	10/60	32B	10/68
D8237	BTH Clayton	-	11/60	34G	03/69
D8238	BTH Clayton	-	12/60	34G	09/68
D8239	BTH Clayton	-	01/61	34G	03/71
D8240	BTH Clayton	-	01/61	34G	10/68
D8241	BTH Clayton	-	02/61	34G	04/68
D8242	BTH Clayton	-	02/61	34G	03/71
D8243	BTH Clayton	-	02/61	34G	02/69

ENGLISH ELECTRIC TYPE 1 Bo-Bo CLASS 20

Class continued from D8199

BR 1957 No.	TOPS No.	Date Re No.	First TOPS Re No.	Date Re No.	Second TOPS Re No.	Date Re No.	Name	Name Date	Built By	Works No.	Date Introduced
D8300	20200	03/74							EE.VF	3681/D1076	03/67
D8301	20201	03/74							EE.VF	3682/D1077	03/67
D8302	20202	02/74							EE.VF	-	03/67
D8303	20203	04/74							EE.VF	3684/D1079	03/67
D8304	20204	02/74							EE.VF	-	03/67
D8305	20205	04/74							EE.VF	3686/D1081	03/67
D8306	20206	03/74							EE.VF	3687/D1082	04/67
D8307	20207	02/74							EE.VF	3688/D1083	04/67
D8308	20208	03/74							EE.VF	3689/D1084	04/67
D8309	20209	03/74	20907*						EE.VF	3690/D1085	04/67
D8310	20210	05/74							EE.VF	3691/D1086	04/67
D8311	20211	02/74							EE.VF	3692/D1087	04/67
D8312	20212	04/74							EE.VF	3693/D1088	04/67
D8313	20213	03/74							EE.VF	3694/D1089	05/67
D8314	20214	04/74							EE.VF	3695/D1090	05/67
D8315	20215	03/74							EE.VF	3696/D1091	05/67
D8316	20216	09/74							EE.VF	3697/D1092	06/67
D8317	20217	09/74							EE.VF	3698/D1093	11/67
D8318	20218	09/74							EE.VF	3699/D1094	11/67
D8319	20219	09/74	20905	08/89			Iona	09/89	EE.VF	3700/D1095	01/68
D8320	20220	09/74							EE.VF	3701/D1096	11/67
D8321	20221	04/74							EE.VF	3702/D1097	12/67
D8322	20222	04/74							EE.VF	3703/D1098	12/67
D8323	20223	04/74							EE.VF	3704/D1099	10/67
D8324	20224	10/73							EE.VF	3705/D1100	11/67
D8325	20225	09/74	20906	08/89			Kilmarnock 400	06/92	EE.VF	3706/D1101	11/67
D8326	20226	09/74							EE.VF	3683/D1078	02/68
D8327	20227	12/73							EE.VF	3685/D1080	02/68

BR 1957 No.	Depot of Final Allocation	Disposal Code	Disposal Detail	Date Cut Up	Notes
D8211	30A	C	BREL Crewe	12/71	
D8212	30A	C	G Cohen, Kettering	04/69	
D8213	30A	C	G Cohen, Kettering	05/69	
D8214	30A	C	BREL Crewe	12/71	
D8215	30A	C	BREL Crewe	11/71	
D8216	30A	C	BREL Crewe	01/72	
D8217	30A	C	G Cohen, Kettering	01/69	
D8218	30A	C	BREL Crewe	11/71	
D8219	30A	C	G Cohen, Kettering	04/69	
D8220	30A	C	BREL Crewe	11/71	
D8221	30A	C	BREL Crewe	12/71	
D8222	30A	C	BREL Crewe	12/71	
D8223	30A	C	G Cohen, Kettering	05/69	
D8224	30A	C	BREL Crewe	12/71	
D8225	30A	C	BREL Crewe	12/71	
D8226	30A	C	BREL Crewe	12/71	
D8227	30A	C	G Cohen, Kettering	05/69	
D8228	30A	C	BREL Crewe	12/71	
D8229	30A	C	BREL Crewe	01/72	
D8230	30A	C	BREL Crewe	11/71	
D8231	30A	C	BREL Crewe	01/72	
D8232	30A	C	BREL Crewe	11/71	Withdrawn: 10/68, R/I: 10/68
D8233	30A	D	To Departmental Stock - 968001	-	
D8234	30A	C	BREL Crewe	11/71	
D8235	30A	C	G Cohen, Kettering	04/69	
D8236	30A	C	G Cohen, Kettering	04/69	
D8237	30A	D	To Departmental Stock - 968002	-	
D8238	30A	C	G Cohen, Kettering	04/69	
D8239	30A	C	BREL Crewe	01/72	
D8240	30A	C	G Cohen, Kettering	07/69	
D8241	30A	C	J Cashmore, Great Bridge	12/68	
D8242	30A	C	BREL Crewe	11/71	
D8243	30A	D	To Departmental Stock - 968000	-	

BR 1957 No.	Depot of First Allocation	Date Withdrawn	Depot of Final Allocation	Disposal Code	Disposal Detail	Date Cut Up	Notes
D8300	50A	08/79	ED	C	BREL Glasgow	09/79	Stored: [U] 06/79
D8301	50A	09/88	IM	C	V Berry, Leicester	03/90	
D8302	50A	03/89	ED	A	BR Toton		
D8303	50A	12/88	ED	C	M C Processors, Glasgow	02/92	
D8304	50A	05/89	ED	C	M C Processors, Glasgow	12/89	Withdrawn: 05/89, R/I: 06/89
D8305	50A	12/89	TE	A	BR Frodingham		
D8306	50A	04/91	TO	A	M C Processors, Glasgow		Stored: [S] 08/89, R/I: 09/89
D8307	50A	07/83	HA	C	BREL Glasgow	06/86	
D8308	50A	02/91	TO	S	RFS Locomotives	-	Stored: [U] 09/89
D8309	50A	11/88	TO	S	Hunslet-Barclay, Kilmarnock		Stored: [U] 01/92
D8310	51L						Stored: [U] 12/82, R/I: 08/83, Stored: [U] 05/89, R/I: 12/89, Stored: [U] 07/90
D8311	51L	07/91	HQ	A	M C Processors, Glasgow		
D8312	51L	07/91	HQ	A	M C Processors, Glasgow		Stored: [U] 05/89, R/I: 07/89, Stored: [S] 09/89, R/I: 12/89, Stored: [U] 07/90
D8313	51L	04/91	TO	C	M C Processors, Glasgow	10/92	Stored: [U] 12/81, R/I: 09/82, Stored: [U] 08/89, R/I: 09/89
D8314	51L						Stored: [S] 03/92
D8315	51L						Stored: [U] 07/83, R/I: 03/85, Stored: [S] 03/92
D8316	64B	11/87	TO	C	V Berry, Leicester	09/88	Stored: [U] 04/87
D8317	64B	07/89	TO	C	M C Processors, Glasgow	02/91	
D8318	64B	10/89	IM	A	M C Processors, Glasgow		Stored: [U] 08/89
D8319	64B	01/89	TO	S	Hunslet-Barclay, Kilmarnock	-	Stored: [U] 03/89, R/I: 08/89
D8320	66A	11/87	HA	A	BR Carlisle Kingmoor		Stored: [U] 05/87
D8321	66A	11/87	HA	A	BR Carlisle Kingmoor		Stored: [U] 05/87
D8322	66A	12/87	HA	C	M C Processors, Glasgow	10/88	
D8323	64B	11/87	HA	A	BR Carlisle Kingmoor		Stored: [U] 05/87
D8324	66A	01/89	TO	C	M C Processors, Glasgow	03/91	
D8325	66A	01/89	TO	S	Hunslet-Barclay, Kilmarnock	-	Named: Georgina 09/89-05/92
D8326	66A	09/88	IM	C	M C Processors, Glasgow	06/92	
D8327	66A	10/90	TO	P #	Class 20 Locomotive Society	-	

Locomotives on hire to RFS Locomotives, to be returned to preservation owners.

NORTH BRITISH TYPE 1 Bo-Bo CLASS 16

BR 1957 No.	Built By	Works No.	Date Introduced	Depot of First Allocation	Date Withdrawn	Depot of Final Allocation	Disposal Code
D8400	NBL	27671	05/58	30A	07/68	30A	C
D8401	NBL	27672	06/58	30A	09/68	30A	C
D8402	NBL	27673	07/58	30A	07/68	30A	C
D8403	NBL	27674	07/58	30A	07/68	30A	C
D8404	NBL	27675	08/58	30A	02/68	30A	C
D8405	NBL	27676	09/58	30A	09/68	30A	C
D8406	NBL	27677	09/58	30A	09/68	30A	C
D8407	NBL	27678	09/58	30A	09/68	30A	C
D8408	NBL	27679	09/58	30A	09/68	30A	C
D8409	NBL	27680	09/58	30A	09/68	30A	C

CLAYTON TYPE 1 Bo-Bo CLASS 17

BR 1957 No.	Built By	Works No.	Date Introduced	Depot of First Allocation	Date Withdrawn	Depot of Final Allocation	Disposal Code
D8500	Clayton	4365U1	09/62	66A	03/71	66A	C
D8501	Clayton	4365U2	10/62	66A	10/68	D10	C
D8502	Clayton	4365U3	10/62	66A	10/71	66A	C
D8503	Clayton	4365U4	10/62	66A	10/71	66A	C
D8504	Clayton	4365U5	10/62	66A	11/71	66A	C
D8505	Clayton	4365U6	10/62	66A	10/71	64B	C
D8506	Clayton	4365U7	10/62	66A	09/71	66A	C
D8507	Clayton	4365U8	10/62	66A	12/71	66A	C
D8508	Clayton	4365U9	10/62	66A	12/71	66A	C
D8509	Clayton	4365U10	11/62	66A	10/68	D10	C
D8510	Clayton	4365U11	11/62	66A	03/71	64B	C
D8511	Clayton	4365U14	11/62	66A	10/68	D10	C
D8512	Clayton	4365U16	12/62	66A	12/68	D10	D
D8513	Clayton	4365U13	12/62	66A	09/71	66A	C
D8514	Clayton	4365U12	07/63	66A	10/68	D10	C
D8515	Clayton	4365U20	08/63	66A	10/71	66A	C
D8516	Clayton	4365U19	04/63	66A	10/71	66A	C
D8517	Clayton	4365U18	06/63	66A	10/68	D10	C
D8518	Clayton	4365U15	04/63	66A	10/68	D10	C
D8519	Clayton	4365U22	07/63	66A	10/68	D10	C
D8520	Clayton	4365U23	05/63	66A	10/68	D10	C
D8521	Clayton	4365U17	07/63	66A	10/68	D10	D
D8522	Clayton	4365U24	06/63	66A	10/68	D10	C
D8523	Clayton	4365U25	07/63	66A	10/68	D10	C
D8524	Clayton	4365U26	07/63	66A	10/68	D10	C
D8525	Clayton	4365U27	08/63	66A	10/71	66A	C
D8526	Clayton	4365U28	08/63	66A	10/68	D10	C
D8527	Clayton	4365U29	08/63	66A	12/68	D10	C
D8528	Clayton	4365U30	07/63	66A	10/71	66A	C
D8529	Clayton	4365U31	05/63	66A	12/71	66A	C
D8530	Clayton	4365U32	04/63	66A	03/71	64B	C
D8531	Clayton	4365U33	05/63	66A	09/71	66A	C
D8532	Clayton	4365U34	07/63	66A	12/68	D10	C
D8533	Clayton	4365U35	08/63	66A	10/68	D10	C
D8534	Clayton	4365U36	06/63	66A	10/68	D10	C
D8535	Clayton	4365U21	07/63	66A	03/71	66A	C
D8536	Clayton	4365U37	08/63	66A	12/71	66A	C
D8537	Clayton	4365U40	06/63	66A	07/68	66A	C
D8538	Clayton	4365U41	06/63	66A	10/71	66A	C
D8539	Clayton	4365U38	06/63	66A	10/71	66A	C
D8540	Clayton	4365U39	08/63	66A	10/71	66A	C
D8541	Clayton	4365U42	06/63	66A	10/71	66A	C
D8542	Clayton	4365U43	08/63	66A	10/71	66A	C
D8543	Clayton	4365U44	09/63	66A	09/71	66A	C
D8544	Clayton	4365U45	09/63	66A	05/69	66A	C
D8545	Clayton	4365U46	09/63	64B	10/71	66A	C
D8546	Clayton	4365U47	09/63	66A	10/71	66A	C
D8547	Clayton	4365U48	09/63	66A	02/69	66A	C
D8548	Clayton	4365U49	09/63	66A	12/71	66A	C
D8549	Clayton	4365U50	10/63	66A	09/71	66A	C
D8550	Clayton	4365U51	11/63	66A	10/71	66A	C
D8551	Clayton	4365U52	10/63	66A	10/71	66A	C
D8552	Clayton	4365U53	10/63	66A	12/71	66A	C
D6553	Clayton	4365U54	10/63	66A	10/68	66A	C
D8554	Clayton	4365U55	10/63	64B	06/69	66A	C
D8555	Clayton	4365U56	10/63	64B	09/71	66A	C
D8556	Clayton	4365U57	10/63	64B	02/69	66A	C
D8557	Clayton	4365U58	11/63	64B	09/71	66A	C
D8558	Clayton	4365U59	11/63	64B	12/71	66A	C

BR 1957 No.	Disposal Detail	Date Cut Up	Notes
D8400	G Cohen, Kettering	11/69	
D8401	G Cohen, Kettering	09/69	
D8402	G Cohen, Kettering	11/69	
D8403	G Cohen, Kettering	04/69	
D8404	Cox & Danks, Park Royal	07/68	
D8405	G Cohen, Kettering	12/69	
D8406	Birds, Long Marston	07/69	
D8407	G Cohen, Kettering	12/69	
D8408	G Cohen, Kettering	05/69	
D8409	G Cohen, Kettering	05/69	

BR 1957 No.	Disposal Detail	Date Cut Up	Notes
D8500	BREL Glasgow	11/71	Withdrawn: 10/68, R/I: 05/69
D8501	BREL Glasgow	12/71	Stored: [U] 01/63, R/I: 04/63
D8502	BREL Glasgow	01/73	Stored: [U] 01/63, R/I: 04/63
D8503	BREL Glasgow	06/72	Stored: [U] 01/63, R/I: 05/63, Stored: [U] 10/68, R/I: 11/69
D8504	J Cashmore, Great Bridge	10/75	Stored: [U] 01/63, R/I: 05/63, Withdrawn: 10/68, R/I: 05/69
D8505	BREL Glasgow	05/74	Stored: [U] 01/63, R/I: 04/63, Withdrawn: 10/68, R/I: 05/69
D8506	BREL Glasgow	12/71	Stored: [U] 01/63, R/I: 04/63, Withdrawn: 10/68, R/I: 05/69, Withdrawn: 05/69, R/I: 11/69
D8507	A King, Norwich	09/75	Stored: [U] 01/63, R/I: 04/63, Withdrawn: 10/68, R/I: 11/69
D8508	A King, Norwich	10/75	Stored: [U] 01/63, R/I: 04/63, Withdrawn: 10/68, R/I: 05/69
D8509	BREL Glasgow	01/72	Stored: [U] 01/63, R/I: 04/63
D8510	BREL Glasgow	11/71	Stored: [U] 01/63, R/I: 04/63, Withdrawn: 12/68, R/I: 11/69
D8511	J McWilliam, Shettleston	08/70	Stored: [U] 01/63, R/I: 04/63
D8512	To Departmental Stock - 8512	-	Stored: [U] 01/63, R/I: 04/63, Withdrawn: 10/68, R/I: 11/68
D8513	BREL Glasgow	12/71	Stored: [U] 01/63, R/I: 04/63, Withdrawn: 10/68, R/I: 05/69
D8514	BREL Glasgow	01/72	Stored: [U] 11/68
D8515	BREL Glasgow	09/72	Stored: [U] 05/69
D8516	A King, Norwich	10/75	Withdrawn: 10/68, R/I: 11/69
D8517	J McWilliam, Shettleston	09/70	
D8518	BREL Glasgow	02/72	
D8519	BREL Glasgow	05/72	
D8520	BREL Glasgow	05/72	
D8521	To Departmental Stock - 18521	-	
D8522	BREL Glasgow	05/72	
D8523	BREL Glasgow	06/72	
D8524	J McWilliam, Shettleston	09/70	
D8525	A King, Norwich	09/75	
D8526	BREL Glasgow	05/72	
D8527	BREL Glasgow	06/72	Stored: [U] 10/68, R/I: 11/68
D8528	BREL Glasgow	07/73	Withdrawn: 10/68, R/I: 11/69
D8529	A King, Norwich	09/75	Withdrawn: 10/68, R/I: 11/68, Withdrawn: 12/68, R/I: 11/69, Withdrawn: 10/71, R/I: 11/71
D8530	BREL Glasgow	10/71	Withdrawn: 10/68, R/I: 05/69, Withdrawn: 06/69, R/I: 10/69
D8531	A King, Norwich	10/75	Withdrawn: 10/68, R/I: 11/69
D8532	BREL Glasgow	06/72	Stored: [U] 10/68, R/I: 11/68
D8533	BREL Glasgow	01/72	
D8534	BREL Glasgow	02/73	
D8535	BREL Glasgow	11/71	
D8536	A King, Norwich	10/75	
D8537	J McWilliam, Shettleston	10/08	
D8538	BREL Glasgow	12/72	
D8539	A King, Norwich	12/75	
D8540	BREL Glasgow	12/72	
D8541	BREL Glasgow	02/73	
D8542	J Cashmore, Great Bridge	09/75	
D8543	BREL Glasgow	06/72	
D8544	J McWilliam, Shettleston	01/71	
D8545	BREL Glasgow	11/73	
D8546	J Cashmore, Great Bridge	10/75	
D8547	Birds, Long Marston	05/69	
D8548	J Cashmore, Great Bridge	10/75	
D8549	BREL Glasgow	06/72	
D8550	J Cashmore, Great Bridge	09/75	
D8551	J Cashmore, Great Bridge	09/75	
D8552	A King, Norwich	10/75	
D8553	J McWilliam, Shettleston	02/69	
D8554	J McWilliam, Shettleston	03/71	
D8555	BREL Glasgow	05/72	
D8556	Birds, Long Marston	05/69	
D8557	J McWilliam, Shettleston	08/75	
D8558	BREL Glasgow	02/73	

BR 1957 No.	Built By	Works No.	Date Introduced	Depot of First Allocation	Date Withdrawn	Depot of Final Allocation	Disposal Code
D8559	Clayton	4365U60	11/63	64B	10/71	66A	C
D8560	Clayton	4365U61	11/63	64B	02/69	66A	C
D8561	Clayton	4365U62	11/63	64B	02/71	64B	C
D8562	Clayton	4365U63	11/63	64B	10/71	66A	C
D8563	Clayton	4365U64	12/63	64B	12/71	66A	C
D8564	Clayton	4365U65	12/63	64B	02/69	66A	C
D8565	Clayton	4365U66	12/63	64B	11/71	64B	C
D8566	Clayton	4365U67	12/63	64B	12/68	64B	C
D8567	Clayton	4365U68	01/64	64B	10/71	66A	C
D8568	Clayton	4365U69	01/64	64B	10/71	66A	P
D8569	Clayton	4365U70	01/64	64B	12/68	66A	C
D8570	Clayton	4365U71	01/64	64B	11/68	66A	C
D8571	Clayton	4365U72	01/64	64B	06/69	66A	C
D8572	Clayton	4365U73	01/64	64B	06/69	66A	C
D8573	Clayton	4365U74	01/64	64B	10/71	66A	C
D8574	Clayton	4365U75	01/64	64B	12/71	66A	C
D8575	Clayton	4365U76	02/64	64B	10/68	64B	C
D8576	Clayton	4365U77	02/64	64B	02/69	66A	C
D8577	Clayton	4365U78	02/64	64B	02/69	64B	C
D8578	Clayton	4365U79	03/64	64B	06/69	66A	C
D8579	Clayton	4365U80	03/64	64B	10/71	64B	C
D8580	Clayton	4365U81	03/64	64B	10/71	64B	C
D8581	Clayton	4365U82	03/64	64B	10/71	64B	C
D8582	Clayton	4365U83	03/64	64B	01/69	64B	C
D8583	Clayton	4365U84	04/64	64B	09/71	64B	C
D8584	Clayton	4365U85	04/64	64B	10/68	64B	C
D8585	Clayton	4365U86	05/64	64B	10/68	64B	C
D8586	Clayton	4365U87	12/64	64B	09/71	64B	C
D8587	Clayton	4365U88	02/65	64B	10/71	64B	C
D8588	B.Peacock	8005	03/64	51L	10/71	64B	C
D8589	B.Peacock	8006	05/64	51L	07/70	52A	C
D8590	B.Peacock	8007	05/64	51L	03/71	52A	C
D8591	B.Peacock	8008	06/64	51L	12/68	52A	C
D8592	B.Peacock	8009	05/64	52A	09/71	64B	C
D8593	B.Peacock	8010	05/64	52A	10/71	64B	C
D8594	B.Peacock	8011	06/64	52A	09/71	64B	C
D8595	B.Peacock	8012	06/64	52A	12/68	52A	C
D8596	B.Peacock	8013	07/64	52A	12/68	52A	C
D8597	B.Peacock	8014	06/64	52A	10/71	64B	C
D8598	B.Peacock	8015	08/64	52A	12/71	66A	D
D8599	B.Peacock	8016	08/64	52A	10/71	64B	C
D8600	B.Peacock	8017	08/64	52A	10/71	64B	C
D8601	B.Peacock	8018	09/64	52A	10/71	64B	C
D8602	B.Peacock	8019	09/64	52A	10/71	64B	C
D8603	B.Peacock	8020	09/64	52A	10/71	64B	C
D8604	B.Peacock	8021	09/64	41A	10/71	64B	C
D8605	B.Peacock	8022	10/64	41A	11/68	51L	C
D8606	B.Peacock	8023	10/64	41A	03/71	64B	C
D8607	B.Peacock	8024	10/64	41A	10/71	66A	C
D8608	B.Peacock	8025	10/64	41A	10/71	66A	C
D8609	B.Peacock	8026	11/64	41A	10/68	64B	C
D8610	B.Peacock	8027	12/64	41A	10/71	66A	C
D8611	B.Peacock	8028	12/64	41A	10/68	64B	C
D8612	B.Peacock	8029	01/65	41A	10/71	66A	C
D8613	B.Peacock	8030	01/65	41A	10/71	66A	C
D8614	B.Peacock	8031	02/65	41A	10/71	66A	C
D8615	B.Peacock	8032	03/65	41A	10/71	66A	C
D8616	B.Peacock	8033	04/65	41E	09/71	66A	C

ENGLISH ELECTRIC TYPE 5 Co-Co CLASS 55

BR 1957 No.	TOPS No.	Date Re No.	Name	Name Date	Built By	Works No.	Date Introduced	Depot of First Allocation	Date Withdrawn
D9000	55022	04/74	Royal Scots Grey	06/62	EE.VF	2905/D557	02/61	64B	01/82
D9001	55001	02/74	St Paddy	07/61	EE.VF	2906/D558	02/61	34G	01/80
D9002	55002	12/73	The King's Own Yorkshire Light Infantry	04/63	EE.VF	2907/D559	03/61	52A	01/82
D9003	55003	02/74	Meld	07/61	EE.VF	2908/D560	03/61	34G	12/80
D9004	55004	05/74	Queen's Own Highlander	05/64	EE.VF	2909/D561	05/61	64B	11/81
D9005	55005	02/74	The Prince of Wales's Own Regiment of Yorkshire	10/63	EE.VF	2910/D562	05/61	52A	02/81
D9006	55006	03/74	The Fife and Forfar Yeomanry	12/64	EE.VF	2911/D563	06/61	64B	02/81
D9007	55007	02/74	Pinza	06/61	EE.VF	2912/D564	06/61	34G	12/81
D9008	55008	02/74	The Green Howards	09/63	EE.VF	2913/D565	07/61	52A	12/81
D9009	55009	01/74	Alycidon	07/61	EE.VF	2914/D566	07/61	34G	01/82
D9010	55010	06/74	The King's Own Scottish Borderer	05/65	EE.VF	2915/D567	07/61	64B	12/81

BR 1957 No.	Disposal Detail	Date Cut Up	Notes
D8559	BREL Glasgow	04/74	
D8560	Birds, Long Marston	05/69	
D8561	BREL Glasgow	07/73	
D8562	BREL Glasgow	07/73	
D8563	J Cashmore, Great Bridge	10/75	
D8564	Birds, Long Marston	05/69	
D8565	BREL Glasgow	05/73	
D8566	J McWilliam, Shettleston	05/69	
D8567	BREL Glasgow	01/73	
D8568	DTG at Chinnor	-	
D8569	J McWilliam, Shettleston	05/69	
D8570	Birds, Long Marston	05/69	
D8571	J McWilliam, Shettleston	07/71	
D8572	J Cashmore, Great Bridge	06/70	
D8573	J McWilliam, Shettleston	08/75	
D8574	A King, Norwich	08/75	Withdrawn: 10/71, R/I: 11/71
D8575	J McWilliam, Shettleston	04/69	
D8576	Birds, Long Marston	06/69	
D8577	Birds, Long Marston	05/69	
D8578	J McWilliam, Shettleston	03/71	
D8579	BREL Glasgow	11/73	
D8580	A King, Norwich	09/75	
D8581	BREL Glasgow	05/74	
D8582	J McWilliam, Shettleston	06/69	
D8583	BREL Glasgow	08/73	
D8584	J McWilliam, Shettleston	05/69	
D8585	J McWilliam, Shettleston	06/69	
D8586	BREL Glasgow	07/73	
D8587	BREL Glasgow	02/74	
D8588	BREL Glasgow	07/73	
D8589	BR Gateshead	04/71	
D8590	BREL Glasgow	05/72	
D8591	J McWilliam, Shettleston	06/69	
D8592	BREL Glasgow	03/72	
D8593	BREL Glasgow	09/73	
D8594	BREL Glasgow	03/72	
D8595	J McWilliam, Shettleston	05/69	
D8596	J McWilliam, Shettleston	06/69	
D8597	BREL Glasgow	07/73	
D8598	To Departmental Stock - 18598	-	Withdrawn: 10/71, R/I: 11/71
D8599	BREL Glasgow	11/71	
D8600	BREL Glasgow	04/73	
D8601	BREL Glasgow	11/73	
D8602	BREL Glasgow	04/72	
D8603	BREL Glasgow	04/72	
D8604	BREL Glasgow	04/72	
D8605	A Draper, Hull	06/70	
D8606	BREL Glasgow	04/72	
D8607	J Cashmore, Great Bridge	08/75	
D8608	J Cashmore, Great Bridge	08/75	
D8609	J Cashmore, Great Bridge	05/69	
D8610	BREL Glasgow	08/73	
D8611	J Cashmore, Great Bridge	06/69	
D8612	J Cashmore, Great Bridge	08/75	
D8613	J Cashmore, Great Bridge	08/75	
D8614	BREL Glasgow	12/72	
D8615	BREL Glasgow	12/73	
D8616	J Cashmore, Great Bridge	08/75	

BR 1957 No.	Depot of Final Allocation	Disposal Code	Disposal Detail	Date Cut Up	Notes
D9000	YK	P	9000 Fund, at BR Old Oak Common	-	
D9001	FP	C	BREL Doncaster	02/80	Stored: [U] 03/78
D9002	YK	P	National Railway Museum, York	-	
D9003	FP	C	BREL Doncaster	03/81	
D9004	YK	C	BREL Doncaster	08/83	Withdrawn: 08/81, R/I: 09/81
D9005	YK	C	BREL Doncaster	02/83	
D9006	YK	C	BREL Doncaster	07/81	
D9007	YK	C	BREL Doncaster	08/82	
D9008	YK	C	BREL Doncaster	08/82	
D9009	YK	P	Deltic Preservation Society	-	
D9010	YK	C	BREL Doncaster	05/82	

BR 1957 No.	TOPS No.	Date Re No.	Name	Name Date	Built By	Works No.	Date Introduced	Depot of First Allocation	Date Withdrawn
D9011	55011	02/74	The Royal Northumberland Fusiliers	05/63	EE.VF	2916/D568	08/61	52A	11/81
D9012	55012	02/74	Crepello	09/61	EE.VF	2917/D569	09/61	34G	05/81
D9013	55013	02/74	The Black Watch	01/63	EE.VF	2918/D570	09/61	64B	12/81
D9014	55014	02/74	The Duke of Wellington's Regiment	10/63	EE.VF	2919/D571	09/61	52A	11/81
D9015	55015	02/74	Tulyar	10/61	EE.VF	2920/D572	10/61	34G	01/82
D9016	55016	03/74	Gordon Highlander	07/64	EE.VF	2921/D573	10/61	64B	12/81
D9017	55017	02/74	The Durham Light Infantry	10/63	EE.VF	2922/D574	11/61	52A	12/81
D9018	55018	02/74	Ballymoss	11/61	EE.VF	2923/D575	11/61	34G	10/81
D9019	55019	11/73	Royal Highland Fusilier	09/65	EE.VF	2924/D576	12/61	64B	12/81
D9020	55020	11/73	Nimbus	02/62	EE.VF	2925/D577	02/62	34G	01/80
D9021	55021	01/74	Argyll and Sutherland Highlander	11/63	EE.VF	2926/D578	05/62	64B	12/81

BR TYPE 1 0-6-0 CLASS 14

BR 1957 No.	Built By	Works No.	Date Introduced	Depot of First Allocation	Date Withdrawn	Depot of Final Allocation	Disposal Code
D9500	BR Swindon	-	07/64	82A	04/69	86A	P
D9501	BR Swindon	-	07/64	86A	03/68	86A	C
D9502	BR Swindon	-	07/64	82A	04/69	86A	P
D9503	BR Swindon	-	07/64	82A	04/68	50B	C
D9504	BR Swindon	-	07/64	82A	04/68	50B	P
D9505	BR Swindon	-	07/64	82A	04/68	50B	E
D9506	BR Swindon	-	08/64	85A	03/68	86A	C
D9507	BR Swindon	-	08/64	85A	04/68	50B	C
D9508	BR Swindon	-	09/64	85A	10/68	87E	C
D9509	BR Swindon	-	09/64	85A	10/68	86A	C
D9510	BR Swindon	-	09/64	86A	04/68	50B	C
D9511	BR Swindon	-	09/64	86A	04/68	50B	C
D9512	BR Swindon	-	09/64	86A	04/68	50B	C
D9513	BR Swindon	-	10/64	86A	03/68	86A	P
D9514	BR Swindon	-	10/64	86A	04/68	86A	C
D9515	BR Swindon	-	10/64	86A	04/68	50B	E
D9516	BR Swindon	-	10/64	86A	04/68	50B	P
D9517	BR Swindon	-	11/64	86A	10/68	86A	C
D9518	BR Swindon	-	10/64	86A	04/69	86A	P
D9519	BR Swindon	-	11/64	86A	10/68	86A	C
D9520	BR Swindon	-	11/64	86A	04/68	50B	P
D9521	BR Swindon	-	11/64	81A	04/69	87E	P
D9522	BR Swindon	-	11/64	81A	12/67	86A	C
D9523	BR Swindon	-	12/64	81A	04/68	50B	P
D9524	BR Swindon	-	12/64	81A	04/69	87E	P
D9525	BR Swindon	-	01/65	82A	04/68	50B	P
D9526	BR Swindon	-	01/65	82A	11/68	86A	P
D9527	BR Swindon	-	01/65	82A	04/69	86A	C
D9528	BR Swindon	-	01/65	82A	03/69	86A	C
D9529	BR Swindon	-	01/65	86A	04/68	50B	P
D9530	BR Swindon	-	02/65	86A	10/68	86A	C
D9531	BR Swindon	-	02/65	86A	12/67	86A	P
D9532	BR Swindon	-	02/65	86A	04/68	50B	C
D9533	BR Swindon	-	02/65	86A	04/68	50B	C
D9534	BR Swindon	-	03/65	86A	04/68	50B	E
D9535	BR Swindon	-	03/65	86A	12/68	86A	C
D9536	BR Swindon	-	03/65	86A	04/69	87E	C
D9537	BR Swindon	-	03/65	86A	04/68	50B	P
D9538	BR Swindon	-	03/65	86A	04/69	87E	C
D9539	BR Swindon	-	04/65	86A	04/68	50B	P
D9540	BR Swindon	-	04/65	86A	04/68	50B	C
D9541	BR Swindon	-	04/65	86A	04/68	50B	C
D9542	BR Swindon	-	05/65	86A	04/68	50B	C
D9543	BR Swindon	-	05/65	86A	04/68	50B	C
D9544	BR Swindon	-	05/65	86A	04/68	50B	C
D9545	BR Swindon	-	06/65	87E	04/68	50B	C
D9546	BR Swindon	-	06/65	86A	04/68	50B	C
D9547	BR Swindon	-	07/65	86A	04/68	50B	C
D9548	BR Swindon	-	07/65	86A	04/68	50B	E
D9549	BR Swindon	-	08/65	86A	04/68	50B	E
D9550	BR Swindon	-	08/65	86A	04/68	50B	C
D9551	BR Swindon	-	09/65	86A	04/68	50B	P
D9552	BR Swindon	-	09/65	86A	04/68	50B	C
D9553	BR Swindon	-	09/65	82A	04/68	50B	P
D9554	BR Swindon	-	10/65	82A	04/68	50B	C
D9555	BR Swindon	-	10/65	82A	04/69	87E	P

BR 1957 No.	Depot of Final Allocation	Disposal Code	Disposal Detail	Date Cut Up	Notes
D9011	YK	C	BREL Doncaster	11/82	
D9012	FP	C	BREL Doncaster	09/81	
D9013	YK	C	BREL Doncaster	12/82	
D9014	YK	C	BREL Doncaster	02/82	
D9015	YK	P	Deltic Preservation Society, MRC	-	
D9016	YK	P	9000 Fund, at BR Old Oak Common	-	
D9017	YK	C	BREL Doncaster	01/83	
D9018	YK	C	BREL Doncaster	01/82	
D9019	YK	P	Deltic Preservation Society, GCR	-	
D9020	FP	C	BREL Doncaster	01/80	
D9021	YK	C	BREL Doncaster	08/82	

BR 1957 No.	Disposal Detail	Date Cut Up	Notes
D9500	South Yorkshire Railway	-	Stored: [U] 12/67, R/I: 02/68, Stored: [U] 04/69, PI
D9501	C F Booth, Rotherham	06/68	Stored: [U] 03/68
D9502	South Yorkshire Railway	-	Stored: [U] 12/67, R/I: 02/68, PI
D9503	BSC Corby	09/80	PI
D9504	Kent & East Sussex Railway	-	PI
D9505	Exported to Belgium [05/75]	-	PI
D9506	Arnott Young, Rawmarsh	07/68	Stored: [U] 02/68
D9507	BSC Corby	12/82	Stored: [U] 05/67, PI
D9508	NCB Ashington	01/84	Stored: [U] 05/67, PI
D9509	G Cohen, Kettering	12/70	
D9510	BSC Corby	09/82	Stored: [U] 05/67, PI
D9511	NCB Ashington	09/79	Stored: [U] 05/67, PI
D9512	BSC Corby	02/82	PI
D9513	Yorkshire Dales Railway	-	Stored: [U] 02/68, PI
D9514	NCB Ashington	12/85	PI
D9515	Exported to Charmartinon, Spain [06/82]	-	PI
D9516	Nene Valley Railway	-	PI
D9517	NCB Ashington Colliery	01/84	PI
D9518	Rutland Railway Museum	-	PI
D9519	G Cohen, Kettering	12/70	Stored: [U] 09/68
D9520	Rutland Railway Museum	-	PI
D9521	Swanage Railway	-	Stored: [U] 03/69, PI
D9522	Arnott Young, Rawmarsh	06/68	
D9523	Nene Valley Railway	-	PI
D9524	Boness & Kinneil Railway	-	Stored: [U] 02/69. PI
D9525	Kent & East Sussex Railway	-	PI
D9526	West Somerset Railway	-	Stored: [U] 10/68, PI
D9527	NCB Ashington	01/84	
D9528	NCB Ashington	12/81	Withdrawn: 09/68, R/I: 12/68, Stored: [U] 02/69, PI
D9529	Nene Valley Railway	-	Stored: [U] 05/69, PI
D9530	NCB Mardy	08/82	PI
D9531	East Lancs Railway	-	PI
D9532	BSC Corby	02/82	Stored: [U] 05/67, R/I: 06/67, PI
D9533	BSC Corby	09/82	Stored: [U] 05/67, R/I: 06/67, PI
D9534	Exported to Bruges, Belgium [05/75]	-	Stored: [U] 06/67, PI
D9535	NCB Ashington	01/84	Stored: [U] 11/68, PI
D9536	NCB Ashington	12/85	PI
D9537	Gloucester Warwickshire Railway	-	Stored: [U] 05/67, R/I: 05/67, PI
D9538	BSC Corby	09/82	PI
D9539	Gloucester Warwickshire Railway	-	Stored: [U] 05/67, R/I: 05/67, PI
D9540	NCB Ashington	01/84	PI
D9541	BSC Corby	09/82	PI
D9542	BSC Corby	09/82	PI
D9543	C F Booth, Rotherham	12/68	Stored: [U] 05/67, R/I: 06/67
D9544	BSC Corby	09/80	Stored: [U] 05/67, R/I: 06/67, PI
D9545	NCB Ashington	07/79	PI
D9546	C F Booth, Rotherham	08/68	
D9547	BSC Corby	09/82	PI
D9548	Exported to Charmartinon, Spain [06/82]	-	PI
D9549	Exported to Charmartinon, Spain [06/82]	-	PI
D9550	C F Booth, Rotherham	11/68	
D9551	West Somerset Railway	-	PI
D9552	BSC Corby	09/80	PI
D9553	Gloucester Warwickshire Railway	-	PI
D9554	BSC Corby	09/82	PI
D9555	Rutland Railway Museum	-	PI

1948-1957 NUMBER CONVERSION TABLES

1948 No. Range	Refer To 1957 No.	1948 No. Range	Refer To 1957 No.
11100	D2200	11181	D2404
11101	D2201	11182	D2405
11102	D2202	11183	D2406
11103	D2203	11184	D2407
11104	-	11185	D2408
11105	D2204	11186	D2409
11106	D2205	11187	D2000
11107	D2206	11188	D2001
11108	D2207	11189	D2002
11109	D2208	11190	D2003
11110	D2209	11191	D2004
11111	D2210	11192	D2005
11112	D2211	11193	D2006
11113	D2212	11194	D2007
11114	D2213	11195	D2008
11115	D2214	11196	D2009
11116	D2500	11197	D2010
11117	D2501	11198	D2011
11118	D2502	11199	D2012
11119	D2503	11200	D2013
11120	D2504	11201	D2014
11121	D2215	11202	D2015
11122	D2216	11203	D2016
11123	D2217	11204	D2017
11124	D2218	11205	D2018
11125	D2219	11206	D2019
11126	D2220	11207	D2020
11127	D2221	11208	D2021
11128	D2222	11209	D2022
11129	D2223	11210	D2023
11130	D2224	11211	D2024
11131	D2225	11212	D2242
11132	D2226	11213	D2243
11133	D2227	11214	D2244
11134	D2228	11215	D2245
11135	D2229	11216	D2246
11136	D2550	11217	D2247
11137	D2551	11218	D2248
11138	D2552	11219	D2249
11139	D2553	11220	D2250
11140	D2554	11221	D2251
11141	D2555	11222	D2252
11142	D2556	11223	D2253
11143	D2557	11224	D2254
11144	D2505	11225	D2255
11145	D2506	11226	D2256
11146	D2507	11227	D2257
11147	D2508	11228	D2258
11148	D2509	11229	D2259
11149	D2230	11500	D2950
11150	D2231	11501	D2951
11151	D2232	11502	D2952
11152	D2233	11503	D2953
11153	D2234	11504	D2954
11154	D2235	11505	D2955
11155	D2236	11506	D2956
11156	D2237	11507	D2957
11157	D2238	11508	D2958
11158	D2239	11700	D2700
11159	D2240	11701	D2701
11160	D2241	11702	D2702
11161	D2558	11703	D2703
11162	D2559	11704	D2704
11163	D2560	11705	D2705
11164	D2561	11706	D2707
11165	D2562	11707	D2707
11166	D2563	11708	D2708
11167	D2564	11709	D2709
11168	D2565	11710	D2710
11169	D2566	11711	D2711
11170	D2567	11712	D2712
11171	D2568	11713	D2713
11172	D2569	11714	D2714
11173	D2570	11715	D2715
11174	D2571	11716	D2716
11175	D2572	11717	D2717
11176	D2573	11719	D2719
11177	D2400	11720	D2710
11178	D2401	13000	D3000
11179	D2402	To	To
11180	D2403	13366	D3366

TOPS NUMBERS AND CONVERSION TABLES, DIESEL

TOPS No. Range	Refer To 1957 No.
Class 01 **D2954-D2955**	
01001	D2954
01002	D2955
Class 02 **D2850-D2869**	
02001	D2851
02002	D2852
02003	D2853
02004	D2856
Class 03 **D2000-D2199/** **D2370-D2399**	
03004	D2004
03005	D2005
03007	D2007
03008	D2008
03009	D2009
03010	D2010
03012	D2012
03013	D2013
03014	D2014
03016	D2016
03017	D2017
03018	D2018
03020	D2020
03021	D2021
03022	D2022
03025	D2025
03026	D2026
03027	D2027
03029	D2029
03034	D2034
03035	D2035
03037	D2037
03044	D2044
03045	D2045
03047	D2047
03050	D2050
03055	D2055
03056	D2056
03058	D2058
03059	D2059
03060	D2060
03061	D2061
03062	D2062
03063	D2063
03064	D2064
03066	D2066
03067	D2067
03068	D2068
03069	D2069
03072	D2072
03073	D2073
03075	D2075
03076	D2076
03078	D2078
03079	D2079
03080	D2080
03081	D2081
03084	D2084
03086	D2086
03089	D2089
03090	D2090
03091	D2091
03092	D2092
03094	D2094
03095	D2095
03096	D2096
03097	D2097
03098	D2098
03099	D2099
03102	D2102
03103	D2103
03104	D2104
03105	D2105
03106	D2106
03107	D2107
03108	D2108
03109	D2109
03110	D2110
03111	D2111
03112	D2112
03113	D2113
03119	D2119
03120	D2120
03121	D2121
03128	D2128
03129	D2129
03134	D2134
03135	D2135
03137	D2137
03141	D2141
03142	D2142
03144	D2144
03145	D2145
03147	D2147
03149	D2149
03151	D2151
03152	D2152
03153	D2153
03154	D2154
03155	D2155
03156	D2156
03157	D2157
03158	D2158
03159	D2159
03160	D2160
03161	D2161
03162	D2162
03163	D2163
03164	D2164
03165	D2165
03166	D2166
03167	D2167
03168	D2168
03169	D2169
03170	D2170
03171	D2171
03172	D2172
03174	D2174
03175	D2175
03179	D2179
03180	D2180
03189	D2189
03196	D2196
03197	D2197
03370	D2370
03371	D2371
03382	D2382
03386	D2386
03389	D2389
03397	D2397
03399	D2399
Class 05 **D2500-D2618**	
05001	D2554
Class 06 **D2410-D2444**	
06001	D2413
06002	D2414
06003	D2420
06004	D2421
06005	D2422
06006	D2423
06007	D2426
06008	D2437
06009	D2440
06010	D2444
Class 07 **D2985-D2998**	
07001	D2985
07002	D2986
07003	D2987
07004	D2988
07005	D2989
07006	D2990
07007	D2991
07008	D2992
07009	D2993
07010	D2994
07011	D2995
07012	D2996
07013	D2997
07014	D2998
Class 08 **D3000-D4192**	
08001	D3004
08002	D3005
08003	D3007
08004	D3008
08005	D3009
08006	D3010
08007	D3012
08008	D3015
08009	D3016
08010	D3017
08011	D3018
08012	D3019
08013	D3020
08014	D3021
08015	D3022
08016	D3023
08017	D3024
08018	D3025
08019	D3027
08020	D3028
08021	D3029
08022	D3030
08023	D3031
08024	D3032
08025	D3033
08026	D3036
08027	D3039
08028	D3040
08029	D3041
08030	D3042
08031	D3043
08032	D3044
08033	D3046
08034	D3047
08035	D3048
08036	D3049
08037	D3050
08038	D3051
08039	D3052
08040	D3053
08041	D3054
08042	D3055
08043	D3056
08044	D3057
08045	D3058
08046	D3059
08047	D3060
08048	D3061
08049	D3062
08050	D3063
08051	D3064
08052	D3065
08053	D3066
08054	D3067
08055	D3068
08056	D3070
08057	D3071
08058	D3072
08059	D3073
08060	D3074
08061	D3075
08062	D3076
08063	D3077
08064	D3079
08065	D3080
08066	D3081
08067	D3082
08070	D3085
08071	D3086
08072	D3087

TOPS No. Range	Refer To 1957 No.	TOPS No. Range	Refer To 1957 No.	TOPS No. Range	Refer To 1957 No.	TOPS No. Range	Refer To 1957 No.	TOPS No. Range	Refer To 1957 No.
08073	D3088	08156	D3224	08239	D3309	08322	D3392	08405	D3520
08074	D3089	08157	D3225	08240	D3310	08323	D3393	08406	D3521
08075	D3090	08158	D3226	08241	D3311	08324	D3394	08407	D3522
08076	D3091	08159	D3227	08242	D3312	08325	D3395	08408	D3523
08077	D3102	08160	D3228	08243	D3313	08326	D3396	08409	D3524
08078	D3103	08161	D3229	08244	D3314	08327	D3397	08410	D3525
08079	D3104	08162	D3230	08245	D3315	08328	D3398	08411	D3526
08080	D3105	08163	D3231	08246	D3316	08329	D3399	08412	D3527
08081	D3106	08164	D3232	08247	D3317	08330	D3400	08413	D3528
08082	D3107	08165	D3233	08248	D3318	08331	D3401	08414	D3529
08083	D3108	08166	D3234	08249	D3319	08332	D3402	08415	D3530
08084	D3109	08167	D3235	08250	D3320	08333	D3403	08416	D3531
08085	D3110	08168	D3236	08251	D3321	08334	D3404	08417	D3532
08086	D3111	08169	D3237	08252	D3322	08335	D3405	08418	D3533
08087	D3112	08170	D3238	08253	D3323	08336	D3406	08419	D3534
08088	D3113	08171	D3239	08254	D3324	08337	D3407	08420	D3535
08089	D3114	08172	D3240	08255	D3325	08338	D3408	08421	D3536
08090	D3115	08173	D3241	08256	D3326	08339	D3409	08422	D3537
08091	D3116	08174	D3242	08257	D3327	08340	D3410	08423	D3538
08092	D3127	08175	D3243	08258	D3328	08341	D3411	08424	D3539
08093	D3128	08176	D3244	08259	D3329	08342	D3412	08425	D3540
08094	D3129	08177	D3245	08260	D3330	08343	D3413	08426	D3541
08095	D3130	08178	D3246	08261	D3331	08344	D3414	08427	D3542
08096	D3131	08179	D3247	08262	D3332	08345	D3415	08428	D3543
08097	D3132	08180	D3248	08263	D3333	08346	D3416	08429	D3544
08098	D3133	08181	D3249	08264	D3334	08347	D3417	08430	D3545
08099	D3134	08182	D3250	08265	D3335	08348	D3418	08431	D3546
08100	D3135	08183	D3251	08266	D3336	08349	D3419	08432	D3547
08101	D3136	08184	D3252	08267	D3337	08350	D3420	08433	D3548
08102	D3167	08185	D3253	08268	D3338	08351	D3421	08434	D3549
08103	D3168	08186	D3254	08269	D3339	08352	D3422	08435	D3550
08104	D3169	08187	D3256	08270	D3340	08353	D3423	08436	D3551
08105	D3170	08188	D3257	08271	D3341	08354	D3424	08437	D3552
08106	D3171	08189	D3258	08272	D3342	08355	D3425	08438	D3553
08107	D3173	08190	D3259	08273	D3343	08356	D3426	08439	D3554
08108	D3174	08191	D3260	08274	D3344	08357	D3427	08440	D3555
08109	D3175	08192	D3262	08275	D3345	08358	D3428	08441	D3556
08110	D3176	08193	D3263	08276	D3346	08359	D3429	08442	D3557
08111	D3177	08194	D3264	08277	D3347	08360	D3430	08443	D3558
08112	D3178	08195	D3265	08278	D3348	08361	D3431	08444	D3559
08113	D3179	08196	D3266	08279	D3349	08362	D3432	08445	D3560
08114	D3180	08197	D3267	08280	D3350	08363	D3433	08446	D3561
08115	D3181	08198	D3268	08281	D3351	08364	D3434	08447	D3562
08116	D3182	08199	D3269	08282	D3352	08365	D3435	08448	D3563
08117	D3184	08200	D3270	08283	D3353	08366	D3436	08449	D3564
08118	D3185	08201	D3271	08284	D3354	08367	D3437	08450	D3565
08119	D3186	08202	D3272	08285	D3355	08368	D3438	08451	D3566
08120	D3187	08203	D3273	08286	D3356	08369	D3454	08452	D3567
08121	D3188	08204	D3274	08287	D3357	08370	D3455	08453	D3568
08122	D3189	08205	D3275	08288	D3358	08371	D3456	08454	D3569
08123	D3190	08206	D3276	08289	D3359	08372	D3457	08455	D3570
08124	D3191	08207	D3277	08290	D3360	08373	D3458	08456	D3571
08125	D3192	08208	D3278	08291	D3361	08374	D3459	08457	D3572
08126	D3194	08209	D3279	08292	D3362	08375	D3460	08458	D3573
08127	D3195	08210	D3280	08293	D3363	08376	D3461	08459	D3574
08128	D3196	08211	D3281	08294	D3364	08377	D3462	08460	D3575
08129	D3197	08212	D3282	08295	D3365	08378	D3463	08461	D3576
08130	D3198	08213	D3283	08296	D3366	08379	D3464	08462	D3577
08131	D3199	08214	D3284	08297	D3367	08380	D3465	08463	D3578
08132	D3200	08215	D3285	08298	D3368	08381	D3466	08464	D3579
08133	D3201	08216	D3286	08299	D3369	08382	D3467	08465	D3580
08134	D3202	08217	D3287	08300	D3370	08383	D3468	08466	D3581
08135	D3203	08218	D3288	08301	D3371	08384	D3469	08467	D3582
08136	D3204	08219	D3289	08302	D3372	08385	D3470	08468	D3583
08137	D3205	08220	D3290	08303	D3373	08386	D3471	08469	D3584
08138	D3206	08221	D3291	08304	D3374	08387	D3472	08470	D3585
08139	D3207	08222	D3292	08305	D3375	08388	D3503	08471	D3586
08140	D3208	08223	D3293	08306	D3376	08389	D3504	08472	D3587
08141	D3209	08224	D3294	08307	D3377	08390	D3505	08473	D3588
08142	D3210	08225	D3295	08308	D3378	08391	D3506	08474	D3589
08143	D3211	08226	D3296	08309	D3379	08392	D3507	08475	D3590
08144	D3212	08227	D3297	08310	D3380	08393	D3508	08476	D3591
08145	D3213	08228	D3298	08311	D3381	08394	D3509	08477	D3592
08146	D3214	08229	D3299	08312	D3382	08395	D3510	08478	D3593
08147	D3215	08230	D3300	08313	D3383	08396	D3511	08479	D3594
08148	D3216	08231	D3301	08314	D3384	08397	D3512	08480	D3595
08149	D3217	08232	D3302	08315	D3385	08398	D3513	08481	D3596
08150	D3218	08233	D3303	08316	D3386	08399	D3514	08482	D3597
08151	D3219	08234	D3304	08317	D3387	08400	D3515	08483	D3598
08152	D3220	08235	D3305	08318	D3388	08401	D3516	08484	D3599
08153	D3221	08236	D3306	08319	D3389	08402	D3517	08485	D3600
08154	D3222	08237	D3307	08320	D3390	08403	D3518	08486	D3601
08155	D3223	08238	D3308	08321	D3391	08404	D3519	08487	D3602

163

TOPS No. Range	Refer To 1957 No.	TOPS No. Range	Refer To 1957 No.	TOPS No. Range	Refer To 1957 No.	TOPS No. Range	Refer To 1957 No.	TOPS No. Range	Refer To 1957 No.
08488	D3603	08571	D3738	08654	D3821	08737	D3905	08820	D3988
08489	D3604	08572	D3739	08655	D3822	08738	D3906	08821	D3989
08490	D3605	08573	D3740	08656	D3823	08739	D3907	08822	D3990
08491	D3606	08574	D3741	08657	D3824	08740	D3908	08823	D3991
08492	D3607	08575	D3742	08658	D3825	08741	D3909	08824	D3992
08493	D3608	08576	D3743	08659	D3826	08742	D3910	08825	D3993
08494	D3609	08577	D3744	08660	D3827	08743	D3911	08826	D3994
08495	D3610	08578	D3745	08661	D3628	08744	D3912	08827	D3995
08496	D3611	08579	D3746	08662	D3829	08745	D3913	08828	D3996
08497	D3652	08580	D3747	08663	D3830	08746	D3914	08829	D3997
08498	D3653	08581	D3748	08664	D3831	08747	D3915	08830	D3998
08499	D3654	08582	D3749	08665	D3832	08748	D3916	08831	D3999
08500	D3655	08583	D3750	08666	D3833	08749	D3917	08832	D4000
08501	D3656	08584	D3751	08667	D3834	08750	D3918	08833	D4001
08502	D3657	08585	D3752	08668	D3835	08751	D3919	08834	D4002
08503	D3658	08586	D3753	08669	D3836	08752	D3920	08835	D4003
08504	D3659	08587	D3754	08670	D3837	08753	D3921	08836	D4004
08505	D3660	08588	D3755	08671	D3838	08754	D3922	08837	D4005
08506	D3661	08589	D3756	08672	D3839	08755	D3923	08838	D4006
08507	D3662	08590	D3757	08673	D3840	08756	D3924	08839	D4007
08508	D3663	08591	D3758	08674	D3841	08757	D3925	08840	D4008
08509	D3664	08592	D3759	08675	D3842	08758	D3926	08841	D4009
08510	D3672	08593	D3760	08676	D3843	08759	D3927	08842	D4010
08511	D3673	08594	D3761	08677	D3844	08760	D3928	08843	D4011
08512	D3674	08595	D3762	08678	D3845	08761	D3929	08844	D4012
08513	D3675	08596	D3763	08679	D3846	08762	D3930	08845	D4013
08514	D3676	08597	D3764	08680	D3847	08763	D3931	08846	D4014
08515	D3677	08598	D3765	08681	D3848	08764	D3932	08847	D4015
08516	D3678	08599	D3766	08682	D3849	08765	D3933	08848	D4016
08517	D3679	08600	D3767	08683	D3850	08766	D3934	08849	D4017
08518	D3680	08601	D3768	08684	D3851	08767	D3935	08850	D4018
08519	D3681	08602	D3769	08685	D3852	08768	D3936	08851	D4019
08520	D3682	08603	D3770	08686	D3853	08769	D3937	08852	D4020
08521	D3683	08604	D3771	08687	D3854	08770	D3938	08853	D4021
08522	D3684	08605	D3772	08688	D3855	08771	D3939	08854	D4022
08523	D3685	08606	D3773	08689	D3856	08772	D3940	08855	D4023
08524	D3686	08607	D3774	08690	D3857	08773	D3941	08856	D4024
08525	D3687	08608	D3775	08691	D3858	08774	D3942	08857	D4025
08526	D3688	08609	D3776	08692	D3859	08775	D3943	08858	D4026
08527	D3689	08610	D3777	08693	D3860	08776	D3944	08859	D4027
08528	D3690	08611	D3778	08694	D3861	08777	D3945	08860	D4028
08529	D3691	08612	D3779	08695	D3862	08778	D3946	08861	D4029
08530	D3692	08613	D3780	08696	D3863	08779	D3947	08862	D4030
08531	D3693	08614	D3781	08697	D3864	08780	D3948	08863	D4031
08532	D3694	08615	D3782	08698	D3865	08781	D3949	08864	D4032
08533	D3695	08616	D3783	08699	D3866	08782	D3950	08865	D4033
08534	D3696	08617	D3784	08700	D3867	08783	D3951	08866	D4034
08535	D3699	08618	D3785	08701	D3868	08784	D3952	08867	D4035
08536	D3700	08619	D3786	08702	D3869	08785	D3953	08868	D4036
08537	D3701	08620	D3787	08703	D3870	08786	D3954	08869	D4037
08538	D3702	08621	D3788	08704	D3871	08787	D3955	08870	D4038
08539	D3703	08622	D3789	08705	D3872	08788	D3956	08871	D4039
08540	D3704	08623	D3790	08706	D3873	08789	D3957	08872	D4040
08541	D3705	08624	D3791	08707	D3874	08790	D3958	08873	D4041
08542	D3706	08625	D3792	08708	D3875	08791	D3959	08874	D4042
08543	D3707	08626	D3793	08709	D3876	08792	D3960	08875	D4043
08544	D3708	08627	D3794	08710	D3877	08793	D3961	08876	D4044
08545	D3709	08628	D3795	08711	D3878	08794	D3962	08877	D4045
08546	D3710	08629	D3796	08712	D3879	08795	D3963	08878	D4046
08547	D3711	08630	D3797	08713	D3880	08796	D3964	08879	D4047
08548	D3712	08631	D3798	08714	D3881	08797	D3965	08880	D4048
08549	D3713	08632	D3799	08715	D3882	08798	D3966	08881	D4095
08550	D3714	08633	D3800	08716	D3883	08799	D3967	08882	D4096
08551	D3715	08634	D3801	08717	D3884	08800	D3968	08883	D4097
08552	D3716	08635	D3802	08718	D3886	08801	D3969	08884	D4098
08553	D3717	08636	D3803	08719	D3887	08802	D3970	08885	D4115
08554	D3718	08637	D3804	08720	D3888	08803	D3971	08886	D4116
08555	D3722	08638	D3805	08721	D3889	08804	D3972	08887	D4117
08556	D3723	08639	D3806	08722	D3890	08805	D3973	08888	D4118
08557	D3724	08640	D3807	08723	D3891	08806	D3974	08889	D4119
08558	D3725	08641	D3808	08724	D3892	08807	D3975	08890	D4120
08559	D3726	08642	D3809	08725	D3893	08808	D3976	08891	D4121
08560	D3727	08643	D3810	08726	D3894	08809	D3977	08892	D4122
08561	D3728	08644	D3811	08727	D3895	08810	D3978	08893	D4123
08562	D3729	08645	D3812	08728	D3896	08811	D3979	08894	D4124
08563	D3730	08646	D3813	08729	D3897	08812	D3980	08895	D4125
08564	D3731	08647	D3814	08730	D3898	08813	D3981	08896	D4126
08565	D3732	08648	D3815	08731	D3899	08814	D3982	08897	D4127
08566	D3733	00049	D3816	08732	D3900	08815	D3983	08898	D4128
08567	D3734	08650	D3817	08733	D3901	08816	D3984	08899	D4129
08568	D3735	08651	D3818	08734	D3902	08817	D3985	08900	D4130
08569	D3736	08652	D3819	08735	D3903	08818	D3986	08901	D4131
08570	D3737	08653	D3820	08736	D3904	08819	D3987	08902	D4132

TOPS No. Range	Refer To 1957 No.	TOPS No. Range	Refer To 1957 No.	TOPS No. Range	Refer To 1957 No.	TOPS No. Range	Refer To 1957 No.	TOPS No. Range	Refer To 1957 No.
08903	D4133	09018	D4106	20031	D8031	20114	D8114	20197	D8197
08904	D4134	09019	D4107	20032	D8032	20115	D8115	20198	D8198
08905	D4135	09020	D4108	20033	D8033	20116	D8116	20199	D8199
08906	D4136	09021	D4109	20034	D8034	20117	D8117	20200	D8300
08907	D4137	09022	D4110	20035	D8035	20118	D8118	20201	D8301
08908	D4138	09023	D4111	20036	D8036	20119	D8119	20202	D8302
08909	D4139	09024	D4112	20037	D8037	20120	D8120	20203	D8303
08910	D4140	09025	D4113	20038	D8038	20121	D8121	20204	D8304
08911	D4141	09026	D4114	20039	D8039	20122	D8122	20205	D8305
08912	D4142	09101	D4001	20040	D8040	20123	D8123	20206	D8306
08913	D4143	09102	D4000	20041	D8041	20124	D8124	20207	D8307
08914	D4144	09103	D3934	20042	D8042	20125	D8125	20208	D8308
08915	D4145	09104	D3917	20043	D8043	20126	D8126	20209	D8309
08916	D4146	09201	D3536	20044	D8044	20127	D8127	20210	D8310
08917	D4147	09202	D3900	20045	D8045	20128	D8050	20211	D8311
08918	D4148	09203	D3949	20046	D8046	20129	D8129	20212	D8312
08919	D4149	09204	D3884	20047	D8047	20130	D8130	20213	D8313
08920	D4150	09205	D3787	20048	D8048	20131	D8131	20214	D8314
08921	D4151			20049	D8049	20132	D8132	20215	D8315
08922	D4152	Class 10		20050	D8000	20133	D8133	20216	D8316
08923	D4153	D3137-D3151/		20051	D8051	20134	D8134	20217	D8317
08924	D4154	D3439-D3453/		20052	D8052	20135	D8135	20218	D8318
08925	D4155	D3473-D3502/		20053	D8053	20136	D8136	20219	D8319
08926	D4156	D3612-D3651/		20054	D8054	20137	D8137	20220	D8320
08927	D4157	D4049-D4094		20055	D8055	20138	D8138	20221	D8321
08928	D4158			20056	D8056	20139	D8139	20222	D8322
08929	D4159	Class 11		20057	D8057	20140	D8140	20223	D8323
08930	D4160	12033-12138		20058	D8058	20141	D8141	20224	D8324
08931	D4161			20059	D8059	20142	D8142	20225	D8325
08932	D4162	Class 12		20060	D8060	20143	D8143	20226	D8326
08933	D4163	15211-15236		20061	D8061	20144	D8144	20227	D8327
08934	D4164			20062	D8062	20145	D8145	20228	D8128
08935	D4165	Class 13		20063	D8063	20146	D8146	20301	D8023
08936	D4166	D4500-D4501		20064	D8064	20147	D8147	20301	D8059
08937	D4167	13001	D4501	20065	D8065	20148	D8148	20303	D8134
08938	D4168	13002	D4502	20066	D8066	20149	D8149	20304	D8168
08939	D4169	13003	D4500	20067	D8067	20150	D8150	20305	D8172
08940	D4170			20068	D8068	20151	D8151	20306	D8173
08941	D4171	Class 14		20069	D8069	20152	D8152	20307	D8194
08942	D4172	D9500-D9555		20070	D8070	20153	D8153	20308	D8196
08943	D4173			20071	D8071	20154	D8154	20901	D8041
08944	D4174	Class 15		20072	D8072	20155	D8155	20902	D8060
08945	D4175	D8200-D8243		20073	D8073	20156	D8156	20903	D8083
08946	D4176			20074	D8074	20157	D8157	20904	D8101
08947	D4177	Class 16		20075	D8075	20158	D8158	20905	D8319
08948	D4178	D8400-D8409		20076	D8076	20159	D8159	20906	D8325
08949	D4179			20077	D8077	20160	D8160	20907	D8309
08950	D4180	Class 17		20078	D8078	20161	D8161		
08951	D4181	D8500-D8616		20079	D8079	20162	D8162	Class 21	
08952	D4182			20080	D8080	20163	D8163	D6100-D6157	
08953	D4183	Class 20		20081	D8081	20164	D8164		
08954	D4184	D8001-D8199/		20082	D8082	20165	D8165	Class 22	
08955	D4185	D8300-D8327		20083	D8083	20166	D8166	D6300-D6357	
08956	D4186	20001	D8001	20084	D8084	20167	D8167		
08957	D4191	20002	D8002	20085	D8085	20168	D8168	Class 23	
08958	D4192	20003	D8003	20086	D8086	20169	D8169	D5900-D5909	
08991	D3273	20004	D8004	20087	D8087	20170	D8170		
08992	D3329	20005	D8005	20088	D8088	20171	D8171	Class 24	
08993	D3759	20006	D8006	20089	D8089	20172	D8172	D5001-D5150	
08994	D3577	20007	D8007	20090	D8090	20173	D8173	24001	D5001
08995	D3854	20008	D8008	20091	D8091	20174	D8174	24002	D5002
		20009	D8009	20092	D8092	20175	D8175	24003	D5003
Class 09		20010	D8010	20093	D8093	20176	D8176	24004	D5004
D3665-D3671/		20011	D8011	20094	D8094	20177	D8177	24005	D5000
D3719-D3721/		20012	D8012	20095	D8095	20178	D8178	24006	D5006
D4099-D4114		20013	D8013	20096	D8096	20179	D8179	24007	D5007
09001	D3665	20014	D8014	20097	D8097	20180	D8180	24008	D5008
09002	D3666	20015	D8015	20098	D8098	20181	D8181	24009	D5009
09003	D3667	20016	D8016	20099	D8099	20182	D8182	24010	D5010
09004	D3668	20017	D8017	20100	D8100	20183	D8183	24011	D5011
09005	D3669	20018	D8018	20101	D8101	20184	D8184	24012	D5012
09006	D3670	20019	D8019	20102	D8102	20185	D8185	24013	D5013
09007	D3671	20020	D8020	20103	D8103	20186	D8186	24014	D5014
09008	D3719	20021	D8021	20104	D8104	20187	D8187	24015	D5015
09009	D3720	20022	D8022	20105	D8105	20188	D8188	24016	D5016
09010	D3721	20023	D8023	20106	D8106	20189	D8189	24017	D5017
09011	D4099	20024	D8024	20107	D8107	20190	D8190	24018	D5018
09012	D4100	20025	D8025	20108	D8108	20191	D8191	24019	D5019
09013	D4101	20026	D8026	20109	D8109	20192	D8192	24020	D5020
09014	D4102	20027	D8027	20110	D8110	20193	D8193	24021	D5021
09015	D4103	20028	D8028	20111	D8111	20194	D8194	24022	D5022
09016	D4104	20029	D8029	20112	D8112	20195	D8195	24023	D5023
09017	D4105	20030	D8030	20113	D8113	20196	D8196	24024	D5024

165

TOPS No. Range	Refer To 1957 No.	TOPS No. Range	Refer To 1957 No.	TOPS No. Range	Refer To 1957 No.	TOPS No. Range	Refer To 1957 No.	TOPS No. Range	Refer To 1957 No.
24025	D5025	24108	D5108	25037	D5187	25120	D5270	25203	D7553
24026	D5026	24109	D5109	25038	D5188	25121	D5271	25204	D7554
24027	D5027	24110	D5110	25039	D5189	25122	D5272	25205	D7555
24028	D5028	24111	D5111	25040	D5190	25123	D5273	25206	D7556
24029	D5029	24112	D5112	25041	D5191	25124	D5274	25207	D7557
24030	D5030	24113	D5113	25042	D5192	25125	D5275	25208	D7558
24031	D5031	24114	D5114	25043	D5193	25126	D5276	25209	D7559
24032	D5032	24115	D5115	25044	D5194	25127	D5277	25210	D7560
24033	D5033	24116	D5116	25045	D5195	25128	D5278	25211	D7561
24034	D5034	24117	D5117	25046	D5196	25129	D5279	25212	D7562
24035	D5035	24118	D5118	25047	D5197	25130	D5280	25213	D7563
24036	D5036	24119	D5119	25048	D5198	25131	D5281	25214	D7564
24037	D5037	24120	D5120	25049	D5199	25132	D5282	25215	D7565
24038	D5038	24121	D5121	25050	D5200	25133	D5283	25216	D7566
24039	D5039	24122	D5122	25051	D5201	25134	D5284	25217	D7567
24040	D5040	24123	D5123	25052	D5202	25135	D5285	25218	D7568
24041	D5041	24124	D5124	25053	D5203	25136	D5286	25219	D7569
24042	D5042	24125	D5125	25054	D5204	25137	D5287	25220	D7570
24043	D5043	24126	D5126	25055	D5205	25138	D5288	25221	D7571
24044	D5044	24127	D5127	25056	D5206	25139	D5289	25222	D7572
24045	D5045	24128	D5128	25057	D5207	25140	D5290	25223	D7573
24046	D5046	24129	D5129	25058	D5208	25141	D5291	25224	D7574
24047	D5047	24130	D5130	25059	D5209	25142	D5292	25225	D7575
24048	D5048	24131	D5131	25060	D5210	25143	D5293	25226	D7576
24049	D5049	24132	D5132	25061	D5211	25144	D5294	25227	D7577
24050	D5050	24133	D5133	25062	D5212	25145	D5295	25228	D7578
24051	D5051	24134	D5134	25063	D5213	25146	D5296	25229	D7579
24052	D5052	24135	D5135	25064	D5214	25147	D5297	25230	D7580
24053	D5053	24136	D5136	25065	D5215	25148	D5298	25231	D7581
24054	D5054	24137	D5137	25066	D5216	25149	D5299	25232	D7582
24055	D5055	24138	D5138	25067	D5217	25150	D7500	25233	D7583
24056	D5056	24139	D5139	25068	D5218	25151	D7501	25234	D7584
24057	D5057	24140	D5140	25069	D5219	25152	D7502	25235	D7585
24058	D5058	24141	D5141	25070	D5220	25153	D7503	25236	D7586
24059	D5059	24142	D5142	25071	D5221	25154	D7504	25237	D7587
24060	D5060	24143	D5143	25072	D5222	25155	D7505	25238	D7588
24061	D5061	24144	D5144	25073	D5223	25156	D7506	25239	D7589
24062	D5062	24145	D5145	25074	D5224	25157	D7507	25240	D7590
24063	D5063	24146	D5146	25075	D5225	25158	D7508	25241	D7591
24064	D5064	24147	D5147	25076	D5226	25159	D7509	25242	D7592
24065	D5065	24148	D5148	25077	D5227	25160	D7510	25243	D7593
24066	D5066	24149	D5149	25078	D5228	25161	D7511	25244	D7594
24067	D5067	24150	D5150	25079	D5229	25162	D7512	25245	D7595
24068	D5068			25080	D5230	25163	D7513	25246	D7596
24069	D5069	**Class 25**		25081	D5231	25164	D7514	25247	D7597
24070	D5070	**D5151-D5299/**		25082	D5232	25165	D7515	25248	D7598
24071	D5071	**D7500-D7677**		25083	D5233	25166	D7516	25249	D7599
24072	D5072	25001	D5151	25084	D5234	25167	D7517	25250	D7600
24073	D5073	25002	D5152	25085	D5235	25168	D7518	25251	D7601
24074	D5074	25003	D5153	25086	D5236	25169	D7519	25252	D7602
24075	D5075	25004	D5154	25087	D5237	25170	D7520	25253	D7603
24076	D5076	25005	D5155	25088	D5238	25171	D7521	25254	D7604
24077	D5077	25006	D5156	25089	D5239	25172	D7522	25256	D7606
24078	D5078	25007	D5157	25090	D5240	25173	D7523	25257	D7607
24079	D5079	25008	D5158	25091	D5241	25174	D7524	25258	D7608
24080	D5080	25009	D5159	25092	D5242	25175	D7525	25259	D7609
24081	D5081	25010	D5160	25093	D5243	25176	D7526	25260	D7610
24082	D5082	25011	D5161	25094	D5244	25177	D7527	25261	D7611
24083	D5083	25012	D5162	25095	D5245	25178	D7528	25262	D7612
24084	D5084	25013	D5163	25096	D5246	25179	D7529	25263	D7613
24085	D5085	25014	D5164	25097	D5247	25180	D7530	25264	D7614
24086	D5086	25015	D5165	25098	D5248	25181	D7531	25265	D7615
24087	D5087	25016	D5166	25099	D5249	25182	D7532	25266	D7616
24088	D5088	25017	D5167	25100	D5250	25183	D7533	25267	D7617
24089	D5089	25018	D5168	25101	D5251	25184	D7534	25268	D7618
24090	D5090	25019	D5169	25102	D5252	25185	D7535	25269	D7619
24091	D5091	25020	D5170	25103	D5253	25186	D7536	25270	D7620
24092	D5092	25021	D5171	25104	D5254	25187	D7537	25271	D7621
24093	D5093	25022	D5172	25105	D5255	25188	D7538	25272	D7622
24094	D5094	25023	D5173	25106	D5256	25189	D7539	25273	D7623
24095	D5095	25024	D5174	25107	D5257	25190	D7540	25274	D7624
24096	D5096	25025	D5175	25108	D5258	25191	D7541	25275	D7625
24097	D5097	25026	D5176	25109	D5259	25192	D7542	25276	D7626
24098	D5098	25027	D5177	25110	D5260	25193	D7543	25277	D7627
24099	D5099	25028	D5178	25111	D5261	25194	D7544	25278	D7628
24100	D5100	25029	D5179	25112	D5262	25195	D7545	25279	D7629
24101	D5101	25030	D5180	25113	D5263	25196	D7546	25280	D7630
24102	D5102	25031	D5181	25114	D5264	25197	D7547	25281	D7631
24103	D5103	25032	D5182	25115	D5265	25198	D7548	25282	D7632
24104	D5104	25033	D5183	25116	D5266	25199	D7549	25283	D7633
24105	D5105	25034	D5184	25117	D5267	25200	D7550	25284	D7634
24106	D5106	25035	D5185	25118	D5268	25201	D7551	25285	D7635
24107	D5107	25036	D5186	25119	D5269	25202	D7552	25286	D7636

TOPS No. Range	Refer To 1957 No.	TOPS No. Range	Refer To 1957 No.	TOPS No. Range	Refer To 1957 No.	TOPS No. Range	Refer To 1957 No.	TOPS No. Range	Refer To 1957 No.
25287	D7637	26028	D5320	27062	D5407	31120	D5538	31203	D5627
25288	D7638	26029	D5329	27063	D5408	31121	D5539	31204	D5628
25289	D7639	26030	D5330	27064	D5409	31122	D5540	31205	D5629
25290	D7640	26031	D5331	27065	D5411	31123	D5541	31206	D5630
25291	D7641	26032	D5332	27066	D5386	31124	D5542	31207	D5631
25292	D7642	26033	D5333	27101	D5374	31125	D5543	31208	D5632
25293	D7643	26034	D5334	27102	D5380	31126	D5544	31209	D5633
25294	D7644	26035	D5335	27103	D5386	31127	D5545	31210	D5634
25295	D7645	26036	D5336	27104	D5387	31128	D5546	31211	D5635
25296	D7646	26037	D5337	27105	D5388	31129	D5547	31212	D5636
25297	D7647	26038	D5338	27106	D5394	31130	D5548	31213	D5637
25298	D7648	26039	D5339	27107	D5395	31131	D5549	31214	D5638
25299	D7649	26040	D5340	27108	D5396	31132	D5550	31215	D5639
25300	D7650	26041	D5341	27109	D5397	31133	D5551	31216	D5641
25301	D7651	26042	D5342	27110	D5399	31134	D5552	31217	D5642
25302	D7652	26043	D5343	27111	D5400	31135	D5553	31218	D5643
25303	D7653	26044	D5344	27112	D5401	31136	D5554	31219	D5644
25304	D7654	26045	D5345	27113	D5404	31137	D5555	31220	D5645
25305	D7655	26046	D5346	27114	D5407	31138	D5556	31221	D5647
25306	D7656			27115	D5408	31139	D5557	31222	D5648
25307	D7657	**Class 27**		27116	D5409	31140	D5558	31223	D5649
25308	D7658	**D5247-D5415**		27117	D5411	31141	D5559	31224	D5650
25309	D7659	27001	D5347	27118	D5413	31142	D5560	31225	D5651
25310	D7660	27002	D5348	27119	D5391	31143	D5561	31226	D5652
25311	D7661	27003	D5349	27120	D5392	31144	D5562	31227	D5653
25312	D7662	27004	D5350	27121	D5393	31145	D5563	31228	D5654
25313	D7663	27005	D5351	27122	D5403	31146	D5564	31229	D5655
25314	D7664	27006	D5352	27123	D5410	31147	D5565	31230	D5657
25315	D7665	27007	D5353	27124	D5412	31148	D5566	31231	D5658
25316	D7666	27008	D5354	27201	D5391	31149	D5567	31232	D5659
25317	D7667	27009	D5355	27202	D5392	31150	D5568	31233	D5660
25318	D7668	27010	D5356	27203	D5393	31151	D5569	31234	D5661
25319	D7669	27011	D5357	27204	D5403	31152	D5570	31235	D5662
25320	D7670	27012	D5358	27205	D5410	31153	D5571	31236	D5663
25321	D7671	27013	D5359	27206	D5412	31154	D5572	31237	D5664
25322	D7672	27014	D5360	27207	D5404	31155	D5573	31238	D5665
25323	D7673	27015	D5361	27208	D5407	31156	D5574	31239	D5666
25324	D7674	27016	D5362	27209	D5408	31157	D5575	31240	D5667
25325	D7675	27017	D5363	27210	D5409	31158	D5576	31241	D5668
25326	D7676	27018	D5364	27211	D5411	31159	D5577	31242	D5670
25327	D7677	27019	D5365	27212	D5386	31160	D5578	31243	D5671
25901	D7612	27020	D5366			31161	D5579	31244	D5672
25902	D7618	27021	D5367	**Class 31**		31162	D5580	31245	D5673
25903	D7626	27022	D5368	**D5500-D5699/**		31163	D5581	31246	D5674
25904	D7633	27023	D5369	**D5800-D5862**		31164	D5582	31247	D5675
25905	D7636	27024	D5370	31001	D5501	31165	D5583	31248	D5676
25906	D7646	27025	D5371	31002	D5502	31166	D5584	31249	D5677
25907	D7647	27026	D5372	31003	D5503	31167	D5585	31250	D5678
25908	D7657	27027	D5373	31004	D5504	31168	D5586	31251	D5679
25909	D7659	27028	D5375	31005	D5505	31169	D5587	31252	D5680
25910	D7665	27029	D5376	31006	D5506	31170	D5588	31253	D5681
25911	D7666	27030	D5377	31007	D5507	31171	D5600	31254	D5682
25912	D7672	27031	D5378	31008	D5508	31172	D5591	31255	D5683
		27032	D5379	31009	D5509	31173	D5593	31256	D5684
Class 26		27033	D5381	31010	D5510	31174	D5594	31257	D5685
D5301-D5346		27034	D5382	31011	D5511	31175	D5595	31258	D5686
26001	D5301	27035	D5384	31012	D5512	31176	D5597	31259	D5687
26002	D5302	27036	D5385	31013	D5513	31177	D5598	31260	D5688
26003	D5303	27037	D5389	31014	D5514	31178	D5599	31261	D5689
26004	D5304	27038	D5390	31015	D5515	31179	D5600	31262	D5690
26005	D5305	27039	D5398	31016	D5516	31180	D5601	31263	D5693
26006	D5306	27040	D5402	31017	D5517	31181	D5602	31264	D5694
26007	D5300	27041	D5405	31018	D5600	31182	D5603	31265	D5695
26008	D5308	27042	D5406	31019	D5519	31183	D5604	31266	D5696
26009	D5309	27043	D5414	31101	D5518	31184	D5607	31267	D5697
26010	D5310	27044	D5415	31102	D5520	31185	D5608	31268	D5698
26011	D5311	27045	D5374	31103	D5521	31186	D5609	31269	D5699
26012	D5312	27046	D5380	31104	D5522	31187	D5610	31270	D5800
26013	D5313	27047	D5413	31105	D5523	31188	D5611	31271	D5801
26014	D5314	27048	D5387	31106	D5524	31189	D5612	31272	D5802
26015	D5315	27049	D5388	31107	D5525	31190	D5613	31273	D5803
26016	D5316	27050	D5394	31108	D5526	31191	D5614	31274	D5804
26017	D5317	27051	D5395	31109	D5527	31192	D5615	31275	D5805
26018	D5318	27052	D5396	31110	D5528	31193	D5617	31276	D5806
26019	D5319	27053	D5397	31111	D5529	31194	D5618	31277	D5807
26020	D5307	27054	D5399	31112	D5530	31195	D5619	31278	D5808
26021	D5321	27055	D5400	31113	D5531	31196	D5620	31279	D5809
26022	D5322	27056	D5401	31114	D5532	31197	D5621	31280	D5810
26023	D5323	27057	D5393	31115	D5533	31198	D5622	31281	D5811
26024	D5324	27058	D5403	31116	D5534	31199	D5623	31282	D5813
26025	D5325	27059	D5410	31117	D5535	31200	D5624	31283	D5815
26026	D5326	27060	D5412	31118	D5536	31201	D5625	31284	D5816
26027	D5327	27061	D5404	31119	D5537	31202	D5626	31285	D5817

TOPS No. Range	Refer To 1957 No.	TOPS No. Range	Refer To 1957 No.	TOPS No. Range	Refer To 1957 No.	TOPS No. Range	Refer To 1957 No.	TOPS No. Range	Refer To 1957 No.
31286	D5818	31441	D5645	33023	D6541	**D6700-D6999**		37083	D6783
31287	D5819	31442	D5679	33024	D6542	37001	D6701	37084	D6784
31288	D5820	31443	D5598	33025	D6543	37002	D6702	37085	D6785
31289	D5821	31444	D5555	33026	D6544	37003	D6703	37086	D6786
31290	D5822	31445	D5833	33027	D6545	37004	D6704	37087	D6787
31291	D5823	31446	D5850	33028	D6546	37005	D6705	37088	D6788
31292	D5825	31447	D5828	33029	D6547	37006	D6706	37089	D6789
31293	D5826	31448	D5566	33030	D6548	37007	D6707	37090	D6790
31294	D5827	31449	D5840	33031	D6549	37008	D6708	37091	D6791
31295	D5828	31450	D5551	33032	D6550	37009	D6709	37092	D6792
31296	D5829	31451	D5852	33033	D6551	37010	D6710	37093	D6793
31297	D5830	31452	D5809	33034	D6552	37011	D6711	37094	D6794
31298	D5831	31453	D5532	33035	D6553	37012	D6712	37095	D6795
31299	D5832	31454	D5654	33036	D6554	37013	D6713	37096	D6796
31300	D5833	31455	D5674	33037	D6555	37014	D6714	37097	D6797
31301	D5834	31456	D5823	33038	D6556	37015	D6715	37098	D6798
31302	D5835	31457	D5587	33039	D6557	37016	D6716	37099	D6799
31303	D5836	31458	D5836	33040	D6558	37017	D6717	37100	D6800
31304	D5837	31459	D5684	33041	D6559	37018	D6718	37101	D6801
31305	D5838	31460	D5696	33042	D6560	37019	D6719	37102	D6802
31306	D5839	31461	D5547	33043	D6561	37020	D6720	37103	D6803
31307	D5840	31462	D5849	33044	D6562	37021	D6721	37104	D6804
31308	D5841	31463	D5830	33045	D6563	37022	D6722	37105	D6805
31309	D5843	31464	D5860	33046	D6564	37023	D6723	37106	D6806
31310	D5844	31465	D5637	33047	D6565	37024	D6724	37107	D6807
31311	D5845	31466	D5533	33048	D6566	37025	D6725	37108	D6808
31312	D5846	31467	D5641	33049	D6567	37026	D6726	37109	D6809
31313	D5847	31468	D5855	33050	D6568	37027	D6727	37110	D6810
31314	D5848	31469	D5807	33051	D6569	37028	D6728	37111	D6811
31315	D5849	31507	D5640	33052	D6570	37029	D6729	37112	D6812
31316	D5850	31511	D5691	33053	D6571	37030	D6730	37113	D6813
31317	D5851	31512	D5692	33054	D6572	37031	D6731	37114	D6814
31318	D5852	31514	D5814	33055	D6573	37032	D6732	37115	D6815
31319	D5853	31516	D5842	33056	D6574	37033	D6733	37116	D6816
31320	D5854	31519	D5697	33057	D6575	37034	D6734	37117	D6817
31321	D5855	31524	D5575	33058	D6577	37035	D6735	37118	D6818
31322	D5857	31526	D5617	33059	D6578	37036	D6736	37119	D6700
31323	D5858	31530	D5695	33060	D6579	37037	D6737	37120	D6820
31324	D5859	31533	D5663	33061	D6581	37038	D6738	37121	D6821
31325	D5860	31537	D5603	33062	D6582	37039	D6739	37122	D6822
31326	D5861	31541	D5645	33063	D6583	37040	D6740	37123	D6823
31327	D5862	31544	D5555	33064	D6584	37041	D6741	37124	D6824
31400	D5579	31545	D5833	33065	D6585	37042	D6742	37125	D6825
31401	D5589	31546	D5850	33101	D6511	37043	D6743	37126	D6826
31402	D5592	31547	D5828	33102	D6513	37044	D6744	37127	D6827
31403	D5596	31548	D5566	33103	D6514	37045	D6745	37128	D6828
31404	D5605	31549	D5840	33104	D6516	37046	D6746	37129	D6829
31405	D5606	31551	D5852	33105	D6517	37047	D6747	37130	D6830
31406	D5616	31552	D5809	33106	D6519	37048	D6748	37131	D6831
31407	D5640	31553	D5532	33107	D6520	37049	D6749	37132	D6832
31408	D5646	31554	D5654	33108	D6521	37050	D6750	37133	D6833
31409	D5656	31555	D5674	33109	D6525	37051	D6751	37134	D6834
31410	D5669	31556	D5823	33110	D6527	37052	D6752	37135	D6835
31411	D5691	31563	D5830	33111	D6528	37053	D6753	37136	D6836
31412	D5692	31565	D5637	33112	D6529	37054	D6754	37137	D6837
31413	D5812	31568	D5855	33113	D6531	37055	D6755	37138	D6838
31414	D5814	31569	D5807	33114	D6532	37056	D6756	37139	D6839
31415	D5824	31970	D5861	33115	D6533	37057	D6757	37140	D6840
31416	D5842			33116	D6535	37058	D6758	37141	D6841
31417	D5856	**Class 33**		33117	D6536	37059	D6759	37142	D6842
31418	D5522	**D6500-D6585**		33118	D6538	37060	D6760	37143	D6843
31419	D5697	33001	D6500	33119	D6580	37061	D6761	37144	D6844
31420	D5591	33002	D6501	33201	D6586	37062	D6762	37145	D6845
31421	D5558	33003	D6503	33202	D6587	37063	D6763	37146	D6846
31422	D5844	33004	D6504	33203	D6588	37064	D6764	37147	D6847
31423	D5621	33005	D6505	33204	D6589	37065	D6765	37148	D6848
31424	D5575	33006	D6506	33205	D6590	37066	D6766	37149	D6849
31425	D5804	33007	D6507	33206	D6591	37067	D6767	37150	D6850
31426	D5617	33008	D6508	33207	D6592	37068	D6768	37151	D6851
31427	D5618	33009	D6509	33208	D6593	37069	D6769	37152	D6852
31428	D5635	33010	D6510	33209	D6594	37070	D6770	37153	D6853
31429	D5699	33011	D6512	33210	D6595	37071	D6771	37154	D6854
31430	D5695	33012	D6515	33211	D6596	37072	D6772	37155	D6855
31431	D5681	33013	D6518	33212	D6597	37073	D6773	37156	D6856
31432	D5571	33014	D6522	33301	D6588	37074	D6774	37157	D6857
31433	D5663	33015	D6523	33302	D6590	37075	D6775	37158	D6858
31434	D5686	33016	D6524	33303	D6591	37076	D6776	37159	D6859
31435	D5600	33017	D6526			37077	D6777	37160	D6860
31436	D5569	33018	D6530	**Class 35**		37078	D6778	37161	D6861
31437	D5603	33019	D6534	**D7000-D7100**		37079	D6779	37162	D6862
31438	D5557	33020	D6537			37080	D6780	37163	D6863
31439	D5666	33021	D6539	**Class 37**		37081	D6781	37164	D6864
31440	D5628	33022	D6540	**D6600-D6608/**		37082	D6782	37165	D6865

TOPS No. Range	Refer To 1957 No.	TOPS No. Range	Refer To 1957 No.	TOPS No. Range	Refer To 1957 No.	TOPS No. Range	Refer To 1957 No.	TOPS No. Range	Refer To 1957 No.
37166	D6866	37249	D6949	37357	D6779	37682	D6936	40013	D213
37167	D6867	37250	D6950	37358	D6791	37683	D6887	40014	D214
37168	D6868	37251	D6951	37359	D6818	37684	D6834	40015	D215
37169	D6869	37252	D6952	37370	D6827	37685	D6934	40016	D216
37170	D6870	37253	D6953	37371	D6847	37686	D6872	40017	D217
37171	D6871	37254	D6954	37372	D6859	37687	D6831	40018	D218
37172	D6872	37255	D6955	37373	D6860	37688	D6905	40019	D219
37173	D6873	37256	D6956	37374	D6865	37689	D6895	40020	D220
37174	D6874	37257	D6957	37375	D6893	37690	D6871	40021	D221
37175	D6875	37258	D6958	37376	D6899	37691	D6879	40022	D222
37176	D6876	37259	D6959	37377	D6900	37692	D6822	40023	D223
37177	D6877	37260	D6960	37378	D6904	37693	D6910	40024	D224
37178	D6878	37261	D6961	37379	D6926	37694	D6892	40025	D225
37179	D6879	37262	D6962	37380	D6959	37695	D6857	40026	D226
37180	D6880	37263	D6963	37381	D6984	37696	D6928	40027	D227
37181	D6881	37264	D6964	37382	D6845	37697	D6943	40028	D228
37182	D6882	37265	D6965	37401	D6968	37698	D6946	40029	D229
37183	D6883	37266	D6966	37402	D6974	37699	D6953	40030	D230
37184	D6884	37267	D6967	37403	D6607	37701	D6730	40031	D231
37185	D6885	37268	D6968	37404	D6986	37702	D6720	40032	D232
37186	D6886	37269	D6969	37405	D6982	37703	D6767	40033	D233
37187	D6887	37270	D6970	37406	D6995	37704	D6734	40034	D234
37188	D6888	37271	D6971 1st	37407	D6605	37705	D6760	40035	D235
37189	D6889	37271	D6603 2nd	37408	D6989	37706	D6716	40036	D236
37190	D6890	37272	D6972 1st	37409	D6970	37707	D6701	40037	D237
37191	D6891	37272	D6604 2nd	37410	D6973	37708	D6789	40038	D238
37192	D6892	37273	D6973 1st	37411	D6990	37709	D6714	40039	D239
37193	D6893	37273	D6606 2nd	37412	D6601	37710	D6744	40040	D240
37194	D6894	37274	D6974 1st	37413	D6976	37711	D6785	40041	D241
37195	D6895	37274	D6608 2nd	37414	D6987	37712	D6802	40042	D242
37196	D6896	37275	D6975	37415	D6977	37713	D6752	40043	D243
37197	D6897	37276	D6976	37416	D6602	37714	D6724	40044	D244
37198	D6898	37277	D6977	37417	D6969	37715	D6721	40045	D245
37199	D6899	37278	D6978	37418	D6971	37716	D6794	40046	D246
37200	D6900	37279	D6979	37419	D6991	37717	D6750	40047	D247
37201	D6901	37280	D6980	37420	D6997	37718	D6784	40048	D248
37202	D6902	37281	D6981	37421	D6967	37719	D6733	40049	D249
37203	D6903	37282	D6982	37422	D6966	37796	D6805	40050	D250
37204	D6904	37283	D6819	37423	D6996	37797	D6781	40051	D251
37205	D6905	37284	D6984	37424	D6979	37798	D6706	40052	D252
37206	D6906	37285	D6985	37425	D6992	37799	D6761	40053	D253
37207	D6907	37286	D6986	37426	D6999	37800	D6843	40054	D254
37208	D6908	37287	D6987	37427	D6988	37801	D6873	40055	D255
37209	D6909	37288	D6988	37428	D6981	37802	D6863	40056	D256
37210	D6910	37289	D6989	37429	D6600	37803	D6908	40057	D257
37211	D6911	37290	D6990	37430	D6965	37883	D6876	40058	D258
37212	D6912	37291	D6991	37431	D6972	37884	D6883	40059	D259
37213	D6913	37292	D6992	37501	D6705	37885	D6877	40060	D260
37214	D6914	37293	D6993	37502	D6782	37886	D6880	40061	D261
37215	D6915	37294	D6994	37503	D6717	37887	D6820	40062	D262
37216	D6916	37295	D6995	37504	D6739	37888	D6835	40063	D263
37217	D6917	37296	D6996	37505	D6728	37889	D6933	40064	D264
37218	D6918	37297	D6997	37506	D6707	37890	D6868	40065	D265
37219	D6919	37298	D6998	37507	D6736	37891	D6800	40066	D266
37220	D6920	37299	D6999	37508	D6790	37892	D6849	40067	D267
37221	D6921	37300	D6600	37509	D6793	37893	D6937	40068	D268
37222	D6922	37301	D6601	37510	D6812	37894	D6824	40069	D269
37223	D6923	37302	D6602	37511	D6803	37895	D6819	40070	D270
37224	D6924	37303	D6603	37512	D6722	37896	D6931	40071	D271
37225	D6925	37304	D6604	37513	D6756	37897	D6855	40072	D272
37226	D6926	37305	D6605	37514	D6815	37898	D6886	40073	D273
37227	D6927	37306	D6606	37515	D6764	37899	D6861	40074	D274
37228	D6928	37307	D6607	37516	D6786	37901	D6850	40075	D275
37229	D6929	37308	D6608	37517	D6718	37902	D6848	40076	D276
37230	D6930	37310	D6852	37518	D6776	37903	D6949	40077	D277
37231	D6931	37311	D6856	37519	D6727	37904	D6825	40078	D278
37232	D6932	37312	D6837	37520	D6741	37905	D6836	40079	D279
37233	D6933	37313	D6845	37521	D6817	37906	D6906	40080	D280
37234	D6934	37314	D6890	37667	D6851			40081	D281
37235	D6935	37320	D6726	37668	D6957	**Class 40**		40082	D282
37236	D6936	37321	D6737	37669	D6829	**D201-D399**		40083	D283
37237	D6937	37322	D6749	37670	D6882	40001	D201	40084	D284
37238	D6938	37323	D6788	37071	D6947	40002	D202	40085	D285
37239	D6939	37324	D6799	37672	D6889	40003	D203	40086	D286
37240	D6940	37325	D6808	37673	D6832	40004	D204	40087	D287
37241	D6941	37326	D6811	37674	D6869	40005	D205	40088	D288
37242	D6942	37350	D6700	37675	D6864	40006	D206	40089	D289
37243	D6943	37351	D6702	37676	D6826	40007	D207	40090	D290
37244	D6944	37352	D6708	37677	D6821	40008	D208	40091	D291
37245	D6945	37353	D6732	37678	D6956	40009	D209	40092	D292
37246	D6946	37354	D6743	37679	D6823	40010	D210	40093	D293
37247	D6947	37355	D6745	37680	D6924	40011	D211	40094	D294
37248	D6948	37356	D6768	37681	D6830	40012	D212	40095	D295

169

TOPS No. Range	Refer To 1957 No.	TOPS No. Range	Refer To 1957 No.	TOPS No. Range	Refer To 1957 No.	TOPS No. Range	Refer To 1957 No.	TOPS No. Range	Refer To 1957 No.
40096	D296	40108	D308	40120	D320	40132	D332	40144	D344
40097	D297	40109	D309	40121	D321	40133	D333	40145	D345
40098	D298	40110	D310	40122	D200	40134	D334	40146	D346
40099	D299	40111	D311	40123	D323	40135	D335	40147	D347
40100	D300	40112	D312	40124	D324	40136	D336	40148	D348
40101	D301	40113	D313	40125	D325	40137	D337	40149	D349
40102	D302	40114	D314	40126	D326	40138	D338	40150	D350
40103	D303	40115	D315	40127	D327	40139	D339	40151	D351
40104	D304	40116	D316	40128	D328	40140	D340	40152	D352
40105	D305	40117	D317	40129	D329	40141	D341	40153	D353
40106	D306	40118	D318	40130	D330	40142	D342	40154	D354
40107	D307	40119	D319	40131	D331	40143	D343	40155	D355

BR HST POWER CARS CLASS 41

Original No.	Revised No.	Date Re No.	Set No.	Built By	Works No.	Date Introduced	Depot of First Allocation	Date Withdrawn	Depot of Final Allocation
41001	43000	05/74	252001	BREL Crewe	-	06/72	RTC	11/76	OO
41002	43001	05/74	252001	BREL Crewe	-	08/72	RTC	11/76	OO

BR HST POWER CARS CLASS 43

BR TOPS No.	Original Set No.	Name	Date Named	Built By	Works No.	Date Introduced	Depot of First Allocation	Date Withdrawn	Depot of Final Allocation
43002	253001	Top of The Pops	08/84	BREL Crewe	-	02/76	PM		
43003	253001			BREL Crewe	-	02/76	PM		
43004	253002	Swan Hunter	11/90	BREL Crewe	-	03/76	OO		
43005	253002			BREL Crewe	-	03/76	OO		
43006	253003			BREL Crewe	-	05/76	PM		
43007	253003			BREL Crewe	-	03/76	PM		
43008	253004			BREL Crewe	-	03/76	OO		
43009	253004			BREL Crewe	-	04/76	OO		
43010	253005	[TSW Today]	10/90-01/93	BREL Crewe	-	04/76	PM		
43011	253005	Reader 125	06/92	BREL Crewe	-	04/76	PM		
43012	253006			BREL Crewe	-	05/76	OO		
43013	253006	[University of Bristol]	10/86-08/89	BREL Crewe	-	05/76	OO		
43014	253007			BREL Crewe	-	05/76	PM		
43015	253007			BREL Crewe	-	05/76	PM		
43016	253008	Gwyl Gerddi Cymru 1992 - Garden Festival Wales 1992	04/92	BREL Crewe	-	06/76	OO		
43017	253008	[HTV West]	05/87-05/89	BREL Crewe	-	06/76	OO		
43018	253009			BREL Crewe	-	06/76	PM		
43019	253009	City of Swansea/ Dinas Abertawe	05/87	BREL Crewe	-	07/76	PM		
43020	253010			BREL Crewe	-	07/76	OO		
43021	253010			BREL Crewe	-	07/76	OO		
43022	253011			BREL Crewe	-	07/76	PM		
43023	253011	County of Cornwall	11/89	BREL Crewe	-	08/76	PM		
43024	253012			BREL Crewe	-	07/76	OO		
43025	253012			BREL Crewe	-	08/76	OO		
43026	253013	City of Westminster	05/85	BREL Crewe	-	08/76	PM		
43027	253013	[Westminster Abbey]	05/85-05/90	BREL Crewe	-	09/76	PM		
43028	253014			BREL Crewe	-	08/76	OO		
43029	253014			BREL Crewe	-	09/76	OO		
43030	253015			BREL Crewe	-	09/76	PM		
43031	253015			BREL Crewe	-	09/76	PM		
43032	253016	The Royal Regiment of Wales	12/89	BREL Crewe	-	10/76	OO		
43033	253016			BREL Crewe	-	10/76	OO		
43034	253017			BREL Crewe	-	11/76	PM		
43035	253017			BREL Crewe	-	11/76	PM		
43036	253018			BREL Crewe	-	11/76	OO		
43037	253018			BREL Crewe	-	11/76	OO		
43038	253019	National Railway Museum The First Ten Years 1975-1985	09/85	BREL Crewe	-	12/76	PM		
43039	253019			BREL Crewe	-	12/76	PM		
43040	253020	Granite City	06/90	BREL Crewe	-	12/76	OO		
43041	253020	City of Discovery	06/90	BREL Crewe	-	12/76	OO		
43042	253021			BREL Crewe	-	02/77	PM		
43043	253021			BREL Crewe	-	02/77	PM		
43044	253022			BREL Crewe	-	03/77	OO		
43045	253022	The Grammar School, Doncaster AD 1350	11/83	BREL Crewe	-	03/77	OO		
43046	253023			BREL Crewe	-	03/77	PM		
43047	253023	Rotherham Enterprise	03/84	BREL Crewe	-	03/77	PM		
43048	253024			BREL Crewe	-	03/77	OO		

TOPS No. Range	Refer To 1957 No.	TOPS No. Range	Refer To 1957 No.	TOPS No. Range	Refer To 1957 No.	TOPS No. Range	Refer To 1957 No.	TOPS No. Range	Refer To 1957 No.
40156	D356	40168	D368	40180	D380	40192	D392	**Class 41**	
40157	D357	40169	D369	40181	D381	40193	D393	**D600-D604**	
40158	D358	40170	D370	40182	D382	40194	D394		
40159	D359	40171	D371	40183	D383	40195	D395	**Class 42**	
40160	D360	40172	D372	40184	D384	40196	D396	**D800-D832/**	
40161	D361	40173	D373	40185	D385	40197	D397	**D866-D870**	
40162	D362	40174	D374	40186	D386	40198	D398		
40163	D363	40175	D375	40187	D387	40199	D399	**Class 43**	
40164	D364	40176	D376	40188	D388			**D833-D865**	
40165	D365	40177	D377	40189	D389				
40166	D366	40178	D378	40190	D390				
40167	D367	40179	D379	40191	D391				

Original No.	Disposal Code	Disposal Detail	Date Cut Up	Notes
41001	D	To Departmental Stock - ADB975812	-	
41002	D	To Departmental Stock - ADB975813	-	

BR TOPS No.	Disposal Code	Disposal Details	Date Cut Up	Notes
43002				Plates off: 05/88-04/91
43003				
43004				
43005				
43006				
43007				
43008				
43009				
43010				
43011				
43012				
43013				
43014				
43015				
43016				
43017				
43018				
43019				Plates off: 06/89-04/91
43020				
43021				
43022				
43023				
43024				
43025				
43026				
43027				
43028				
43029				
43030				
43031				
43032				
43033				
43034				
43035				
43036				
43037				
43038				Plates off: 09/89-02/91
43039				
43040				
43041				
43042				
43043				
43044				
43045				Plates off: 07/89-06/91
43046				
43047				Plates off: 06/89-03/91
43048				

BR TOPS No.	Original Set No.	Name	Date Named	Built By	Works No.	Date Introduced	Depot of First Allocation	Date Withdrawn	Depot of Final Allocation
43049	253024	Neville Hill	01/84	BREL Crewe	-	04/77	OO		
43050	253025			BREL Crewe	-	05/77	PM		
43051	253025	The Duke and Duchess of York	07/87	BREL Crewe	-	04/77	PM		
43052	253026	City of Peterborough	05/84	BREL Crewe	-	04/77	OO		
43053	253026	County of Humberside	04/84	BREL Crewe	-	05/77	OO		
43054	253027			BREL Crewe	-	05/77	PM		
43055	253027			BREL Crewe	-	06/77	PM		
43056	254001	University of Bradford	11/83	BREL Crewe	-	07/77	BN		
43057	254001	Bounds Green	03/84	BREL Crewe	-	07/77	BN		
43058	254002			BREL Crewe	-	07/77	HT		
43059	254002			BREL Crewe	-	08/77	HT		
43060	254003	County of Leicestershire	03/85	BREL Crewe	-	08/77	NL		
43061	254003	City of Lincoln	05/84	BREL Crewe	-	08/77	NL		
43062	254004			BREL Crewe	-	09/77	BN		
43063	254004			BREL Crewe	-	09/77	BN		
43064	254005	City of York	09/83	BREL Crewe	-	10/77	HT		
43065	254005			BREL Crewe	-	10/77	HT		
43066	254006			BREL Crewe	-	10/77	BN		
43067	254006			BREL Crewe	-	10/77	BN		
43068	254007			BREL Crewe	-	11/77	NL		
43069	254007			BREL Crewe	-	11/77	NL		
43070	254008			BREL Crewe	-	11/77	NL		
43071	254008			BREL Crewe	-	11/77	NL		
43072	254009			BREL Crewe	-	12/77	NL		
43073	254009			BREL Crewe	-	12/77	NL		
43074	254010			BREL Crewe	-	12/77	BN		
43075	254010			BREL Crewe	-	12/77	BN		
43076	254011	BBC East Midlands Today	01/91	BREL Crewe	-	12/77	NL		
43077	254011	County of Nottingham	09/84	BREL Crewe	-	12/77	NL		
43078	254012	[Shildon - County Durham]	09/83-06/90	BREL Crewe	-	01/78	HT		
43079	254012			BREL Crewe	-	01/78	HT		
43080	254013			BREL Crewe	-	01/78	EC		
43081	254013			BREL Crewe	-	02/78	EC		
43082	254014			BREL Crewe	-	02/78	HT		
43083	254014			BREL Crewe	-	02/78	HT		
43084	254015	County of Derbyshire	07/86	BREL Crewe	-	02/78	NL		
43085	254015	[City of Bradford]	06/83-11/89	BREL Crewe	-	03/78	NL		
43086	254016			BREL Crewe	-	04/78	HT		
43087	254016			BREL Crewe	-	05/78	HT		
43088	254017	XIIIth Commonwealth Games Scotland 1986	03/85	BREL Crewe	-	05/78	EC		
43089	254017			BREL Crewe	-	05/78	EC		
43090	254018			BREL Crewe	-	05/78	EC		
43091	254018	[Edinburgh Military Tattoo]	08/85-11/88	BREL Crewe	-	05/78	EC		
43092	254019	[Highland Chieftain]	05/84-01/88	BREL Crewe	-	06/78	EC		
43093	254019	York Festival '88	02/88	BREL Crewe	-	06/78	EC		
43094	254020			BREL Crewe	-	06/78	HT		
43095	254020	[Heaton]	02/84-04/91	BREL Crewe	-	06/78	HT		
43096	254021	The Queens Own Hussars	05/85	BREL Crewe	-	07/78	HT		
43097	254021	[The Light Infantry]	11/83-05/90	BREL Crewe	-	07/78	HT		
43098	254022	[Tyne and Wear Metropolitan County]	09/85-12/87	BREL Crewe	-	07/78	EC		
43099	254022			BREL Crewe	-	08/78	EC		
43100	254023	Craigentinny	03/84	BREL Crewe	-	08/78	NL		
43101	254023	[Edinburgh International Festival]	08/84-10/87	BREL Crewe	-	10/78	NL		
43102	254024	[City of Wakefield]	06/84-10/87	BREL Crewe	-	10/78	NL		
43103	254024	John Wesley	05/88	BREL Crewe	-	10/78	NL		
43104	254025	[County of Cleveland]	04/85-08/89	BREL Crewe	-	10/78	NL		
43105	254025	[Hartlepool]	07/84-03/88	BREL Crewe	-	10/78	NL		
43106	254026	Songs of Praise	06/89	BREL Crewe	-	10/78	BN		
43107	254026	[City of Derby]	05/86-10/88	BREL Crewe	-	11/78	BN		
43108	254027	[BBC Television Railwatch]	02/89-03/89	BREL Crewe	-	11/78	NL		
43109	254027	Yorkshire Evening Press	05/89	BREL Crewe	-	01/79	NL		
43110	254028	Darlington	05/84	BREL Crewe	-	01/79	HT		
43111	254028			BREL Crewe	-	02/79	HT		
43112	254029			BREL Crewe	-	02/79	NL		
43113	254029	[City of Newcastle upon Tyne]	04/83-02/89	BREL Crewe	-	02/79	NL		
43114	254030	National Garden Festival Gateshead 1990	08/89	BREL Crewe	-	03/79	BN		
43115	254030	Yorkshire Cricket Academy	06/89	BREL Crewe	-	03/79	BN		
43116	254031	City of Kingston upon Hull	05/83	BREL Crewe	-	03/79	NL		
43117	254031			BREL Crewe	-	03/79	NL		
43118	254032	Charles Wesley	05/88	BREL Crewe	-	03/79	NL		
43119	254032			BREL Crewe	-	04/79	NL		
43120	S			BREL Crewe	-	09/77	SPARE		
43121	S	[West Yorkshire Metropolitan County]	09/84-09/88	BREL Crewe	-	09/77	SPARE		
43122	S	South Yorkshire Metropolitan County	01/85	BREL Crewe	-	04/79	NL		

BR TOPS No.	Disposal Code	Disposal Details	Date Cut Up	Notes
43049				
43050				
43051				Plates off: 11/89-05/91
43052				Plates off: 10/90-03/91
43053				Plates off: 07/90-02/91
43054				
43055				
43056				Plates off: 05/90-12/92
43057				Plates off: 10/89-08/91
43058				
43059				
43060				Plates off: 05/90-03/91
43061				
43062				
43063				
43064				Plates off: 10/89-03/91
43065				
43066				
43067				
43068				
43069				
43070				
43071				
43072				
43073				
43074				
43075				
43076				
43077				Plates off: 05/89-06/91
43078				
43079				
43080				
43081				
43082				
43083				
43084				
43085				
43086				
43087				
43088				Plates off: 02/98-03/92
43089				
43090				
43091				
43092				
43093				
43094				
43095				
43096				Plates off: 12/87-12/92
43097				
43098				
43099				
43100				Plates off: 12/87-06/91
43101				
43102				
43103				
43104				
43105				
43106				
43107				
43108				
43109				
43110				Plates off: 12/88-01/91
43111				
43112				
43113				
43114				
43115				
43116				Plates off: 08/89-06/91
43117				
43118				
43119				
43120				
43121				
43122				Plates off: 08/89-03/91

BR TOPS No.	Original Set No.	Name	Date Named	Built By	Works No.	Date Introduced	Depot of First Allocation	Date Withdrawn	Depot of Final Allocation
43123	S			BREL Crewe	-	04/79	BN		
43124	S	[BBC Points West]	09/86-08/89	BREL Crewe	-	06/81	PM		
43125	253028	[Merchant Venturer]	04/85-05/89	BREL Crewe	-	04/79	OO		
43126	253028	City of Bristol	04/85	BREL Crewe	-	05/79	OO		
43127	253029			BREL Crewe	-	05/79	PM		
43128	253029			BREL Crewe	-	06/79	PM		
43129	253030			BREL Crewe	-	06/79	OO		
43130	253030	Sulis Minerva	06/92	BREL Crewe	-	07/79	OO		
43131	253031	[Sir Felix Pole]	08/85-11/87	BREL Crewe	-	07/79	PM		
43132	253031	Worshipful Company of Carmen	10/87	BREL Crewe	-	07/79	PM		
43133	253032			BREL Crewe	-	07/79	OO		
43134	253032	County of Somerset	07/92	BREL Crewe	-	08/79	OO		
43135	253033			BREL Crewe	-	08/79	PM		
43136	253033			BREL Crewe	-	08/79	PM		
43137	253034			BREL Crewe	-	09/79	OO		
43138	253034			BREL Crewe	-	09/79	OO		
43139	253035			BREL Crewe	-	03/80	PM		
43140	253035			BREL Crewe	-	03/80	PM		
43141	253036			BREL Crewe	-	04/80	OO		
43142	253036	[St Marys Hospital Paddington]	11/86-11/88	BREL Crewe	-	05/80	OO		
43143	253037			BREL Crewe	-	02/81	PM		
43144	253037			BREL Crewe	-	02/81	PM		
43145	253038			BREL Crewe	-	03/81	OO		
43146	253038			BREL Crewe	-	03/81	OO		
43147	253039	The Red Cross	02/91	BREL Crewe	-	05/81	PM		
43148	253039			BREL Crewe	-	05/81	PM		
43149	253040	BBC Wales Today	09/88	BREL Crewe	-	06/81	OO		
43150	253040	Bristol Evening Post	10/88	BREL Crewe	-	06/81	OO		
43151	253041	[Blue Peter II]	05/87-11/89	BREL Crewe	-	06/81	OO		
43152	253041	St Peters School York AD 627	11/84	BREL Crewe	-	06/81	OO		
43153	254033	University of Durham	07/83	BREL Crewe	-	01/81	NL		
43154	254033			BREL Crewe	-	01/81	NL		
43155	254034	BBC Look North	06/85	BREL Crewe	-	01/81	NL		
43156	254034			BREL Crewe	-	01/81	NL		
43157	254035	Yorkshire Evening Post	01/84	BREL Crewe	-	04/81	NL		
43158	254035			BREL Crewe	-	04/81	NL		
43159	254036			BREL Crewe	-	04/81	NL		
43160	254036	Storm Force	04/91	BREL Crewe	-	04/81	NL		
43161	254037	Reading Evening Post	04/91	BREL Crewe	-	06/81	NL		
43162	254037	Borough of Stevenage	03/84	BREL Crewe	-	06/81	NL		
43163	253041			BREL Crewe	-	07/81	LA		
43164	253041			BREL Crewe	-	07/81	LA		
43165	253042			BREL Crewe	-	09/81	LA		
43166	253042			BREL Crewe	-	09/81	LA		
43167	253043			BREL Crewe	-	09/81	LA		
43168	253043			BREL Crewe	-	09/81	LA		
43169	253044	The National Trust	07/89	BREL Crewe	-	09/81	LA		
43170	253044			BREL Crewe	-	09/81	LA		
43171	253045			BREL Crewe	-	09/81	LA		
43172	253045			BREL Crewe	-	09/81	LA		
43173	253046			BREL Crewe	-	09/81	LA		
43174	253046			BREL Crewe	-	09/81	LA		
43175	253047			BREL Crewe	-	10/81	LA		
43176	253047			BREL Crewe	-	10/81	LA		
43177	253048			BREL Crewe	-	11/81	LA		
43178	253048			BREL Crewe	-	11/81	LA		
43179	253049	Pride of Laira	09/91	BREL Crewe	-	11/81	LA		
43180	253049			BREL Crewe	-	11/81	LA		
43181	253050			BREL Crewe	-	11/81	NL		
43182	253050			BREL Crewe	-	11/81	NL		
43183	253051			BREL Crewe	-	02/82	LA		
43184	253051			BREL Crewe	-	02/82	LA		
43185	253052	Great Western	05/92	BREL Crewe	-	03/82	LA		
43186	253052	Sir Francis Drake	07/88	BREL Crewe	-	03/82	LA		
43187	253053			BREL Crewe	-	04/82	LA		
43188	253053	City of Plymouth	05/86	BREL Crewe	-	04/82	LA		
43189	253054			BREL Crewe	-	04/82	LA		
43190	253054			BREL Crewe	-	04/82	LA		
43191	253055	Seahawk	10/88	BREL Crewe	-	05/82	LA		
43192	253055	City of Truro	10/88	BREL Crewe	-	05/82	LA		
43193	253056	Yorkshire Post	12/83	BREL Crewe	-	06/82	LA		
43194	253056	[Royal Signals]	10/85-08/89	BREL Crewe	-	06/82	LA		
43195	253057			BREL Crewe	-	07/82	LA		
43196	253057	The Newspaper Society Founded 1836	04/86	BREL Crewe	-	07/82	LA		
43197	253058			BREL Crewe	-	08/82	LA		
43198	253058			BREL Crewe	-	08/82	LA		

BR TOPS No.	Disposal Code	Disposal Details	Date Cut Up	Notes
43123				
43124				
43125				
43126				Plates off: 09/87-09/91
43127				
43128				
43129				
43130				
43131				
43132				
43133				
43134				
43135				
43136				
43137				
43138				
43139				
43140				
43141				
43142				
43143				
43144				
43145				
43146				
43147				Plates read: 'Red Cross' 05/88-02/91
43148				
43149				
43150				
43151				
43152				Plates off: 05/89-06/91
43153				Plates off: 07/89-06/91
43154				
43155				Plates off: 03/90-05/91
43156				
43157				Plates off: 05/90-05/91
43158				
43159				
43160				
43161				
43162				Plates off: 09/90-01/91
43163				
43164				
43165				
43166				
43167				
43168				
43169				
43170				
43171				
43172				
43173				
43174				
43175				
43176				
43177				
43178				
43179				
43180				
43181				
43182				
43183				
43184				
43185				
43186				
43187				
43188				
43189				
43190				
43191				
43192				
43193				
43194				
43195				
43196				Plates off: 02/89-02/91
43197				
43198				

TOPS NUMBERS AND CONVERSION TABLES CONTINUED

TOPS No. Range	Refer To 1957 No.	TOPS No. Range	Refer To 1957 No.	TOPS No. Range	Refer To 1957 No.	TOPS No. Range	Refer To 1957 No.	TOPS No. Range	Refer To 1957 No.
Class 44		45067	D115	46018	D155	47040	D1621	47121	D1710
D1-D10		45068	D118	46019	D156	47041	D1622	47122	D1711
44001	D1	45069	D121	46020	D157	47042	D1623	47123	D1712
44002	D2	45070	D122	46021	D158	47043	D1624	47124	D1714
44003	D3	45071	D125	46022	D159	47044	D1625	47125	D1715
44004	D4	45072	D127	46023	D160	47045	D1626	47126	D1717
44005	D5	45073	D129	46024	D161	47046	D1628	47127	D1718
44006	D6	45074	D131	46025	D162	47047	D1629	47128	D1719
44007	D7	45075	D132	46026	D163	47048	D1630	47129	D1720
44008	D8	45076	D134	46027	D164	47049	D1631	47130	D1721
44009	D9	45077	D136	46028	D165	47050	D1632	47131	D1722
44010	D10	45101	D96	46029	D166	47051	D1633	47132	D1723
		45102	D51	46030	D167	47052	D1634	47133	D1724
Class 45		45103	D116	46031	D168	47053	D1635	47134	D1726
D1-D137		45104	D59	46032	D169	47054	D1638	47135	D1727
45001	D13	45105	D86	46033	D170	47055	D1639	47136	D1728
45002	D29	45106	D106	46034	D171	47056	D1640	47137	D1729
45003	D133	45107	D43	46035	D172	47057	D1641	47138	D1730
45004	D77	45108	D120	46036	D173	47058	D1642	47139	D1731
45005	D79	45109	D85	46037	D174	47059	D1643	47140	D1732
45006	D89	45110	D73	46038	D175	47060	D1644	47141	D1733
45007	D119	45111	D65	46039	D176	47061	D1645	47142	D1735
45008	D90	45112	D61	46040	D177	47062	D1646	47143	D1736
45009	D37	45113	D80	46041	D178	47063	D1647	47144	D1737
45010	D112	45114	D94	46042	D179	47064	D1648	47145	D1738
45011	D12	45115	D81	46043	D180	47065	D1649	47146	D1739
45012	D108	45116	D47	46044	D181	47066	D1650	47147	D1740
45013	D20	45117	D35	46045	D182	47067	D1651	47148	D1741
45014	D137	45118	D67	46046	D183	47068	D1652	47149	D1742
45015	D14	45119	D33	46047	D184	47069	D1653	47150	D1743
45016	D16	45120	D107	46048	D185	47070	D1654	47151	D1744
45017	D23	45121	D18	46049	D186	47071	D1655	47152	D1745
45018	D15	45122	D11	46050	D187	47072	D1656	47153	D1746
45019	D33	45123	D52	46051	D188	47073	D1657	47154	D1747
45020	D26	45124	D28	46052	D189	47074	D1658	47155	D1748
45021	D25	45125	D123	46053	D190	47075	D1659	47156	D1749
45022	D60	45126	D32	46054	D191	47076	D1660	47157	D1750
45023	D54	45127	D87	46055	D192	47077	D1661	47158	D1751
45024	D17	45128	D113	46056	D193	47078	D1663	47159	D1752
45025	D19	45129	D111			47079	D1664	47160	D1754
45026	D21	45130	D117	**Class 47**		47080	D1665	47161	D1755
45027	D24	45131	D36	**D1100-D1999**		47081	D1666	47162	D1756
45028	D27	45132	D22	47001	D1521	47082	D1667	47163	D1757
45029	D30	45133	D39	47002	D1522	47083	D1668	47164	D1758
45030	D31	45134	D126	47003	D1523	47084	D1669	47165	D1759
45031	D36	45135	D99	47004	D1524	47085	D1670	47166	D1761
45032	D38	45136	D88	47005	D1526	47086	D1672	47167	D1762
45033	D39	45137	D56	47006	D1528	47087	D1673	47168	D1763
45034	D42	45138	D92	47007	D1529	47088	D1674	47169	D1764
45035	D44	45139	D109	47008	D1530	47089	D1675	47170	D1765
45036	D45	45140	D102	47009	D1532	47090	D1676	47171	D1766
45037	D46	45141	D82	47010	D1537	47091	D1677	47172	D1767
45038	D48	45142	D83	47011	D1538	47092	D1678	47173	D1768
45039	D49	45143	D62	47012	D1539	47093	D1679	47174	D1769
45040	D50	45144	D55	47013	D1540	47094	D1680	47175	D1770
45041	D53	45145	D128	47014	D1543	47095	D1681	47176	D1771
45042	D57	45146	D66	47015	D1544	47096	D1682	47177	D1772
45043	D58	45147	D41	47016	D1546	47097	D1684	47178	D1773
45044	D63	45148	D130	47017	D1570	47098	D1685	47179	D1774
45045	D64	45149	D135	47018	D1572	47099	D1686	47180	D1775
45046	D68	45150	D78	47019	D1573	47100	D1687	47181	D1776
45047	D69			47020	D1583	47101	D1688	47182	D1777
45048	D70	**Class 46**		47021	D1584	47102	D1690	47183	D1778
45049	D71	**D138-D197**		47022	D1585	47103	D1691	47184	D1779
45050	D72	46001	D138	47023	D1588	47104	D1692	47185	D1780
45051	D74	46002	D139	47024	D1591	47105	D1693	47186	D1781
45052	D75	46003	D140	47025	D1592	47106	D1694	47187	D1837
45053	D76	46004	D141	47026	D1597	47107	D1695	47188	D1838
45054	D95	46005	D142	47027	D1599	47108	D1696	47189	D1839
45055	D84	46006	D143	47028	D1605	47109	D1697	47190	D1840
45056	D91	46007	D144	47029	D1606	47110	D1698	47191	D1841
45057	D93	46008	D145	47030	D1609	47111	D1699	47192	D1842
45058	D97	46009	D146	47031	D1610	47112	D1700	47193	D1843
45059	D98	46010	D147	47032	D1611	47113	D1701	47194	D1844
45060	D100	46011	D148	47033	D1613	47114	D1702	47195	D1845
45061	D101	46012	D149	47034	D1614	47115	D1703	47196	D1846
45062	D103	46013	D150	47035	D1615	47116	D1704	47197	D1847
45063	D104	46014	D151	47036	D1617	47117	D1705	47198	D1848
45064	D105	46015	D152	47037	D1618	47118	D1706	47199	D1849
45065	D110	46016	D153	47038	D1619	47119	D1708	47200	D1850
45066	D114	46017	D154	47039	D1620	47120	D1709	47201	D1851

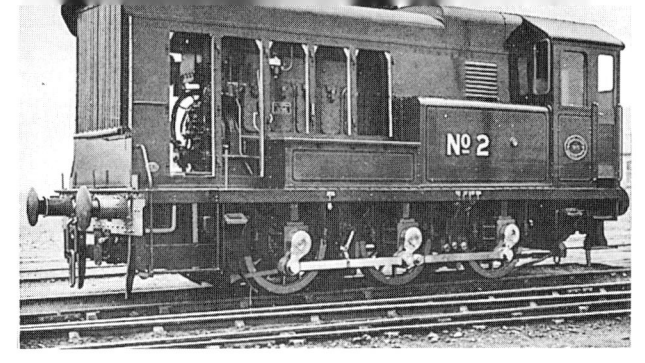

1. LMS 'prototype' 0-6-0 diesel shunter No. 1831.
2. GWR 0-6-0 diesel-electric shunter No. 2
3. BR 0-6-0 diesel-electric shunter No. 15101.
4. LMS designed main line prototypes Nos 10000 and 10001. 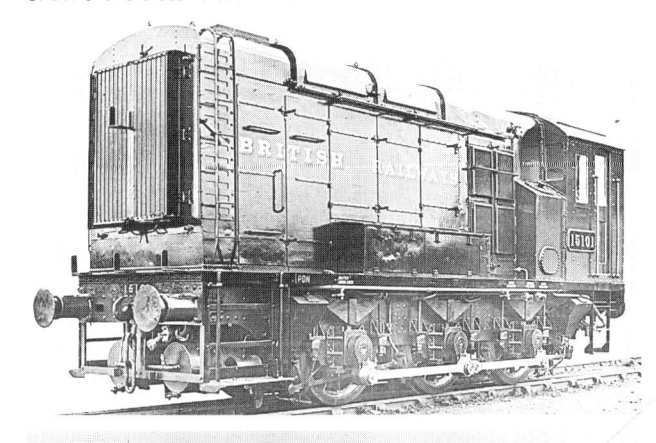   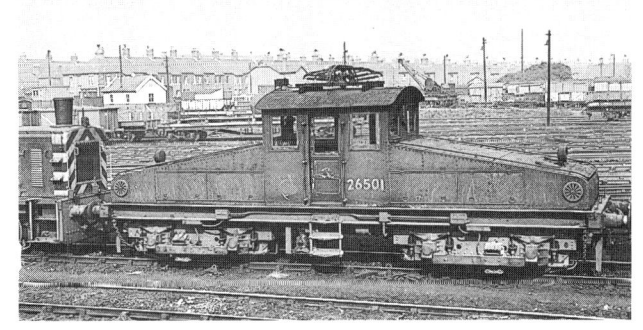
5. North Eastern Railway electric locomotive No. 3.
6. North Eastern Bo-Bo electric as BR No. 26501.  
7. Brush prototype Type 5 No. HS4000 *Kestrel*.
8. The AEI/BRCW/Sulzer Type 4 prototype No. D0260 *Lion*.
9. Brush prototype No. D0280 *Falcon*.
10. The prototype 'Fell' locomotive, No. 10100.

11. Brown-Boveri gas-turbine prototype No. 18000.
12. Motor Rail (Simplex) 4-wheel petrol locomotive No. 15099.
13. North British-built diesel-hydraulic shunter No. D2903.
14. Hunslet-built diesel shunter No. D2950.

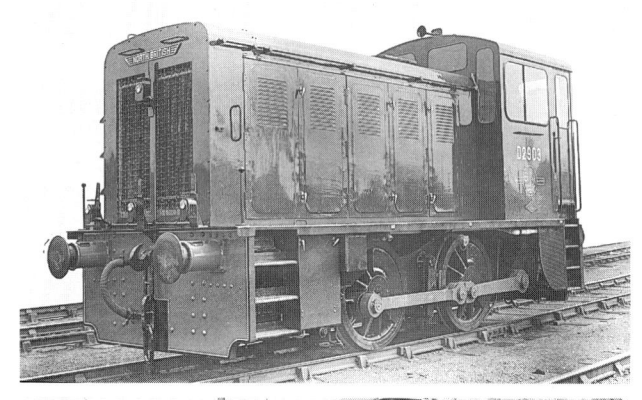

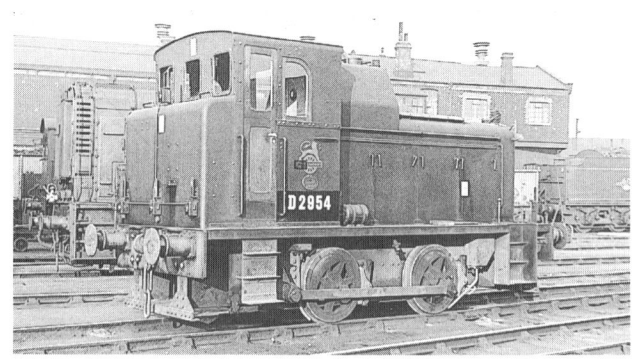

15. North British 0-4-0 diesel-hydraulic No. 11706.
16. Class 01 No. D2954.

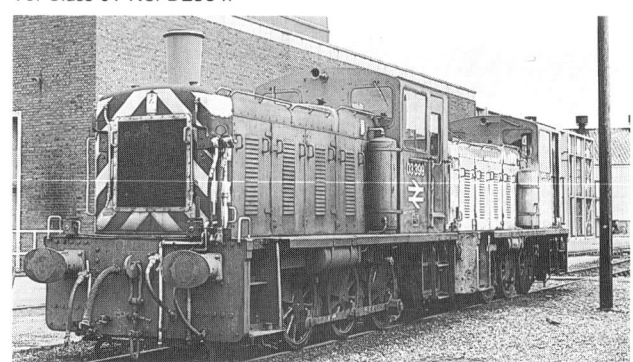

17. Class 02 No. D2852.
18. Class 03 No. 03399.
19. Class 04 No. 11105.
20. Class 05 No. D2583.

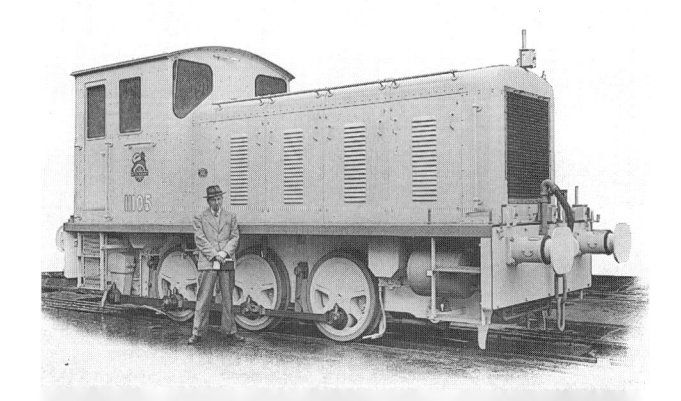

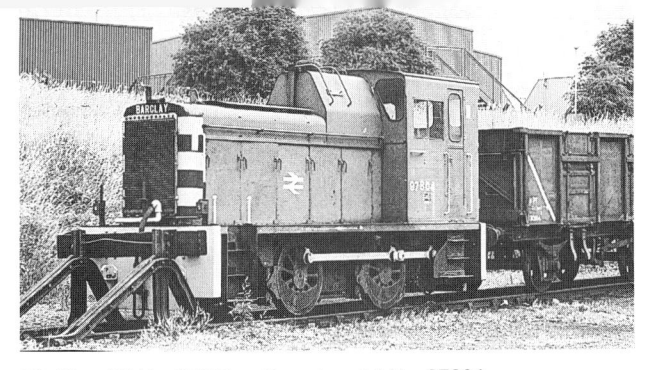

21. Class 06 No. 06003 as Departmental No. 97804.
22. Class 07 No. 07011.
23. LMS Hawthorn Leslie-built 0-6-0 diesel-electric No. 7073.
24. LNER/Brush design 0-6-0 diesel-electric No. 15004.

25. Class 08 No. 08642/D3809. 
26. Class 08 No. D3232 as RFS Industries No. 002 *Prudence*.

27. Class 12 No. 15228.
28. Class 11 No. 12100.
29. Class 13 No. D4502.
30. Class 14 No. D9500.

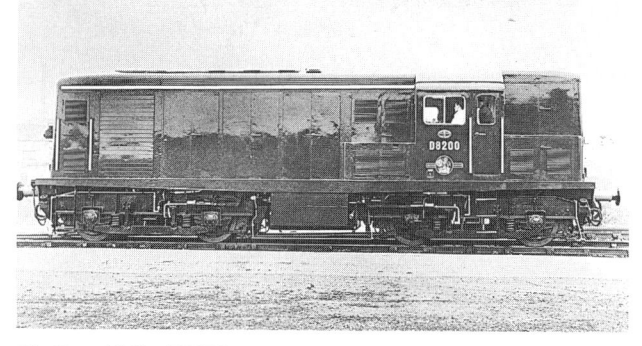

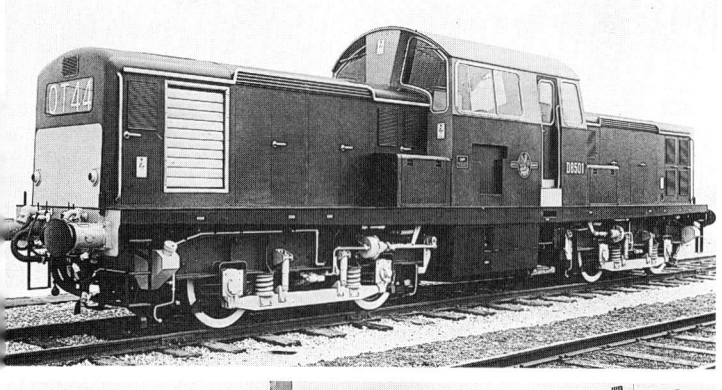

31. Class 15 No. D8200.
32. Class 16 No. D8402.
33. Class 17 No. D8501.
34. Class 20 No. 20215.  
35. Class 29 No. D6130. 
36. Class 22 No. D6300. 
37. Class 23 No. D5903.
39. Class 25 No. D7672 *Tamworth Castle*.  
38. Class 24 No. D5030.
40. Class 26 Nos D5301 (26001) *Eastfield* and D5300 (26007).

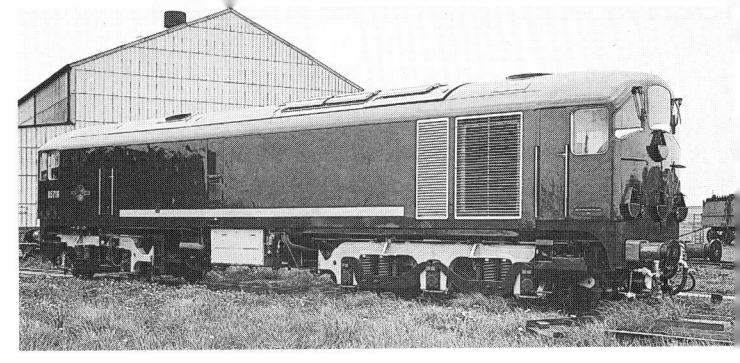

41. Class 27 No. 27008.
42. Class 28 No. D5716.
43. Class 31 No. 31412.
44. Class 33 No. 33114 *Ashford 150*.

45. Class 35 No. 35017 (D7017) *Williton*. 
46. Class 37 No. 37893.

47. Class 40 No. 40129.
48. Class 41 'Warship' No. D600 *Active*.
49. Class 42 'Warship' No. 823 *Hermes*.
50. Class 44 'Peak' No. D1 *Scafell Pike*.

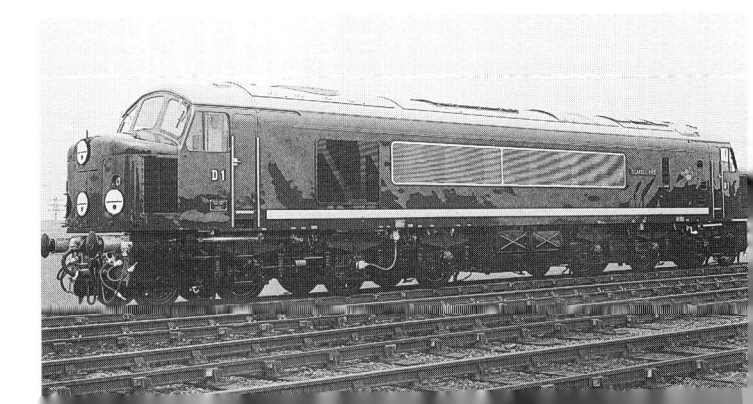

51. Class 45 'Peak' No. 45057.
52. Class 46 'Peak' No. 46035 as Departmental No. 97403.
53. Class 47 No. D1733.
54. Class 47 No. 47569 *The Gloucestershire Regiment*.   
55. Class 50 No. 50046 *Ajax*. 
56. Class 50 No. 50033 *Glorious*.  
57. Class 52 No. D1013 *Western Ranger*.
58. Class 55 No. D9010 *The King's Own Scottish Borderer*.
59. Class 56 No. 56072.
60. Class 58 Nos 58048 and 58020.

61. Class 59 No. 59004 *Yeoman Challenger*.
62. Class 60 No. 60006 *Great Gable*.
63. SR/BR Co-Co electric locomotive No. 20003.
64. Class 71 No. 71003. 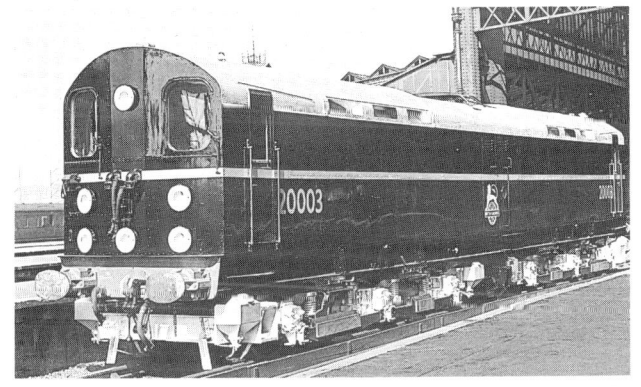  
65. Class 73 No. 73101 *Brighton Evening Argus*. 
66. Class 74 No. 74010. 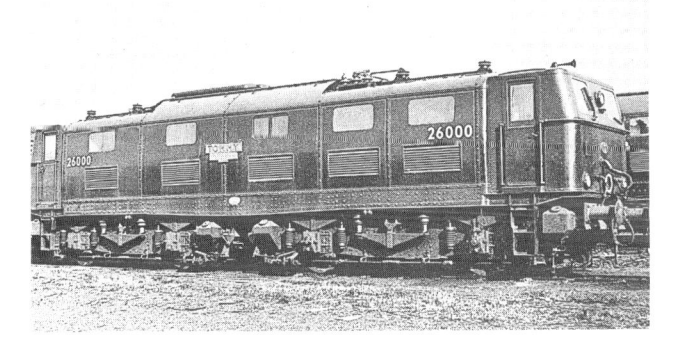 
67. Class 76 No. 26000 *Tommy*.
68. Class 77 No. 27000 *Electra*.
69. Electric locomotive No. E1000, ex-gas-turbine No. 18100.
70. Class 81 No. E3001.

71. Class 82 No. E3047.
72. Class 83 No. E3026.
73. Class 84 No. E3036.
74. Class 85 No. E3070.

75. Class 86 No. 86246 *Royal Anglian Regiment*. 
76. Class 87 No. 87010 *King Arthur*.

77. Class 89 No. 89001.
78. Class 90 No. 90010 *275 Railway Squadron (Volunteers)*.
79. Class 91 No. 91006.
80. Class 43 No. 43159.

TOPS No. Range	Refer To 1957 No.	TOPS No. Range	Refer To 1957 No.	TOPS No. Range	Refer To 1957 No.	TOPS No. Range	Refer To 1957 No.	TOPS No. Range	Refer To 1957 No.
47202	D1852	47283	D1985	47364	D1883	47464	D1587	47546	D1747
47203	D1853	47284	D1986	47365	D1884	47465	D1589	47547	D1642
47204	D1854	47285	D1987	47366	D1885	47466	D1590	47548	D1715
47205	D1855	47286	D1988	47367	D1886	47467	D1593	47549	D1724
47206	D1856	47287	D1989	47368	D1887	47468	D1594	47550	D1731
47207	D1857	47288	D1990	47369	D1888	47469	D1595	47551	D1746
47208	D1858	47289	D1991	47370	D1889	47470	D1596	47552	D1950
47209	D1859	47290	D1992	47371	D1890	47471	D1598	47553	D1956
47210	D1860	47291	D1993	47372	D1891	47472	D1600	47554	D1957
47211	D1861	47292	D1994	47373	D1892	47473	D1601	47555	D1717
47212	D1862	47293	D1995	47374	D1893	47474	D1602	47556	D1583
47213	D1863	47294	D1996	47375	D1894	47475	D1603	47557	D1591
47214	D1864	47295	D1997	47376	D1895	47476	D1604	47558	D1599
47215	D1865	47296	D1998	47377	D1896	47477	D1607	47559	D1605
47216	D1866	47297	D1999	47378	D1897	47478	D1608	47560	D1610
47217	D1867	47298	D1100	47379	D1898	47479	D1612	47561	D1614
47218	D1868	47299	D1866	47380	D1899	47481	D1627	47562	D1617
47219	D1869	47300	D1594	47381	D1900	47482	D1636	47563	D1618
47220	D1870	47301	D1782	47401	D1500	47483	D1637	47564	D1619
47221	D1871	47302	D1783	47402	D1501	47484	D1662	47565	D1620
47222	D1872	47303	D1784	47403	D1502	47485	D1683	47566	D1624
47223	D1873	47304	D1785	47404	D1503	47486	D1689	47567	D1625
47224	D1874	47305	D1786	47405	D1504	47487	D1707	47568	D1626
47225	D1901	47306	D1787	47406	D1505	47488	D1713	47569	D1629
47226	D1902	47307	D1788	47407	D1506	47489	D1716	47570	D1630
47227	D1903	47308	D1789	47408	D1507	47490	D1725	47571	D1750
47228	D1904	47309	D1790	47409	D1508	47491	D1753	47572	D1763
47229	D1905	47310	D1791	47410	D1509	47492	D1760	47573	D1768
47230	D1906	47311	D1792	47411	D1510	47493	D1932	47574	D1769
47231	D1907	47312	D1793	47412	D1511	47494	D1936	47575	D1770
47232	D1909	47313	D1794	47413	D1512	47495	D1937	47576	D1771
47233	D1910	47314	D1795	47414	D1513	47496	D1939	47577	D1774
47234	D1911	47315	D1796	47415	D1514	47497	D1940	47578	D1776
47235	D1912	47316	D1797	47416	D1515	47498	D1941	47579	D1778
47236	D1913	47317	D1798	47417	D1516	47499	D1942	47580	D1762
47237	D1914	47318	D1799	47418	D1517	47500	D1943	47581	D1764
47238	D1915	47319	D1800	47419	D1518	47501	D1944	47582	D1765
47239	D1916	47320	D1801	47420	D1519	47502	D1945	47583	D1767
47240	D1917	47321	D1802	47421	D1520	47503	D1946	47584	D1775
47241	D1918	47322	D1803	47422	D1525	47504	D1947	47585	D1779
47242	D1919	47323	D1804	47423	D1527	47505	D1948	47586	D1623
47243	D1920	47324	D1805	47424	D1531	47506	D1949	47587	D1963
47244	D1921	47325	D1806	47425	D1533	47507	D1951	47588	D1773
47245	D1922	47326	D1807	47426	D1534	47508	D1952	47589	D1928
47246	D1923	47327	D1808	47427	D1535	47509	D1953	47590	D1759
47247	D1924	47328	D1809	47428	D1536	47510	D1954	47591	D1965
47248	D1925	47329	D1810	47429	D1541	47511	D1955	47592	D1766
47249	D1926	47330	D1811	47430	D1542	47512	D1958	47593	D1973
47250	D1927	47331	D1812	47431	D1545	47513	D1950	47594	D1615
47251	D1928	47332	D1813	47432	D1547	47514	D1960	47595	D1969
47252	D1929	47333	D1814	47433	D1548	47515	D1961	47596	D1933
47253	D1930	47334	D1815	47434	D1549	47516	D1968	47597	D1597
47254	D1931	47335	D1816	47435	D1550	47517	D1975	47598	D1777
47255	D1933	47336	D1817	47436	D1552	47518	D1101	47599	D1772
47256	D1934	47337	D1818	47437	D1553	47519	D1102	47600	D1927
47257	D1935	47338	D1819	47438	D1554	47520	D1103	47601	D1628
47258	D1938	47339	D1820	47439	D1555	47521	D1104	47602	D1780
47259	D1950	47340	D1821	47440	D1556	47522	D1105	47603	D1967
47260	D1956	47341	D1822	47441	D1557	47523	D1106	47604	D1972
47261	D1957	47342	D1823	47442	D1558	47524	D1107	47605	D1754
47262	D1962	47343	D1824	47443	D1559	47525	D1108	47606	D1666
47263	D1963	47344	D1825	47444	D1560	47526	D1109	47607	D1730
47264	D1964	47345	D1826	47445	D1561	47527	D1110	47608	D1962
47265	D1965	47346	D1827	47446	D1563	47528	D1111	47609	D1656
47266	D1966	47347	D1828	47447	D1564	47529	D1551	47610	D1757
47267	D1967	47348	D1829	47448	D1565	47530	D1930	47611	D1761
47268	D1969	47349	D1830	47449	D1566	47531	D1584	47612	D1665
47269	D1970	47350	D1831	47450	D1567	47532	D1641	47613	D1661
47270	D1971	47351	D1832	47451	D1568	47533	D1651	47614	D1723
47271	D1972	47352	D1833	47452	D1569	47534	D1678	47615	D1929
47272	D1973	47353	D1834	47453	D1571	47535	D1649	47616	D1925
47273	D1974	47354	D1835	47454	D1574	47536	D1655	47617	D1742
47274	D1976	47355	D1836	47455	D1575	47537	D1657	47618	D1609
47275	D1977	47356	D1875	47456	D1576	47538	D1669	47619	D1964
47276	D1978	47357	D1876	47457	D1577	47539	D1718	47620	D1654
47277	D1979	47358	D1877	47458	D1578	47540	D1723	47621	D1728
47278	D1980	47359	D1878	47459	D1579	47541	D1755	47622	D1726
47279	D1981	47360	D1879	47460	D1580	47542	D1585	47623	D1676
47280	D1982	47361	D1880	47461	D1581	47543	D1588	47624	D1673
47281	D1983	47362	D1881	47462	D1582	47544	D1592	47625	D1660
47282	D1984	47363	D1882	47463	D1586	47545	D1646	47626	D1667

177

TOPS No. Range	Refer To 1957 No.	TOPS No. Range	Refer To 1957 No.	TOPS No. Range	Refer To 1957 No.	TOPS No. Range	Refer To 1957 No.	TOPS No. Range	Refer To 1957 No.
47627	D1974	47648	D1744	47674	D1972	47801	D1746	47822	D1758
47628	D1663	47649	D1645	47675	D1969	47802	D1950	47823	D1757
47629	D1966	47650	D1935	47676	D1623	47803	D1956	47824	D1780
47630	D1622	47651	D1931	47677	D1742	47804	D1965	47825	D1759
47631	D1643	47652	D1639	47701	D1932	47805	D1935	47826	D1976
47632	D1652	47653	D1674	47702	D1947	47806	D1931	47827	D1928
47633	D1668	47654	D1640	47703	D1960	47807	D1639	47828	D1966
47634	D1751	47655	D1924	47704	D1937	47808	D1674	47829	D1964
47635	D1606	47656	D1719	47705	D1957	47809	D1640	47830	D1645
47636	D1920	47657	D1916	47706	D1936	47810	D1924	47831	D1618
47637	D1976	47658	D1720	47707	D1949	47811	D1719	47832	D1610
47638	D1653	47659	D1919	47708	D1968	47812	D1916	47833	D1962
47639	D1648	47660	D1748	47709	D1942	47813	D1720	47834	D1656
47640	D1921	47661	D1650	47710	D1939	47814	D1919	47835	D1654
47641	D1672	47662	D1611	47711	D1941	47815	D1748	47836	D1609
47642	D1621	47663	D1917	47712	D1948	47816	D1650	47837	D1761
47643	D1970	47664	D1727	47713	D1954	47817	D1611	47838	D1665
47644	D1923	47665	D1909	47714	D1955	47818	D1917	47839	D1728
47645	D1659	47671	D1925	47715	D1945	47819	D1727	47840	D1661
47646	D1658	47672	D1617	47716	D1951	47820	D1909	47841	D1726
47647	D1677	47673	D1973	47717	D1940	47821	D1730	47842	D1666

BR/BRUSH TYPE 5 Co-Co CLASS 56

TOPS No.	Name	Date Named	Built By	Works No.	Date Introduced	Depot of First Allocation	Date Withdrawn	Depot of Final Allocation
56001	Whatley	10/87	Electroputere	-	02/77	TI		
56002			Electroputere	-	02/77	TI	05/92	TO
56003			Electroputere	-	02/77	TI		
56004			Electroputere	-	02/77	TI		
56005			Electroputere	-	03/77	TI		
56006			Electroputere	-	02/77	TI		
56007			Electroputere	-	04/77	TI		
56008			Electroputere	-	04/77	TI		
56009			Electroputere	-	05/77	TI		
56010			Electroputere	-	07/77	TI		
56011			Electroputere	-	06/77	TI		
56012	[Maltby Colliery]	06/89-05/92	Electroputere	-	04/77	TI		
56013			Electroputere	-	08/77	TI		
56014			Electroputere	-	03/77	TI		
56015			Electroputere	-	02/77	TI		
56016			Electroputere	-	05/77	TI		
56017			Electroputere	-	05/77	TI	05/92	TO
56018			Electroputere	-	08/77	TI		
56019			Electroputere	-	05/77	TI		
56020			Electroputere	-	05/77	TI		
56021			Electroputere	-	06/77	TI		
56022			Electroputere	-	05/77	TI		
56023			Electroputere	-	07/77	TI		
56024			Electroputere	-	06/77	TI		
56025			Electroputere	-	07/77	TI		
56026			Electroputere	-	07/77	TI		
56027			Electroputere	-	09/77	TI		
56028	[West Burton Power Station]	09/88-10/92	Electroputere	-	09/77	TI		
56029			Electroputere	-	09/77	TI		
56030	Eggborough Power Station	09/89	Electroputere	-	10/77	TI		
56031	Merehead	09/83	BREL Doncaster	-	05/77	TO		
56032	Sir De Morgannwg/County of South Glamorgan	10/83	BREL Doncaster	-	07/77	TO		
56033			BREL Doncaster	-	08/77	TO		
56034	Castell Ogwr/Ogmore Castle	06/85	BREL Doncaster	-	08/77	TO		
56035	[Taff Merthyr]	11/81-06/89	BREL Doncaster	-	10/77	TO		
56036			BREL Doncaster	-	01/78	TO		
56037	Richard Trevithick	07/81	BREL Doncaster	-	01/78	TO		
56038	Western Mail	06/81	BREL Doncaster	-	02/78	TO		
56039			BREL Doncaster	-	01/78	TO		
56040	Oystermouth	03/82	BREL Doncaster	-	02/78	TO		
56041			BREL Doncaster	-	02/78	TO		
56042			BREL Doncaster	-	05/79	TO	08/91	TO
56043			BREL Doncaster	-	03/78	TO		
56044	Cardiff Canton	08/91	BREL Doncaster	-	05/78	TO		
56045			BREL Doncaster	-	06/78	TO		
56046			BREL Doncaster	-	07/78	TO		
56047			BREL Doncaster	-	07/78	TO		
56048			BREL Doncaster	-	09/78	TO		
56049			BREL Doncaster	-	10/78	TO		
56050			BREL Doncaster	-	10/78	TO		
56051			BREL Doncaster	-	11/78	TO		
56052			BREL Doncaster	-	12/78	TO		

TOPS No. Range	Refer To 1957 No.	TOPS No. Range	Refer To 1957 No.	TOPS No. Range	Refer To 1957 No.	TOPS No. Range	Refer To 1957 No.	TOPS No. Range	Refer To 1957 No.
47843	D1676	**Class 50**		50020	D420	50041	D441	55002	D9002
47844	D1583	**D401-D449**		50021	D421	50042	D442	55003	D9003
47845	D1653	50001	D401	50022	D422	50043	D443	55004	D9004
47846	D1677	50002	D402	50023	D423	50044	D444	55005	D9005
47847	D1677	50003	D403	50024	D424	50045	D445	55006	D9006
47848	D1652	50004	D404	50025	D425	50046	D446	55007	D9007
47849	D1630	50005	D405	50026	D426	50047	D447	55008	D9008
47850	D1744	50006	D406	50027	D427	50048	D448	55009	D9009
47851	D1648	50007	D407	50028	D428	50049	D449	55010	D9010
47852	D1621	50008	D408	50029	D429	50050	D400	55011	D9011
47853	D1733	50009	D409	50030	D430	50149	D449	55012	D9012
47901	D1628	50010	D410	50031	D431			55013	D9013
47971	D1616	50011	D411	50032	D432	**Class 52**		55014	D9014
47972	D1646	50012	D412	50033	D433	**D1000-D1073**		55015	D9015
47973	D1614	50013	D413	50034	D434			55016	D9016
47974	D1584	50014	D414	50035	D435	**Class 53**		55017	D9017
47975	D1723	50015	D415	50036	D436	**D0280/D1200**		55018	D9018
47976	D1747	50016	D416	50037	D437			55019	D9019
		50017	D417	50038	D438	**Class 55**		55020	D9020
		50018	D418	50039	D439	**D9001-D9021**		55021	D9021
		50019	D419	50040	D440	55001	D9001	55022	D9000

TOPS No.	Disposal Code	Disposal Detail	Date Cut Up	Notes
56001				
56002	A	BR Doncaster		Stored: [U] 08/91
56003				Stored: [U] 06/90, R/I: 07/90
56004				
56005				
56006				Stored: [U] 02/90
56007				
56008				Stored: [U] 09/92
56009				
56010				
56011				
56012				
56013				
56014				
56015				
56016				Stored: [U] 06/92
56017	A	BR Toton		Stored: [U] 12/91
56018				
56019				
56020				Stored: [U] 09/92
56021				
56022				Stored: [S] 09/92, R/I: 11/92, Stored: [U] 12/92, R/I: 12/92
56023				
56024				Stored: [U] 06/92
56025				
56026				
56027				
56028				
56029				
56030				
56031				
56032				
56033				
56034				
56035				
56036				Stored: [U] 06/92
56037				
56038				
56039				
56040				Names off 12/90 - 01/92
56041				
56042	A	BR Toton		Stored: [U] 02/90
56043				
56044				
56045				
56046				
56047				
56048				
56049				
56050				
56051				
56052				

TOPS No.	Name	Date Named	Built By	Works No.	Date Introduced	Depot of First Allocation	Date Withdrawn	Depot of Final Allocation
56053	County of Mid Glamorgan/ Sir Morgannwg Ganol	03/86	BREL Doncaster	-	12/78	TO		
56054			BREL Doncaster	-	01/79	TO		
56055			BREL Doncaster	-	02/79	TO		
56056			BREL Doncaster	-	03/79	TO		
56057			BREL Doncaster	-	03/79	TO		
56058			BREL Doncaster	-	04/79	TO		
56059			BREL Doncaster	-	05/79	TO		
56060	The Cardiff Rod Mill	12/92	BREL Doncaster	-	06/79	TO		
56061			BREL Doncaster	-	07/79	TO		
56062	Mountsorrel	03/89	BREL Doncaster	-	08/79	TO		
56063	Bardon Hill	10/86	BREL Doncaster	-	08/79	TO		
56064			BREL Doncaster	-	09/79	TO		
56065			BREL Doncaster	-	10/79	TO		
56066			BREL Doncaster	-	10/79	TO		
56067			BREL Doncaster	-	11/79	TO		
56068			BREL Doncaster	-	11/79	TO		
56069			BREL Doncaster	-	12/79	TO		
56070			BREL Doncaster	-	12/79	TO		
56071			BREL Doncaster	-	12/79	TO		
56072			BREL Doncaster	-	01/80	TO		
56073			BREL Doncaster	-	02/80	TI		
56074	Kellingley Colliery	06/82	BREL Doncaster	-	03/80	TI		
56075	West Yorkshire Enterprise	07/85	BREL Doncaster	-	04/80	TI		
56076	[Blyth Power]	09/82-10/86	BREL Doncaster	-	04/80	TI		
56077	Thorpe Marsh Power Station	09/90	BREL Doncaster	-	05/80	TI		
56078			BREL Doncaster	-	05/80	TI		
56079			BREL Doncaster	-	06/80	TI		
56080	Selby Coalfield	10/89	BREL Doncaster	-	07/80	TI		
56081			BREL Doncaster	-	07/80	TI		
56082			BREL Doncaster	-	08/80	TI		
56083			BREL Doncaster	-	09/80	TI		
56084			BREL Doncaster	-	10/80	TI		
56085			BREL Doncaster	-	11/80	TI		
56086			BREL Doncaster	-	12/80	TI		
56087			BREL Doncaster	-	12/80	TI		
56088			BREL Doncaster	-	01/81	TI		
56089	Ferrybridge C Power Station	09/91	BREL Doncaster	-	01/81	TI		
56090			BREL Doncaster	-	02/81	TI		
56091	Castle Donington Power Station	06/88	BREL Doncaster	-	05/81	TI		
56092			BREL Doncaster	-	06/81	TI		
56093	Institution of Mining Engineers	11/89	BREL Doncaster	-	06/81	TI		
56094			BREL Doncaster	-	07/81	TI		
56095	Harworth Colliery	10/87	BREL Doncaster	-	08/81	TI		
56096			BREL Doncaster	-	08/81	TI		
56097			BREL Doncaster	-	09/81	TI		
56098			BREL Doncaster	-	09/81	TI		
56099	Fiddlers Ferry Power Station	07/89	BREL Doncaster	-	10/81	TI		
56100			BREL Doncaster	-	11/81	TI		
56101	Mutual Improvement	04/90	BREL Doncaster	-	11/81	TI		
56102	Scunthorpe Steel Centenary	03/90	BREL Doncaster	-	12/81	TI		
56103			BREL Doncaster	-	12/81	TI		
56104			BREL Doncaster	-	02/82	TI		
56105			BREL Doncaster	-	03/82	TI		
56106			BREL Doncaster	-	04/82	TI		
56107			BREL Doncaster	-	05/82	TI		
56108			BREL Doncaster	-	06/82	HM		
56109			BREL Doncaster	-	08/82	TI		
56110	Croft	09/92	BREL Doncaster	-	09/82	HM		
56111			BREL Doncaster	-	10/82	HM		
56112			BREL Doncaster	-	11/82	TI		
56113			BREL Doncaster	-	12/82	HM		
56114	Maltby Colliery	07/92	BREL Doncaster	-	01/83	HM		
56115			BREL Doncaster	-	01/83	HM		
56116			BREL Crewe	-	03/83	HM		
56117	Wilton-Coalpower	09/92	BREL Crewe	-	03/83	HM		
56118			BREL Crewe	-	04/83	TI		
56119			BREL Crewe	-	05/83	TI		
56120			BREL Crewe	-	05/83	TI		
56121			BREL Crewe	-	06/83	TI		
56122	[Wilton-Coalpower]	04/88-03/92	BREL Crewe	-	07/83	TI	09/92	TO
56123	Drax Power Station	05/88	BREL Crewe	-	07/83	TI		
56124	[Blue Circle Cement]	10/83-09/89	BREL Crewe	-	09/83	TI		
56125			BREL Crewe	-	10/83	TI		
56126			BREL Crewe	-	11/83	TI		
56127			BREL Crewe	-	12/83	TI		
56128			BREL Crewe	-	12/83	TI		
56129			BREL Crewe	-	01/84	TI		
56130	Wardley Opencast	11/90	BREL Crewe	-	04/84	GD		
56131	Ellington Colliery	08/87	BREL Crewe	-	04/84	GD		
56132	[Fina Energy]	10/86-03/89	BREL Crewe	-	06/84	GD		
56133	Crewe Locomotive Works	06/84	BREL Crewe	-	07/84	GD		

TOPS No.	Disposal Code	Disposal Detail	Date Cut Up	Notes
56053				
56054				
56055				
56056				
56057				
56058				
56059				
56060				
56061				
56062				
56063				
56064				
56065				
56066				
56067				
56068				
56069				
56070				
56071				
56072				
56073				
56074				
56075				
56076				
56077				
56078				
56079				
56080				
56081				
56082				
56083				
56084				
56085				
56086				
56087				
56088				
56089				
56090				
56091				
56092				
56093				
56094				
56095				
56096				
56097				
56098				
56099				
56100				
56101				
56102				
56103				Stored: [U] 02/90
56104				
56105				
56106				
56107				
56108				
56109				
56110				
56111				
56112				
56113				
56114				
56115				
56116				
56117				
56118				
56119				
56120				
56121				
56122	A	BR Toton		Stored: [U] 03/92
56123				
56124				
56125				
56126				
56127				
56128				
56129				
56130				
56131				
56132				
56133				

TOPS No.	Name	Date Named	Built By	Works No.	Date Introduced	Depot of First Allocation	Date Withdrawn	Depot of Final Allocation
56134	Blyth Power	11/86	BREL Crewe	-	09/84	GD		
56135	Port of Tyne Authority	10/85	BREL Crewe	-	11/84	TI		

BR TYPE 5 Co-Co CLASS 58

TOPS No.	Name	Date Named	Built By	Works No.	Date Introduced	Depot of First Allocation	Date Withdrawn	Depot of Final Allocation
58001			BREL Doncaster	-	05/83	TO		
58002	Daw Mill Colliery	03/88	BREL Doncaster	-	05/83	TO		
58003	Markham Colliery	11/88	BREL Doncaster	-	07/83	TO		
58004			BREL Doncaster	-	09/83	TO		
58005			BREL Doncaster	-	10/83	TO		
58006			BREL Doncaster	-	10/83	TO		
58007	Drakelow Power Station	08/90	BREL Doncaster	-	11/83	TO		
58008			BREL Doncaster	-	12/83	TO		
58009			BREL Doncaster	-	01/84	TO		
58010			BREL Doncaster	-	02/84	TO		
58011			BREL Doncaster	-	03/84	TO		
58012			BREL Doncaster	-	03/84	TO		
58013			BREL Doncaster	-	03/84	TO		
58014	Didcot Power Station	06/88	BREL Doncaster	-	04/84	TO		
58015			BREL Doncaster	-	09/84	TO		
58016			BREL Doncaster	-	10/84	TO		
58017			BREL Doncaster	-	10/84	TO		
58018	High Marnham Power Station	05/88	BREL Doncaster	-	10/84	TO		
58019	Shirebrook Colliery	10/89	BREL Doncaster	-	11/84	TO		
58020	Doncaster Works	11/84	BREL Doncaster	-	11/84	TO		
58021			BREL Doncaster	-	12/84	TO		
58022			BREL Doncaster	-	12/84	TO		
58023			BREL Doncaster	-	12/84	TO		
58024			BREL Doncaster	-	12/84	TO		
58025			BREL Doncaster	-	01/85	TO		
58026			BREL Doncaster	-	03/85	TO		
58027			BREL Doncaster	-	03/85	TO		
58028			BREL Doncaster	-	03/85	TO		
58029			BREL Doncaster	-	03/85	TO		
58030			BREL Doncaster	-	06/85	TO		
58031			BREL Doncaster	-	09/85	TO		
58032			BREL Doncaster	-	09/85	TO		
58033			BREL Doncaster	-	09/85	TO		
58034	Bassetlaw	12/85	BREL Doncaster	-	11/85	TO		
58035			BREL Doncaster	-	01/86	TO		
58036			BREL Doncaster	-	02/86	TO		
58037			BREL Doncaster	-	02/86	TO		
58038			BREL Doncaster	-	02/86	TO		
58039	Rugeley Power Station	09/86	BREL Doncaster	-	03/86	TO		
58040	Cottam Power Station	09/86	BREL Doncaster	-	03/86	TO		
58041	Ratcliffe Power Station	09/86	BREL Doncaster	-	03/86	TO		
58042	Ironbridge Power Station	09/86	BREL Doncaster	-	05/86	TO		
58043			BREL Doncaster	-	07/86	TO		
58044	Oxcroft Opencast	05/92	BREL Doncaster	-	08/86	TO		
58045			BREL Doncaster	-	09/86	TO		
58046	Thorseby Colliery	06/91	BREL Doncaster	-	10/86	TO		
58047	Manton Colliery	05/92	BREL Doncaster	-	10/86	TO		
58048	Coventry Colliery	05/91	BREL Doncaster	-	11/86	TO		
58049	Littleton Colliery	03/87	BREL Doncaster	-	12/86	TO		
58050	Toton Traction Depot	05/87	BREL Doncaster	-	03/87	TO		

GENERAL MOTORS TYPE 5 Co-Co CLASS 59

TOPS No.	Name	Date Named	Built By	Works No.	Date Introduced	Depot of Allocation and Owner	Date Withdrawn
59001	Yeoman Endeavour	06/86	GM EMD (USA)	848002-1	01/86	F.Yeoman, Merehead	
59002	Yeoman Enterprise	06/86	GM EMD (USA)	848002-2	01/86	F.Yeoman, Merehead	
59003	Yeoman Highlander	06/86	GM EMD (USA)	848002-3	01/86	F.Yeoman, Merehead	
59004	Yeoman Challenger	06/86	GM EMD (USA)	848002-4	01/86	F.Yeoman, Merehead	
59005	Kenneth J Painter	06/89	GM EMD (USA)	878039-1	06/89	F.Yeoman, Merehead	
59101	Village of Whatley	05/92	GM EMD (CANADA)	878029-1	10/90	ARC, Whatley	
59102	Village of Chantry	09/91	GM EMD (CANADA)	878029-2	10/90	ARC, Whatley	
59103	Village of Mells	08/91	GM EMD (CANADA)	878029-3	10/90	ARC, Whatley	
59104	Village of Great Elm	09/91	GM EMD (CANADA)	878029-4	10/90	ARC, Whatley	

TOPS No.	Disposal Code	Disposal Detail	Date Cut Up	Notes
56134				
56135				

TOPS No.	Disposal Code	Disposal Detail	Date Cut Up	Notes
58001				
58002				
58003				
58004				
58005				
58006				
58007				
58008				
58009				
58010				
58011				
58012				
58013				
58014				
58015				
58016				
58017				
58018				
58019				
58020				Plate read 'Doncaster Works BRE' 11/84-08/86
58021				
58022				
58023				
58024				
58025				
58026				
58027				
58028				
58029				
58030				
58031				
58032				
58033				
58034				
58035				
58036				
58037				
58038				
58039				
58040				
58041				
58042				
58043				
58044				
58045				
58046				
58047				
58048				
58049				
58050				

TOPS No.	Depot of Final Allocation	Disposal Code	Disposal Detail	Date Cut Up	Notes
59001					
59002					
59003					
59004					
59005					
59101					
59102					
59103					
59104					

BR/BRUSH TYPE 5 Co-Co CLASS 60

TOPS No.	Name	Date Named	Built By	Works No.	Date Ex Brush	Date Introduced	Depot of First Allocation
60001	Steadfast	06/89	Brush	903	06/89	09/91	SL
60002	Capability Brown	07/89	Brush	904	10/89	12/92	IM
60003	Christopher Wren	12/89	Brush	905	12/89	01/92	IM
60004	Lochnagar	10/89	Brush	906	10/89	09/91	TE
60005	Skiddaw	10/89	Brush	907	10/89	09/91	IM
60006	Great Gable	12/89	Brush	908	12/89	09/91	TO
60007	Robert Adam	12/89	Brush	909	12/89	02/93	TE
60008	Moel Fammau	01/90	Brush	910	01/90	12/92	IM
60009	Carnedd Dafydd	01/90	Brush	911	01/90	02/93	SL
60010	Plynlimon/Pumlumon	02/90	Brush	912	02/90	01/91	CF
60011	Cader Idris	02/90	Brush	913	02/90	10/91	TO
60012	Glyder Fawr	03/90	Brush	914	03/90	11/91	TO
60013	Robert Boyle	03/90	Brush	915	03/90	02/93	IM
60014	Alexander Fleming	04/90	Brush	916	04/90	01/93	IM
60015	Bow Fell	07/90	Brush	917	07/90	03/93	TE
60016	Langdale Pikes	07/90	Brush	918	07/90	02/93	SL
60017	Arenig Fawr	09/90	Brush	919	09/90	10/91	SL
60018	Moel Siabod	10/90	Brush	920	09/90	10/90	TE
60019	Wild Boar Fell	10/90	Brush	921	09/90	12/90	IM
60020	Great Whernside	10/90	Brush	922	10/90	01/91	TE
60021	Pen-y-Ghent	11/90	Brush	923	11/90	12/90	TE
60022	Ingleborough	09/90	Brush	924	11/90	01/91	TE
60023	The Cheviot	08/90	Brush	925	09/90	11/90	TE
60024	Elizabeth Fry	08/90	Brush	926	11/90	12/90	CF
60025	Joseph Lister	09/90	Brush	927	11/90	12/90	TE
60026	William Caxton	10/90	Brush	928	11/90	12/90	CF
60027	Joseph Banks	11/90	Brush	929	11/90	02/91	IM
60028	John Flamsteed	10/90	Brush	930	10/90	11/90	TE
60029	Ben Nevis	10/90	Brush	931	10/90	11/90	TE
60030	Cir Mhor	10/90	Brush	932	11/90	11/90	TE
60031	Ben Lui	04/91	Brush	933	04/91	09/91	TE
60032	William Booth	11/90	Brush	934	11/90	12/90	IM
60033	Anthony Ashley Cooper	11/90	Brush	935	12/90	02/91	CF
60034	Carnedd Llewelyn	11/90	Brush	936	11/90	12/90	TE
60035	Florence Nightingale	04/91	Brush	937	03/91	09/91	TO
60036	Sgurr Na Ciche	01/91	Brush	938	06/91	09/91	TE
60037	Helvellyn	01/91	Brush	939	01/91	09/91	TE
60038	Bidean Nam Bian	01/91	Brush	940	01/91	03/91	TE
60039	Glastonbury Tor	04/91	Brush	941	04/91	09/91	CF
60040	Brecon Beacons	06/91	Brush	942	07/91	02/92	SL
60041	High Willhays	04/91	Brush	943	05/91	06/91	TO
60042	Dunkery Beacon	05/91	Brush	944	05/91	06/91	SL
60043	Yes Tor	06/91	Brush	945	06/91	06/91	SL
60044	Ailsa Craig	06/91	Brush	946	06/91	09/91	TO
60045	Josephine Butler	02/91	Brush	947	02/91	03/91	SL
60046	William Wilberforce	02/91	Brush	948	02/91	09/91	SL
60047	Robert Owen	02/91	Brush	949	02/91	07/91	SL
60048	Saddleback	04/91	Brush	950	04/91	06/91	TO
60049	Scafell	05/91	Brush	951	04/91	09/91	IM
60050	Roseberry Topping	02/91	Brush	952	02/91	03/91	TE
60051	Mary Somerville	03/91	Brush	953	03/91	09/91	CF
60052	Goat Fell	04/91	Brush	954	04/91	09/91	IM
60053	John Reith	03/91	Brush	955	04/91	06/91	TO
60054	Charles Babbage	04/91	Brush	956	04/91	09/91	IM
60055	Thomas Barnardo	05/91	Brush	957	05/91	09/91	TO
60056	William Beveridge	05/91	Brush	958	05/91	09/91	IM
60057	Adam Smith	05/91	Brush	959	05/91	09/91	TO
60058	John Howard	06/91	Brush	960	05/91	09/91	IM
60059	Samuel Plimsoll	06/91	Brush	961	05/91	09/91	TE
60060	James Watt	06/91	Brush	962	06/91	09/91	TO
60061	Alexander Graham Bell	06/91	Brush	963	05/91	09/91	TO
60062	Samuel Johnson	06/91	Brush	964	06/91	06/91	CF
60063	James Murray	06/91	Brush	965	06/91	09/91	CF
60064	Back Tor	06/91	Brush	966	06/91	09/91	IM
60065	Kinder Low	06/91	Brush	967	06/91	09/91	CF
60066	John Logie Baird	08/91	Brush	968	07/91	09/91	IM
60067	James Clerk-Maxwell	07/91	Brush	969	07/91	09/91	IM
60068	Charles Darwin	07/91	Brush	970	07/91	10/91	TO
60069	Humphry Davy	08/91	Brush	971	08/91	09/91	TO
60070	John Loudon McAdam	08/91	Brush	972	08/91	10/91	TO
60071	Dorothy Garrod	08/91	Brush	973	08/91	09/91	TO
60072	Cairn Toul	09/91	Brush	974	09/91	10/91	TO
60073	Cairn Gorm	10/91	Brush	975	10/91	11/91	TO
60074	Braeriach	10/91	Brush	976	10/91	11/91	TO
60075	Liathach	11/91	Brush	977	11/91	12/91	TO
60076	Suilven	10/91	Brush	978	10/91	11/91	TO
60077	Canisp	10/91	Brush	979	10/91	11/91	TO
60078	Stac Pollaidh	10/91	Brush	980	10/91	11/91	IM
60079	Foinaven	11/91	Brush	981	11/91	01/92	CF
60080	Kinder Scout	11/91	Brush	982	11/91	11/91	IM

TOPS No.	Depot of Final Allocation	Disposal Code	Disposal Detail	Date Cut Up	Notes
60001					
60002					
60003					
60004					
60005					
60006					
60007					
60008					
60009					
60010					
60011					
60012					
60013					
60014					
60015					
60016					
60017					
60018					
60019					
60020					
60021					
60022					
60023					
60024					
60025					
60026					
60027					
60028					
60029					
60030					
60031					
60032					
60033					
60034					
60035					
60036					
60037					
60038					
60039					
60040					
60041					
60042					
60043					
60044					
60045					
60046					
60047					
60048					
60049					
60050					
60051					
60052					
60053					
60054					
60055					
60056					
60057					
60058					
60059					
60060					
60061					
60062					
60063					
60064					
60065					
60066					
60067					
60068					
60069					
60070					
60071					
60072					
60073					
60074					
60075					
60076					
60077					
60078					
60079					
60080					

TOPS No.	Name	Date Named	Built By	Works No.	Date Ex Brush	Date Introduced	Depot of First Allocation
60081	Bleaklow Hill	11/91	Brush	983	11/91	12/91	IM
60082	Mam Tor	11/91	Brush	984	11/91	12/91	IM
60083	Shining Tor	11/91	Brush	985	11/91	03/92	TO
60084	Cross Fell	11/91	Brush	986	11/91	01/93	IM
60085	Axe Edge	11/91	Brush	987	11/91	12/91	IM
60086	Schiehallion	12/91	Brush	988	12/91	01/92	SL
60087	Slioch	12/91	Brush	989	12/91	12/91	TO
60088	Buachaille Etive Mor	12/91	Brush	990	12/91	01/92	TO
60089	Arcuil	12/91	Brush	991	12/91	01/92	TO
60090	Quinag	12/91	Brush	992	12/91	02/92	CF
60091	An Teallach	01/92	Brush	993	01/92	02/92	TO
60092	Reginald Munns	09/92	Brush	994	01/92	01/92	TO
60093	Jack Stirk	09/92	Brush	995	02/92	02/92	TO
60094	Tryfan	02/92	Brush	996	02/92	02/92	SL
60095	Crib Goch	02/92	Brush	997	02/92	03/92	IM
60096	Ben Macdui	03/92	Brush	998	03/92	05/92	IM
60097	Pillar	11/92	Brush	999	11/92	12/92	IM
60098	Charles Francil Brush	11/92	Brush	1000	11/92	12/92	IM
60099	Ben More Assynt	11/92	Brush	1001	11/92	12/92	TO
60100	Boar of Badenoch	11/92	Brush	1002	11/92	12/92	TO

ELECTRIC LOCOMOTIVES
PRE-NATIONALISATION NUMBERS

NER Bo-Bo ELECTRIC LOCOMOTIVES

Original NER No.	LNER 1946 No.	Date Re No.	BR 1948 No.	Date Re No.	Built By	Works No.	Date Introduced	Depot of First Allocation	Date Withdrawn
1	6480	06/46	26500	05/48	Brush/BTH	-	12/03	52B	09/64
2	6481	06/46	26501	05/48	Brush/BTH	-	12/03	52B	09/64

NER Bo-Bo ELECTRIC LOCOMOTIVES

Original NER No.	LNER 1946 No.	Date Re No.	BR 1948 No.	Date Re No.	Built By	Works No.	Date Introduced	Depot of First Allocation	Date Withdrawn	Depot of Final Allocation
3	6490	05/46	26502	07/49	NER Darlington	999	06/15	Shildon	08/50	51A
4	6491	05/46	26503	07/49	NER Darlington	1000	06/15	Shildon	08/50	51A
5	6492*	-	26504*	-	NER Darlington	1001	06/15	Shildon	08/50	51A
6	6493	05/46	26505	08/49	NER Darlington	1002	06/15	Shildon	08/50	51A
7	6494	05/46	26506	07/49	NER Darlington	1003	06/15	Shildon	08/50	51A
8	6495	06/46	26507	09/49	NER Darlington	1004	06/15	Shildon	08/50	51A
9	6496	06/46	26508	07/49	NER Darlington	1005	06/15	Shildon	08/50	51A
10	6497	05/46	26509	08/49	NER Darlington	1006	06/15	Shildon	08/50	51A
11	6498	05/46	26510	08/49	NER Darlington	1007	12/15	Shildon	01/59	51A
12	6499	05/46	26511	07/49	NER Darlington	1008	05/20	Shildon	08/50	51A

NER PROTOTYPE MAIN LINE ELECTRIC LOCOMOTIVE

Original NER No.	LNER 1946 No.	Date Re No.	Original BR 1948 No.	Date Re No.	Second BR 1948 No.	Date Re No.	Built By	Works No.	Date Introduced	Depot of First Allocation	Date Withdrawn
13	6999	05/46	26999*	-	26600	09/48	NER Darlington	1169	05/22	Shildon	08/50

SR/BR Co-Co ELECTRIC LOCOMOTIVES

Original SR No.	BR 1948 No.	Date Re No.	Built By	Works No.	Date Introduced	Depot of First Allocation	Date Withdrawn	Depot of Final Allocation
CC1	20001	12/48	SR Ashford	-	07/41	75A	01/69	75A
CC2	20002	02/49	SR Ashford	-	09/45	75A	12/68	75A
CC3*	20003	-	SR Brighton	-	09/48	75A	10/68	75A

TOPS No.	Depot of Final Allocation	Disposal Code	Disposal Detail	Date Cut Up	Notes
60081					
60082					
60083					
60084					
60085					
60086					
60087					
60088					
60089					
60090					
60091					
60092					
60093					
60094					
60095					
60096					
60097					
60098					
60099					
60100					

Original NER No.	Depot of Final Allocation	Disposal Code	Disposal Detail	Date Cut Up	Notes
1	52J	P	National Railway Museum, York	-	No. 4075 carried 09/44 10/44, Stored: [U] 03/64
2	52J	C	W Willoughby, Choppington	07/66	

Original NER No.	Disposal Code	Disposal Detail	Date Cut Up	Notes
3	C	Wanty & Co, Catcliffe	06/51	Stored: [S] 01/35
4	C	Wanty & Co, Catcliffe	04/51	Stored: [S] 01/35
5	C	BR Darlington Works	12/50	Stored: [S] 01/35
6	C	Wanty & Co, Catcliffe	08/51	Stored: [S] 01/35
7	C	Wanty & Co, Catcliffe	05/51	Stored: [S] 01/35
8	C	Wanty & Co, Catcliffe	07/51	Stored: [S] 01/35
9	C	Wanty & Co, Catcliffe	04/51	Stored: [S] 01/35
10	C	Wanty & Co, Catcliffe	08/51	Stored: [S] 01/35
11	D	To Departmental Stock - 100	-	Stored: [S] 01/35
12	C	Wanty & Co, Catcliffe	04/51	Stored: [S] 01/35

Original NER No.	Depot of Final Allocation	Disposal Code	Disposal Detail	Date Cut Up	Notes
13	51A	C	Wanty & Co, Catcliffe	07/51	Stored for many years prior to withdrawal

Original SR No.	Disposal Code	Disposal Detail	Date Cut Up	Notes
CC1	C	J Cashmore, Newport	09/69	Withdrawn: 12/68, R/I: 01/69
CC2	C	J Cashmore, Newport	09/69	
CC3*	C	G Cohen, Kettering	11/69	Stored: [U] 09/68

6000 See BR 1948 No. E26000
6480-6481 See NER Nos. 1-2
6490-6499 See NER Nos. 3-12
6700 See BR 1948 No. E26000
6999 See NER No. 13

20001-20003 See SR numbers CC1-CC3

BRITISH RAILWAYS 1948 NUMBERS-ELECTRIC
BR 1500v DC ELECTRIC LOCOMOTIVES CLASS 76

BR 1948 No.	First TOPS No.	Date Re No.	Second TOPS No.	Date Re No.	Name	Name Date	Built By	Original Works No.	Amended Works No.	Date Introduced	Depot of First Allocation
26000	-	-	-	-	Tommy	06/52	LNER Doncaster	-	-	03/41	39A
26001	76001	02/74	-	-	-	-	BR Gorton	1004	1008	10/50	39A
26002	76002	04/74	-	-	-	-	BR Gorton	1005	1009	10/50	39A
26003	76003	03/72	76036	10/76	-	-	BR Gorton	1006	1010	11/50	39A
26004	76004	02/74	-	-	-	-	BR Gorton	1007	1011	02/51	39A
26005	-	-	-	-	-	-	BR Gorton	1008	1012	01/51	39A
26006	76006	01/74	-	-	-	-	BR Gorton	1009	1013	01/51	39A
26007	76007	02/74	-	-	-	-	BR Gorton	1010	1014	02/51	39A
26008	76008	11/72	-	-	-	-	BR Gorton	1011	1015	03/51	39A
26009	76009	04/74	-	-	-	-	BR Gorton	1012	1016	03/51	39A
26010	76010	02/73	-	-	-	-	BR Gorton	1013	1017	03/51	39A
26011	76011	07/73	-	-	-	-	BR Gorton	1014	1018	05/51	39A
26012	76012	04/73	-	-	-	-	BR Gorton	1015	1019	05/51	39A
26013	76013	09/72	-	-	-	-	BR Gorton	1016	1020	05/51	39A
26014	76014	02/74	-	-	-	-	BR Gorton	1017	1021	05/51	39A
26015	76015	11/72	-	-	-	-	BR Gorton	1018	1022	07/51	39A
26016	76016	03/73	-	-	-	-	BR Gorton	1019	1023	07/51	39A
26017	-	-	-	-	-	-	BR Gorton	1020	1024	07/51	39A
26018	76018	02/74	76035	06/76	-	-	BR Gorton	1021	1025	09/51	39A
26019	-	-	-	-	-	-	BR Gorton	1022	1026	09/51	39A
26020	76020	02/74	-	-	-	-	BR Gorton	1023	1027	02/51	39A
26021	76021	06/73	-	-	-	-	BR Gorton	1024	1028	10/51	39A
26022	76022	04/72	-	-	-	-	BR Gorton	1025	1029	09/51	39A
26023	76023	09/73	-	-	-	-	BR Gorton	1026	1030	09/51	39A
26024	76024	11/73	-	-	-	-	BR Gorton	1027	1031	09/51	39A
26025	76025	08/72	-	-	-	-	BR Gorton	1032	-	01/52	39A
26026	76026	02/74	-	-	-	-	BR Gorton	1033	-	01/52	39A
26027	76027	01/74	-	-	-	-	BR Gorton	1034	-	01/52	39A
26028	76028	11/73	-	-	-	-	BR Gorton	1035	-	01/52	39A
26029	76029	02/74	-	-	-	-	BR Gorton	1036	1032	12/51	39A
26030	76030	12/72	-	-	-	-	BR Gorton	1037	1033	12/51	39A
26031	-	-	-	-	-	-	BR Gorton	1038	-	01/52	39A
26032	76032	01/72	-	-	-	-	BR Gorton	1039	-	01/52	39A
26033	76033	01/74	-	-	-	-	BR Gorton	1040	-	01/52	39A
26034	76034	10/73	-	-	-	-	BR Gorton	1041	-	01/52	39A
26035	-	-	-	-	-	-	BR Groton	1042	-	01/52	39A
26036	76036	01/74	76003	10/76	-	-	BR Gorton	1043	-	02/52	39A
26037	76037	03/72	-	-	-	-	BR Gorton	1044	-	02/52	39A
26038	76038	10/72	76050	10/76	-	-	BR Gorton	1045	-	04/52	39A
26039	76039	03/74	76048	10/76	-	-	BR Gorton	1046	-	04/52	36B
26040	76040	02/74	-	-	-	-	BR Gorton	1047	-	04/52	36B
26041	76041	03/74	-	-	-	-	BR Gorton	1048	-	04/52	36B
26042	-	-	-	-	-	-	BR Gorton	1049	-	05/52	36B
26043	76043	04/72	-	-	-	-	BR Gorton	1050	-	05/52	36B
26044	76044	03/72	76031	03/76	-	-	BR Gorton	1051	-	06/52	36B
26045	-	-	-	-	-	-	BR Gorton	1052	-	06/52	36B
26046	76046	02/74	-	-	Archimedes	05/59	BR Gorton	1053	-	08/52	36B
26047	76047	02/74	-	-	Diomedes	09/60	BR Gorton	1054	-	08/52	36B
26048	76048	05/72	76039	10/76	Hector	03/60	BR Gorton	1055	-	09/52	36B
26049	76049	08/72	-	-	Jason	08/60	BR Gorton	1056	-	10/52	36B
26050	76050	11/71	76038	11/76	Stentor	08/60	BR Gorton	1057	-	11/52	36B
26051	76051	02/74	-	-	Mentor	06/59	BR Gorton	1058	-	01/53	36B
26052	76052	02/74	-	-	Nestor	08/61	BR Gorton	1059	-	01/53	39A
26053	76053	03/74	-	-	Perseus	10/60	BR Gorton	1060	-	03/53	39A
26054	76054	12/73	-	-	Pluto	04/61	BR Gorton	1061	-	04/53	39A
26055	76055	02/74	-	-	Prometheus	06/59	BR Gorton	1062	-	06/53	39A
26056	76056	02/74	-	-	Triton	07/59	BR Gorton	1063	-	07/53	39A
26057	76057	07/72	-	-	Ulysses	04/60	BR Gorton	1064	-	08/53	39A

26500-26501 See NER No. 1-2
26502-26512 See NER No. 3-12
26600 See NER No. 13

BR 1948 No.	Date Withdrawn	Depot of Final Allocation	Disposal Code	Disposal Detail	Date Cut Up	Notes
26000	03/70	9C	C	BREL Crewe	11/72	Loaned to Netherlands Railway 09/47-03/52, Numbered: 6701 03/41-06/46, 6000: 06/46-04/52
26001	11/80	RS	C	C F Booth, Rotherham	05/83	Stored: [U] 07/68, R/I: 09/68
26002	06/78	RS	C	C F Booth, Rotherham	12/83	Stored: [U] 07/68, R/I: 09/68, Stored: [U] 07/77
26003	07/81	RS	C	C F Booth, Rotherham	05/83	Stored: [U] 08/80
26004	06/78	RS	C	C F Booth, Rotherham	01/84	Stored: [U] 07/68, R/I: 09/68, Withdrawn: 02/77, R/I: 08/77
26005	03/70	RS	C	BREL Crewe	08/71	Stored: [U] 07/68, R/I: 09/68
26006	07/81	RS	C	C F Booth, Rotherham	05/83	
26007	07/81	RS	C	C F Booth, Rotherham	06/83	
26008	07/81	RS	C	C F Booth, Rotherham	06/83	
26009	07/81	RS	C	C F Booth, Rotherham	07/83	
26010	07/81	RS	C	C F Booth, Rotherham	06/83	
26011	07/81	RS	C	C F Booth, Rotherham	06/83	
26012	07/81	RS	C	C F Booth, Rotherham	07/83	
26013	07/81	RS	C	C F Booth, Rotherham	05/83	
26014	07/81	RS	C	C F Booth, Rotherham	05/83	
26015	07/81	RS	C	C F Booth, Rotherham	03/83	
26016	07/81	RS	C	C F Booth, Rotherham	04/83	
26017	03/70	RS	C	BR Reddish, by J Cashmore	10/71	
26018	07/81	RS	C	C F Booth, Rotherham	06/83	
26019	10/71	RS	C	BREL Crewe	05/72	Stored: [U] 02/70
26020	08/77	RS	P	National Railway Museum, York	-	Stored: [U] 07/77
26021	07/81	RS	C	C F Booth, Rotherham	03/83	
26022	07/81	RS	C	C F Booth, Rotherham	05/83	
26023	07/81	RS	C	C F Booth, Rotherham	04/83	
26024	07/81	RS	C	C F Booth, Rotherham	06/83	
26025	07/81	RS	C	C F Booth, Rotherham	03/83	
26026	07/81	RS	C	C F Booth, Rotherham	07/83	
26027	07/81	RS	C	C F Booth, Rotherham	03/83	
26028	07/81	RS	C	C F Booth, Rotherham	06/83	
26029	07/81	RS	C	Coopers Metals, Brightside	03/83	
26030	07/81	RS	C	C F Booth, Rotherham	04/83	
26031	10/71	RS	C	BREL Crewe	05/72	
26032	07/81	RS	C	Coopers Metals, Brightside	03/83	
26033	07/81	RS	C	Coopers Metals, Brightside	03/83	Stored: [U] 06/80, R/I: 09/80
26034	07/81	RS	C	C F Booth, Rotherham	03/83	
26035	03/70	RS	C	BR Reddish, by J Cashmore	10/71	Stored: [U] 06/68, R/I: 09/68
26036	07/81	RS	C	V Berry, Leicester	05/83	Stored: [U] 08/80
26037	07/81	RS	C	V Berry, Leicester	04/83	Withdrawn: 10/71, R/I: 11/71
26038	02/77	RS	C	C F Booth, Rotherham	02/84	
26039	02/77	RS	C	C F Booth, Rotherham	02/84	
26040	07/81	RS	C	V Berry, Leicester	05/83	Stored: [S] 08/80, R/I: 11/80
26041	11/80	RS	C	C F Booth, Rotherham	03/83	Stored: [U] 04/80
26042	03/70	RS	C	BR Reddish, by J Cashmore	10/71	Stored: [U] 05/68
26043	06/78	RS	C	C F Booth, Rotherham	02/84	Stored: [U] 07/77
26044	07/81	RS	C	Coopers Metals, Brightside	02/84	Stored: [U] 06/80, R/I: 09/80
26045	11/71	RS	C	BREL Crewe	04/72	
26046	11/80	RS	C	C F Booth, Rotherham	03/83	Stored: [U] 08/80
26047	11/80	RS	C	C F Booth, Rotherham	05/83	Stored: [U] 09/80
26048	07/81	RS	C	C F Booth, Rotherham	06/83	1 Cab preserved at Liverpool Road, Manchester
26049	11/80	RS	C	C F Booth, Rotherham	03/83	Stored: [S] 08/80, Stored: [U] 09/80
26050	07/81	RS	C	C F Booth, Rotherham	05/83	Stored: [U] 06/80, R/I: 09/80
26051	07/81	RS	C	C F Booth, Rotherham	03/83	Stored: [S] 06/80, R/I: 08/80
26052	06/78	RS	C	C F Booth, Rotherham	02/84	Stored: [U] 07/77
26053	11/80	RS	C	C F Booth, Rotherham	03/83	Stored: [U] 06/80, R/I: 09/80
26054	07/81	RS	C	C F Booth, Rotherham	05/83	Stored: [S] 07/77, R/I: 03/78, Stored: [S] 07/80, R/I: 09/80
26055	02/77	RS	C	C F Booth, Rotherham	02/84	
26056	06/78	RS	C	BR Reddish, by C F Booth	03/83	Stored: [U] 03/78
26057	02/77	RS	C	BR Reddish, by C F Booth	03/83	

BR 1500v DC ELECTRIC LOCOMOTIVES CLASS 77

BR 1948 No.	Name	Name Date	Built By	Works No.	Date Introduced	Depot of First Allocation	Date Withdrawn	Depot of Final Allocation
E27000	Electra	08/59	MV/BR Gorton	1065	12/53	36B	09/68	9C
E27001	Ariadne	10/59	MV/BR Gorton	1066	03/54	36B	09/68	9C
E27002	Aurora	06/59	MV/BR Gorton	1067	05/54	36B	09/68	9C
E27003	Diana	01/61	MV/BR Gorton	1068	08/54	39A	09/68	9C
E27004	Juno	06/59	MV/BR Gorton	1069	09/54	39A	09/68	9C
E27005	Minerva	05/59	MV/BR Gorton	1070	12/54	39A	09/68	9C
E27006	Pandora	05/59	MV/BR Gorton	1071	12/54	39A	09/68	9C

BRITISH RAILWAYS 1957 NUMBERS-ELECTRIC

E1000 See BR 1948 numbers (Diesel) 18100
E2001 See BR 1948 numbers (Diesel) 18100

BIRMINGHAM RCW 25kv Bo-Bo CLASS 81

BR 1957 No.	TOPS No.	Date Re No.	Built By	Works No.	Date Introduced	Depot of First Allocation	Date Withdrawn	Depot of Final Allocation
E3001	81001	04/73	BRCW	1083	11/59	9A	07/84	GW
E3002	-	-	BRCW	1084	01/60	9A	11/68	ACL
E3003	81002	06/74	BRCW	1085	02/60	9A	10/90	WN
E3004	81003	06/73	BRCW	1086	04/60	9A	03/88	GW
E3005	81004	02/75	BRCW	1087	05/60	9A	04/90	WN
E3006	81005	10/74	BRCW	1088	07/60	9A	02/89	GW
E3007	81006	09/74	BRCW	1089	08/60	9A	02/89	GW
E3008	81007	07/74	BRCW	1090	09/60	9A	11/89	GW
E3009	-	-	BRCW	1091	10/60	ACL	08/68	ACL
E3010	81008	09/74	BRCW	1092	10/60	ACL	03/88	GW
E3011	81009	01/75	BRCW	1093	11/60	ACL	02/90	WN
E3012	81010	08/75	BRCW	1094	11/60	ACL	05/90	WN
E3013	81011	05/74	BRCW	1095	12/60	ACL	04/89	GW
E3014	81012	08/73	BRCW	1096	12/60	ACL	08/91	WN
E3015	81013	08/73	BRCW	1097	12/60	ACL	11/89	GW
E3016	81014	10/73	BRCW	1098	03/61	ACL	03/88	GW
E3017	81015	05/73	BRCW	1099	05/61	ACL	12/84	GW
E3018	81016	04/74	BRCW	1100	03/61	ACL	07/83	GW
E3019	-	-	BRCW	1101	04/61	ACL	07/71	ACL
E3020	81017	11/74	BRCW	1102	04/61	ACL	07/91	WN
E3021	81018	10/73	BRCW	1103	06/61	ACL	01/86	GW
E3022	81019	05/73	BRCW	1104	09/61	ACL	01/89	GW
E3023	81020	05/75	BRCW	1105	09/61	ACL	07/87	GW

Class continued from E3096

ENGLISH ELECTRIC 25kv Bo-Bo CLASS 83

BR 1957 No.	TOPS No.	Date Re No.	Built By	Works No.	Date Introduced	Depot of First Allocation	Date Withdrawn	Depot of Final Allocation
E3024	83001	01/73	EE.VF	2928/E264	07/60	9A	08/83	LG
E3025	83002	08/72	EE.VF	2929/E265	07/60	9A	08/83	LG
E3026	83003	01/73	EE.VF	2930/E266	08/60	9A	05/75	LG
E3027	83004	05/72	EE.VF	2931/E267	09/60	9A	01/78	LG
E3028	83005	04/72	EE.VF	2932/E268	09/60	ACL	08/83	LG
E3029	83006	08/72	EE.VF	2933/E269	10/60	ACL	08/83	LG
E3030	83007	12/72	EE.VF	2934/E270	10/60	ACL	08/83	LG
E3031	83008	11/72	EE.VF	2935/E271	11/60	ACL	08/83	LG
E3032	83009	02/72	EE.VF	2936/E272	11/60	ACL	03/89	WN
E3033	83010	03/72	EE.VF	2937/E273	12/60	ACL	08/83	LG
E3034	83011	02/72	EE.VF	2938/E274	02/61	ACL	08/83	LG
E3035	83012	07/72	EE.VF	2941/E277	07/61	ACL	03/89	WN

Class continued from E3098

BR 1948 No.	Disposal Code	Disposal Detail	Date Cut Up	Notes
E27000	P	EM2 Locomotive Society at Ilford	-	Exported to NS in 09/69 and renumbered 1502
E27001	P	Greater Manchester Museum of Transport	-	Exported to NS in 09/69 and renumbered 1505
E27002	E	Exported to NS in 09/69, renumbered 1506	-	
E27003	P	Werkgroep 1501	-	Exported to NS in 09/69 and renumbered 1501
E27004	E	Exported to NS in 09/69, renumbered 1503	-	
E27005	E	Exported to NS in 09/69, to provide spare parts	01/70	
E27006	E	Exported to NS in 09/69, renumbered 1504	-	

BR 1957 No.	Disposal Code	Disposal Detail	Date Cut Up	Notes
E3001	C	BREL Crewe	09/86	Stored: [U] 11/83
E3002	C	BR Workshops, Crewe	01/69	
E3003	P	The Railway Age, Crewe	-	Stored: [U] 01/90
E3004	C	Coopers Metals, Sheffield	01/92	
E3005	C	M C Processors, Glasgow	05/92	
E3006	C	Coopers Metals, Sheffield	01/92	
E3007	C	Coopers Metals, Sheffield	01/92	
E3008	C	Coopers Metals, Sheffield	01/92	
E3009	C	BR Workshops, Crewe	08/68	
E3010	C	Coopers Metals, Sheffield	11/91	
E3011	C	Coopers Metals, Sheffield	02/92	
E3012	C	Coopers Metals, Sheffield	01/92	
E3013	C	Coopers Metals, Sheffield	12/91	
E3014	C	Coopers Metals, Sheffield	02/92	
E3015	C	Coopers Metals, Sheffield	12/91	Stored: [U] 08/89
E3016	C	Coopers Metals, Sheffield	12/91	Stored: [U] 01/61, R/I: 03/61
E3017	C	M C Processors, Glasgow	05/92	
E3018	C	BREL Crewe	02/85	Stored: [U] 03/83
E3019	C	BREL Crewe	10/71	
E3020	C	Coopers Metals, Sheffield	01/92	
E3021	C	M C Processors, Glasgow	07/92	Stored: [U] 09/85
E3022	C	Coopers Metals, Sheffield	11/91	
E3023	C	Coopers Metals, Sheffield	12/91	

BR 1957 No	Disposal Code	Disposal Detail	Date Cut Up	Notes
E3024	C	V Berry, Leicester	09/84	Stored: [U] 06/68, R/I: 01/73, Stored: [U] 10/82
E3025	C	V Berry, Leicester	09/84	Stored: [U] 05/69, R/I: 08/72, Stored: [U] 10/81
E3026	C	BREL Crewe	07/75	Stored: [U] 05/68, R/I: 01/73
E3027	C	BR Willesden	02/78	Stored: [U] 05/69, R/I: 05/72
E3028	C	V Berry, Leicester	09/84	Stored: [U] 05/69, R/I: 04/72, Stored: [U] 10/82
E3029	C	V Berry, Leicester	09/84	Stored: [U] 05/69, R/I: 08/72, Stored: [U] 10/82
E3030	C	V Berry, Leicester	10/84	Stored: [U] 05/68, R/I: 12/72, Stored: [U] 10/82
E3031	C	V Berry, Leicester	09/84	Stored: [U] 05/69, R/I: 11/72, Stored: [U] 12/81
E3032	A	BR Crewe		Stored: [U] 03/68, R/I: 02/72, Withdrawn: 08/83, R/I: 09/85
E3033	C	V Berry, Leicester	10/84	Stored: [U] 03/68, R/I: 02/72, Stored: [U] 10/82
E3034	C	V Berry, Leicester	09/84	Stored: [U] 03/68, R/I: 02/72, Stored: [U] 10/82
E3035	P	P Waterman at M C Processors	-	Stored: [U] 03/69, R/I: 07/72

NORTH BRITISH 25kv Bo-Bo CLASS 84

BR 1957 No.	TOPS No.	Date Re No.	Built By	Works No.	Date Introduced	Depot of First Allocation	Date Withdrawn	Depot of Final Allocation
E3036	84001	12/72	NBL	27793	03/60	9A	01/79	CE
E3037	84002	09/72	NBL	27794	05/60	9A	09/80	CE
E3038	84003	05/72	NBL	27795	06/60	9A	11/80	CE
E3039	84004	06/72	NBL	27796	07/60	9A	11/77	CE
E3040	84005	07/72	NBL	27797	08/60	9A	04/77	CE
E3041	84006	10/72	NBL	27798	09/60	ACL	01/78	CE
E3042	84007	05/72	NBL	27799	10/60	ACL	04/77	CE
E3043	84008	08/72	NBL	27800	11/60	ACL	10/79	CE
E3044	84009	11/72	NBL	27801	12/60	ACL	08/78	CE
E3045	84010	10/72	NBL	27802	03/61	ACL	11/80	CE

BEYER PEACOCK 25 Kv Bo-Bo CLASS 82

BR 1957 No.	TOPS No.	Date Re No.	Built By	Works No.	Date Introduced	Depot of First Allocation	Date Withdrawn	Depot of Final Allocation
E3046	-	-	B.Peacock	1021/7884	05/60	9A	01/71	ACL
E3047	82001	04/74	B.Peacock	1022/7885	07/60	9A	08/83	LG
E3048	82002	02/74	B.Peacock	1023/7886	08/60	9A	08/83	LG
E3049	82003	05/74	B.Peacock	1024/7887	08/60	9A	08/83	LG
E3050	82004	04/74	B.Peacock	1025/7888	09/60	ACL	10/83	LG
E3051	82005	02/74	B.Peacock	1026/7889	10/60	ACL	10/87	WN
E3052	82006	04/74	B.Peacock	1027/7890	12/60	ACL	08/83	LG
E3053	82007	04/74	B.Peacock	1028/7891	01/62	ACL	08/83	LG
E3054	82008	02/74	B.Peacock	1029/7892	11/61	ACL	12/87	WN
E3055	-	-	B.Peacock	1030/7893	04/62	ACL	09/69	ACL

BR 25kv Bo-Bo CLASS 85

BR 1957 No.	TOPS No.	Date Re No.	TOPS Re No.	Date Re No.	Built By	Works No.	Date Introduced	Depot of First Allocation	Date Withdrawn	Depot of Final Allocation
E3056	85001	06/74			BR Doncaster	-	08/61	ACL	10/85	CE
E3057	85002	01/74			BR Doncaster	-	06/61	ACL	05/89	CE
E3058	85003	08/73	85113	10/90	BR Doncaster	-	06/61	ACL	11/91	CE
E3059	85004	09/74	85111	11/89	BR Doncaster	-	07/61	ACL	03/90	CE
E3060	85005	03/74			BR Doncaster	-	07/61	ACL	05/90	CE
E3061	85006	11/74	85101	06/89	BR Doncaster	-	12/61	ACL	11/92	CE
E3062	85007	11/73	85112	04/90	BR Doncaster	-	12/61	ACL	07/91	CE
E3063	85008	11/74			BR Doncaster	-	10/61	ACL	09/90	CE
E3064	85009	07/74	85102	06/89	BR Doncaster	-	12/61	ACL	05/91	CE
E3065	85010	04/73	85103	07/89	BR Doncaster	-	12/61	ACL	05/91	CE
E3066	85011	10/74	85114	10/90	BR Doncaster	-	04/62	ACL	07/91	CE
E3067	85012	10/74	85104	06/89	BR Doncaster	-	01/62	ACL	07/91	CE
E3068	85013	09/73			BR Doncaster	-	05/62	ACL	10/90	CE
E3069	85014	05/74			BR Doncaster	-	05/62	ACL	10/89	CE
E3070	85015	02/75			BR Doncaster	-	07/62	ACL	10/90	CE
E3071	85016	03/75	85105	07/89	BR Doncaster	-	10/62	ACL	07/91	CE
E3072	85017	03/74			BR Doncaster	-	07/62	ACL	08/87	CE
E3073	85018	05/73			BR Doncaster	-	12/62	ACL	10/91	LG
E3074	85019	09/74			BR Doncaster	-	12/62	ACL	12/89	CE
E3075	85020	02/74			BR Doncaster	-	01/63	ACL	10/90	CE
E3076	85021	10/73	85106	07/89	BR Doncaster	-	04/63	ACL	10/90	CE
E3077	85022	02/74			BR Doncaster	-	03/63	ACL	02/89	CE
E3078	85023	06/74			BR Doncaster	-	03/63	ACL	04/90	CE
E3079	85024	11/74	85107	06/89	BR Doncaster	-	09/63	ACL	05/90	CE
E3080	85025	08/74			BR Doncaster	-	03/63	ACL	01/90	CE
E3081	85026	02/74			BR Doncaster	-	06/63	ACL	05/90	CE
E3082	85027	07/74			BR Doncaster	-	06/63	ACL	05/83	CE
E3083	85028	12/74			BR Doncaster	-	10/63	ACL	01/90	CE
E3084	85029	10/74			BR Doncaster	-	05/63	ACL	05/88	CE
E3085	85030	07/73			BR Doncaster	-	07/64	ACL	09/90	CE
E3086	85031	11/74			BR Doncaster	-	02/62	ACL	06/90	CE
E3087	85032	04/75	85108	06/89	BR Doncaster	-	10/62	ACL	07/91	CE
E3088	85033	06/73			BR Doncaster	-	02/63	ACL	07/84	CE
E3089	85034	05/73			BR Doncaster	-	06/63	ACL	10/90	CE
E3090	85035	11/73	85109	06/89	BR Doncaster	-	11/63	ACL	07/91	CE
E3091	85036	08/74	85110	06/89	BR Doncaster	-	10/63	ACL	10/91	LG
E3092	85037	12/73			BR Doncaster	-	02/64	ACL	09/90	CE
E3093	85038	01/74			BR Doncaster	-	11/63	ACL	01/90	CE
E3094	85039	03/73			BR Doncaster	-	02/64	ACL	03/87	CE
E3095	85040	04/74			BR Doncaster	-	12/64	ACL	11/91	WN

BR 1957 No.	Disposal Code	Disposal Detail	Date Cut Up	Notes
E3036	P	National Railway Museum, York	-	Stored: [U] 10/67, R/I: 05/72, Stored: [U] 12/78
E3037	C	Texas Metals, Hyde	12/82	Stored: [U] 11/67, R/I: 07/71
E3038	C	V Berry, Leicester	01/86	Stored: [U] 10/67, R/I: 07/70
E3039	C	Birds, Long Marston	03/85	Stored: [U] 11/67, R/I: 09/71
E3040	C	Birds, Long Marston	03/85	Stored: [U] 11/67, R/I: 11/71
E3041	C	BR Crewe Gresty Lane, by J Cashmore	10/79	Stored: [U] 11/67, R/I: 07/70
E3042	C	BR Crewe Gresty Lane, by J Cashmore	10/79	Stored: [U] 10/67, R/I: 07/70
E3043	C	BREL Crewe, by A Hampton	11/88	Stored: [U] 11/67, R/I: 05/68
E3044	D	To Departmental Stock - 968021	-	Stored: [U] 10/67, R/I: 04/69
E3045	C	Texas Metals, Hyde	12/82	Stored: [U] 10/67, R/I: 02/72

BR 1957 No.	Disposal Code	Disposal Detail	Date Cut Up	Notes
E3046	C	BREL Crewe	06/71	
E3047	C	V Berry, Leicester	02/85	Stored: [U] 05/82
E3048	C	V Berry, Leicester	02/85	Stored: [U] 10/81
E3049	A	BR Crewe		Stored: [U] 08/82
E3050	C	V Berry, Leicester	09/84	Stored: [U] 09/82, R/I: 08/83
E3051	A	BR Crewe		Stored: [U] 11/82, R/I: 04/83
E3052	C	V Berry, Leicester	09/84	Stored: [U] 06/82
E3053	C	V Berry, Leicester	09/84	Stored: [U] 09/82
E3054	P	P Waterman at M C Processors	-	Stored: [U] 11/82, R/I: 04/83
E3055	C	BREL Crewe	08/70	Stored: [U] 09/66

BR 1957 No.	Disposal Code	Disposal Detail	Date Cut Up	Notes
E3056	C	M C Processors, Glasgow	04/89	
E3057	C	M C Processors, Glasgow	10/92	
E3058	C	M C Processors, Glasgow	01/93	Stored: [U] 12/90
E3059	C	M C Processors, Glasgow	09/92	
E3060	C	M C Processors, Glasgow	01/93	Stored: [U] 04/90
E3061	A	BR Crewe		Withdrawn: 11/91, R/I: 06/92, Stored: [U] 06/92
E3062	C	M C Processors, Glasgow	01/93	Stored: [U] 03/83, R/I: 09/83
E3063	C	M C Processors, Glasgow	01/93	Stored: [U] 06/90
E3064	C	M C Processors, Glasgow	10/92	
E3065	C	M C Processors, Glasgow	09/92	
E3066	C	M C Processors, Glasgow	02/93	
E3067	C	M C Processors, Glasgow	01/93	Stored: [U] 02/90, R/I: 04/90, Sold for preservation 11/92, sold back for scrap 01/93
E3068	A	BR Crewe		Stored: [U] 03/89
E3069	C	M C Processors, Glasgow	10/92	Stored: [U] 08/89
E3070	C	M C Processors, Glasgow	10/92	
E3071	C	M C Processors, Glasgow	10/92	
E3072	A	BR Crewe		Stored: [U] 05/87
E3073	C	M C Processors, Glasgow	10/92	Stored: [U] 05/86, R/I: 07/86, Withdrawn: 10/90, R/I: 10/90
E3074	C	V Berry, Leicester	08/90	
E3075	A	BR Crewe		Stored: [U] 03/89
E3076	C	M C Processors, Glasgow	09/92	
E3077	A	BR Crewe		Stored: [U] 12/88
E3078	C	M C Processors, Glasgow	11/92	Stored: [U] 07/89, R/I: 07/89. Stored: [U] 02/90
E3079	A	BR Crewe		
E3080	C	V Berry, Leicester	08/90	Stored: [U] 09/89
E3081	C	M C Processors, Glasgow	01/93	
E3082	C	BREL Crewe	03/85	Stored: [U] 05/83
E3083	A	BR Crewe		Stored: [U] 05/86, R/I: 08/86
E3084	A	BR Crewe		
E3085	C	M C Processors, Glasgow	10/92	
E3086	C	M C Processors, Glasgow	09/92	
E3087	C	M C Processors, Glasgow	09/92	Stored: [U] 07/89, R/I: 09/89
E3088	C	BREL Crewe	03/85	Stored: [U] 04/83, R/I: 09/83
E3089	C	M C Processors, Glasgow	01/93	Withdrawn: 10/90, R/I: 10/90
E3090	C	M C Processors, Glasgow	10/92	
E3091	C	M C Processors, Glasgow	10/92	
E3092	C	M C Processors, Glasgow	10/92	
E3093	C	M C Processors, Glasgow	10/92	
E3094	C	M C Processors, Glasgow	04/89	
E3095	C	M C Processors, Glasgow	01/93	Withdrawn: 10/90, R/I: 10/90

BIRMINGHAM RCW 25kv Bo-Bo Class 81

Continued from E3023

Revised 1957 No.	Original 1957 No.	Date Re No.	TOPS No.	Date Re No.	Built By	Works No.	Date Introduced	Depot of First Allocation	Date Withdrawn	Depot of Final Allocation
E3096	E3301	06/63	81021	05/74	BRCW	1106	04/62	5H	05/87	GW
E3097	E3302*	-	81022	12/73	BRCW	1107	02/64	5H	07/87	GW

ENGLISH ELECTRIC 25kv Bo-Bo CLASS 83

Continued from E3035

Revised 1957 No.	Original 1957 No.	Date Re No.	TOPS No.	Date Re No.	Built By	Works No.	Date Introduced	Depot of First Allocation	Date Withdrawn	Depot of Final Allocation
E3098	E3303	09/62	83013	03/72	EE.VF	2939/E275	03/61	ACL	07/83	LG
E3099	E3304	11/62	83014	10/72	EE.VF	2940/E276	05/61	ACL	07/83	LG
E3100	E3305*	-	83015	10/73	EE.VF	2942/E278	06/62	ACL	02/89	WN

BR/EE 25kv Bo-Bo Class 86

BR 1957 No.	First TOPS No.	Date Re No.	Second TOPS No.	Date Re No.	Third TOPS No.	Date Re No.	Fourth TOPS No.	Date Re No.	Fifth TOPS No.	Date Re No.	Name	Name Date
E3101	86252	05/74									The Liverpool Daily Post	11/80
E3102	86009	10/73	86409	11/86	86609	06/89						
E3103	86004	05/73	86404	01/86	86604	11/90						
E3104	86010	07/73	86410	08/86	86610	07/90						
E3105	86030	02/74	86430	02/87							[Scottish National Orchestra]	06/87-11/91
E3106	86214	08/73									Sans Pereil	04/80
E3107	86248	04/74									Sir Clwyd/County of Clwyd	03/81
E3108	86038	05/74	86438	02/87	86638	11/90						
E3109	86016	09/73	86316	01/82	86416	03/87	86616	09/89	86416	05/92	[Wigan Pier]	09/84-10/92
E3110	86027	01/74	86327	06/80	86427	05/85	86627	09/89			The Industrial Society	07/85
E3111	86024	12/73	86324	04/80	86424	05/86						
E3112	86006	06/73	86406	05/86	86606	11/90						
E3113	86232	01/74									Norwich Festival	10/90
E3114	86020	09/73	86320	04/80	86420	12/84	86620	05/89				
E3115	86003	05/73	86403	06/86	86603	11/90						
E3116	86238	02/74									European Community	04/86
E3117	86227	12/73									Sir Henry Johnson	03/81
E3118	86041	10/73	86261	08/75							[Driver John Axon GC]	02/81-06/92
E3119	86229	01/74									Sir John Betjeman	06/83
E3120	86019	09/73	86319	04/81	86419	03/86					Post Haste - 150 Years of Travelling Post Offices	07/90
E3121	86241	03/74	86508	02/89	86241	06/89					Glenfiddich	03/79
E3122	86012	06/73	86312	02/81	86412	10/85	86612	10/90			Elizabeth Garrett Anderson	10/83
E3123	86015	07/73	86315	08/80	86415	06/86	86615	10/90	86415	05/92	Rotary International	06/84
E3124	86035	04/74	86435	04/86	86635	09/89						
E3125	86209	07/73									City of Coventry	02/79
E3126	86231	01/74									Starlight Express	10/84
E3127	86240	03/74									Bishop Eric Treacy	04/79
E3128	86013	08/73	86313	11/80	86413	03/85	86613	11/89			County of Lancashire	04/85
E3129	86205	06/73	86503	09/88	86205	11/89					City of Lancaster	10/79
E3130	86037	05/74	86437	05/86	86637	11/90						
E3131	86222	11/73	86502	06/88	86222	10/89					Lloyds List 250th Anniversary	06/87
E3132	86221	11/73									BBC Look East	05/87
E3133	86236	02/74									Josiah Wedgwood Master Potter 1730-1795	11/80
E3134	86224	11/73									Caledonian	07/79
E3135	86040	11/73	86256	03/75							Pebble Mill	11/81
E3136	86044	02/74	86253	01/75							The Manchester Guardian	11/80
E3137	86045	01/74	86259	05/75							Peter Pan	10/79
E3138	86242	03/74									James Kennedy GC	11/81
E3139	86043	12/73	86257	05/75							Snowdon	01/81
E3140	86046	03/74	86258	05/75	86501	05/88	86258	11/89			Talyllyn - The First Preserved Railway	04/84
E3141	86208	08/73									City of Chester	03/79
E3142	86047	01/74	86254	12/74							[William Webb Ellis]	10/80-08/92
E3143	86203	09/72	86103	07/74							Andre Chapelon	01/81
E3144	86048	03/74	86260	06/75							Driver Wallace Oakes GC	02/81
E3145	86014	08/73	86314	09/81	86414	07/86	86614	12/90	86414	05/92	Frank Hornby	09/86
E3146	86017	08/73	86317	02/81	86417	02/85					The Kingsman	07/85
E3147	86211	08/73									[City of Milton Keynes]	05/82-10/87
E3148	86032	03/74	86432	01/85	86632	08/89					Brookside	08/87
E3149	86246	04/74	86505	10/88	86246	09/89					Royal Anglian Regiment	05/85
E3150	86202	12/72	86102	07/74							Robert A Riddles	05/81
E3151	86212	09/73									Preston Guild - 1328 - 1992	05/92
E3152	86023	12/73	86323	08/80	86423	11/86	86623	12/90				

BR 1957 No.	Disposal Code	Disposal Detail	Date Cut Up	Notes
E3301	A	M C Processors, Glasgow		
E3302*	C	BREL Crewe, by A Hampton	11/88	

BR 1957 No.	Disposal Code	Disposal Detail	Date Cut Up	Notes
E3303	C	V Berry, Leicester	09/84	Stored: [U] 09/68, R/I: 03/72, Stored: [U] 10/81, R/I: 05/82, Stored: [U] 10/82
E3304	C	V Berry, Leicester	09/84	Stored: [U] 04/69, R/I: 10/72, Stored: [U] 03/81
E3305	A	BR Crewe		Stored: [U] 03/69, R/I: 12/69, Stored: [U] 01/82, Withdrawn: 07/83, R/I: 11/83

BR 1957 No.	Built By	Works No.	Date Introduced	Depot of First Allocation	Date Withdrawn	Depot of Final Allocation	Disposal Code	Disposal Detail	Date Cut Up	Notes
E3101	BR Doncaster	-	08/65	5H						
E3102	BR Doncaster	-	08/65	5H						
E3103	BR Doncaster	-	08/65	5H						
E3104	BR Doncaster	-	10/65							
E3105	BR Doncaster	-	06/65	5H						
E3106	BR Doncaster	-	06/65	5H						
E3107	BR Doncaster	-	10/65	5H						
E3108	BR Doncaster	-	06/65	5H						
E3109	BR Doncaster	-	06/65	5H						
E3110	BR Doncaster	-	06/65	5H						
E3111	BR Doncaster	-	06/65	5H						
E3112	BR Doncaster	-	08/65	5H						
E3113	BR Doncaster	-	08/65	5H						Named: Harold Macmillan 10/79-08/90
E3114	BR Doncaster	-	10/65	5H						
E3115	BR Doncaster	-	10/65	5H						
E3116	BR Doncaster	-	10/65	5H						
E3117	BR Doncaster	-	10/65	5H						
E3118	BR Doncaster	-	09/65	5H						
E3119	BR Doncaster	-	00/65	5H						
E3120	BR Doncaster	-	09/65	5H						
E3121	BR Doncaster	-	09/65	5H						
E3122	BR Doncaster	-	10/65	5H						
E3123	BR Doncaster	-	10/65	5H						
E3124	BR Doncaster	-	10/65	5H						
E3125	BR Doncaster	-	11/65	5H						
E3126	BR Doncaster	-	11/65	5H						
E3127	BR Doncaster	-	10/65	5H						
E3128	BR Doncaster	-	11/65	5H						
E3129	BR Doncaster	-	11/65	5H						
E3130	BR Doncaster	-	12/65	5H						
E3131	BR Doncaster	-	01/66	ACL						Named: Fury 04/79-05/87
E3132	BR Doncaster	-	12/65	5H						Named: Vesta 05/79-04/87
E3133	BR Doncaster	-	12/65	5H						
E3134	BR Doncaster	-	12/65	5H						
E3135	BR Doncaster	-	01/66	ACL						
E3136	BR Doncaster	-	12/65	5H						
E3137	BR Doncaster	-	01/66	ACL						
E3138	BR Doncaster	-	01/66	ACL						
E3139	BR Doncaster	-	02/66	ACL						
E3140	BR Doncaster	-	03/66	ACL						
E3141	EE.VF	3722/E382	02/66	ACL						
E3142	EE.VF	3723/E383	02/66	ACL						
E3143	EE.VF	3724/E384	03/66	ACL						
E3144	EE.VF	3725/E385	03/66	ACL						
E3145	EE.VF	3726/E386	03/66	ACL						
E3146	EE.VF	3727/E387	04/66	ACL						
E3147	EE.VF	3728/E388	04/66	ACL	11/86	WN	C	BREL Crewe	11/86	Stored: [U] 10/86
E3148	EE.VF	3729/E389	04/66	ACL						
E3149	EE.VF	3730/E390	04/66	ACL						
E3150	EE.VF	3731/E391	04/66	ACL						
E3151	EE.VF	3732/E392	04/66	ACL						Named: Preston Guild 05/79-05/92
E3152	EE.VF	3733/E393	05/66	ACL						

BR 1957 No.	First TOPS No.	Date Re No.	Second TOPS No.	Date Re No.	Third TOPS No.	Date Re No.	Fourth TOPS No.	Date Re No.	Fifth TOPS No.	Date Re No.	Name	Name Date
E3153	86039	10/74	86439	10/86	86639	12/90						
E3154	86042	12/73	86255	04/75								
E3155	86234	01/74									Penrith Beacon	11/80
E3156	86220	11/73									J B Priestley OM	12/80
E3157	86021	09/73	86321	05/80	86421	06/85	86621	06/89			The Round Tabler	05/87
E3158	86223	11/73									London School of Economics	10/85
E3159	86028	01/74	86328	05/80	86428	02/86	86628	10/90	86428	05/92	Norwich Union	12/87
E3160	86036	04/74	86436	08/85	86636	08/89					Aldaniti	03/84
E3161	86249	04/74										
E3162	86226	11/73									County of Merseyside	09/81
E3163	86018	08/73	86318	11/81	86418	08/85	86618	07/90			Royal Mail Midlands	01/85
E3164	86225	11/73										
E3165	86215	09/73									Hardwicke	10/80
E3166	86216	10/73									Joseph Chamberlain	04/81
E3167	86228	12/73									Meteor	08/79
E3168	86230	01/74									Vulcan Heritage	03/80
E3169	86239	02/74	86507	02/89	86239	07/89					The Duke of Wellington	06/81
E3170	86002	05/73	86402	01/85	86602	10/89					L S Lowry	10/80
E3171	86011	08/73	86311	10/80	86411	07/86	86611	09/90	86411	05/92	Airey Neave	05/83
E3172	86233	01/74	86506	02/89	86233	04/89					Laurence Olivier	06/80
E3173	86204	06/73									City of Carlisle	12/78
E3174	86022	11/73	86322	05/80	86422	11/85	86622	04/90				
E3175	86218	10/73									Planet	06/79
E3176	86007	07/73	86407	05/87	86607	08/89					Institution of Electrical Engineers	07/87
E3177	86217	09/73	86504	10/88	86217	10/89					Halley's Comet	11/85
E3178	86244	03/74									The Royal British Legion	11/81
E3179	86207	06/73									City of Lichfield	03/81
E3180	86008	06/73	86408	11/85	86608	09/89					St John Ambulance	11/87
E3181	86243	03/74									The Boys' Brigade	04/83
E3182	86245	04/74									Dudley Castle	05/84
E3183	86251	05/74									The Birmingham Post	02/80
E3184	86206	07/73									City of Stoke on Trent	12/78
E3185	86005	06/73	86405	03/86	86605	12/90	86405	05/92			Intercontainer	06/91
E3186	86025	11/73	86325	04/80	86425	02/86						
E3187	86034	03/74	86434	09/86	86634	04/90					University of London	04/86
E3188	86031	02/74	86431	04/86	86631	11/90	86431	05/92				
E3189	86250	05/74									The Glasgow Herald	09/80
E3190	86210	08/73									City of Edinburgh	02/79
E3191	86201	08/72	86101	06/74							Sir William A. Stanier FRS	10/78
E3192	86247	04/74									Abraham Derby	10/81
E3193	86213	10/73									Lancashire Witch	03/81
E3194	86235	01/74									Crown Point	10/92
E3195	86026	12/73	86326	03/80	86426	12/84						
E3196	86219	10/73										
E3197	86237	02/74									Phoenix	08/79
E3198	86033	05/74	86433	04/85	86633	09/89					Sir Charles Halle	11/83
E3199	86001	05/73	86401	12/86							Wulfruna	06/85
E3200	86029	02/74	86329	05/80	86429	03/85					[Northampton Town]	05/89-10/91
											[The Times]	06/85-10/86

BR 750v DC LOCOMOTIVES CLASS 71

Original 1957 No.	Revised 1957 No.	Date Re No.	TOPS No.	Date Re No.	Built By	Works No.	Date Introduced	Depot of First Allocation	Date Withdrawn	Depot of Final Allocation
E5000	E5024	12/62	-	-	BR Doncaster	-	12/58	73A	10/66	73F
E5001	-	-	71001	01/74	BR Doncaster	-	02/59	73A	11/77	AF
E5002	-	-	71002	12/75	BR Doncaster	-	02/59	73A	11/77	AF
E5003	-	-	-	-	BR Doncaster	-	03/59	73A	02/67	73F
E5004	-	-	71004	01/74	BR Doncaster	-	04/59	73A	11/77	AF
E5005	-	-	-	-	BR Doncaster	-	05/59	73A	03/67	73F
E5006	-	-	-	-	BR Doncaster	-	05/59	73A	06/66	73D
E5007	-	-	71007	01/74	BR Doncaster	-	06/59	73A	11/77	AF
E5008	-	-	71008	12/73	BR Doncaster	-	07/59	73A	11/77	AF
E5009	-	-	71009	01/74	BR Doncaster	-	08/59	73A	11/77	AF
E5010	-	-	71010	12/73	BR Doncaster	-	10/59	73A	11/77	AF
E5011	-	-	71011	12/73	BR Doncaster	-	10/59	73A	11/77	AF
E5012	-	-	71012	01/74	BR Doncaster	-	11/59	73A	11/77	AF
E5013	-	-	71013	01/74	BR Doncaster	-	12/59	73A	11/77	AF
E5014	-	-	71014	12/73	BR Doncaster	-	02/60	73A	11/77	AF
E5015	-	-	-	-	BR Doncaster	-	02/60	73A	03/66	75D
E5016	-	-	-	-	BR Doncaster	-	04/60	73A	10/66	73F
E5017	-	-	-	-	BR Doncaster	-	03/60	73A	04/67	73F
E5018	E5003	12/68	71003	12/73	BR Doncaster	-	04/60	73A	11/77	AF
E5019	-	-	-	-	BR Doncaster	-	06/60	73A	10/66	73A
E5020	E5005	10/68	71005	12/73	BR Doncaster	-	06/60	73A	11/77	AF
E5021	-	-	-	-	BR Doncaster	-	08/60	73A	03/67	73F
E5022	E5006	10/68	71006	12/73	BR Doncaster	-	09/60	73A	11/77	AF
E5023	-	-	-	-	BR Doncaster	-	11/60	73A	01/67	73F
E5024 - See E5000										

BR 1957 No.	Built By	Works No.	Date Introduced	Depot of First Allocation	Date Withdrawn	Depot of Final Allocation	Disposal Code	Disposal Detail	Date Cut Up	Notes
E3153	EE.VF	3734/E394	05/66	ACL						
E3154	EE.VF	3735/E395	05/66	ACL						
E3155	EE.VF	3736/E396	04/66	ACL						
E3156	EE.VF	3737/E397	07/66	ACL						Named: Goliath 08/79-04/87
E3157	EE.VF	3738/E398	07/66	ACL						
E3158	EE.VF	3739/E399	07/66	ACL						Named: Hector 07/79-11/86
E3159	EE.VF	3740/E400	07/66	ACL						
E3160	EE.VF	3741/E401	10/66	ACL						
E3161	EE.VF	3453/E299	10/65	5H						
E3162	EE.VF	3454/E300	08/65	5H						Named: Mail 07/79-01/84
E3163	EE.VF	3455/E301	08/65	5H						
E3164	EE.VF	3456/E302	08/65	5H						
E3165	EE.VF	3457/E303	08/65	5H						
E3166	EE.VF	3458/E304	10/65	5H						
E3167	EE.VF	3459/E305	08/65	5H						
E3168	EE.VF	3460/E306	06/65	5H						
E3169	EE.VF	3461/E307	06/65	5H						
E3170	EE.VF	3462/E308	06/65	5H						
E3171	EE.VF	3463/E309	10/65	5H						
E3172	EE.VF	3464/E310	06/65	5H						
E3173	EE.VF	3465/E311	08/65	5H						
E3174	EE.VF	3466/E312	08/65	5H						
E3175	EE.VF	3467/E313	10/65	5H						
E3176	EE.VF	3468/E314	08/65	5H						
E3177	EE.VF	3469/E315	08/65	5H						Named: Comet 10/80-11/85
E3178	EE.VF	3470/E316	08/65	5H						
E3179	EE.VF	3471/E317	10/65	5H						
E3180	EE.VF	3472/E318	10/65	5H						
E3181	EE.VF	3473/E319	10/65	5H						
E3182	EE.VF	3474/E320	09/65	5H						
E3183	EE.VF	3475/E321	10/65	5H						
E3184	EE.VF	3476/E322	10/65	5H						
E3185	EE.VF	3477/E323	10/65	5H						
E3186	EE.VF	3478/E324	10/65	5H						
E3187	EE.VF	3479/E325	10/65	5H						
E3188	EE.VF	3480/E326	10/65	5H						
E3189	EE.VF	3481/E327	11/65	5H						
E3190	EE.VF	3482/E328	11/65	5H						
E3191	EE.VF	3483/E329	11/65	5H						
E3192	EE.VF	3484/E330	12/65	5H						
E3193	EE.VF	3485/E331	12/65	5H						
E3194	EE.VF	3486/E332	01/66	ACL						Named: Novelty 06/79-08/90, and Harold Macmillan 10/90-10/92
E3195	EE.VF	3487/E333	12/65	5H						
E3196	EE.VF	3488/E334	12/65	5H						
E3197	EE.VF	3489/E335	01/66	ACL						
E3198	EE.VF	3490/E336	01/66	ACL						
E3199	EE.VF	3491/E337	02/66	ACL						
E3200	EE.VF	3492/E338	02/66	ACL	11/86	WN	C	BREL Crewe	11/86	Stored: [U] 10/86

Original 1957 No.	Disposal Code	Disposal Detail	Date Cut Up	Notes
E5000	R	Rebuilt as Class 74 No. E6104	-	
E5001	P●	National Railway Museum, York	-	Stored: [U] 10/76, ●Authorised for main line running
E5002	C	J Cashmore, Newport	08/78	Stored: [U] 10/76
E5003	R	Rebuilt as Class 74 No. E6107	-	
E5004	C	BREL Doncaster	12/79	Stored: [U] 10/76
E5005	R	Rebuilt as Class 74 No. E6108	-	
E5006	R	Rebuilt as Class 74 No. E6103	-	
E5007	C	J Cashmore, Newport	08/78	Stored: [U] 10/76
E5008	C	J Cashmore, Newport	08/78	Stored: [U] 10/76
E5009	C	BREL Doncaster	09/79	Stored: [U] 10/76
E5010	C	BREL Doncaster	08/79	Stored: [U] 10/76
E5011	C	BREL Doncaster	11/79	Stored: [U] 10/76
E5012	C	J Cashmore, Newport	08/78	Stored: [U] 10/76
E5013	C	BREL Doncaster	11/79	Stored: [U] 10/76
E5014	C	BREL Doncaster	09/79	Stored: [U] 10/76
E5015	R	Rebuilt as Class 74 No. E6101	-	
E5016	R	Rebuilt as Class 74 No. E6102	-	
E5017	R	Rebuilt as Class 74 No. E6109	-	
E5018	C	BREL Doncaster	03/80	Stored: [U] 10/76
E5019	R	Rebuilt as Class 74 No. E6105	-	
E5020	C	J Cashmore, Newport	08/78	Stored: [U] 10/76
E5021	R	Rebuilt as Class 74 No. E6110	-	
E5022	C	J Cashmore, Newport	08/78	Stored: [U] 10/76
E5023	R	Rebuilt as Class 74 No. E6106	-	

E5024 - See E5000

BR/ENGLISH ELECTRIC ELECTRO DIESELS CLASS 73

BR 1957 No.	TOPS No.	Date Re No.	First TOPS Re No.	Date Re No.	Name	Name Date	Built By	Works No.	Date Introduced	Depot of First Allocation	Date Withdrawn
E6001	73001	02/74					BR Eastleigh	-	02/62	73A	
E6002	73002	02/74					BR Eastleigh	-	03/62	73A	
E6003	73003	02/74					BR Eastleigh	-	04/62	73A	
E6004	73004	02/74			[The Bluebell Railway]	09/87-09/90	BR Eastleigh	-	07/62	75D	03/91
E6005	73005	02/74			Mid Hants, Watercress Line	09/88	BR Eastleigh	-	07/62	75D	
E6006	73006	02/74					BR Eastleigh	-	11/62	75D	
E6007	73101	01/74			The Royal Alex'	05/92	EE.VF	3569/E339	10/65	75D	
E6008	73102	02/74	73212	02/88	Airtour Suisse	04/85	EE.VF	3570/E340	10/65	75D	
E6009	73103	02/74					EE.VF	3571/E341	11/65	75D	
E6010	73104	03/74					EE.VF	3572/E342	11/65	75D	
E6011	73105	01/74			[Quadrant]	11/87-08/89	EE.VF	3573/E343	12/65	75D	
E6012	73106	01/74					EE.VF	3574/E344	12/65	75D	
E6013	73107	02/74					EE.VF	3575/E345	12/65	75D	
E6014	73108	01/74					EE.VF	3576/E346	01/66	75D	
E6015	73109	01/74			Battle of Britain 50th Anniversary	09/90	EE.VF	3577/E347	01/66	75D	
E6016	73110	02/74					EE.VF	3578/E348	01/66	75D	
E6017	73111	02/74					EE.VF	3579/E349	01/66	75D	05/91
E6018	73112	03/74			University of Kent at Canterbury	04/90	EE.VF	3580/E350	02/66	75D	
E6019	73113	01/74	73211	02/88	[County of West Sussex]	07/86-04/91	EE.VF	3581/E351	02/66	75D	
E6020	73114	02/74					EE.VF	3582/E352	02/66	75D	
E6021	73115	02/74					EE.VF	3583/E353	02/66	75D	04/82
E6022	73116	02/74	73210	02/88	Selhurst	09/86	EE.VF	3584/E354	03/66	75D	
E6023	73117	01/74			University of Surrey	07/87	EE.VF	3585/E355	03/66	75D	
E6024	73118	02/74			The Romney Hythe and Dymchurch Railway	05/87	EE.VF	3586/E356	03/66	75D	
E6025	73119	02/74			Kentish Mercury	08/86	EE.VF	3587/E357	03/66	75D	
E6026	73120	04/74	73209	02/88			EE.VF	3588/E358	04/66	75D	
E6027	-	-					EE.VF	3589/E359	04/66	75D	07/72
E6028	73121	01/74	73208	02/88	Croydon 1883-1983	09/83	EE.VF	3590/E360	04/66	75D	
E6029	73122	01/74	73207	02/88	County of East Sussex	07/85	EE.VF	3591/E361	04/66	75D	
E6030	73123	01/74	73206	02/88	Gatwick Express	05/84	EE.VF	3592/E362	05/66	75D	
E6031	73124	04/74	73205	02/88	London Chamber of Commerce	06/87	EE.VF	3593/E363	05/66	75D	
E6032	73125	02/74	73204	02/88	Stewarts Lane 1860-1985	09/85	EE.VF	3594/E364	05/66	75D	
E6033	73126	01/74			Kent & East Sussex Railway	05/91	EE.VF	3595/E365	05/66	75D	
E6034	73127	01/74	73203	02/88			EE.VF	3596/E366	05/66	75D	
E6035	73128	02/74			O.V.S Bulleid CBE	09/91	EE.VF	3597/E367	06/66	75D	
E6036	73129	02/74			City of Winchester	12/82	EE.VF	3598/E368	06/66	75D	
E6037	73130	02/74			City of Portsmouth	07/88	EE.VF	3709/E369	07/66	75D	
E6038	73131	01/74			County of Surrey	03/88	EE.VF	3710/E370	07/66	75D	
E6039	73132	01/74					EE.VF	3711/E371	07/66	75D	
E6040	73133	01/74			The Bluebell Railway	09/90	EE.VF	3712/E372	08/66	75D	
E6041	73134	02/74			Woking Homes 1885-1985	10/85	EE.VF	3713/E373	08/66	75D	
E6042	73135	02/74	73235	04/91			EE.VF	3714/E374	08/66	75D	
E6043	73136	02/74			Kent Youth Music	05/92	EE.VF	3715/E375	09/66	75D	
E6044	73137	02/74	73202	02/88	Royal Observer Corps	10/85	EE.VF	3716/E376	09/66	75D	
E6045	73138	02/74			[Post Haste - 150 Years of Travelling Post Offices]	05/88-06/90	EE.VF	3717/E377	10/66	75D	
E6046	73139	03/74					EE.VF	3718/E378	10/66	75D	
E6047	73140	01/74					EE.VF	3719/E379	10/66	75D	
E6048	73141	02/74					EE.VF	3720/E380	12/66	75D	
E6049	73142	01/74	73201	02/88	Broadlands	09/80	EE.VF	3721/E381	01/67	75D	

BR ELECTRO-DIESELS CLASS 74

BR 1957 No.	TOPS No.	Date Re No.	Original Class 71 No.	Rebuilt By	Works No.	Date Introduced	Depot of First Allocation	Date Withdrawn	Depot of Final Allocation
E6101	74001	12/73	E5015	BR Crewe	-	03/68	73A	12/77	EH
E6102	74002	02/74	E5016	BR Crewe	-	11/67	73A	06/77	EH
E6103	74003	02/74	E5006	BR Crewe	-	01/68	73A	12/77	EH
E6104	74004	12/73	E5024	BR Crewe	-	03/68	73A	12/77	EH
E6105	74005	12/73	E5019	BR Crewe	-	03/68	73A	12/77	EH
E6106	74006	02/74	E5023	BR Crewe	-	04/68	73A	06/76	EH
E6107	74007	02/74	E5003	BR Crewe	-	04/68	73A	12/77	EH
E6108	74008	02/74	E5005	BR Crewe	-	05/68	73A	12/77	EH
E6109	74009	01/74	E5017	BR Crewe	-	05/68	73A	12/77	EH
E6110	74010	02/74	E5021	BR Crewe	-	06/68	73A	12/77	EH

BR 1957 No.	Depot of Final Allocation	Disposal Code	Disposal Detail	Date Cut Up	Notes
E6001					
E6002					
E6003					Stored: [U] 03/90, R/I: 04/90
E6004	SL	A	BR Stewarts Lane		Stored: [U] 03/90, R/I: 04/90
E6005					
E6006					Stored: [U] 11/90, R/I: 08/91
E6007					Numbered: 73100 12/80-01/81, Numbered: 73801 07/89-07/89 Named 'Brighton Evening Argus' 12/80 - 04/92
E6008					
E6009					
E6010					
E6011					
E6012					
E6013					
E6014					
E6015					
E6016					
E6017	SL	A	BR Stewarts Lane		
E6018					
E6019					
E6020					
E6021	SL	C	BR Selhurst	08/82	
E6022					
E6023					
E6024					Stored: [S] 08/81, R/I: 11/81
E6025					
E6026					
E6027	75D	C	BR Slade Green	02/73	
E6028					
E6029					Stored: [S] 08/81, R/I: 11/81
E6030					Stored: [S] 08/81, R/I: 11/81
E6031					
E6032					
E6033					Stored: [S] 07/81, R/I: 11/81
E6034					
E6035					
E6036					
E6037					
E6038					
E6039					
E6040					
E6041					Stored: [S] 08/81, R/I: 01/82
E6042					
E6043					Stored: [S] 08/81, R/I: 01/82
E6044					
E6045					
E6046					
E6047					
E6048					
E6049					

BR 1957 No.	Disposal Code	Disposal Detail	Date Cut Up	Notes
E6101	C	Birds, Long Marston	08/78	
E6102	C	J Cashmore, Newport	12/77	
E6103	C	J Cashmore, Newport	12/80	
E6104	C	Birds, Long Marston	08/78	
E6105	C	BR Fratton, by Pounds	01/81	
E6106	C	G Cohen, Kettering	06/77	
E6107	C	Birds, Long Marston	08/78	
E6108	C	Birds, Long Marston	08/78	
E6109	C	Birds, Long Marston	08/78	
E6110	C	BREL Doncaster	10/79	

TOPS NUMBERS-ELECTRIC

TOPS No. Range	Refer To 1957 No.	TOPS No. Range	Refer To 1957 No.	TOPS No. Range	Refer To 1957 No.	TOPS No. Range	Refer To 1957 No.	TOPS No. Range	Refer To 1957 No.
Class 71 **E5001-E5024**		73123	E6030	**Class 76** **E26000-E26057**		76048	E26048 1st	**Class 83** **E3024-E3035/** **E3098-E3100**	
71001	E5001	73124	E6031	76001	E26001	76048	E26039 2nd		
71002	E5002	73125	E6032	76002	E26002	76050	E26050 1st	83001	E3024
71003	E5018/E5003	73126	E6033	76003	E26003 1st	76050	E26038 2nd	83002	E3025
71004	E5004	73127	E6034	76003	E26036 2nd	76051	E26051	83003	E3026
71005	E5020/E5005	73128	E6035	76004	E26004	76052	E26052	83004	E3027
71006	E5022/E5006	73129	E6036	76006	E26006	76053	E26053	83005	E3028
71007	E5007	73130	E6037	76007	E26007	76054	E26054	83006	E3029
71008	E5008	73131	E6038	76008	E26008	76055	E26055	83007	E3030
71009	E5009	73132	E6039	76009	E26009	76056	E26056	83008	E3031
71010	E5010	73133	E6040	76010	E26010	76057	E26057	83009	E3032
71011	E5011	73134	E6041	76011	E26011			83010	E3033
71012	E5012	73135	E6042	76012	E26012	**Class 81** **E3001-E3023/** **E3096-E3097**		83011	E3034
71013	E5013	73136	E6043	76013	E26013			83012	E3035
71014	E5014	73137	E6044	76014	E26014	81001	E3001	83013	E3098
		73138	E6045	76015	E26015	81002	E3003	83014	E3099
		73139	E6046	76016	E26016	81003	E3004	83015	E3100
Class 73 **E6001-E6049**		73140	E6047	76018	E26018	81004	E3005		
		73141	E6048	76020	E26020	81005	E3006	**Class 84** **E3036-E3045**	
73001	E6001	73142	E6049	76021	E26021	81006	E3007		
73002	E6002	73201	E6049	76022	E26022	81007	E3008	84001	E3036
73003	E6003	73202	E6044	76023	E26023	81008	E3010	84002	E3037
73004	E6004	73203	E6034	76024	E26024	81009	E3011	84003	E3038
73005	E6005	73204	E6032	76025	E26025	81010	E3012	84004	E3039
73006	E6006	73205	E6031	76026	E26026	81011	E3013	84005	E3040
73100	E6007	73206	E6030	76027	E26027	81012	E3014	84006	E3041
73101	E6007	73207	E6029	76028	E26028	81013	E3015	84007	E3042
73102	E6008	73208	E6028	76029	E26029	81014	E3016	84008	E3043
73103	E6009	73209	E6026	76030	E26030	81015	E3017	84009	E3044
73104	E6010	73210	E6022	76031	E26044	81016	E3018	84010	E3045
73105	E6011	73211	E6019	76032	E26032	81017	E3020		
73106	E6012	73212	E6008	76033	E26033	81018	E3021	**Class 85** **E3056-E3095**	
73107	E6013	73235	E6042	76034	E26034	81019	E3022		
73108	E6014	73801	E6007	76035	E26018	81020	E3023	85001	E3056
73109	E6015			76036	E26036 1st	81021	E3096	85002	E3057
73110	E6016	**Class 74** **E6101-E6110**		76036	E26003 2nd	81022	E3097	85003	E3058
73111	E6017			76037	E26037			85004	E3059
73112	E6018	74001	E6101	76038	E26038 1st	**Class 82** **E3046-E3055**		85005	E3060
73113	E6019	74002	E6102	76038	E26050 2nd			85006	E3061
73114	E6020	74003	E6103	76039	E26039 1st	82001	E3047	85007	E3062
73115	E6021	74004	E6104	76039	E26048 2nd	82002	E3048	85008	E3063
73116	E6022	74005	E6105	76040	E26040	82003	E3049	85009	E3064
73117	E6023	74006	E6106	76041	E26041	82004	E3050	85010	E3065
73118	E6024	74007	E6107	76043	E26043	82005	E3051	85011	E3066
73119	E6025	74008	E6108	76044	E26044	82006	E3052	85012	E3067
73120	E6026	74009	E6109	76046	E26046	82007	E3053	85013	E3068
73121	E6028	74010	E6110	76047	E26047	82008	E3054	85014	E3069
73122	E6029								

TOPS No. Range	Refer To 1957 No.	TOPS No. Range	Refer To 1957 No.	TOPS No. Range	Refer To 1957 No.	TOPS No. Range	Refer To 1957 No.	TOPS No. Range	Refer To 1957 No.
85015	E3070	86006	E3112	86203	E3143	86251	E3183	86420	E3114
85016	E3071	86007	E3176	86204	E3173	86252	E3101	86421	E3157
85017	E3072	86008	E3180	86205	E3129	86253	E3136	86422	E3174
85018	E3073	86009	E3102	86206	E3184	86254	E3142	86423	E3152
85019	E3074	86010	E3104	86207	E3179	86255	E3154	86424	E3111
85020	E3075	86011	E3171	86208	E3141	86256	E3135	86425	E3186
85021	E3076	86012	E3122	86209	E3125	86257	E3139	86426	E3195
85022	E3077	86013	E3128	86210	E3190	86258	E3140	86427	E3110
85023	E3078	86014	E3145	86211	E3147	86259	E3137	86428	E3159
85024	E3079	86015	E3123	86212	E3151	86260	E3144	86429	E3200
85025	E3080	86016	E3109	86213	E3193	86311	E3171	86430	E3105
85026	E3081	86017	E3146	86214	E3106	86312	E3122	86431	E3188
85027	E3082	86018	E3163	86215	E3165	86313	E3128	86432	E3148
85028	E3083	86019	E3120	86216	E3166	86314	E3145	86433	E3198
85029	E3084	86020	E3114	86217	E3177	86315	E3123	86434	E3187
85030	E3085	86021	E3157	86218	E3175	86316	E3109	86435	E3124
85031	E3086	86022	E3174	86219	E3196	86317	E3146	86436	E3160
85032	E3087	86023	E3152	86220	E3156	86318	E3163	86437	E3180
85033	E3088	86024	E3111	86221	E3132	86319	E3120	86438	E3108
85034	E3089	86025	E3186	86222	E3131	86320	E3114	86439	E3153
85035	E3090	86026	E3195	86223	E3158	86321	E3157	86501	E3140
85036	E3091	86027	E3110	86224	E3134	86322	E3174	86502	E3131
85037	E3092	86028	E3159	86225	E3164	86323	E3152	86503	E3129
85038	E3093	86029	E3200	86226	E3162	86324	E3111	86504	E3177
85039	E3094	86030	E3105	86227	E3117	86325	E3186	86505	E3149
85040	E3095	86031	E3188	86228	E3167	86326	E3195	86506	E3172
85101	E3061	86032	E3148	86229	E3119	86327	E3110	86507	E3169
85102	E3064	86033	E3198	86230	E3168	86328	E3159	86508	E3121
85103	E3065	86034	E3187	86231	E3126	86329	E3200	86509	E3180
85104	E3067	86035	E3124	86232	E3113	86401	E3199	86602	E3170
85105	E3071	86036	E3160	86233	E3172	86402	E3170	86607	E3176
85106	E3076	86037	E3130	86234	E3155	86403	E3115	86608	E3180
85107	E3079	86038	E3108	86235	E3194	86404	E3103	86609	E3102
85108	E3087	86039	E3153	86236	E3133	86405	E3185	86613	E3128
85109	E3090	86040	E3185	86237	E3197	86406	E3112	86616	E3109
85110	E3091	86041	E3118	86238	E3116	86407	E3176	86620	E3114
85111	E3059	86042	E3154	86239	E3169	86408	E3180	86621	E3157
85112	E3062	86043	E3139	86240	E3127	86409	E3102	86627	E3110
85113	E3058	86044	E3136	86241	E3121	86410	E3104	86632	E3148
85114	E3066	86045	E3137	86242	E3138	86411	E3171	86633	E3198
		86046	E3140	86243	E3181	86412	E3122	86635	E3124
Class 86		86047	E3142	86244	E3178	86413	E3128	86636	E3160
E3101-E3200		86048	E3144	86245	E3182	86414	E3145		
86001	E3199	86101	E3191	86246	E3149	86415	E3123		
86002	E3170	86102	E3150	86247	E3192	86416	E3109		
86003	E3115	86103	E3143	86248	E3107	86417	E3146		
86004	E3103	86201	E3191	86249	E3161	86418	E3163		
86005	E3185	86202	E3150	86250	E3189	86419	E3120		

BR 25kv Bo-Bo CLASS 87

TOPS No.	Name	Date Named	Built By	Works No.	Date Introduced	Depot of First Allocation	Date Withdrawn	Depot of Final Allocation
87001	Royal Scot	07/77	BREL Crewe	-	06/73	WN		
87002	Royal Sovereign	07/78	BREL Crewe	-	06/73	WN		
87003	Patriot	06/78	BREL Crewe	-	07/73	WN		
87004	Britannia	04/78	BREL Crewe	-	07/73	WN		
87005	City of London	11/77	BREL Crewe	-	08/73	WN		
87006	City of Glasgow	12/77	BREL Crewe	-	11/73	WN		
87007	City of Manchester	10/77	BREL Crewe	-	10/73	WN		
87008	City of Liverpool	11/77	BREL Crewe	-	11/73	WN		
87009	City of Birmingham	11/77	BREL Crewe	-	11/73	WN		
87010	King Arthur	06/78	BREL Crewe	-	12/73	WN		
87011	The Black Prince	05/78	BREL Crewe	-	01/74	WN		
87012	Royal Bank of Scotland	11/88	BREL Crewe	-	01/74	WN		
87013	John o' Gaunt	03/78	BREL Crewe	-	02/74	WN		
87014	Knight of the Thistle	05/78	BREL Crewe	-	01/74	WN		
87015	Howard of Effingham	05/78	BREL Crewe	-	02/74	WN		
87016	Willesden Intercity Depot	08/92	BREL Crewe	-	03/74	WN		
87017	Iron Duke	05/78	BREL Crewe	-	03/74	WN		
87018	Lord Nelson	03/78	BREL Crewe	-	05/74	WN		
87019	Sir Winston Churchill	05/78	BREL Crewe	-	03/74	WN		
87020	North Briton	05/78	BREL Crewe	-	03/74	WN		
87021	Robert the Bruce	06/78	BREL Crewe	-	04/74	WN		
87022	Cock o' the North	06/78	BREL Crewe	-	04/74	WN		
87023	Velocity	10/85	BREL Crewe	-	04/74	WN		
87024	Lord of the Isles	05/78	BREL Crewe	-	04/74	WN		
87025	County of Cheshire	11/82	BREL Crewe	-	04/74	WN		
87026	Sir Richard Arkwright	10/82	BREL Crewe	-	05/74	WN		
87027	Wolf of Badenoch	05/78	BREL Crewe	-	05/74	WN		
87028	Lord President	05/78	BREL Crewe	-	05/74	WN		
87029	Earl Marischal	06/78	BREL Crewe	-	06/74	WN		
87030	Black Douglas	07/78	BREL Crewe	-	06/74	WN		
87031	Hal o' the Wynd	06/78	BREL Crewe	-	07/74	WN		
87032	Kenilworth	05/78	BREL Crewe	-	07/74	WN		
87033	Thane of Fife	05/78	BREL Crewe	-	08/74	WN		
87034	William Shakespeare	05/78	BREL Crewe	-	09/74	WN		
87035	Robert Burns	04/78	BREL Crewe	-	10/74	WN		
87101	Stephenson	10/77	BREL Crewe	-	03/75	RTC		

BRUSH PROTOYPE 25kv Co-Co CLASS 89

TOPS No.	Name	Date Named	Built By	Works No.	Date Introduced	Depot of First Allocation	Date Withdrawn	Depot of Final Allocation
89001	[Avocet]	01/89-06/92	BREL Crewe/Brush	875	10/86	RTC	07/92	BN

BR 25kv Bo-Bo CLASS 90

Original TOPS No.	TOPS Re No.	Date Re No.	Name	Date Named	Built By	Works No.	Date Introduced	Depot of First Allocation	Date Withdrawn	Depot of Final Allocation
90001			BBC Midlands Today	09/89	BREL Crewe	-	04/88	WN		
90002					BREL Crewe	-	04/88	WN		
90003					BREL Crewe	-	04/88	WN		
90004			D'Oyle Carte Opera Company	03/91	BREL Crewe	-	04/88	WN		
90005			Financial Times	03/88	BREL Crewe	-	03/88	WN		
90006			High Sheriff	05/92	BREL Crewe	-	09/88	WN		
90007			Lord Stamp	04/92	BREL Crewe	-	04/88	WN		
90008			The Birmingham Royal Ballet	08/90	BREL Crewe	-	05/88	WN		
90009			Royal Show	05/89	BREL Crewe	-	09/88	WN		
90010			275 Railway Squadron (Volunteers)	11/89	BREL Crewe	-	10/88	WN		
90011			The Chartered Institute of Transport	10/88	BREL Crewe	-	09/88	WN		
90012			Glasgow 1990 Cultural Capital of Europe	06/90	BREL Crewe	-	11/88	WN		
90013			The Law Society	10/92	BREL Crewe	-	11/88	WN		
90014			'The Liverpool Phil'	03/90	BREL Crewe	-	10/88	WN		
90015			BBC North West	10/89	BREL Crewe	-	11/88	WN		
90016					BREL Crewe	-	12/88	WN		
90017					BREL Crewe	-	12/88	WN		
90018					BREL Crewe	-	12/88	WN		
90019			Penny Black	05/90	BREL Crewe	-	01/89	WN		
90020			Colonel Bill Cockburn CBE TD	10/91	BREL Crewe	-	01/89	WN		
90021					BREL Crewe	-	01/89	WN		
90022			Freightconnection	09/92	BREL Crewe	-	01/89	WN		
90023					BREL Crewe	-	02/89	WN		
90024					BREL Crewe	-	02/89	WN		

TOPS No.	Disposal Code	Disposal Detail	Date Cut Up	Notes
87001				Named: Stephenson 01/76-07/77
87002				
87003				
87004				
87005				
87006				Named: Glasgow Garden Festival 04/87-01/89
87007				
87008				
87009				
87010				
87011				
87012				Named: Coeur de Lion 05/78-11/84
87013				
87014				
87015				
87016				Named: Sir Francis Drake 04/78-06/88
87017				
87018				
87019				
87020				
87021				
87022				
87023				Named: Highland Chieftain 06/78-05/84
87024				
87025				Named: Borderer 06/78-11/82
87026				Named: Redgauntlet 05/78-10/82
87027				
87028				
87029				
87030				
87031				
87032				
87033				
87034				
87035				
87101				Entered service 08/76

TOPS No.	Disposal Code	Disposal Detail	Date Cut Up	Notes
89001	P	Midland Railway Centre	-	Stored: [U] 07/91

TOPS No	Disposal Code	Disposal Detail	Date Cut Up	Notes
90001				
90002				
90003				
90004				
90005				
90006				
90007				
90008				
90009				
90010				
90011				
90012				
90013				
90014				
90015				
90016				
90017				
90018				
90019				
90020				
90021				
90022				
90023				
90024				

Original TOPS No.	TOPS Re No.	Date Re No.	Name	Date Named	Built By	Works No.	Date Introduced	Depot of First Allocation	Date Withdrawn	Depot of Final Allocation
90025					BREL Crewe	-	03/89	WN		
90026	90126	07/91			BREL Crewe	-	03/89	WN		
90027	90127	07/91			BREL Crewe	-	04/89	WN		
90028	90128	07/91	Vrachtverbinding	09/92	BREL Crewe	-	03/89	WN		
90029	90129	08/91	Frachtverbindungen	09/92	BREL Crewe	-	04/89	WN		
90030	90130	07/91	Fretconnection	09/92	BREL Crewe	-	05/89	WN		
90031	90131	07/91			BREL Crewe	-	05/89	WN		
90032	90132	07/91			BREL Crewe	-	09/89	WN		
90033	90133	08/91			BREL Crewe	-	09/89	WN		
90034	90134	08/91			BREL Crewe	-	09/89	WN		
90035	90135	07/91			BREL Crewe	-	09/89	WN		
90036	90136	07/91			BREL Crewe	-	05/90	WN		
90037	90137	07/91			BREL Crewe	-	05/90	WN		
90038	90138	08/91			BREL Crewe	-	05/90	CE		
90039	90139	07/91			BREL Crewe	-	05/90	CE		
90040	90140	08/91			BREL Crewe	-	05/90	CE		
90041	90141	07/91			BREL Crewe	-	05/90	CE		
90042	90142	07/91			BREL Crewe	-	06/90	CE		
90043	90143	07/91			BREL Crewe	-	06/90	CE		
90044	90144	07/91			BREL Crewe	-	06/90	CE		
90045	90145	07/91			BREL Crewe	-	06/90	CE		
90046	90146	08/91			BREL Crewe	-	06/90	CE		
90047	90147	08/91			BREL Crewe	-	06/90	CE		
90048	90148	07/91			BREL Crewe	-	07/90	CE		
90049	90149	07/91			BREL Crewe	-	09/90	CE		
90050	90150	07/91			BREL Crewe	-	09/90	CE		

GEC/BR 25kv Bo-Bo CLASS 91

TOPS No.	Name	Date Named	Built By	Works No.	Date Introduced	Depot of First Allocation	Date Withdrawn	Depot of Final Allocation
91001	Swallow	09/89	GEC/BREL Crewe	-	04/88	BN		
91002			GEC/BREL Crewe	-	04/88	BN		
91003			GEC/BREL Crewe	-	04/88	BN		
91004	The Red Arrows	11/89	GEC/BREL Crewe	-	06/88	BN		
91005	Royal Air Force Regiment	02/92	GEC/BREL Crewe	-	05/88	BN		
91006			GEC/BREL Crewe	-	06/88	BN		
91007	Ian Allan	10/92	GEC/BREL Crewe	-	07/88	BN		
91008	Thomas Cook	10/91	GEC/BREL Crewe	-	07/88	BN		
91009	Saint Nicholas	12/92	GEC/BREL Crewe	-	08/88	BN		
91010			GEC/BREL Crewe	-	04/89	BN		
91011	Terence Cuneo	03/90	GEC/BREL Crewe	-	02/90	BN		
91012			GEC/BREL Crewe	-	03/90	BN		
91013	Michael Faraday	11/91	GEC/BREL Crewe	-	04/90	BN		
91014	Northern Electric	06/92	GEC/BREL Crewe	-	05/90	BN		
91015			GEC/BREL Crewe	-	06/90	BN		
91016			GEC/BREL Crewe	-	06/90	BN		
91017			GEC/BREL Crewe	-	07/90	BN		
91018			GEC/BREL Crewe	-	08/90	BN		
91019	Scottish Enterprise	09/90	GEC/BREL Crewe	-	08/90	BN		
91020			GEC/BREL Crewe	-	09/90	BN		
91021			GEC/BREL Crewe	-	09/90	BN		
91022			GEC/BREL Crewe	-	09/90	BN		
91023			GEC/BREL Crewe	-	09/90	BN		
91024			GEC/BREL Crewe	-	10/90	BN		
91025	BBC Radio One FM	09/92	GEC/BREL Crewe	-	11/90	BN		
91026			GEC/BREL Crewe	-	11/90	BN		
91027			GEC/BREL Crewe	-	12/90	BN		
91028	Guide Dog	07/91	GEC/BREL Crewe	-	12/90	BN		
91029	Queen Elizabeth II	06/91	GEC/BREL Crewe	-	01/91	BN		
91030	Palace of Holyroodhouse	06/91	GEC/BREL Crewe	-	01/91	BN		
91031	Sir Henry Royce	06/91	GEC/BREL Crewe	-	02/91	BN		

TOPS No.	Disposal Code	Disposal Detail	Date Cut Up	Notes
90025				
90026				
90027				
90028				
90029				
90030				
90031				
90032				
90033				
90034				
90035				
90036				
90037				
90038				
90039				
90040				
90041				
90042				
90043				
90044				
90045				
90046				
90047				
90048				
90049				
90050				

TOPS No.	Disposal Code	Disposal Detail	Date Cut Up	Notes
91001				
91002				
91003				
91004				
91005				
91006				
91007				
91008				
91009				
91010				
91011				
91012				
91013				
91014				
91015				
91016				
91017				
91018				
91019				
91020				
91021				
91022				
91023				
91024				
91025				
91026				
91027				
91028				
91029				
91030				
91031				

BRUSH DUAL VOLTAGE CHANNEL TUNNEL LOCOMOTIVES CLASS 92

TOPS No.	Name	Date Named	Built By	Works No.	Date Introduced	Depot of First Allocation	Date Withdrawn	Depot of Final Allocation
92001			Brush					
92002			Brush					
92003			Brush					
92004			Brush					
92005			Brush					
92006			Brush					
92007			Brush					
92008			Brush					
92009			Brush					
92010			Brush					
92011			Brush					
92012			Brush					
92013			Brush					
92014			Brush					
92015			Brush					
92016			Brush					
92017			Brush					
92018			Brush					
92019			Brush					
92020			Brush					
92021			Brush					
92022			Brush					
92023			Brush					
92024			Brush					
92025			Brush					
92026			Brush					
92027			Brush					
92028			Brush					
92029			Brush					
92030			Brush					
92031			Brush					
92032			Brush					
92033			Brush					
92034			Brush					
92035			Brush					
92036			Brush					
92037			Brush					
92038			Brush					
92039			Brush					
92040			Brush					
92041			Brush					
92042			Brush					
92043			Brush					
92044			Brush					
92045			Brush					
92046			Brush					
92047			Brush					

DEPARTMENTAL STOCK

PRE-GROUPING COMPANY DEPARTMENTAL STOCK

Original Departmental No.	Subsequent Departmental No.	Date Re No.	Owning Company	Built By	Works No.	Date Introduced	Depot of First Allocation	Date Withdrawn	Depot of Final Allocation
1	-	-	L & Y	Horwich	-	/12	Bury	/20	Bury
1	-	-	L & Y	M Rail	1947	/20	Horwich	11/30	Edge Hill
2	-	-	L & Y	Horwich	-	/16	Clifton Jn	07/33	Clifton Jn
2	-	-	L & Y	M Rail	2022	/20	Horwich	11/30	Edge Hill
3	-	-	L & Y	M Rail	2023	/20	Horwich	11/30	Edge Hill
74S	DS74*	-	L & SWR	Nine Elms	1899	-	Waterloo	07/65	Wimbledon
75S	DS75	09/49	L & SWR	Siemens	-	1898	Waterloo	10/68	Waterloo
1550	BEL 1	06/51	Midland	Derby	-	/14	Poplar	09/64	Poplar
BEL 2	-	-	N Staffs	N Staffs	-	/17	Oakamoor	06/63	Oakamoor

LMS DEPARTMENTAL STOCK

LMSR Departmental No.	Date Re No.	BR Departmental No.	Built By	Works No.	Date Introduced to Departmental	First Allocation	Date Withdrawn	Final Allocation	Use
5519	-	ZM9	H.Clarke	D563	/30	Crewe Works	10/57	Horwich Works	Shunter
2	-	ED1	Fowler	JF21048	08/35	Beeston	06/62	Derby	Shunter

206

TOPS No.	Disposal Code	Disposal Detail	Date Cut Up	Notes
92001				
92002				
92003				
92004				
92005				
92006				
92007				
92008				
92009				
92010				
92011				
92012				
92013				
92014				
92015				
92016				
92017				
92018				
92019				
92020				
92021				
92022				
92023				
92024				
92025				
92026				
92027				
92028				
92029				
92030				
92031				
92032				
92033				
92034				
92035				
92036				
92037				
92038				
92039				
92040				
92041				
92042				
92043				
92044				
92045				
92046				
92047				

Departmental No.	Disposal Code	Disposal Detail	Date Cut Up	Notes
1	C	Horwich Works		Based on 2-4-2T steam locomotive
1	P	Chacewater RPS	-	
2	C	LMS Workshops, Derby	c1946	
2	C	Arnott Young, Parkgate	c1963	
3	C	Synthetite Mould	c1965	
74S	C	Cox & Danks, Park Royal	c1965	
75S	P	National Railway Museum, York	-	Stored: [U] 05/68
1550	C	A King, Norwich	c1964	Allocated No. 41550
BEL 2	P	Staffordshire Industrial Museum	-	

LMSR Departmental No.	Disposal Code	Disposal Detail	Date Cut Up	Notes
5519	C	BR Workshops, Horwich	c1958	1ft 6in gauge
2	C	BR Workshops, Derby	07/62	

SR DEPARTMENTAL STOCK

SR Departmental No.	BR Departmental No.	Date Re No.	Built By	Works No.	Date Introduced	First Allocation	Date Withdrawn	Final Allocation	Use
49S	DS49*	-	Exmouth Jn	-	06/40	Exmouth	09/59	Broad Clyst	Shunter
343S	DS343*	-	AEC	-	/30	Eastleigh Wks	10/52	Eastleigh Wks	Shunter
346S	DS346*	-	LBSCR/Drewry	615	/15	LBSCR	06/49	Norwood	Saloon
400S	DS400	--/49	Fowler	22934	11/46	Southampton Dks	06/57	Southampton Dks	Shunter
499S	DS499*	-	Lancing	-	/35	Lancing C&W	03/65	Lancing C&W	Shunter
600S	DS600	--/49	Fowler	22997	12/47	Eastleigh	01/63	Eastleigh	Shunter

GWR DEPARTMENTAL STOCK

GWR Departmental No.	Date Re No.	BR Departmental No.	Built By	Works No.	Date Introduced	First Allocation	Date Withdrawn	Final Allocation	Use
1	-	-	Fowler	19451	04/33	Swindon	03/40	Swindon	Shunter
15	-	-	M Rail	2138	03/23	Wolverhampton	02/51	Taunton	Shunter
22	-	-	M Rail	5031	--/30	Banbury	/52	Theale	Shunter
23	-	-	M Rail	3731	05/25	Bridgwater	06/60	Swindon	Shunter
24	-	-	M Rail	3821	01/26	Taunton	11/60	Swindon	Shunter
25	-	-	M Rail	5074	--/30	St Austell	/51	Hayes	Shunter
26	-	-	M Rail	4178	04/27	Didcot	07/60	Swindon	Shunter
27	-	-	M Rail	3820	01/26	Reading	07/60	Swindon	Shunter

BR DEPARTMENTAL STOCK

Departmental No.	Date Re No.	Former No.	Name	Built By	Works No.	Date Introduced to Departmental Stock	First Location	Date Withdrawn
1	See GWR	No. 1						
ED1	See LMSR	No. 2						
ED2	-	-	-	Fowler	4200041	03/49	Derby	06/65
ED3	-	-	-	Fowler	4200042	10/49	Castleton	09/67
ED4	-	-	-	Fowler	4200043	11/49	Northampton	02/64
ED5	-	-	-	Fowler	4200044	11/49	Beeston	06/65
ED6	-	-	-	Fowler	4200045	07/49	Castleton	09/67
ED7	-	-	-	Fowler	22891	07/55	Fazackerley	02/64
ZM9	See LMSR 5519							
ED10	-	-	-	R.Hornsby	411322	07/58	Beeston	02/65
15	See GWR	No. 15						
20	See BR	No. 97020						
22	See GWR	No. 22						
23	See GWR	No. 23						
24	See GWR	No. 24						
25	See GWR	No. 25						
26	See GWR	No. 26						
27	See GWR	No. 27						
PO1	05/78	08173	-	BR Darlington	-	05/78	Polmadie	09/84
ZM32	-	-	-	R.Hornsby	416214	10/57	Horwich Wks	03/64
DS49	See SR	No. 49S						
52	04/53	11104	-	Hibberd	3466	/50	Hartlepool	03/67
56	-	-	-	R.Hornsby	338424	07/55	Hull	05/70
DS74	See LSWR	No. 74S						
DS75	See LSWR	No. 75S						
81	-	-	-	A.Barclay	424	/58	Peterborough	07/67
82	-	-	-	R.Hornsby	425485	/59	Dinsdale	05/70
83	-	-	-	R.Hornsby	432477	/59	Low Fell	05/70
84	-	-	-	R.Hornsby	432478	/59	York	05/70
85	-	-	-	R.Hornsby	432489	/59	Hunslet	05/70
86	-	-	-	R.Hornsby	463151	/61	Hartlepool	05/70
87	-	-	-	R.Hornsby	463152	/61	Darlington	05/70
88	02/61	D2612	-	Hunslet	5661	02/61	Barassie	12/67
89	01/64	D2615	-	Hunslet	5664	01/64	Dinsdale	11/67
91	-	-	-	BR Swindon	-	/58	Chesterton	07/67
92	-	-	-	BR Swindon	-	/58	Chesterton	07/67
100	08/49	26510	-	NER Darlington	1007	01/59	Stratford	04/63
DS209*	-	-	-	Secmafer	-	10/66	Shalford	04/68
DS343	See SR	No. 343S						
DS346	See SR	No. 346S						
DS400	See SR	No. 400S						
DS600	See SR	No. 600S						
PWM650	See BR	No. 97650						
PWM651	See BR	No. 97651						
PWM652	See BR	No. 97652						
PWM653	See BR	No. 97653						
PWM654	See BR	No. 97654						
DS1169	-	-	-	R.Hornsby	237923	12/48	Folkestone	10/72

SR Departmental No.	Disposal Code	Disposal Detail	Date Cut Up	Notes
49S	C	T W Ward, Briton Ferry		
343S	C	BR Workshops, Eastleigh		
346S	C			
400S	C			
499S	C	BR Workshops, Lancing		
600S	C	Birds, Long Marston	07/69	

GWR Departmental No.	Disposal Code	Disposal Detail	Date Cut Up	Notes
1	S	Ministry of Supply	-	Built as demonstrator
15	C	BR Workshops, Swindon	03/51	
22	C			2ft 0in gauge, Also No. PWM1780
23	C	BR Workshops, Swindon	10/60	
24	C	BR Workshops, Swindon	11/60	
25	C			2ft 0in gauge, Also No. PWM1779
26	C	BR Workshops, Swindon	10/60	
27	C	BR Workshops, Swindon	10/60	

Departmental No.	Final Location	Use	Disposal Code	Disposal Detail	Date Cut Up	Notes
ED2	Beeston	CCE Shunter	C	BR Workshops, Derby	09/67	
ED3	Bedford	CCE Shunter	C	G Cohen, Kettering	06/68	
ED4	Northampton	CCE Shunter	C	J Cashmore, Great Bridge	08/67	
ED5	Derby Works	CCE Shunter	C	J Cashmore, Great Bridge	08/67	
ED6	Holyhead	CCE Shunter	C	Valley Goods, Anglesey	07/68	
ED7	Derby Works	CCE Shunter	C	BR Workshops, Derby	04/64	Built 1940
ED10	Beeston	CCE Shunter	P	Church Farm, Fen Stanton	-	3ft 0in gauge
PO1	Polmadie	M&EE Shunter	C	BR Thornton, by V Berry	03/87	
ZM32	Horwich Wks	Works Shunter	P	NGR Centre, Gloddfa Ganol	-	
52	Woking	CCE Shunter	C	J Cashmore, Newport	03/67	
50	Thornaby	CCE Shunter	C	T J Thomson, Stockton	/80	
81	Cambridge	CCE Shunter	I	To Capital stock as D2956 [2nd loco]	-	
82	Darlington	CCE Shunter	C	T J Thomson, Stockton	/80	
83	Low Fell	CCE Shunter	C	W Hestlewood, Attercliffe	08/70	
84	York	CCE Shunter	C	Arnott Young, Parkgate	07/70	
85	Crofton	CCE Shunter	C	Arnott Young, Parkgate	07/70	
86	York	CCE Shunter	C	Arnott Young, Parkgate	07/70	
87	Darlington	CCE Shunter	C	T J Thomson, Stockton	/80	
88	Darlington	Works Shunter	C	Argosy Salvage, Shettleston	12/67	
89	Dinsdale	Works Shunter	I	To Capital stock as D2615	-	
91	Chesterton	CCE Shunter	I	To Capital stock as D2370	-	
92	Chesterton	CCE Shunter	I	To Capital stock as D2371	-	
100	Ilford	Pilot	C	BR Workshops, Doncaster	05/64	
DS209	Theale	CCE Shunter	E	Returned to France	-	
DS1169	Yeovil	CCE Shunter	C	G Cohen, Kettering	07/73	

Departmental No.	Date Re No.	Former No.	Name	Built By	Works No.	Date Introduced to Departmental Stock	First Location	Date Withdrawn
DS1173	-	-	-	Drewry	2217/D46	10/48	Hither Green	03/67
D2554 / 05001	See BR	No. 97803						
D2991	-	-	-	R.Hornsby	480692	05/73	Eastleigh	04/86
D5705	See BR	No. 15705						
D5901	-	-	-	EE.VF	2378/D418	12/69	RTC Derby	11/75
D7076	-	-	-	B.Peacock	7980	08/74	RTC Derby	11/82
D7096	-	-	-	B.Peacock	8000	08/74	RTC Derby	11/82
D8512	-	-	-	Clayton	4365U16	07/69	RTC Derby	01/72
D8521	-	-	-	Clayton	4365U17	07/69	RTC Derby	10/78
D8598	-	-	-	B.Peacock	8015	03/72	RTC Derby	10/78
15705	12/68	D5705	-	M.Vickers	-	12/68	RTC Derby	09/77
97020	05/80	20	-	R.Hornsby	408493	08/57	Reading	04/81
97201	08/79	24061	Experiment	BR Crewe	-	11/75	RTC Derby	01/88
97202	01/84	25131	-	BR Derby	-	10/83	Toton	01/84
97203	07/86	31298	-	Brush	367	07/86	RTC Derby	05/87
97204	05/87	31326	-	Brush	397	05/87	RTC Derby	-
97250	05/83	25310	Ethel 1	BR Derby	-	05/83	Eastfield	10/87
97251	08/83	25305	Ethel 2	B.Peacock	8065	08/83	Eastfield	
97252	07/83	25314	Ethel 3	BR Derby	-	07/83	Eastfield	
97401*	-	46009	-	BR Derby	-	10/83	RTC Derby	07/84
97402*	-	46023	-	BR Derby	-	12/83	RTC Derby	10/84
97403	12/84	46035	Ixion	BR Derby	-	12/84	RTC Derby	03/91
97404	01/85	46045	-	BR Derby	-	12/84	RTC Derby	09/90
97405	04/85	40060	-	EE.VF	2782/D476	04/85	Crewe	03/87
97406	05/85	40135	-	EE.VF	3081/D631	05/85	Crewe	12/86
97407	05/85	40012	-	EE.VF	2668/D429	05/85	Crewe	04/86
97408	05/85	40118	-	EE.RSH	2853/8148	05/85	Crewe	02/86
97409	10/87	45022	-	BR Crewe	-	10/87	Gateshead	07/88
97410	10/87	45029	-	BR Derby	-	10/87	Gateshead	08/88
97411	10/87	45034	-	BR Derby	-	10/87	Gateshead	07/88
97412	10/87	45040	-	BR Crewe	-	10/87	Gateshead	08/88
97413	10/87	45066	-	BR Crewe	-	10/87	Gateshead	08/88
97472	See Class 47 No. 47472							
97480	See Class 47 No. 47480							
97545	See Class 47 No. 47545							
97561	See Class 47 No. 47561							
97650	09/79	PWM650	-	R.Hornsby	312990	02/53	Swindon	04/87
97651	09/79	PWM651	-	R.Hornsby	431758	08/59	Cardiff	
97652	04/80	PWM652	-	R.Hornsby	431759	08/59	Bristol	04/87
97653	09/83	PWM653	-	R.Hornsby	431760	08/59	Reading	08/92
97654	11/80	PWM654	-	R.Hornsby	431761	08/59	Worcester	
97800	06/79	08600	Ivor	BR Derby	-	05/79	Slade Green	-
97801	11/79	08267	Pluto	BR Derby	-	03/78	RTC Derby	08/81
97802*	-	08070	-	BR Derby	-	02/79	Polmadie	01/80
97803	01/81	05001	-	Hunslet	4870	01/81	Ryde	09/83
97804	05/81	06003	-	A.Barclay	435	03/81	Reading	09/84
97805*	-	03079	-	BR Doncaster	-	04/84	Ryde	-
97806	10/87	09017	-	BR Horwich	-	10/87	Severn T Jn	
97807*		03179	-	BR Swindon	-	01/89	Ryde	
97880	See No. 97800							
ADB966506	12/73	D3078	-	BR Darlington	-	09/73	Doncaster	12/78
ADB966507	12/73	D3006	-	BR Derby	-	10/73	Lincoln	06/79
ADB966508	01/74	D3035	-	BR Derby	-	01/74	Lincoln	01/79
ADB966509	04/75	D3069	-	BR Darlington	-	07/74	Thornaby	06/69
ADB966510	11/74	D3037	-	BR Derby	-	07/74	Grantham	12/78
ADB966511	See No. ADB968011							
ADB966512	See No. ADB968012							
ADB966513	See No. ADB968010							
ADB966514	See No. ADB968021							
ADB968000	09/69	D8243	-	BTH	-	09/69	Stratford	10/87
ADB968001	09/69	D8233	-	BTH	1131	09/69	Finsbury Pk	12/82
ADB968002	07/69	D8237	-	BTH	-	07/69	Bradford	09/82
ADB968003	11/69	D8203	-	BTH	1034	11/69	Hornsey	07/81
TDB968004*	Allocated to conversion of D7055 to coach heating vehicle, but cancelled							
TDB968005*		D7089	-	B.Peacock	7993	02/75	Laira	11/75
TDB968006	See	No. 15705						
RDB968007	See	No. 97201						
TDB968008	08/76	24054	-	BR Crewe	-	08/76	Exeter	10/82
TDB968009	08/76	24142	-	BR Derby	-	08/76	Exeter	09/82
ADB968010	08/77	08117	-	BR Derby	-	06/77	Tinsley	01/79
ADB968011	08/77	08119	-	BR Derby	-	06/77	Wath	06/79
ADB968012	08/77	08111	-	BR Derby	-	06/77	Frodingham	04/79
ADB968013	03/79	31013	-	Brush	84	03/79	Stratford	01/83
ADB968014	07/80	31002	-	Brush	73	07/80	Stratford	10/82
ADB968015	07/77	31014	-	Brush	85	05/77	Stratford	10/82
ADB968016	02/81	31008	-	Brush	79	02/81	Stratford	10/82
RDB968020	See No. 97801							
ADB968021	10/79	84009	-	NBL	27801	09/78	Derby RTC	
ADB968024	07/85	45017	-	BR Crewe	-	07/85	Toton	03/88
ADB968025	11/86	27207	-	BRCW	DEL247	11/86	Eastfield	07/87
ADB968026*	-	25908	-	B.Peacock	8067	10/86	Toton	02/88
ADB968027*	-	25912	-	BR Derby	-	03/87	Holbeck	-

Departmental No.	Final Location	Use	Disposal Code	Disposal Detail	Date Cut Up	Notes
DS1173	Hither Green	CCE Shunter	I	To Capital stock - D2341	-	
D2991	Eastleigh	Generator	P	Eastleigh Preservation Society	-	
D5901	RTC Derby	Traction unit	C	BREL Doncaster	02/77	
D7076	RTC Derby	Dead Weight	P	East Lancs Railway	-	
D7096	RTC Derby	Dead Weight	C	Marple & Gillott, Attercliffe	02/86	
D8512	RTC Derby	Traction unit	C	BREL Glasgow	01/73	
D8521	RTC Derby	Generator	C	BREL Glasgow	04/79	
D8598	RTC Derby	Traction unit	C	BREL Glasgow	03/79	
15705	Bristol	Traction/Heat unit	P	Peak Railway, Matlock	-	Carried No. TDB968006 from 01/75
97020	Reading	CS&TE Shunter	C	BR S & T Works Reading by Cartwrights of Tipton	08/82	
97201	RTC Derby	Traction unit	P	Stephenson Railway Museum, Tyne & Wear	-	Carried No. RDB968007 11/75-08/79
97202	Toton	Training loco	C	V Berry, Leicester	06/87	
97203	RTC Derby	Traction unit	C	Booth-Roe, Rotherham	01/90	
97204	-	Traction unit	I	Returned to Capital stock - 31970 [31326]	-	
97250	Marylebone	Train Heat unit	A	BR Inverness		
97251		Train Heat unit				
97252		Train Heat unit				
97401*	RTC Derby	CEGB collision loco	C	BR Old Dalby	10/84	
97402*	RTC Derby	Traction unit	A	BR Crewe		
97403	RTC Derby	Test locomotive	P	Pete Waterman (site to be advised)	-	Stored: [U] 07/91
97404			P	D4 preservation group, at MRC	-	Stored: [U] 04/86
97405	Crewe	CCE Locomotive	C	V Berry, Leicester	03/88	
97406	Crewe	CCE Locomotive	P	Class 40 Preservation Society, Bury	-	
97407	Crewe	CCE Locomotive	P	Class 40 Appeal, Butterley	-	
97408	Crewe	CCE Locomotive	P	Birmingham Railway Museum, Tyseley	-	
97409	Tinsley	CCE Locomotive	C	M C Processors, Glasgow	10/91	
97410	Tinsley	CCE Locomotive	A	M C Processors, Glasgow		
97411	Tinsley	CCE Locomotive	C	M C Processors, Glasgow	07/92	
97412	Tinsley	CCE Locomotive	C	M C Processors, Glasgow	11/91	
97413	Tinsley	CCE Locomotive	C	M C Processors, Glasgow	11/91	
97650	Reading	CCE Shunter	P	Private in Lincoln		
97651		CCE Shunter				
97652	Taunton	CCE Shunter	C	BR Laira	10/90	
97653	Cardiff	CCE Shunter	A	BR Reading		Withdrawn: 04/87, R/I: 06/87
97654	CCE Shunter					
97800	-	M&EE Shunter	I	Returned to Capital Stock - 08600	-	
97801	RTC Derby	Research loco	C	V Berry, Leicester	08/85	Carried No. RDB968020 03/78-12/79
97802*	Polmadie	M&EE Shunter	C	BREL Glasgow	04/80	
97803	Ryde	CCE Shunter	P	Isle of Wight Railway Centre	-	
97804	Reading	CS&TE Shunter	P	South Yorkshire Railway	-	
97805*	-	CCE Shunter	I	Returned to Capital Stock - 03079	-	
97806	M&EE Shunter					
97807*	-	CCE Shunter	I	Returned to Capital Stock - 03179	-	
ADB966506	Doncaster	Snowplough	C	BREL Doncaster	02/79	
ADB966507	March	Snowplough	C	BREL Doncaster	08/79	
ADB966508	Lincoln	Snowplough	C	BREL Doncaster	05/79	
ADB966509	Thornaby	Snowplough	C	BR Thornaby	03/80	
ADB966510	Grantham	Snowplough	C	BREL Doncaster	03/79	
ADB968000	Marylebone	Heating vehicle	C	V Berry, Leicester	03/91	
ADB968001	Stratford	Heating vehicle	P	Mangapps Farm, Burnham on Crouch	-	
ADB968002	Finsbury Pk	Heating vehicle	C	Marple & Gillott, Attercliffe	04/85	
ADB968003	Colchester	Heating vehicle	C	BR Colchester, by V Berry	10/81	
TDB968005*	Laira	Heating vehicle	C	T J Thomson, Stockton	12/75	
TDB968008	Newton Abbot	Heating vehicle	P	East Lancs Railway	-	
TDB968009	Stratford	Heating vehicle	C	Coopers Metals, Sheffield	05/84	
ADB968010	Tinsley	Snowplough	C	BREL Doncaster	03/79	Numbered ADB966513 06/77-08/77
ADB968011	Doncaster	Snowplough	C	BREL Doncaster	02/80	Numbered ADB966511 06/77-08/77
ADB968012	Frodingham	Snowplough	C	BREL Doncaster	04/79	
ADB968013	Stratford	Heating vehicle	C	BREL Doncaster	08/83	
ADB968014	Stratford	Heating vehicle	C	BREL Crewe	05/84	
ADB968015	Stratford	Heating vehicle	C	BREL Doncaster	06/83	
ADB968016	Stratford	Heating vehicle	C	BREL Crewe	06/85	
ADB968021	Load Bank					
ADB968024	Tinsley	Training loco	A	M C Processors, Glasgow		
ADB968025	Eastfield	Training loco	C	M C Processors, Glasgow	03/88	
ADB968026*	Toton	Training loco	C	V Berry, Leicester	08/88	
ADB968027*	-	Training loco	I	Returned to Capital Stock 09/90	-	

Departmental No.	Date Re No.	Former No.	Name	Built By	Works No.	Date Introduced to Departmental Stock	First Location	Date Withdrawn
ADB968028*	-	27024	-	BRCW	DEL213	06/89	Eastfield	12/89
TDB968030	07/89	33018	-	BRCW	DEL122	07/89	M on Lugg	
ADB975812	10/79	43000	-	BREL Crewe	-	10/79	RTC Derby	01/82
RDB975813	10/79	43001	-	BREL Crewe	-	10/79	RTC Derby	12/88

LOCOMOTIVE REBUILDS FROM EMU STOCK

BR TOPS No.	Original Departmental No.	Original Class 501 No.	Rebuilt By	Date Introduced	Depot of First Allocation	Date Withdrawn	Depot of Final Allocation
97701	DB975178	M61136	BREL Wolverton	05/74	BD		
97702	DB975179	M61139	BREL Wolverton	05/74	BD		
97703		M61182	BREL Doncaster	01/80	CW		
97704		M61185	BREL Doncaster	01/80	CW		
97705		M61194	BREL Doncaster	05/80	CW		
97706		M61189	BREL Doncaster	05/80	CW		
97707	LDB975407	M61166	BREL Doncaster	06/75	GW		
97708	LDB975408	M61173	BREL Doncaster	06/75	GW		
97709	LDB975409	M61172	BREL Doncaster	09/75	HE	07/87	HE
97710	LDB975410	M61175	BREL Doncaster	09/75	HE	07/87	HE

Class 73/0 No. 73003 as restored to original green livery as No. E6003, outside Selhurst depot on 27th March 1993.   *Colin J. Marsden*

Departmental No.	Final Location	Use	Disposal Code	Disposal Detail	Date Cut Up	Notes
ADB968028*	Eastfield	Training loco	P	Brechin Railway	-	
TDB968030		MOD Locomotive				
ADB975812	RTC Derby	Traction unit	P	National Railway Museum, York	-	
RDB975813	RTC Derby	Dead Weight vehicle	C	Booth-Roe, Rotherham	12/90	

BR TOPS No.	Disposal Code	Disposal Detail	Date Cut Up	Notes
97701				
97702				
97703				
97704				
97705				
97706				
97707				
97708				
97709	A	BR Hornsey		Stored: [U] 07/85
97710	A	BR Hornsey		Stored: [U] 07/85

Class 60s Nos 60042 *Dunkery Beacon,* 60043 *Yes Tor* and 60041 *High Willhays* at Old Oak Common depot on 20th March 1993.

Colin J. Marsden